Tip-Based Nanofabrication

Ampere A. Tseng
Editor

Tip-Based Nanofabrication

Fundamentals and Applications

Editor
Ampere A. Tseng
School for Engineering of Matter,
Transport, and Energy
Arizona State University
Tempe, AZ 85287-6106, USA
ampere.tseng@asu.edu

ISBN 978-1-4899-9566-7 ISBN 978-1-4419-9899-6 (eBook)
DOI 10.1007/978-1-4419-9899-6
Springer New York Dordrecht Heidelberg London

Printed on acid-free paper

Springer is part of Springer Science+Business Media (www.springer.com)

Preface

Many optoelectronic devices and systems used in modern industry are becoming progressively smaller and have reached the nanoscale domain. Nanofabrication is critical to the realization of the potential benefits of these devices and systems for society. An important enabling technology in nanofabrication is Tip-Based Nanofabrication (TBN), which makes use of functionalized probes consisting of microscale cantilevers (or tip holders) with attached nanoscale tips. These tip-based probes, evolved in essence from scanning probe microscopy, can perform all types of manufacturing activities, from material removal and material modification to material deposition and material manipulation, all in the nanoscale. Not only can TBN create nanostructures through a conventional top-down approach, it can also build nano-components from the bottom-up. Moreover, this technology can fully integrate with stations in a semiconductor production line, as well as be performed in an ordinary chemistry or physics laboratory.

This monograph consists of twelve chapters with subjects ranging from the basic principles of TBN to recent advances in several major TBN technologies related to atomic force microscopy (AFM), scanning tunneling microscopy (STM), and dip-pen nanolithography (DPN). Two of the twelve chapters are devoted to a single material, one with a specific focus on graphene, and the other with a more general discussion of diamondoid. The former topic is particularly timely given that the 2010 Nobel Prize in Physics was awarded to Geim and Novoselov for their efforts in extracting graphene. The remaining ten chapters address a wide variety of materials, from metals and semiconductors to polymers and ceramics. This monograph is the first book of its kind dedicated solely to examining the technology of TBN and is designed both to disseminate scientific knowledge and technical information from recent findings, as well as to expand on the needs and challenges facing the TBN community.

This is an exciting moment for TBN, not least because of the enormous growth of the field in the past few years. The major advancements in TBN can be found in three categories: capability (manipulability), repeatability (reliability), and productivity (throughput). Techniques for capability enhancement presented in this monograph include AFM oxidation using dynamic force mode and double-layer approach. Eventually, the most attractive approach for capability enhancement will be a hybrid approach, such as one where the tip is loaded with a dual- or multi-energy source, or one where a bottom-up scheme is integrated with a top-down procedure. The chapters reviewing thermochemical nanolithography and electric-field-assisted nanolithography provide good examples of using dual-source tips, while the chapters on nanomanipulation and nanografting involve the mixing of a bottom-up scheme with a top-down procedure. Approaches for improving repeatability, such as the development of automated equipment and expert software, are outlined nicely in the first and last chapters. Finally, increases in throughput, or productivity, through the use of parallel processing, control strategies for increasing speed, and micro/macro tips are addressed in the chapters on the high-throughput control technique and its accompanying constraints and challenges. In the near future, we will all likely bear witness to these new advances dominating research in the area of nanofabrication, and TBN playing a key role in bridging and communicating between the nanoscopic world and our macroscopic world.

Each chapter in this monograph has been authored by world-class researchers, to whom I am grateful for their contribution. I am also indebted to a large number of reviewers whose critiques have ensured that each chapter is of the highest quality. Members of this reviewing committee are Susanne Dröscher of Swiss Federal Institute of Technology Zurich, Jayne Garno of Louisiana State University, Shao-Kang Hung of National Chiao Tung University (Taiwan), Shyankay Jou of National Taiwan University of Science and Technology, Andres La Rosa of Portland State University, Zhuang Li of Chinese Academy of Sciences, Heh-Nan Lin of National Tsinghua University (Taiwan), Hui-Hsin Lu of National Taiwan University, Andrea Notargiacomo of CNR-IFN (Italy), Luca Pellegrino of CNR-INFM-LAMIA (Italy), Debin Wang of Lawrence Berkeley National Laboratory, and Guoliang Yang of Drexel University. I hope that readers will find this book both stimulating and useful.

Tempe, Arizona, USA Ampere A. Tseng

Contents

Contributors

Bernhard Basnar Center for Micro- and Nanostructures, Vienna University of Technology, 1040 Vienna, Austria, basnar@tuwien.ac.at; basnar@fkeserver.fke.tuwien.ac.at

Jennifer E. Curtis School of Physics, Petit Institute for Bioengineering and Bioscience, Georgia Institute of Technology, Atlanta, GA 30332-0430, USA, jennifer.curtis@physics.gatech.edu

John A. Dagata National Institute of Standards and Technology, Gaithersburg, MD 20899-8212, USA, john.dagata@nist.gov

Klaus Ensslin Solid State Physics Laboratory, ETH Zurich, 8093 Zurich, Switzerland, ensslin@phys.ethz.ch

Robert A. Freitas Jr. Institute for Molecular Manufacturing, Palo Alto, CA 94301, USA, rfreitas@rfreitas.com

Jayne C. Garno Department of Chemistry, Louisiana State University, Baton Rouge, LA 70803, USA, jgarno@lsu.edu

Urszula Gasser Solid State Physics Laboratory, ETH Zurich, 8093 Zurich, Switzerland, szerer@phys.ethz.ch

Simon Gustavsson Solid State Physics Laboratory, ETH Zurich, 8093 Zurich, Switzerland; Research Laboratory of Electronics, Massachusetts Institute of Technology, Cambridge, MA, USA, simongus@phys.ethz.ch

Thomas Ihn Solid State Physics Laboratory, ETH Zurich, 8093 Zurich, Switzerland, ihn@phys.ethz.ch

Vamsi K. Kodali Department of Biophysical Chemistry, School of Physics, Georgia Institute of Technology, University of Heidelberg, Atlanta, GA 30332-0430, USA, vamsi.kodali@physics.gatech.edu

Hiromi Kuramochi MANA, NIMS, Tsukuba 305-0044, Japan; NRI, AIST, Tsukuba 305-8568, Japan, kuramochi.hiromi@nims.go.jp

Andres La Rosa Department of Physics, Portland State University, Portland, OR 97201, USA, andres@pdx.edu

Zorabel M. LeJeune Department of Chemistry, Louisiana State University, Baton Rouge, LA 70803, USA, zmallo1@lsu.edu

Pasqualantonio Pingue Laboratorio NEST, Scuola Normale Superiore, I-56127 Pisa, Italy, p.pingue@sns.it

Elisa Riedo School of Physics, Georgia Institute of Technology, Atlanta, GA 30332-0430, USA, elisa.riedo@physics.gatech.edu

Wilson K. Serem Department of Chemistry, Louisiana State University, Baton Rouge, LA 70803, USA; Department of Physical Sciences, Masinde Muliro University of Science and Technology, Kakamega 50100, Kenya, wserem1@lsu.edu

Martin Sigrist Solid State Physics Laboratory, ETH Zurich, 8093 Zurich, Switzerland, msigrist@phys.ethz.ch

Tian Tian Department of Chemistry, Louisiana State University, Baton Rouge, LA 70803, USA, ttian2@lsu.edu

Ampere A. Tseng School for Engineering of Matter, Transport, and Energy, Arizona State University, Tempe, AZ 85287-6106, USA, ampere.tseng@asu.edu

Debin Wang The Molecular Foundry, Lawrence Berkeley National Laboratory, Berkeley, CA 94720, USA, debinwang@lbl.gov

Haiming Wang Department of Mechanical and Aerospace Engineering, Rutgers, The State University of New Jersey, Piscataway, NJ 08854, USA, haimingw@eden.rutgers.edu

Mingdi Yan Department of Chemistry, Portland State University, Portland, OR 97201, USA, yanm@pdx.edu

Guoliang Yang Department of Physics, Drexel University, Philadelphia, PA 19104, USA, gyang@drexel.edu

Jing-Jiang Yu Nanotechnology Measurements Division, Agilent Technologies, Inc., Chandler, AZ 85226, USA, jing-jiang_yu@agilent.com

Qingze Zou Department of Mechanical and Aerospace Engineering, Rutgers, The State University of New Jersey, Piscataway, NJ 08854, USA, qzzou@rci.rutgers.edu

Chapter 1
Nanoscale Scratching with Single and Dual Sources Using Atomic Force Microscopes

Ampere A. Tseng

Abstract AFM (atomic force microscope) scratching is a simple yet versatile material removing technique for nanofabrication. It has evolved from a purely mechanical process to one in which the tip can be loaded by additional energy sources, such as thermal, electric, or chemical. In this chapter, scratching techniques using tips with both single and dual sources are reviewed with an emphasis on associated material removing behavior. Recent developments in scratching systems equipped with automated stages or platforms using both single tip and multiple tips are assessed. The characteristics of various approaches for scratching different types of materials, including polymers, metals, and semiconductors, are presented and evaluated. The effects of the major scratching parameters on the final nanostructures are reviewed with the goal of providing quantitative information for guiding the scratching process. Advances in several techniques using dual sources for AFM scratching are then studied with a focus on their versatility and potential for different applications. Finally, following a section on the applications of AFM scratching for fabricating a fairly wide range of nanoscale devices and systems, concluding remarks are presented to recommend subjects for future technological improvement and research emphasis, as well as to provide the author's perspective on future challenges in the field of AFM scratching.

Keywords Atomic force microscope/microscopy · Chip forming · Contact force · Dual sources · Groove · Heated tip · Machinability · Machining · Metals · Multiple scratches · Multiple probes · Nanofabrication · Nanolithography · Nanostructure · Polymer · Protuberance · Scanning probe microscopy · Semiconductor · Scratchability · Scratch direction · Scratching · Scratch ratio · Scratch speed · Threshold force · Wear coefficient

Abbreviations

AE Acoustic emission
AFM Atomic force microscope/microscopy
CNT Carbon nanotubes

A.A. Tseng (✉)
School for Engineering of Matter, Transport, and Energy, Arizona State University, Tempe, AZ 85287-6106, USA
e-mail: ampere.tseng@asu.edu

A.A. Tseng (ed.), *Tip-Based Nanofabrication*, DOI 10.1007/978-1-4419-9899-6_1,

DA	Diels–Alder
EBD	Electron beam induced deposition
ECM	Electrochemical nanomachining
FET	Field effect transistor
FWHM	Full width at half maximum
GO	Graphene oxide
KOH	Potassium hydroxide
LAO	Local anodic oxidation
PC	Polycarbonate or personal computer
PDMS	Polydimethylsiloxane
PGMA	Polyglycidyl-methacrylate
PMMA	Polymethylmethacrylate
PNBA	Poly(n-butyl acrylate)
PNIPAM	Poly(n-isopropylacrylamide)
PS	Polystyrene
R^2	Coefficient of determination
SAD	Self-amplified depolymerization
SAM	Self-assembled monolayer
SD	Standard deviation
SEM	Scanning electron microscope
SIMS	Secondary ion mass spectrometry
SNOM	Scanning near-field optical microscopy
SOI	Silicon on insulator
SPDT	Single point diamond tools
SPM	Scanning probe microscopy
SQUID	Superconducting quantum interference device
SR	Scratch ratio
STM	Scanning tunneling microscopy
TCNL	Thermochemical nanolithography
TEM	Transmission electron microscopy
TMNL	Thermomechanical nanolithography
USD	Unit scratch depth
1DES	One-dimensional electron system
2DEG	Two-dimensional electron gas

1.1 Introduction

Many devices and systems used in modern industry are becoming progressively smaller and have reached the nanoscale domain. Nanofabrication is the central theme in the realization of the potential benefits of these modern devices and systems. An atomic force microscopy (AFM) based scratching technique, also known as AFM machining, is one way to enable technology to nanofabricate a small quantity of product. This scratching technique employs functionalized microcantilevers with nanoscale tips to create nanostructures with nanometer precision. AFM, developed by Binnig et al. [1] in 1986, evolved from scanning tunneling microscopy (STM) invented in 1982. AFM operates by measuring the attractive or repulsive forces between a tip and a sample, which vary according to the distance between the two. Since the tip is located at the free-end of a deformable cantilever, the attractive or repulsive forces cause the cantilever to deflect. Typically, the deflection is measured using a laser spot reflected from the back-top of the cantilever into an array of photodetector, as shown in Fig. 1.1. AFM has a much broader potential and range of applications compared to STM because AFM can be performed under either conducting or nonconducting surfaces at room environment [2]. By contrast, STM is normally to be performed with a conducting substrate in a vacuum environment.

In the past two decades, owing to its low cost and unique atomic-level manipulation precision, AFM has undergone enormous development and has evolved from a powerful imaging instrument for atomic and molecular analyses to a major tool for nanoscale component and device fabrication [2, 3]. Many AFM fabrication techniques have been developed with different degrees of similarities and success. This chapter will focus on one of the most versatile processes: mechanical scratching, also known as AFM machining, which constitutes all the processes or techniques involved with nanoscale material removal through mechanical means by AFM. In other words, the materials are directly extracted or removed by tip scratching or plowing. Kim and Lieber [4] were the first to apply an AFM tip to mechanically scratch a thin MoO_3 crystal layer that was grown on a MoS_2 surface. In scratching, a certain amount of force is applied on the tip by controlling the cantilever deflection during scanning. The scratched depth is usually controlled by the applied normal force, which is kept constant using feedback control from a piezo scanner (Fig. 1.1). Trenches or grooves with widths from tens to hundreds of nm and depths from a few to tens of nm have been scratched on many hard surfaces of metals, oxides, and semiconductors, in addition to various soft materials [5]. In

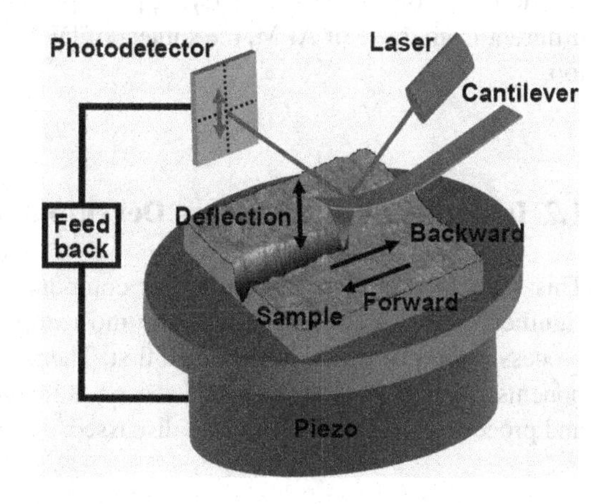

Fig. 1.1 Schematic of nanoscale scratching using atomic force microscopy

order to form various scratched profiles with desired dimensions, the grooves or trenches can be overlapped by repeated scratching or scanning. Many two- and three-dimensional (2D/3D) nanostructures have thus been fabricated by different repeating and overlapping techniques [6, 7].

In this chapter, following a general introduction of AFM scratching, an overview on advances in instrument and process development for scratching is presented. Observations and guidelines in scratching soft and hard materials are provided and elaborated. The major material and processing parameters and their significance to the AFM scratching are analyzed and quantified. Techniques where the AFM tip is charged with additional energy other than the mechanical forces, such as thermal, electrical, or chemical energy, are reviewed and discussed with an emphasis on the coupling effects and the advantages for fabricating various types of nanostructures, in which structures or grooves with higher aspect ratios may be created with a higher speed as compared with those scratched by solely mechanical means. Finally, the applications of the scratching technique for fabricating different devices and systems are selectively presented in order to illustrate the versatility of the scratching process considered.

As mentioned earlier, the scratching results based on AFM tips charged with sources other than mechanical energy will be presented. However, to limit the scope of this chapter, the techniques, in which the mechanical scratching does not play a major role in the involved material removing process, will not be considered in this chapter. For example, the AFM tip can be loaded by an optical source and the associated material removing mechanism can be dominated by the optical or photonic energy. In fact, the AFM essentially becomes a scanning near-field optical microscopy (SNOM), where the sharp tip acts as a nanoscale light source–collector or as a scatterer and materials are removed mainly by photonic excitation in near field [8]. As a result, the activities by a SNOM probe will not be discussed any further. Furthermore, if the cantilevered tip is imposed on some chemical or biological substances, the AFM can function as a dip pen to perform nanofabrication activities, also known as dip pen nanolithography (DPN), and the tip is not used for mechanical energy transfer but rather for coating material transfer for chemical or biological reactions [3]. Because the underlying principle and configuration of DPN are quite different from those of AFM, the subject of DPN will not be covered in this chapter too.

1.2 Instrument and Process Developments for Scratching

This section will focus on the major components of AFM equipment and their significances to AFM scratching. The most important component that affects the process is the probe and is presented first. Then, the presentation on the other components, such as piezoscanner and sensor, is included. The advances in equipment and process developments are also discussed.

1.2.1 AFM Probes for Scratching

An AFM probe, consisting of a microscale cantilever and a nanoscale tip, is distinguished by its stiffness or spring constant and resonant frequency. The tip is integrated with or attached to the cantilever and is characterized by its material properties and dimensions, including apex radius, cross-sectional shape, height, aspect ratio, hardness, and stiffness. The tip and cantilever are typically fabricated from the same material when mass production or integrated fabrication is required. Both the tip and the cantilever can be realized in distinct ways: direct fabrication by etching and indirect fabrication by molding. As indicated by Santschi et al. [9], material deposition and milling by focused ion beams have also been used to fabricate different types of tips. A variety of Si- and Si_3N_4-based probes are commercially available with spring constants ranging from 0.01 to 50 N/m and resonant frequencies ranging from 5 to 300 kHz. With its high aspect ratio, carbon nanotube (CNT) has also been used as a scanning tip for imaging or patterning different surfaces.

In scratching, the microscale cantilever is frequently under relatively larger applied loads (several hundreds nN or larger) than those for imaging or other AFM-based fabrication process. As compared to imaging probes, the probes used for scratching normally possess higher spring constants and higher resonance with a high hardness or wear-resistance tips. The corresponding quality factor (Q) is also higher. AFM scratching probes are preferred to be operated under contact mode. Although tapping and dynamic modes have been used for scratching, they suffer from the inherent lower tip force than that of the contact mode, which limits the depth of the scratched grooves to a few nanometers [10]. Thus, the tapping mode has only been adopted for scratching soft substrates. In this mode, in addition to the tip force, the shape or profile of the groove is dependent on the tapping drive amplitude and frequency, which makes the scratch process too complex to be precisely controlled [11].

Major limitations on direct scratching include the shallowness of scratch depth and tool wear. Often, with a conventional AFM setup, several hundred scratching repetitions are needed to scratch a required deep groove or curved surface, which is not only time-consuming but can also cause undesirable tip deterioration with unacceptable scratch precision. To cope with these limitations, tips coated by or made of hard materials, including diamond and diamond-like carbon, have been adopted for scratching. These types of tips can be made by postcoating or postassembling processes. Although these high-hardness tips can alleviate the tool wear problem, the associated probe stiffness should be increased to have a better depth control for producing deeper grooves at relatively large scratch loads. Ashida et al. [12] developed a diamond-tip cantilever with a spring constant of 820 N/m, which is about 1,000 times greater than that of a typical cantilever used for imaging. Kawasegi et al. [13] also fabricated several diamond-tip cantilevers with a spring constant on the order of 500 N/m. These cantilevers can allow the normal load on the order of 500 μN and the depths up to 100 nm for scratching a Si surface.

To improve the throughput, efforts have been dedicated to increase the AFM writing or scanning speeds. Several studies have reported that fast AFM can be operated at a rate of 30 frames per second [14–16]. This rate gives the AFM to have real-time panning and zooming capabilities and to be comparable to a typical e-beam lithography system. CNT based probes have also been developed to have well-defined tip shapes at sub-nm precision with extraordinary mechanical and electrical properties. The CNT tips can be assembled or catalytically grown from the cantilevers [17, 18]. Extremely fine patterns can be written by such probes without noticeable wearing. Although, the improvement in scanning speeds and tip hardness (or increasing tip life) is certainly helpful, a new strategy should also be explored. For example, with its high aspect ratio and low stiffness, the CNT tip is seldom used in scratching hard surfaces and a scheme to make a CNT probe with high stiffness should be studied. More information on the enhancement of throughput will be provided in later sections.

1.2.2 Multiple Probes

One of the major challenges in the development of AFM nanofabrication is to increase its throughput. Extensive efforts have been made on using multiple probe arrays for parallel-processing nanostructures. Approaches ranging from individual multifunction probes to independently activated array probes have been evaluated [19, 20]. Minne et al. [21] developed a system to operate an array of 50 cantilevered probes for local oxidation patterning in parallel at high speed. IBM has applied the multiple-tip concept, also called Millipede, for data-storage applications in which an array of heated tips is used to scratch (write), to image (read), and to melt (erase) nanoscale holes (data) on very thin polymer films. To demonstrate the potential of Millipede for ultrahigh storage density, a 64×64 cantilever/tip array for Write/Read/Erase functionality has been developed [22, 23].

In the fabrication of multiple probes, not only the height and shape of each tip are important, but also the dimensions, which affect the bending and torsion of the cantilever, play crucial roles. These geometric factors have consequences in particular for the variation of the probe properties such as deflection, compliance, and resonant frequency. For example, during scratching, the approach angle for a cantilevered single probe to a planar substrate is not a critical issue. However, in the case of a multiple parallel probe arrangement, the approach angle becomes critical in the plane parallel to the surface. The alignment of the cantilevers and the approach of the array determine which tip comes in contact with the surface first. Often, the two outermost probes are intentionally made longer to serve as adjustment or guide probes [9].

1.2.3 AFM Probe Sensing

Many types of sensing probes including acoustic emission [24], magnetic stress [25, 26], thermal [27] and electrochemical sensors [28], and scanning Hall probes

[29, 30] have also been developed for sensing tip-surface interactions. These sensors could be adopted for process control or monitoring tool for AFM scratching or nanofabrication.

Ahn and Lee [24] mounted an acoustic-emission (AE) sensor on a specially designed AFM cantilever to monitor the scratching process to determine the scratching characteristics. AE refers to the transient elastic stress waves, normally at low-intensity, high-frequency, generated due to the rapid release of strain energy within the workpiece by scratching. AE signals had been used for either on-line monitoring or off-line diagnosing of various nano or micro scale machining processes [31, 32]. By scratching Si wafers with a diamond coated Si tip, Ahn and Lee reported that their sensor-embedded AFM probe could distinguish the chip forming (shearing) stage from the protuberant (ploughing) stage and found the minimum scratch depth with chip formation. The probe used has a spring constant of 35 N/m while the tip radius is 100 nm. As reported by Zhang and Tanaka [33] and Tseng et al. [34], by increasing the scratching forces or more specifically the scratch depth, three distinct scratching phases: adhering, ploughing (or protuberance forming), and shearing (or chip forming), were observed. In adhering, no noticeable plastic deformation can be observed but atomic mass transfer from scratching surface to the tip can occur, especially for scratching soft materials [35]. In ploughing, protuberances or ridges are formed outside the scratched groove. Similar to those phenomena observed in nano-indentation of materials, protuberances are mainly the material plastically squeezed or deformed by the stress generated by the tip during scratching and the height of the squeezed protuberances can be as large as the depth of the indentation [36, 37]. As the scratching depth increases further (such as depth increasing from 4 to 8 nm), materials can be found are mainly removed by shearing and cutting chips are formed. In shearing, the height of the protuberance formation is normally smaller than those observed in the ploughing phase, but the related AE signals are stronger, which can be used as an index to gauge the phase transition. In scratching, shearing or chip forming is more desirable, since smoother surfaces and deeper grooves can be formed, if the chips have no difficulty to be removed or cleaned out from the surface afterward.

For multiple sensing probes, Minne et al. [21] have presented arrays with 10, 32 and 50 cantilevers arranged in a single line. The piezoresistive sensors provide a vertical resolution of 3.5 nm and have a bandwidth of 20 kHz. A typical image size recorded using a conventional AFM configuration is on the order of $100 \times 100\ \mu m^2$. McNamara et al. [39] fabricated a 1×8 probe array in order to combine individual thermal images of a commercial IC into an image covering $750 \times 200\ \mu m^2$ total surface. The scan speed of an individual probe was 200 μm/s resulting in an apparent total scan speed of 1,600 μm/s with a lateral resolution of 2 μm.

In magnetic stress sensing, the stress dependence of magnetic properties of materials, also known as Joule and Villari effects, is utilized to measure the deflection of microscopic cantilevers [40]. Takezaki et al. [26] and Mamin et al. [40] used the properties of spin-valve sensors in order to detect the deflection of commercially available Si_xN_y cantilevers for AFM force measurement. The magnetic stress sensor has also been adopted for measuring the magnetostriction coefficient of ferromagnetic films [25]. Information on other types of AFM based sensors for both single or multiple probes can be found in a probe review article by Santschi et al. [9].

1.2.4 AFM Process Development

Gozen and Ozdoganlar [41] developed an AFM probe, in which the probe tip could act as a single-tooth milling cutter and be rotated at high speeds to remove materials in form of long curled chips. The tip was directly driven by a three-axis piezoelectric actuator to perform both in-plan and out-of-plan rotations, while the feeding motions and depth prescription were provided by a nano-positioning stage. With limited results, Gozen and Ozdoganlar claimed that scratching by rotating tips or by nanomilling had the potential to yield lower forces, reduced tool wear, and improved feature quality.

In AFM, a raster scan is normally used to onstruct the images or scratching the required patterns. Recently, AFM, in which its tip can be moved in spiral scans, has been studied and developed. Mahmood and Moheimani [42] reported that a spiral scan could be performed by applying single frequency cosine and sine signals with slowly varying amplitudes to the x-axis and y-axis, respectively, with an AFM scanner. The use of the single tone input signals permitted the scanner to move at high speeds without exciting the mechanical resonance of the scanner and with relatively small control efforts. Experimental results obtained by Mahmood and Moheimani indicated that high-quality images could be generated at scan frequencies well beyond the raster scans. Hung [43] found that the time to complete an imaging cycle could be reduced from 800 to 314 s by using spiral scans instead of the line-by-line raster scan, without sacrificing the image resolution. Since the spiral AFM can be directly applied for performing scratching as well as other AFM based activities, it is expected that the scratching speed or AFM throughput can be improved by using spiral AFM.

Kim et al. [44] developed a tip-based tool that used multi-arrayed diamond tips driven by a high-speed air turbine spindle to scratch soft materials for patterning nanostructures with a nanoscale surface quality. The tips on the scratching tool were uniform in shape and size, and were located in the same orientation so that all the tool paths could be prescribed and the roughness of the scratched surface could be controlled within 50 nm with the number of tip arrays from 3×3 to 10×10. Some of their analyses are expected to be used for the design of the future multiple tips, especially for the scratching process.

1.2.5 AFM Three-Dimensional (3D) Patterning

In scratching curved surfaces or 3D structures, many researchers have integrated an AFM piezo tube scanner with a commercially available multi-axis stage to enhance the manipulability of workpiece movements. Thus, the workpiece can move or rotate in 3D and a curvilinear pattern of grooves can be singly or overlapped scratched. However, to handle relative large dimensions, on the order of mm, the resolution of the stage is normally one to two orders of magnitude higher than that of an AFM piezo scanner and thus the integrated scratching system may be limited to create microscale instead of nanoscale 3D patterns. In an AFM scanner, the vertical accuracy (in the normal direction between the tip and workpiece) is normally less than 1 nm while the horizontal accuracy is on the order of a few or tens nm. Certainly, the horizontal accuracy of the AFM scanner can be improved if an appropriate closed-loop control system is installed to tune the horizontal positions.

Yan et al. [6] coupled an AFM with a commercial piezo-driven stage to perform 3D scratching of a 1-μm thick copper film, which was deposited on a Si substrate. The coupled system was equipped with a capacitive based feedback sensor to improve the accuracy of stage movements. A few of 3D structures were scratched layer-by-layer (or slice by slice). Depending on the accuracy required, the 3D profile of the object was sliced into a good number of layers, and each layer was treated as a 2D contour. Figure 1.2a was presented by Yan et al. [6] showing a circular Cu contour with a diameter of 17 μm scratched by a diamond tip with a normal load of 70 μN and a scratching speed of 60 μm/s. The apex radius of the tip used was 50 nm while the average spring constant of the stainless steel cantilever adopted was 250 N/m. The depth and width of each groove were mainly controlled by the applied normal load. The tip (tool) paths of the overlapped grooves scratched are shown in Fig. 1.2b, in which the feed marked as "f", also known as pitch (or pixel pitch), is the distance between two adjacent parallel-grooves scratched and equal to 60 nm. In scratching a smooth or curved surface, the pitch size is extremely critical and dictates the smoothness and depth of the surface scratched. Theoretically, it can be found that the ratio of the pitch to the curvature radius of the tip should be smaller

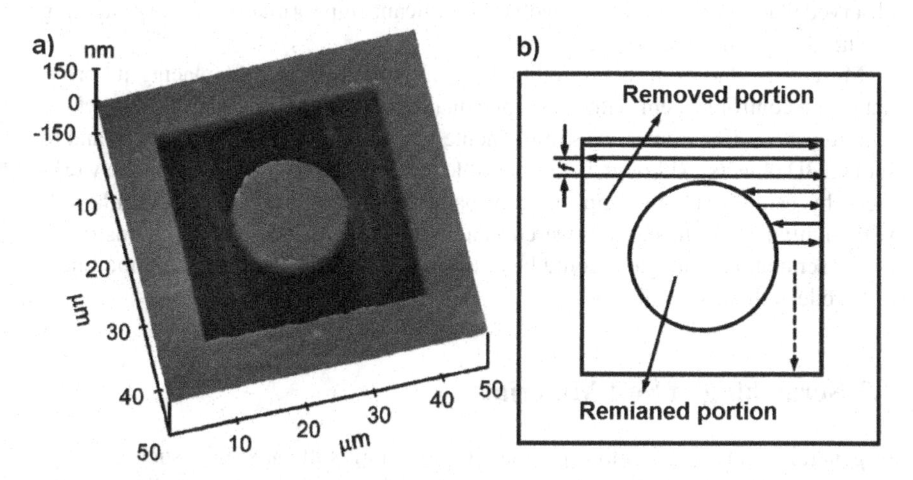

Fig. 1.2 AFM scratching of circular microdisk: (**a**) AFM image of scratched disk, (**b**) scan or tip path used for scratching (Reprinted with permission from [6] by Elsevier)

than or equal to 1.274. Yan et al. [6] reported that the scratch depth increased 30% for the feed or pitch reducing 33%. Also, it is well-known that, in layer manufacturing or rapid prototyping, the geometric inaccuracy or the size of "stair-step" errors is on the same order of the size of the pitch or feed [45].

Since the depth of scratching is dictated by the applied tip force (for more discussions, see Section 1.5.2), if the tip force can be arbitrarily changed, a nanostructure with an arbitrarily 3D profile can be scratched. Recently, a number of studies have been reported by using this changing tip force approach to generate 3D polymeric nanostructures. For example, Knoll et al. [7] used a heated tip to activate certain chemical reactions to modify an organic material structure, which could be easier to be scratched. As a result, the scratch depth was a function of both the tip force and temperature. Materials can also be scratched out layer by layer with fixed depths. Pires et al. [46] demonstrated that a 3D pattern could be written by simultaneously applying a force and temperature pulse for several μs and a microscale replica of the Matterhorm Mountain with a resolution of 15 nm into a 100-nm thick molecular glass film was created. The replica was created by 120 steps of layer-by-layer scratches, resulting in a 25-nm tall structure. More detailed information can be found in Section 1.6.1.

Efforts have also been made on the modification of AFM equipment for scratching 3D or curvilinear patterns. Bourne, Kapoor, and DeVor [47] assembled an AFM based scratching system, which could hold an AFM probe at varying angles relative to a workpiece and permitted deflections of the AFM cantilever to be measured via a displacement sensor. This system or assembly combined an AFM probe with a five-axis microscale stage, which had a resolution of 20 nm in order to achieve high scratching speeds. Since the workpiece could be rotated by 360°, the assembly had a capability of scratching curvilinear patterns of grooves. Grooves with a length of 82 mm but with depths of only a few hundred nm, using a single tool pass at scratching speeds as high at 417 μm/s, were demonstrated. The authors also observed that groove formation involved significant chip formation, while ploughing occurred particularly at low load levels [47].

Moreover, Mao et al. [48] modified a commercial AFM by replacing its original probe control system with a PC (personal computer) based controller, in which a multiaxes motion control was implemented to perform trajectory planning and to scratch 3D objects. Although it was possible to create 3D nano-profiles on the workpiece by using a specified tip and tool path, significant geometrical irregularities with highly unsmooth surfaces were found in those scratched objects demonstrated. Better scratching strategies should be developed to improve the surface smoothness or to reduce those irregularities.

1.3 Scratching of Soft Materials

In general, scratching is more popular in patterning soft materials, such as polymers, biomaterials, resists, and mica, than that of hard materials, including metals, ceramics, and semiconductors. For scratching soft materials, a regular Si or Si_3N_4

tip is usually used with a reasonable tip lifetime. As compared to imaging probes, the probes used for scratching normally possess higher spring constants (higher stiffness) and high resonance. The observation and guidelines in scratching soft materials are presented and elaborated in this section, while the subject related to scratching hard materials is reserved for the subsequent section.

Trenches or grooves can be scratched by AFM tips under contact and intermittent-contact modes. In the former, the groove is continuously carved by plowing forces, while, in the latter, the groove is formed by overlapping a series of indented holes. Because the geometry scratched by contact mode is relatively easier to be controlled, the contact mode is preferred in AFM scratching.

1.3.1 Scratching Polymers

Jung et al. [49] applied a regular SiN_4 tip under a load of 100 nN to plough a groove of 10-nm deep and 70-nm wide on a PC surface. Jin and Unertl [50] used a normal load of 490 nN to scribe grooves with widths less than 120 nm and depths of about 5 nm on a PI (polyimide) sample, where the tip used was a Si_3N_4 pyramid with an apex radius less than 40 nm on a 0.2-mm long triangular cantilever with a force constant of 0.37 N/m. Yamamoto, Yamada, and Tokumoto [51] mechanically scratched grooves on a negative resist film for e-beam lithography, called PGMA (polyglycidyl-methacrylate) using a Si_3N_4 pyramid tip under normal forces of the order of 100 nN. The scratched PGMA film was used as a mask for subsequent wet etching of a SiO_2 substrate. Li et al. [52] applied a Si tip loaded with two normal forces of 5 and 10 μN to scribe a polymeric photoresist for pattern transferring to the Si substrate underneath by subsequent wet etching. Kunze and Klehn [53] used a Si tip with an apex radius less than 10 nm for dynamic plowing of a PMMA (polymethylmethacrylate) resist and the plowed resist patterns were transferred into SiO_2, Si, GaAs, Ti, and Au substrates by wet-chemical etching. An AFM Si tip coated with a diamond layer was utilized by Choi et al. [54] as a machining tool to repeatedly scratch a grating structure of 100 μm $\times$ 150 μm on PC surfaces. The period and the depth of the grating were 500 nm and 50 nm, respectively. Light with a wavelength of 632.8 nm was well diffracted on the grated PC surfaces.

Normally, the scratched polymer surfaces can form large ridges or protuberances outside the grooves. Figure 1.3a and 1.3b show the amorphousness of the scratched surfaces for the PMMA (provided by Aldrich, M.W. 230,000) and polyimide (PI, provided by Kapton VN) films, respectively. The PMMA film is prepared by depositing droplets of a 0.2-g solution of PMMA in 10 μL of acetone on a mica surface and letting the acetone slowly evaporate over a few days. The PMMA trench shown in Fig. 1.3a was scratched by a Si tip with a normal force of 17.5 μN in a contact mode. It possesses a V-shaped cross-section with protuberances or bulges observed along the trench banks. Except from the two end regions, the depth of the groove gradually increased along the scratch direction, the arrow direction shown in Fig. 1.3a. The groove was 4-μm long and the depths were 150.0, 190.6, and

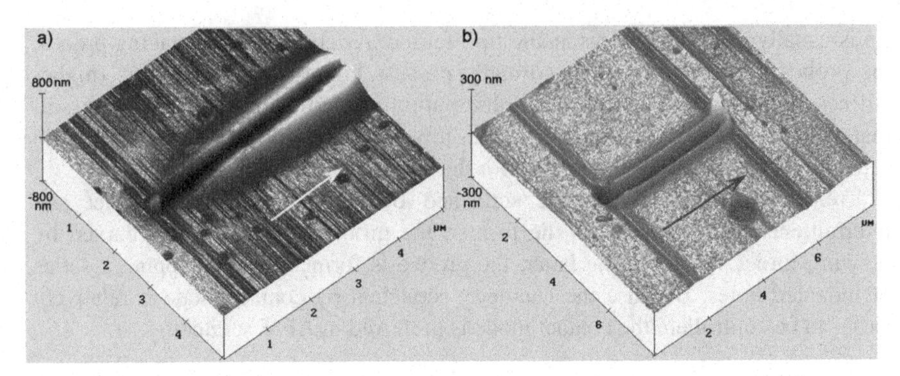

Fig. 1.3 AFM images of grooves on polymer films by AFM using Si tip at contact mode: (**a**) single scratch on PMMA film with normal force of 17.5 μN, (**b**) single scratch on PI film with normal force of 8.5 μN

225.0 nm measured at the locations of approximately 1, 2, and 3 μm from the starting scratching position, respectively. The cantilever used is 110 μm long with a spring constant of 17.5 N/m and a resonant frequency of 210 kHz. The trench on the PI film shown in Fig. 1.3b was scratched by the same Si tip used for PMMA scratching but with a normal force of 8.5 μN. The depths of the PI trench were rather consistent and were 140.2 and 138.9 nm measured at the locations about 0.5 μm from the two ends. Similar to the PMMA trench, the PI trench also had sizable side protuberances. Also it was observed that at relatively high scratching forces, the chip debris can be formed and cover the trench, which could not only greatly deteriorate the uniformity of the trench profile but also reduced the material removing rate.

1.3.2 Scratching Self-Assembled Monolayers

Grooves and pits with lateral sizes down to a few tens of nanometers could also be obtained by scratching in self-assembled monolayers (SAMs). Sugihara, Takahara, and Kajiyama [55] applied a Si tip in order to dig pits with a minimum diameter of ~20 nm in a lignoceric acid SAM in an ambient environment. The pits were made by a tip with a radius of 10-nm, a scanning rate of 2 Hz, and a tip force of 0.3 nN. The pits could be artificially distributed with different sizes and surface area densities in the organic monolayer. Zhang, Balaur, and Schmuki [56] used a diamond-coated tip with a cantilever spring constant of 17 N/m and a scratch speed of 4 μm/s to scratch through an organic monolayer covered Si(111) surface in contact mode. The scratched monolayer (1-octadecene) was used as an insulating mask for patterning a Cu-based nanostructure on the Si surface by a subsequent immersion plating process. The 1-octadecene ($C_{18}H_{36}$) layer was covalently attached to a hydrogen-terminated Si(111) surface.

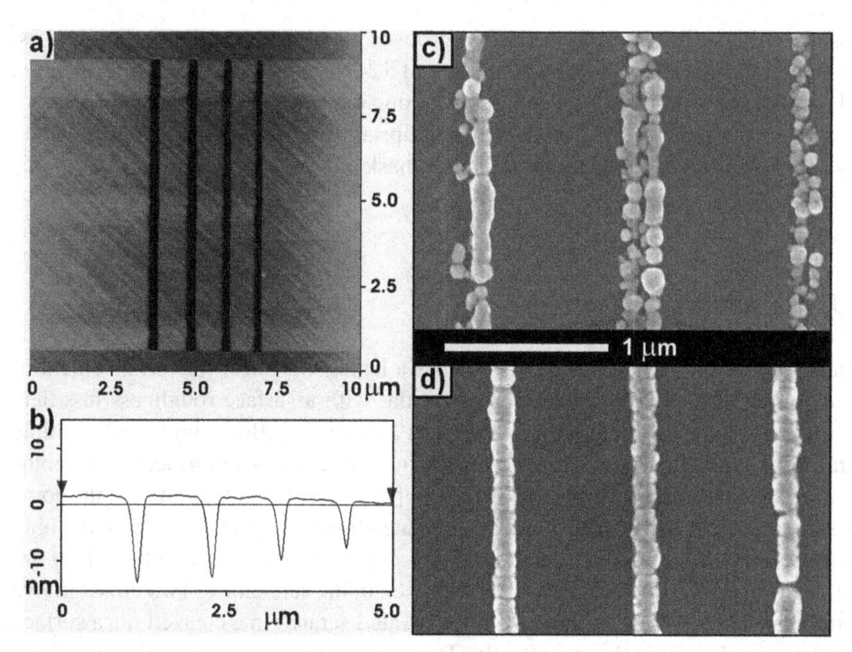

Fig. 1.4 Nanopatterning of organic monolayer covered Si surfaces by AFM scratching: (**a**) AFM image (5 μm × 5 μm) of four grooves (from left to right) scratched with the normal tip loads decreasing from 30 to 10 μN under contact mode on 1-octadecene ($C_{18}H_{36}$) coated Si surface, (**b**) corresponding AFM measured groove profile, in which the distance between two neighboring grooves is about 1 μm, (**c**) SEM image of Cu deposited in scratched grooves by immerse plating in 0.05 M $CuSO_4$ + 1% HF solution with immersion time of 10 s, (**d**) of 15 s (courtesy of Professor Patrik Schmuki of University of Erlangen-Nuremberg, Germany & Dr. Eugeniu Balaur of La Trobe University, Australia)

Figure 1.4a shows an AFM image of four scratched grooves that were spaced by 1 μm on the 1-octadecene covered Si surface, where the four grooves shown from left to right were scratched with normal tip loads decreasing from 30 to 10 μN. Figure 1.4b is the corresponding AFM measured groove profiles and show a uniform V-shape with the depth range from 15 to 7 nm and width range from 300 to 200 nm as the normal load decreases from 30 to 10 μN. Immediately after scratching, the Cu was deposited on the scratched grooves in 0.05 M $CuSO_4$ + 1% HF solution by immersion plating. Different immersion times were used to test the selectivity of the deposition. Figure 1.4c, d show two SEM images that were obtained after different immersion times, 10 and 15 s, respectively. Clear effects of the immersion time on the copper deposit morphology were found: For deposition time up to 10 s single copper nuclei form selectively in the scratch, at approximately 10–15 s these nuclei coalesce to coherent lines. It is also apparent that the longer the immersion time, the wider the deposit lines. The line width increases from 150 to 440 nm within 10–25 s. In Fig. 1.4c, it is clear that the initial step of the copper deposition is the formation of globular nuclei. It was observed that for a deposition time longer than 20 s, Cu was not only deposited on the scratched grooves but also randomly on the surfaces. As shown in Fig. 1.4d, an immersion time of 12–15 s can give the best results, i.e. coherently deposited lines but no deposits outside the scratched region. As concluded by the authors [56], under an appropriate immersion time, the scratched organic layer can be used as an effective mask for patterning Cu nanostructures on Si.

1.3.3 Scratching Mica

Mica is a group of silicate minerals and has a hexagonal sheet-like arrangement of its atoms, which make its cleavage perfect flat with a surface roughness less than 1 nm. It is a soft material and just as soft as a fingernail. Both single and multiple scratches have been used to scratch trenches or grooves on mica surfaces with depths up to 1 μm. As demonstrated by Müller et al. [57], grooves with mouth widths down to 3-nm in a cleaved mica layer can be generated by repeated mechanical scratching using an AFM tip in contact mode. A V-shaped Si_3N_4 tip with a radius of 30 nm with a normal force of several 100 nN was used in the scratching. Two cross-groove patterns generated by a single and a five-repeated scratch in a cleaved mica surface are shown in Fig. 1.5a, b, respectively. The grooves with widths of 80 and 300 nm and with depths of 7 and 25 nm were scratched with the number of scratches equal to 1 and 5, respectively, by a Si tip in contact mode with a normal force of 5 μN under ambient conditions. The tip used was pyramidal shaped with a tip height of 13 μm having a spring constant of approximately 60 N/m and a resonant frequency of 260 kHz. For a freshly cleaved mica surface, the normal force to have a noticeable scratch should be larger than 100-nN, which is the typical threshold force observed by other investigators [57, 58]. No debris was found outside the trenches, indicating

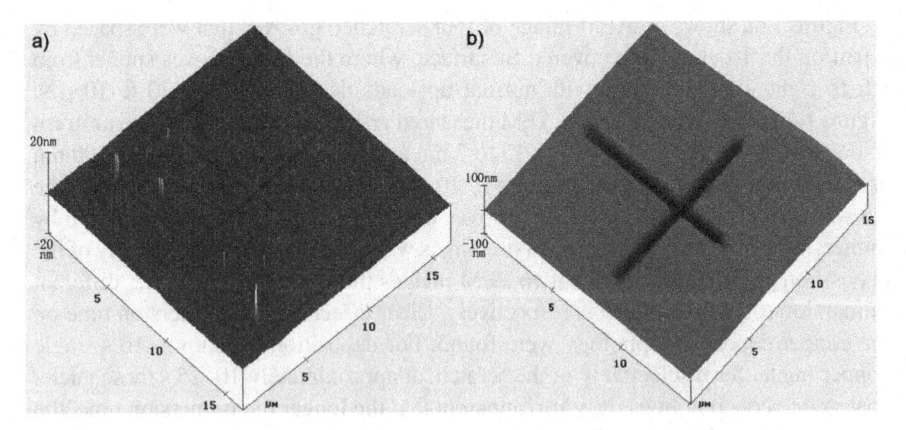

Fig. 1.5 AFM images of grooves scratched on mica substrate by AFM using Si tip at contact mode: (**a**) single scratch with a normal force of 5 nN, (**b**) five repeated scratches a normal force of 5 μN

that the material removing mechanism for multiple scratches is also dominated by atomic-scale abrasive wear due to sliding friction as those observed in single scratch [58]. During mica surface scratching, a top 0.2-nm thick layer, which is about the size of the molecular layer, is first removed. Then, the mica surface is removed layer by layer, sequentially, in which the layer thickness is about 1 nm. With a proper design of the tip profile, nano- or micro-scale groove or channel with controlled profiles and patterns can be produced.

1.3.4 Scratching Bio or Other Soft Materials

Firtel et al. [59] used a Si tip for nanosurgery to mechanically remove large patches of outer cell walls after appropriate medical treatment, which typically left the bacteria alive. Their study can open a direction of using AFM scratching for biological applications.

Muir et al. [60] created nanoscale biologically active protein patterns by scratching the top low-fouling surface coating of a bilayer film, which consists of a 4 nm-thick low-fouling DGpp (diethylene glycol dimethyl ether plasmapolymer) coating and a protein-fouling HApp (heptylamine plasma-polymer) non-specific-binding base. The material of the top DGpp coating can be selectively scratched to form a desired pattern so that the underlying HApp base, which can provide a protein-adsorbing surface, can be exposed. Incubation in a rabbit IgG protein solution leads to the adsorption of proteins onto the exposed protein-fouling patterns, while the regions with the low-fouling top coating resist protein adsorption. For example, subsequent exposure to a fluorescently labeled anti-IgG antibody results in the selective binding of the antibody to the surface-bound proteins, resulting in fluorescently active protein patterns which can be readily imaged by fluorescence as demonstrated in Fig. 1.6. Figure 1.6a shows the AFM images of a scratched groove pattern with a typical groove depth of 7 nm and a mouth width of 300-nm, while Fig. 1.6b is the AFM image after sequential adsorption of an IgG protein and a fluorescently labeled anti-IgG protein antibody. The fluorescence microscopy image of Fig. 1.6b is shown in Fig. 1.6c. The schematic diagrams illustrating the steps for protein patterning are also shown in Fig. 1.6, where Fig. 1.6d is showing the DGpp surface being scratched by an AFM tip to expose the underlying HApp substrate; Fig. 1.6e depicts the scratched surface incubated with a rabbit IgG protein solution; Fig. 1.6f exhibits that a fluorescently labeled anti-IgG protein that selectively binds to the rabbit IgG molecules being incubated with the surface; Fig. 1.6g displays that, after rinsing, immobilized protein remains on the exposed HApp scratched regions.

Park et al. [61] used a multi-wall CNT tip for drilling sub-100-nm holes on pyrolytic graphite surfaces. Gnecco, Bennewitz, and Meyer [62] studied the AFM induced abrasive wear or multiple scratch on a soft optical substrate, KBr(001), and explained AFM scratching as the result of the removal and rearrangement of single ion pairs. The debris was reorganized in regular terraces with the same periodicity and orientation as the unscratched surface as in local epitaxial growth. The applied

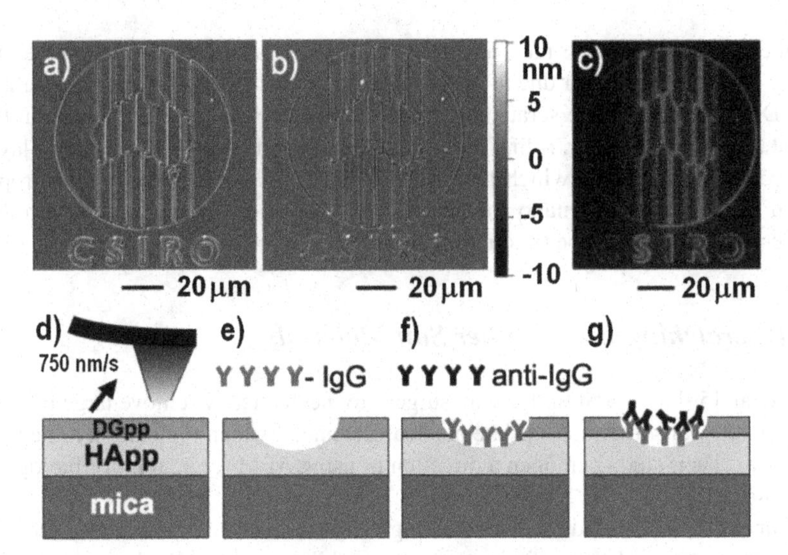

Fig. 1.6 Biological protein patterning by AFM scratching: (**a**) AFM image of a scratched groove pattern with a typical groove depth of 7 nm and a mouth width of 300-nm, (**b**) AFM image after sequential adsorption of an IgG protein and a fluorescently labeled anti-IgG protein antibody, (**c**) fluorescence microscopy image of b, (**d**) DGpp surface being scratched by an AFM tip to expose the underlying HApp substrate, (**e**) scratched surface incubated with a rabbit IgG protein solution, (**f**) a fluorescently labeled anti-IgG protein that selectively binds to the rabbit IgG molecules being incubated with the surface, (**g**) after rinsing, immobilized protein remains on the exposed HApp scratched regions (Reproduced with permission from [60] by Wiley-VCH Verlag GmbH & Co.KGaA)

tip load had a strong influence on this abrasive or scratch process, whereas the scan speed was less dominant. KBr is a sort of soft materials similar to mica and is much softer than a bronze coin. More discussions on the effects of the applied tip load and scan speed on the scratched geometry will be presented in Section 1.5.

1.4 Scratching of Hard Materials

In scratching hard materials, relatively high applied loads (several hundreds nN or larger) are required as compared to those required for imaging or scratching soft materials. Although, a regular Si or Si_3N_4 tip has been used to scribe hard materials, the tool wear problem has limited the adoption of these regular tips in scratching hard materials. To have a longer tip life, diamond or diamond-coated tips, which have the best hardness or wear resistance, are preferred for scratching hard materials. In this section, scratching hard materials, including metals, semiconductors, and ceramics as well as the carbon-based nanomaterials, will be reviewed and related issues will be discussed.

1.4.1 Scratching Metals

Rank et al. [63] used a carbon-coated Si_3N_4 tip to plow lines with widths less than 80-nm in Al and AuPd substrates under both contact (17–51 nN tip force) and non-contact (10–100 μN tip force) modes. Sumomogi et al. [64] used a diamond tip to repeatedly scratch Au, Cu, and Ni surfaces to a depth of 100 nm. Watanabe et al. [65] applied a Cr and Au coated Si tip to scratch a 1.5-nm thick Au film to fabricate Au nanowires 70–110 nm wide and 4–7 nm high. The AFM probe used has a spring constant of 0.6 N/m with a tip radius of 50-nm at a scan speed of 5.61 μm/s. Li et al. [66] used an AFM to scratch various types of cavities or channel on Au nanowires with a height of 160 nm, width of 350 nm, and length of 5 μm, which were made by e-beam lithography. A Veeco Dimension-3100 AFM equipped with a stiff stainless steel probe with a diamond tip of 15 nm in radius was used in scratching.

Fang, Chang and Weng [67] experimentally and numerically studied the AFM scratch characteristics of gold and platinum thin films. Their results indicated that at the same scratching conditions considered, the depths of the scratched Au grooves were larger than that of Pt grooves. Filho et al. [68] equipped an AFM with a diamond tip under contact mode to scratch an Al layer for patterning a Si-based mask. The diamond tip used had an apex radius of 25-nm and the scratch depth grew from 20 to 80-nm as the applied tip force increased from 15 to 30 μN. Fang, Weng, and Chang [69] applied a diamond tip to scribe an Al surface and found that the surface roughness improves as the scratch speed or the number of scratches increases. Notargiacomo et al. [70] ploughed 60-nm gaps in Al stripes and fabricated nanogap electrodes to be used for molecular devices and single-electron transistors. Tseng et al. [71] used a diamond coated tip to study the scratch properties of a NiFe-based alloy for making a nanoconstriction for a magnetic device. Kawasegi et al. [72] applied a high-stiffness probe with a diamond tip to machine a $Zr_{55}Cu_{39}Al_{10}Ni_5$ metallic glass and discovered that the metallic glass is more difficult than Si to be scratched. As indicated in the above cited studies, AFM has been indeed applied to a wide range of metals for various purposes.

1.4.2 Scratching Semiconductors

AFM diamond tips have been popular for scratching semiconductors, because of their high wear resistance. Santinacci et al. [73] applied an AFM equipped with a diamond tip to scratch through a 10-nm thick oxide layer onto a p-type Si (100) substrate. Kawasegi et al. [13] used a high-stiffness diamond-tip probe to study the scratching behavior of undoped Si(100) wafer, while Ogino, Nishimura, and Shirakashi [74] and Tseng et al. [5, 71] applied a diamond coated tip to scratch p-type Si(100) wafer. It was found that the depth and width of the grooves scratched on Si surface increase logarithmically with the normal force applied [5, 71] and the dimensional increase follows the power-law with the number of scratches or repeated cycles [37]. Recently, Brousseau et al. [75] used Si tips (Nanosensors PPP-NCH) to scratch both Si and brass ($CuZn_{39}Pb_3$) surface for microinjection mold making. By applying these scratched molds for molding polypropylenes they discovered that the brass mold has longer mold life than that of the Si one and is more appropriate for micromolding applications. The effects of the applied normal force, scratching speed, number of scratches, and pitch (or feed) on the depth and roughness of the scratched brass mold were also reported.

Figure 1.7 shows AFM images of three groove patterns on a NiFe coated Si substrate created with a single scratch or scan. Figure 1.7a depicts a four-groove array scratched with a pitch (distance between two adjacent grooves) equal to 80 nm while Fig. 1.7b reveals a ten-groove pattern machined with a pitch of 35 nm. All grooves

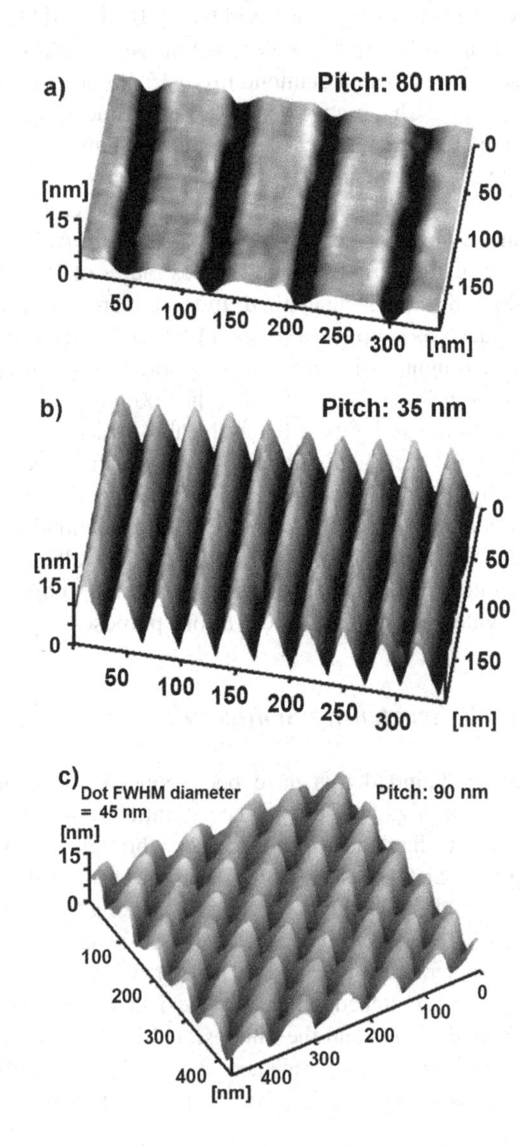

Fig. 1.7 AFM images of single-scratch patterns by AFM using diamond-coated tip at contact mode: (**a**) four-parallel grooves scratched with a pitch of 80 nm at a normal force of 9 μN, (**b**) ten-parallel grooves scratched with a pitch of 35 nm at a normal force of 9 μN, (**c**) nanodots with a diameter of 45 nm created by cross scratching with a pitch of 90 nm at a normal force of 9 μN (courtesy of Professor Ampere A. Tseng of Arizona State University)

were scratched by an applied normal force (F_n) of 9 μN and a speed of 100 nm/s. A diamond coated tip that had an apex radius of 120-nm attached on a Si cantilever with a spring constant of 42 N/m was used. These single-scratch patterns can be used as a stamp or mold for nanoimprinting different waveguide devices, which normally have equal grating elements [76]. The parallel groove pattern can also be used to create quasi-3D patterns by adding parallel groove scratches in the perpendicular direction. As illustrated in Fig. 1.7c, the hemispheric dot pattern was constructed by double perpendicular scratches with a pitch of 90 nm. The FWHM (full width at half maximum) dot diameter is approximately 45 nm and corresponding dot density can be as high as 2.6×10^8 per mm^2. Since backward or orthogonal scratching was used, the formation of protuberances or ridges along the groove sides was minimized. Further details on the effects of the scratch direction on the protuberance creation will be presented in Section 1.5.1.

Other Si surfaces had also been scratched by a diamond tip with a much large apex radius and under a much higher force. By applying a normal force from a few to several tens of μN, Ashida, Morita, and Yoshida [12] and Miyake and Kim [77] studied the scratching behavior of single-crystal Si by a diamond tip and found that the Si on the scratched area was not removed but was protruding to a height of 1–2 nm. The authors [77] believed that the scratching or sliding enhanced the reaction of Si with environmental moisture and oxygen to form Si oxide or Si hydroxide that made the area protruded. However, Park et al. [78] and Kawasegi et al. [79] used transmission electron microscopy (TEM) and secondary ion mass spectrometry (SIMS) analyses to study the protruded area and found that the area scratched by a diamond tip at a normal force of 350 μN had a 15–20-nm thick amorphous structure because of a pressure induced phase transformation. Numerous overlaps of the collision cascade are generally required to render the crystal amorphous. While etched in KOH (potassium hydroxide) solution, the amorphous layer that formed on the scratched area could be used as a mask and could withstand the etching while the non-scratched area was etched. As a result, protruding nanostructures with a height of several tens to several hundreds of nm can be obtained.

1.4.3 Scratching Ceramics and Carbon-Based Nanomaterials

In addition to metals and semiconductor, AFM has also been used for scratching materials with very high hardness, including glasses and carbon-based nanomaterials. The carbon-based nanomaterials considered here are carbon or carbon-like thin films, CNT, and graphenes, and normally have very unique material properties including extremely high hardness.

Normally, glass or ceramics is brittle and is not easy to be machined to have nanoscale quality surfaces and patterns, because of the occurrence of the fractures or more specifically, nano- and micro-scale cracks. Yan et al. [80] equipped a high-precision stage with an AFM diamond tip to scratch various shapes of cavities on curved glass surfaces. Although Yan et al. did not study the ductile and brittle modes in scratching glasses, for a brittle material like glass, ductile mode scratching is desirable to avoid brittle fracture damage of surface layer and to have smooth and controllable surface quality.

In ductile mode scratching, the scratch depth should be less than the critical or threshold depth of cut or scratch, d_c. Bifano et al. [81] developed an empirical equation to estimate d_c,

$$d_c = 0.15\ (E/H)\ (K_c/H)^2 \tag{1}$$

where E, H, and K_c are the Young's modulus, hardness, and fracture toughness of the scratched film or substrate, respectively. For most of brittle materials, the values of d_c can vary from 10 to 100 nm. In AFM scratching, the round shape of the tip possesses a negative rake-angle face, which can produce sufficient hydrostatic compressive stress in the scratching zone. This compressive stress can provide more favorable environment for the material transferring from a brittle to a ductile mode as compared with conventional machining, which normally has a positive rake angle. As a result, the critical depth predicted by Eq. (1) may be underestimated. As reported by Young et al. [82], the value of d_c is in the range of 20–40 nm by AFM scratching on a Si substrate, which is somewhat greater than that estimated by Eq. (1). In fact, using single point diamond tools (SPDTs), similar to AFM diamond tips to have a negative rake angle, are one of the emerging machining technologies for making optical glasses with nm surface finishes, which is much better than that of others optical manufacturing processes [82]. The findings from AFM scratching of glass should be able to be implemented for making optical components and devices.

In scratching fused silica (amorphous SiO_2), Park et al. [83] used an AFM equipped with a Berkovich diamond tip with a apex radius of 40 nm under normal loads in a range from a few mN to tens mN and found that both protruding and depressed patterns could be generated for the area under scratching similar to the scratch behavior mentioned earlier in scratching Si substrates at high normal loads. The scratched area or pattern, which was under both normal and shear deformations, has been used as an etching mask against HF (hydrofluoric) solution for the subsequent etching process [83]. However, little information on the specific conditions that can create a protruding or depressed pattern was presented. Also, no convinced explanation has been reported on why the scratched or deformed area can resist HF etching. Further research in this area is needed to better apply this technique for making etching mask or other applications.

In carbon-based nanomaterials, Prioli, Chhowalla, and Freire [84] applied AFM scratching with a diamond tip to scratch nanostructured carbon (ns-C) and tetrahedral amorphous carbon (ta-C) thin films with a focus on finding the relationship of the scratch depth as a function of the normal scratch load. The ta-C thin films containing 80% sp^3 bonding have a hardness of 60–70 GPa and Young's modulus of 300–400 GPa while the Young's modulus of ns-C thin films measured by a surface acoustic wave method is approximately 500 GPa. The ns-C films have strongly interacting graphene planes, which yield a unique property to have higher hardness and elastic properties than those of ta-C. Both types of C films have been widely used as protective overcoats for magnetic recording media and storage devices and as sliding parts of MEMS [85].

By applying a diamond tip, Tsuchitani et al. [86] scratched amorphous carbon (a-C) films, which were deposited on a Si substrate by an ECR (electron cyclotron resonance) plasma sputtering process. It is known that the hardness of this a-C film is higher than that of the RF sputtered C films and is close to that of bulk diamond. The effect of the humidity on the scratch depth was evaluated and it was found that the higher the humidity larger the depth scratched, although some contradicted behavior in a few cases was also observed [86]. It is expected that mass transfer of carbon atoms between the diamond tip and a-C film could occur during scratching and the details of the mass transfer effect on the scratch behavior remains to be explored.

By using a diamond tip to scratch carbon nitride (CN_x) overcoats with thickness from 1 to 10 nm, Bai et al. [87] observed that the scratch behavior was changed, where the nanoscale material removal rate increases as the overcoat thickness decreases. For example, at the scratch normal force equal to 45 μN with 20 scratch cycles, the scratch depth increased from 4 to 13.5 nm as the overcoat thickness decreased from 10 to 1 nm while the corresponding material removal rate grew from 0.2×10^{-4} to 0.8×10^{-4} mm^3/nm. The AFM tip used was a diamond tip with radius less than 100 nm. Furthermore, non-contact mode imaging of scratched surfaces showed that the scratch mode was gradually changed from the ploughing (protuberance or plastic deformation dominated) to shearing (chip or debris dominated) mode as the overcoat thickness decreased from 10 to 1 nm. This scratch mode change may associate with the cutting phase change, i.e., from brittle (at thinner overcoats) to ductile (at thicker overcoats), which is also dependent on the ratio of the scratch depth to the substrate or film thickness, as investigated by Fang et al. [88]. The thickness effects of coated film or substrate on the scratch or wear mechanism are extremely interesting and more research along this direction should be encouraged.

An AFM Si tip was used by Lu et al. [89] to scratch graphene oxide (GO) films, resulting in GO-free trenches or grooves and various single-layer GO patterns such as gaps, ribbons, squares, and triangles can be fabricated. By using the GO patterns as templates, hybrid GO-Ag nanoparticle patterns can be created and can have the potential for making graphene material-based devices. The GO film was synthesized using the modified Hummers scheme from graphite powder [90] and assembled onto a Si/SiO_2 substrate using the Langmuir-Blodgett technique. Kim, Koo, and Kim [91] found that multiwalled CNTs and graphite on a Si substrate can be cut or broken by a conducting AFM tip imposed with a negative bias. It is believed that the field-emission current from the negatively biased AFM tip may provide the activation energy to break or to cut the atomic bonds in CNTs (or graphite) [91]. Scratching caused by other energy sources will be presented and discussed in later sections.

1.5 Controlling Parameters in Scratching

The characteristics of material removing by scratching at the nanometer scale depend strongly on the major scratching parameters. Without proper understanding of those parameters, desirable nanostructures cannot be appropriately patterned and high-precision nano-product cannot be efficiently fabricated. In this section, the recent progress on the understanding of these controlling parameters will be studied and discussed.

There are many ways to study the controlling parameters of AFM scratching. The most simple and controllable scheme should be using a single scratch or scan to assess the effects of the major scratch parameters, including the scratch direction, normal tip force and scratch speed, on the geometry and dimensions of the scratched structures. The results based on single groove scratch will be presented, while other effects, including multiple scratches, scratchability, and wear coefficients, will also be assessed.

1.5.1 Effects of Scratch Directions and Protuberance Formation

To study the geometric effects of the tip or scratch direction, a pyramidal diamond coated Si tip was selected to scratch a $Ni_{80}Fe_{20}$ coated Si substrate in four different directions. The NiFe thin film is well-known for its high magnetic permeability, low coercivity, near zero magnetostriction, and significant anisotropic magnetoresistance [92]. The NiFe thin film was deposited on a Si substrate by an e-beam deposition process, which yielded a nearly isotropic Ni-Fe thin film material, which could make the influence of changing scratch direction caused by the substrate to be negligible. As shown in Fig. 1.8, the 10-μm tall tip with a three-sided pyramidal geometry is attached a 125-μm long Si cantilever with a rectangular cross section of 4 μm × 30 μm, provided by Nanosensor (DT-NCHR). The probe has a spring constant of 42 N/m and a resonant frequency of 330 kHz. As shown in Fig. 1.8b, c, the tip radius of curvature is approximately 120 nm and the vertex angle of the triangular pyramid tip in the last 200 nm is tapered to a half cone angle of 10° at the very end of the tip.

In scratching, the AFM tip was moved by a vector scan method in contact mode and loaded normal to the sample surface with a constant tip force, F_n, of 9μN. As shown in Fig. 1.9a, four grooves on the $Ni_{80}Fe_{20}$ surfaces were scratched in four different directions, where the forward and backward directions are parallel to the longitudinal direction of the cantilever while the upward and downward directions are perpendicular to the cantilever as defined in Fig. 1.8a. Each groove was scratched with a speed of 100 nm/s at low relative humidity in the range from 20 to 30% to alleviate the influence of adhesive force. The corresponding groove profiles, which were taken in ten different locations along the respective groove scratched, are shown in Fig. 1.9b, where the cross-sectional profiles are V-shaped and protuberances or ridges have been observed along the banks, the two sides near the

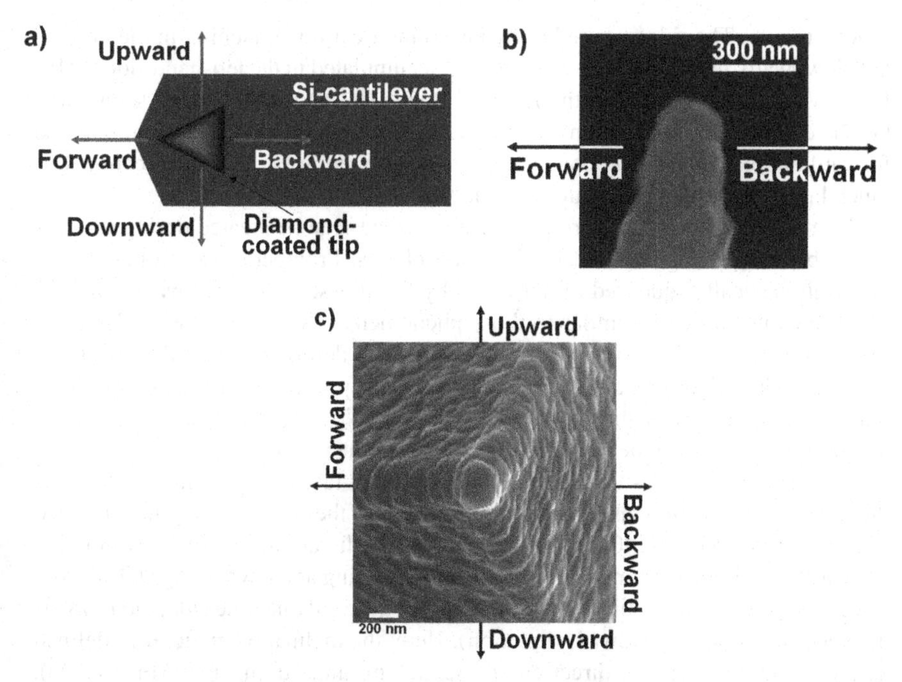

Fig. 1.8 Diamond-coated AFM probe: (**a**) rectangular microcantilever with triangular pyramid tip, (**b**) SEM image of tip side view, (**c**) SEM image of tip top view

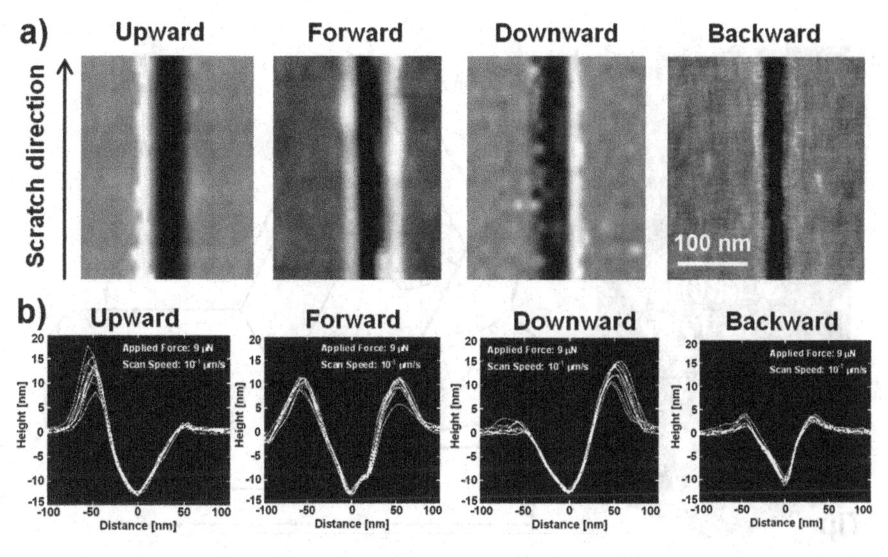

Fig. 1.9 Grooves scratched by AFM with triangular pyramid diamond-coated tip on $Ni_{80}Fe_{20}$ surface with a normal force of 9 μN: (**a**) AFM image of groove scratched at upward, forward, downward, and backward directions (from left to right), (**b**) cross-sectional profiles measured by AFM at 10 different groove locations for each groove scratched (courtesy of Professor Ampere A. Tseng of Arizona State University)

groove mouth. The ridges were highly unsymmetrical for scratching in the upward and downward directions; they were mainly accumulated in the left-hand side (LHS) for upward scratching and in the right-hand side (RHS) for downward scratching. On the contrary, the protuberances were basically symmetrical for scratching in the forward and backward directions, but their magnitude in the former direction was much larger than that of the latter direction [37].

As mentioned earlier, the protuberances or ridges can be created during scratching in both the ploughing and the shearing phases. Protuberances are mainly the material plastically squeezed or deformed by the stresses generated by the tip during scratching, which is similar to those phenomena observed in nano-indentation of materials [36, 93]. The protuberances caused by deformation are difficult to be removed. The ridges can also be the adhered pile-up of scratched debris or chips, which are the materials sheared from the groove [37]. As shown in Fig. 1.9b, the height of the ridges can be as large as the depth of the grooves.

The pyramidal tip has three scratching or cutting faces. If scratching is towards the downward direction, as illustrated in Fig. 1.10a, the tip scratching face is tilted with an inclination angle (θ) of 60° with the scratch direction, which is in a situation known as the oblique cutting in conventional machining as shown in Fig. 1.10b. As a result, the protuberances tilt at an angle and are squeezed onto one side, the RHS, as depicted in Fig. 1.9 (downward direction). Here, the inclination angle, θ, is defined as the angle between the directions of scratching and cutting face (Fig. 1.10b). Scratching in the upward direction is almost identical to that of the downward direction, except the protuberances are generated in the opposite side, the LHS.

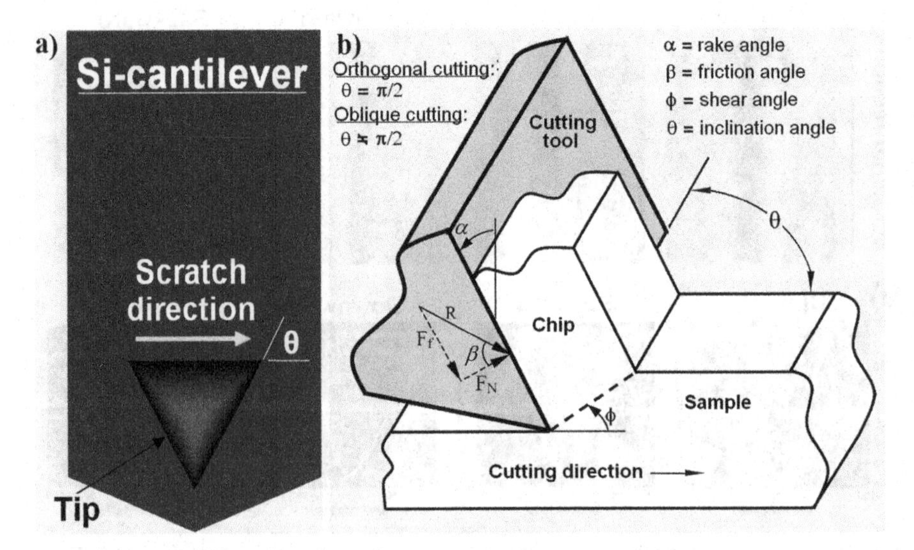

Fig. 1.10 Schematic of oblique cutting: (**a**) inclination angle, θ, defined in AFM scratching with triangular pyramid tip, (**b**) inclination angle defined in conventional machining (courtesy of Professor Ampere A. Tseng of Arizona State University)

In the case of the backward direction, the tip scratch face is perpendicular to the scratch direction, i.e., the inclination angle, θ, equals 90°, which satisfies the requirement to become orthogonal cutting. The protuberances are created evenly along two sides of the grooves. On the other hand, if scratching is in the forward direction, the "V"-shaped scratching face is composed of two inclination angles, i.e., one is –30° and the other is 30°. As a result, the protuberances are squeezed evenly onto the two sides of the groove scratched. Since the 30° or –30° inclination angle provides much more favorable stress states to squeeze the materials onto the two sides as compared with that of the 90° inclination angle, the protuberances created in forward scratching is more or larger than that of the backward direction but is less or smaller than the larger one of the upward or downward direction. This can clearly observed in Fig. 1.9b by comparing the protuberances generated by the different directions.

Consequently, to minimize the protuberances or ridges, the backward scratching should be used in creating high-precision and high quality nanostructures. The groove patterns shown in Fig. 1.7 were scratched based on the backward scratching and very little protuberances were generated.

In addition to the changes on the formation of protuberance, Yan et al. [94] also observed that both the groove depth and roughness were affected by varying the scratch direction. However, only limited data were reported and no conclusive results were presented. To have a better control of the scratched dimensions and surface quality, the effects of the scratch direction on the formation of protuberances and the roughness of scratched surfaces in more complicated scratching conditions should be understood. The related research to quantify these effects should be encouraged.

1.5.2 Effects of Scratch Forces on Scratch Geometry

Tseng et al. [5, 34, 71] have studied the characteristic of scratching Si, Ni, and $Ni_{80}Fe_{20}$ thin films by changing the scratch force (F_n) from 1 to 9 μN and they found that the depth, d, and the width, w_f, of the grooves increase with F_n following a logarithmic relationship:

$$d(F_n) = \alpha_1 \mathrm{Ln}(F_n/F_{t1}) \tag{2a}$$

and

$$w_f(F_n) = \alpha_2 \mathrm{Ln}(F_n/F_{t2}) \tag{2b}$$

where α_1 and α_2 are the scratch parameters called the scratch penetration depth and penetration width, respectively; F_{t1} and F_{t2} are the threshold forces based on the depth and width data, respectively. The parameter of α is a measure of the machining efficiency of the tip to a specific material, while the threshold force (F_t) can be considered as the minimum applied normal force to machine or scratch any measurable grooves or no scratch can be observed at $F_n = F_t$. The groove depth, d,

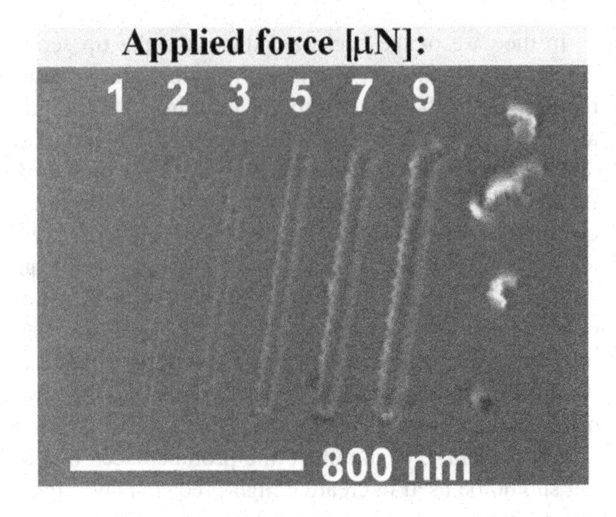

Fig. 1.11 SEM image of six grooves scratched on $Ni_{80}Fe_{20}$ surface at normal force (F_n) equal to 1, 2, 3, 5, 7, and 9 μN

is defined as the distance between the original surface and the lowest point in the groove, while w_f is the FWHM width. By comparing the measurement data with the correlated values for several types of metals, Tseng et al. [5] concluded that the logarithmic relationship between the scratched geometry and normal force is accurate and reliable.

Figure 1.11 shows the SEM image of six grooves scratched in the backward direction on the NiFe thin film with a normal force (F_n) varying from 1 to 9 μN (from left to right) at a speed of 100 nm/s. The scratching was performed in all four scratch directions, similar to those indicated in Fig. 1.9. As mentioned earlier, the behavior of scratching in the upward and downward are physically identical. It has been found that the differences of the scratch data between the upward (or downward) case and the forward case were relatively small and indistinguishable, i.e, the differences were smaller than their associated standard deviation (SD). Consequently only the upward and backward cases are discussed here. The measurement data and correlation for both upward and backward cases are plotted in Fig. 1.12 and shows that the measurement data of the groove depth, d, and width, w_f, fit the logarithmic relationship of Eq. (2) very well. Both d and w_f were based on ten measurements at ten randomly-selected locations along the groove. Only their means were plotted. Their SDs were calculated and are less than 4% of their respective means. The associated coefficients of determination (R^2) for the depth and width data are 0.971 and 0.975, respectively, for the upward scratching, and are 0.983 and 0.993, respectively, for the backward scratching. Based on the R^2 values obtained, it can be expected that the mean deviation between the correlation and the measurement data is less that 2% for the backward case and is less than 3% for the upward case.

In the data fitting depicted in Fig. 1.12, if d and w_f are in [nm], the values of α_1 and α_2 are 5.250 nm and 17.43 nm, respectively, for the upward scratching and are 4.964 nm and 14.50 nm, respectively, for the backward scratching. The depth scratched in the upward direction, is approximately 8% deeper than that of

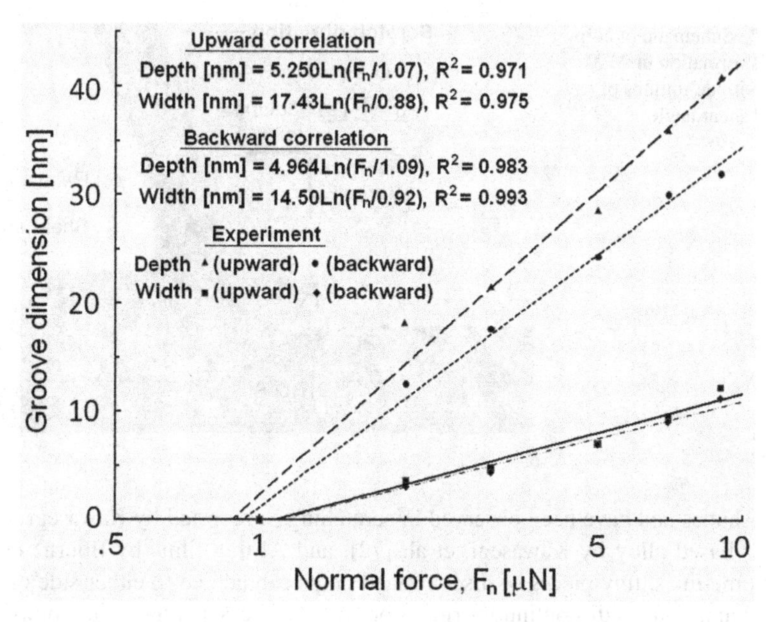

Fig. 1.12 Experimental data correlation of depth (d) and width (w_f) of groove scratched on NiFe surface with applied normal force (F_n) at two different scratching directions

the backward direction. The threshold forces F_{t1} and F_{t2}, can be found to be 1.07 and 0.88 μN, respectively, for the upward scratching and 1.09 and 0.92 μN, respectively, for the backward scratching. Since all measurement data indicated that the scratched depth equals zero at $F_n = 1.0$ μN, i.e., the threshold forces (F_t) for both cases should be higher than 1.0 μN. Based on the depth measurements for both cases, the correlations predict that $F_{t1} = 1.07$ or 1.09 μN, which are consistent with the experiment results. However, based on the width measurements, the correlations predict that $F_{t2} = 0.88$ or 0.92 μN, which are approximately 10% underestimated if $F_t = 1.0$ μN. This 10% underestimation can be caused by the inaccuracy in measuring the groove width. By comparing and judging the measurement data with the correlated values for several types of metals, Tseng et al. [5, 34, 37, 71] selected F_{t1} as the scratch threshold force (F_t) to avoid its double definition and to achieve better correlation accuracy and reliability. Thus, it was recommended the correlation based on only the depth data to be adopted.

As shown in Fig. 1.11, the chip or debris can be observed at the load being higher than, say 5 μN. Basically, the scratched chips are created ahead of the tip by shearing the material along the shear plane as shown in Fig. 1.11. The applied normal force should be large enough to produce a shear stress along the shear plane to overcome the shear strength to sever or cleave the material to form the chips. Both continuous and discontinuous chips can be developed. The scratched chip climbs up along the curved face of the semispherical portion of the tip and eventually moves up to the face of the usually flat or triangular pyramid section of the tip. Very often, the scratched chips can be in several or tens micrometers long in either spiral or

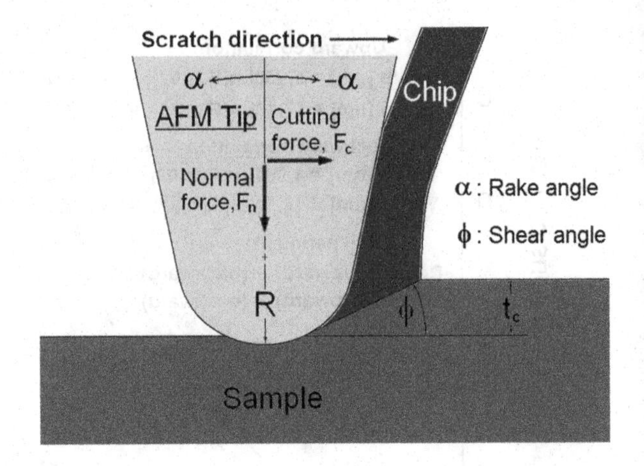

Fig. 1.13 Schematic of chip or debris formation in AFM scratch with definitions of rake and shear angle

helical shapes and have been observed in scratching Si reported by Kawasegi et al. [13], Zr based alloy by Kawasegi, et al. [72], and Al thin films by Bourne et al. [47]. Sometimes, tiny pieces of discontinued chips can adhere to either side of the groove bank to form discontinuous ridges or protuberances symmetrically or asymmetrically. This situation is similar to an angle snow-plough blade, which throws the snow sideways.

Furthermore, in Fig. 1.11, the height of the protuberance increases with the applied load, F_n, To have a better view on the profiles of these protuberances or grooves, the SEM image in Fig. 1.11 can also be compared with the cross-sectional profiles measured by AFM for $F_n = 9\ \mu N$ at four different scratch directions displayed in Fig. 1.9. Note that the characteristics of AFM scratching are quite different from the conventional machining process. Not only the scratching is in a much smaller scale but also the cutting angle of the tip is different. For example, as shown in Fig. 1.13, the rake angle using a typical AFM tip is highly negative while it is normally positive in a conventional machining indicated in Fig. 1.10b. Mechanically, it is very inefficient to have a negative rake angle in machining because higher machining force and energy is needed as comparing with that of machining with a positive rake angle. By adjusting the probe mounting angle, Bourne et al. [47] studied the effect of the rake angle (still in the negative range) on the scratched geometry and found that the scratched groove depth increases with the normal load and the magnitude of this increase is affected by changing the rake angle.

1.5.3 Scratch Ratio and Scratchability

From Eq. (2), the parameter, α, can be defined as d/Ln (F_n/F_{t1}), which is a ratio of the groove depth to the logarithm of the normalized applied force. It can be considered as a measure of the scratch efficiency of the tip to a specific material. It can be expected that the higher the α value the easier the material to be scratched. The threshold force (F_t) is also used to gauge the ease of a material that can be scratched. However, its effect on scratching is opposite to α_i, because the higher the F_{ti}, the harder the material to be scratched. Since the scratch behavior can be reliably represented by the logarithmic equation of Eq. (2a), which is governed by the two parameters of α and F_t, a ratio of these two parameters, called the scratch ratio (SR) or α/F_t (or, more precisely, α_1/F_{t1}), should be able to appropriately characterize the scratch behavior. The SR should also be considered as a material property that dictates the ease or difficulty with which the material can be scratched using an AFM tip. The scratch ratio in the AFM scratching process is similar to those material properties used to quantify the machinability in conventional machining process [95].

As mentioned earlier, Kawasegi et al. [13] developed a diamond-tip probe with a spring constant of 239 N/m to scratch grooves on an undoped Si(110) wafer to a depth up to 97 nm using a normal load of 556 μN. Their scratching results are plotted in Fig. 1.14, where the applied load varies from 79.5 to 556.2 μN and the corresponding depth of the grooves changes from 20.4 to 96.7 nm. The data can also be well correlated with the logarithmic form of Eq. (2a) and the parameters of α_1 and F_{t1} can be found to be 38.65 nm and 52.65 μN, respectively, with the associated R^2 equal to 0.97. The corresponding scratch ratio can be calculated to be 0.734 nm/μN, which is only 3% smaller than the value obtained from the results, i.e. $\alpha_1/F_{t1} = 0.756$ nm/μN, reported by Tseng [5], in which a diamond coated tip that had an apex radius of 120-nm attached on a Si cantilever with a spring constant of 42 N/m was used to perform the single-scratch experiment. For the sake of comparison, the experiment results obtained by Tseng [5] are also plotted in Fig. 1.14. The probe used by Tseng is similar to that shown in Fig. 1.8 and is much

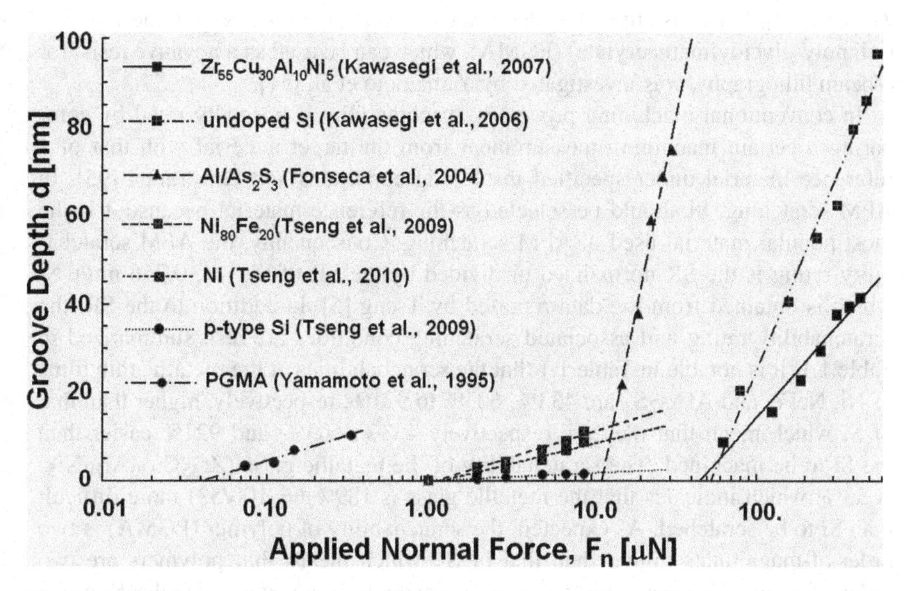

Fig. 1.14 Correlation of groove depth (d) with applied normal force (F_n) for various materials (Reproduced with permission from [5] by Elsevier B.V.)

smaller than those used by Kawasegi et al. [13]. More specifically, the tip material and shape, the cantilever stiffness and resonant frequency, the AFM instrument, the applied force, and scratch speed of the two Si scratching cases compared are extremely different. For example, the probe spring constant, the maximum force applied, and the depth of the groove scratched by Kawasegi et al. are 5.7, 61.8, and 60.1 times larger than those adopted by the experiments presented by Tseng et al. [5]. Since these two cases cover a wide range of processing conditions used for AFM scratching, the agreement reached by these two cases should be applicable for a wide range of AFM scratching conditions and therefore, the scratch ratio should be considered as the nanoscale scratchability property to gauge the ease of the material to be scratched by AFM.

Normally, α_1 is the slope of the logarithmic correlation and represents the unit amount of material to be removed by an AFM tip, while the threshold force, F_t, is dictated by the radius of the tip and the loaded force. The scratch ratio is α divided by F_t and therefore can make the ratio or the scratchability less sensitive to the change of the scratch tool and the applied load. Furthermore, since the effects of the scratch speed on the scratched geometry have been shown to be negligible, the ratio of α/F_t instead of a rate quantity, such as the volume removal rate, is selected for representing the scratchability. The speed effect will be presented in Section 1.5.8.

To obtain the scratch ratio (SR) for other materials, five sets of scratch data for five different materials, including three metallic thin films, one metallic glass, and one polymer, are also logarithmically correlated and plotted in Fig. 1.14. The data for the three metallic thin films were reported by Tseng et al. [34] for scratching a nickel thin film, by Filho et al. [68] dealing an Al/As_2S_3 bilayer thin film, and by Tseng et al. [71] for machining a permalloy. The metallic glass, $Zr_{55}Cu_{30}Al_{10}Ni_5$, was studied by Kawasegi et al. [72], while the polymeric material, poly(glycidylmethacrylate) (PGMA), which can be used as a negative resist for e-beam lithography, was investigated by Yamamoto et al. [51].

In conventional machining processes, machinability is normally rated by comparing a certain machining measurement from the target material with that of a reference material under specified machining conditions and constraints [95]. In AFM scratching, Si should be selected as the reference material because it is the most popular material used in AFM scratching. Consequently, the AFM scratchability rating is the SR normalized or divided by the SR of Si, i.e., 0.756 nm/μN, which is obtained from the data reported by Tseng [5]. In addition to the SR, the scratchability rating and associated scratching conditions are also summarized in Table 1.1. It is notable in Table 1.1 that the scratchabilities of the metallic thin films of Ni, NeFe, and $AlAs_2S_3$ are 433%, 603% to 921%, respectively, higher than that of Si which mean that they are respectively 433%, 603%, and 921% easier than the Si to be machined. The scratchability of the metallic glass ($Zr_{55}Cu_{30}Al_{10}Ni_5$) is 53%, which indicates that the metallic glass is 189% (= 100/53) more difficult than Si to be scratched. As expected, the scratchability of polymer (PGMA) is two order-of-magnitudes higher than that of Si, which means that polymers are two order-of-magnitudes easier than Si to be scratched. As a result, a regular Si probe

Table 1.1 Comparison of scratch properties (*with respect to Si properties)

Parameter	p-type Si	Undopped Si	$Ni_{80}Fe_{20}$ thin film	Ni thin film	Al/As_2S_3 thin film	$Zr_{55}Cu_{30}Al_{10}Ni_5$	PGMA
Spring const. [N/m]	42.0	239	42.0	42.0	165	239	0.68
Scratch speed [μm/s]	0.10	30	0.10	0.10	5.0	30	10
Scratch coeff. α_1 [nm]	0.765	38.65	4.96	3.94	80.51	18.09	4.538
Threshold force, F_{t1} [μN]	1.01	52.66	1.09	1.20	11.57	45.35	0.037
Coefficient R^2	0.96	0.97	0.98	0.95	0.98	0.96	0.96
Scratch ratio, α_1/F_{t1} [nm/μN]	0.756	0.734	4.555	3.272	6.962	0.399	123.44
Scratch ratio rating [%]*	100	97	603	433	921	53	16,328
Hardness, H [GPa]	10	10	8.0	7.35	1.00	4.90	–
Rating based on hardness*	1.00	1.00	1.25	1.36	10.0	2.04	–
Data source	[5]	[13]	[71]	[34]	[68]	[72]	[51]

with a spring constant of 0.68 N/m, which is typically applied for topographical imaging of surfaces, could be used for scratching PGMA [51].

1.5.4 Scratch Cycle Number and Multiple Scratches

Repeated scratches were also conducted in scratching experiments to study the effects of the number of the scratch or scan cycle on material removing behavior. It is expected that multiple or repeating scratch cycles along the same scratch path can enlarge the groove size. Tseng et al., [34] found that the depth, d, and the width, w_f, of the grooves increase with the number of scratch cycles (N_o) following a power-law relationship:

$$d(N_o) = M_1(N_o)^{n_1} \tag{3a}$$

and

$$w_f(N_o) = M_2(N_o)^{n_2} \tag{3b}$$

where M_i and n_i are the multiple scratch coefficient and multiple scratch exponent, respectively, and both n and N_o are dimensionless. Here, the coefficient M has the same dimension of the depth or width and is equal to $\alpha Ln(F_n/F_t)$, where α and F_t are the scratch parameter and threshold force, respectively, defined in Eq. (2). The d and w_f data depicted in Fig. 1.15 $Ni_{80}Fe_{20}$ thin film were obtained by Tseng et al. [34] where the correlation values of M_1, M_2, n_1, and n_2 can be found to be

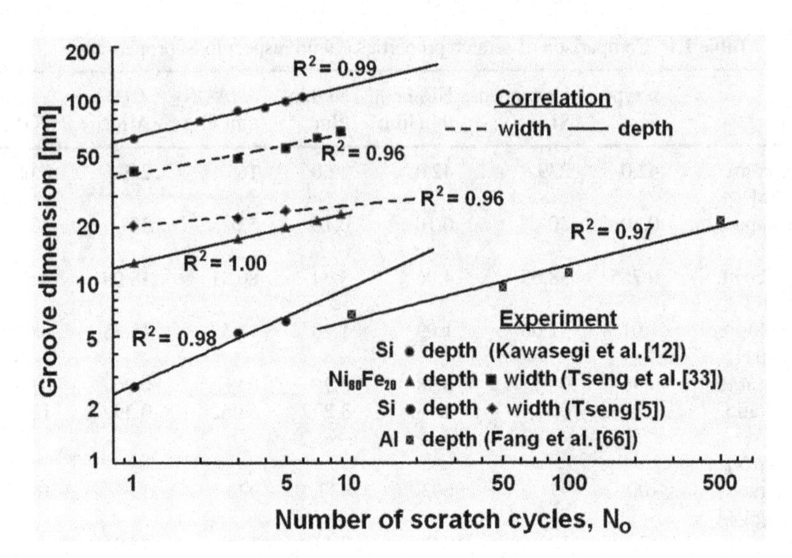

Fig. 1.15 Correlation of dimensions of grooves with number of scratching cycle (N_o) for various materials (Reproduced with permission from [5] by Elsevier B.V.)

12.711 nm, 39.61 nm, 0.279, and 0.227, respectively. The associated coefficient of determination (R^2) is 1.00 for d and 0.96 for w_f, indicating that the power-law correlation fits the depth data perfectly and width measurements reasonably well. In addition, the power-law correlation can also satisfy the physical requirement, i.e. predicting no scratching at $N_o = 0$.

The experimental data for scratching a p-type Si(100) substrate along the forward direction were obtained by Tseng [5] using a normal load of 9 μN for the scratch cycle equal to 1, 3, and 5 with the correlated values of M_1, M_2, n_1, and n_2 can be found to be 2.694 nm, 20.34 nm, 0.538, and 0.112, respectively, while the coefficient R^2 for the depth and width correlations are 0.98 and 0.96, respectively. This also indicates that the data fits the power-law correlation very well. The mean values of the depth, d, and width, w_f, based on ten measurements are also plotted in Fig. 1.15. The associated standard deviation, which is not shown, is roughly twice as large as those based on the single cycle cases shown in Fig. 1.14. Since no feedback correction is applied to the movement in the horizontal (x and y) directions, the larger SD should be caused by the hysteresis-creep of the piezo-scanner used in AFM [96]. Furthermore, the depth data under a normal load of 318 μN reported by Kawasegi et al. [13] for scratching an undoped Si(100) wafer at cycle number equal to 1–6 are depicted in Fig. 1.15. As shown, these multiple scratching data are correlated with the power-law very well, where M_1, n_1, and R^2 are equal to 63.06 nm, 0.285, and 0.99, respectively. Kawasegi et al. used their high-stiffness diamond-tip cantilever to scratch the Si surface with a normal load of 318 μN at a speed of 30 μm/s.

The results reported by Fang, Weng, and Chang [69] for scribing a 500-nm thick Al thin film using a diamond tip has also been found to be correlated well with the power-law. The depth correlation for the cycle number varying from 10 to 500 under a tip force of 8 μN is depicted in Fig. 1.15 and the corresponding correlation factors M_l, N_l and R^2 are 3.039, 0.303, and 0.97, respectively. On the other hand, the results reported by Tsuchitani, Kaneko, and Hiron [86] for scratching carbon thin films under different environments and by Ogino, Nishimura, and Shirakashi [74] for scribing a Si substrate indicate that the groove depths increase somewhat linearly with N_o. Both of the experiments were conducted using diamond-coated tips. Although Tsuchitani et al. also found that in some cases, the wear depths could correlate the power law relationship better than the linear one; the reason to cause this change is still unclear. To better understand the scratch cycle effects on the scratched sizes, more efforts are needed to resolve this inconsistence. Note that, the linear correlation cannot satisfy the physical requirement, i.e. predicting no scratching at $N_o = 0$ whereas the power-law not only meets the physical requirement but also fits the data very well in a wide range of scratching conditions, as shown in Fig. 1.15. The power law is reliable to use for correlating the scratch cycle to the scratched geometry. For the sake of convenience for comparison, the correlated M_l and n_l values of the four cases considered are summarized in Table 1.2.

In some aspects, multiple scratching, is similar to the wear test in evaluating specific wear coefficient or wear behavior [62, 84, 97]. Gnecco, Bennewitz, and Meyer [62] investigated the abrasive wear of KBr(001) by a Si AFM tip in ultra-high vacuum to avoid the environmental humidity effect. Multiple scratches with scratch cycles up to 5,120 were used to study the formation and morphology of scratched debris. An AFM equipped with a soft cantilever having a spring constant of 0.87 N/m and a tip apex radius less than 15 nm was used to ensure that atomic scale wear mechanism could be discovered. Multiple scratches with scratch cycles up to 1,024 and normal forces up to 100 μN were applied by Prioli, Chhowalla, and Freire [84] to evaluate the wear and friction properties of nanostructured and amorphous carbon thin films. A standard AFM with a diamond-tipped cantilever was used. Kato, Sakairi, and Takahashi [97] had applied an AFM with a Si tip to scratch Al surfaces to understand the changes of the formation of grooves scratched in different solutions. Multiple scratches at 1,600 cycles under a load of up to 20 μN were performed.

Table 1.2 Comparison of parameters related to wear coefficient correlation

Parameter	p-type Si	Undopped Si	$Ni_{80}Fe_{20}$ thin film	Ni thin film	Al thin film
Multiple scratch coefficient, M_l	2.694	63.06	12.71	–	3.039
Multiple scratch exponent, n_l	0.538	0.285	0.279	–	0.303
R^2 for correlation of M_l and n_l	0.98	0.99	1.00	–	0.97
Parameter β_w defined in Eq. (10)	0.014	0.115	0.117	0.107	–
Threshold force in Eq. (10), F_{tw} [μN]	0.760	29.31	1.014	1.013	–
R^2 for correlation of β_w and F_{tw}	0.81	0.96	0.99	0.99	–
Data source	[5]	[13]	[71]	[34]	[69]

As indicated in the above cited references in the assessment of wear behavior, the number of scratch cycles is from hundreds to thousands or even millions. The wear properties are normally evaluated based on a significant number of scratches or sliding and the effects from each individual scratch are not specifically distinguishable. On the other hand, the AFM scratching discussed in this article deals with fabrication that makes high-quality and high-precision nanostructures. The key to achieve this goal is to operate an AFM tip that has proper sizes and material properties with appropriate normal loads, scratch cycles, and others to remove the required amount of material from a pre-defined location in a controllable manner. As a result, in AFM scratching, each scratch should be conducted under a controllable condition and should be distinguished from other scratches so that the nanoscale precision can be achieved and desirable surface quality can be maintained. Only a limited number of scratches should be performed in AFM scratching experiments, otherwise the scratch experiment becomes too time consuming and cumbersome to be performed.

1.5.5 Threshold Forces and Contact Stresses

The threshold force (F_t) discussed in Sections 1.5.2 and 1.5.3 is re-examined in this subsection by analyzing its corresponding stresses and predicting its theoretical value using fundamental material properties. If an isotropic and homogeneous substrate is scratched at the threshold force, no scratch (d = 0) occurs, which implies no plastic deformation or permanent indentation on the substrate. The diamond coated tip and substrate should be under an elastic contact. The contact mechanics states that the contact area, A_c, under a normal contact force F_c, can be expressed as:

$$A_c = C_c\left(\frac{3F_cR}{4E_c}\right)^{2/3} \tag{4}$$

where R is the radius of tip curvature, and E_c is the contact elastic modulus, which is defined as

$$\frac{1}{E_c} = \frac{1-\upsilon_t^2}{E_t} + \frac{1-\upsilon_s^2}{E_s} \tag{5}$$

where E is Young's modulus; υ is Poisson's ratio; the subscripts t and s denote the material property of the tip and substrate, respectively. Here, C_c is a contact constant and equal to π based on the Hertz contact theory [98]. However, in nanoscale contact, such as AFM scratching, the long- and short- range adhesive and surface forces can play an important role in predicting the contact or interfacial force between the two contacted materials. Since the Hertz theory assumes no adhesive and friction forces acting between the two materials, the contact area predicted from the theory is highly underestimated, especially for the case of single scratching. Based on a study done by Carpick and Salmeron [99], the contact area by using the JKR (Johnson, Kendall, & Robert) theory should be more than double of the value by the Hertz theory for the range of the contact load considered. Consequently, C_c should be fine-tuned to be 2π.

Since the maximum contact stress σ_c can be found as $1.5F_c/A_c$, the contact force, F_c can be expressed by

$$F_c = \frac{2\sigma_c A_c}{3} = \frac{4\pi^3 R^2 \sigma_c^3}{3E_c^2} \tag{6}$$

If σ_c reaches the yield strength of the substrate, Y, the contact force, F_c, becomes the calculated threshold force, F_{tc}. The above equation can then be rewritten as

$$F_{tc} = F_c = \frac{4\pi^3 R^2 Y^3}{3E_c^2} \approx \frac{1.388R^2 H^3}{E_c^2} \tag{7}$$

where H is the hardness of the substrate and is assumed to be 3.1Y, based on a plasticity analysis [100].

By providing the properties of E_t, E_s, υ_t, υ_s and H (or Y) as well as the tip radius, R, the threshold force, F_t can be evaluated. If a diamond coated tip is used to scratch a Ni thin film, E_t, E_s, υ_t, υ_s and H can be found as 1,143 GPa, 190 GPa, 0.07, 0.24, and 7.35 GPa, respectively. The values of the material properties for the diamond tip were reported by Klein [101], while the properties for Ni are obtained from Ref. [102]. Using the above material properties with the tip radius equal to 120 nm and based on Eq. (7), F_{tc} can be calculated to be 285 nN, which is approximately three times less than the threshold force (F_t) of 1.0 μN obtained from the scratching experiment conducted by Tseng et al. [34]. As reported by Tseng et al. [34], if the calculated threshold force, F_{tc} is based on that the contact stress, F_c, reaches its hardness, H, instead of the yield stress, Y, the F_{tc} can be 372% higher than that predicted by Eq. (7). Thus, F_{tc} agrees very well with the threshold force correlated, i.e., $F_{tc} \approx F_t$ ($\approx$ 1.0 μN).

A wide range of materials is selected to further evaluate the usefulness of Eq. (7) for predicting the threshold force (F_t), which is one of the major processing parameters needed to be known prior to performing the scratching. Table 1.3 summarizes the results of the calculated threshold force, F_{tc}, in which the materials selected are all scratched by a diamond tip, where the material properties including E, υ, and H used in the calculation are all reported in the table. The material properties for the thin films of Al, Au, Cu, NiFe, Si, steel, Ti, W, metal glass ($Zr_{55}Cu_{30}Al_{10}Ni_5$), glass, quartz ($SiO_2$), and PGMA (Micposit S1813 by Shipley) are reported by Zhou & Yao [103], Mulloni et al. [104], Volinsky et al. [105], Wu et al. [106], Madou [107], Zhou & Yao [103], Kuruvilla et al. [108], Madou [107], Kawasegi et al. [72], Yan et al. [80], Golovin et al. [109], and Calabri et al. [110], respectively. Since the hardness and yield strength of PGMA cannot be found, the values reported for a similar material, Micposit S1813 by made Shipley were used for the calculated

Table 1.3 Calculation of contact threshold force, F_{tc}, for various materials

Materials	Young's mod.[GPa]	Poisson's ratio, υ	Yield streng. Y [MPa]	Hardness H [GPa]	Radius, R [nm]	Cal. Thresh. force, F_{tc}[μN]	Mea.Thresh. Force, F_t [μN]	References
Diamond	1,143	0.07	53,000	164	120	–	–	Klein [101]
Al	69	0.33	170	1.0	100	0.000570	–	Zhou et al. [103]
Au	80	0.42	900	2.4	100	0.023922	–	Mulloni et al. [104]
Cu	120	0.35	400	1.3	100	0.002042	–	Volinsky et al. [105]
Ni	190	0.24	873	7.35	120	0.284975	1.20	Tseng et al. [34]
NiFe	215	0.26	1,000	8.0	120	0.541982	1.09	Wu et al. [106]
p-type Si	169	0.35	7,000	10	120	0.734707	1.01	Zhou et al. [103]
Undoped Si	169	0.35	7,000	10	918	26.73471	52.7	Zhou et al. [103]
Steel	207	0.27	200	1.5	100	0.001341	–	Madou [107]
Ti	120	0.36	900	3.0	100	0.024734	–	Kuruvilla et al. [108]
W	410	0.17	400	9.0	100	0.106164	–	Madou [107]
ZrCuAlNi	90	0.35	–	4.9	918	15.52249	45.3	Kawasegi et al. [13]
Glass	65	0.17	–	9.0	100	2.529379	–	Yan et al. [80]
SiO_2	76	0.19	1,500	9.5	100	2.186065	–	Golovin et al. [109]
PGMA	3.0	0.37	41	0.13	170	0.017961	0.037	Calabri et al. [110]

[110]. The elastic properties of NiFe thin films reported by Zhou et al. [111] were also used in the calculation. Note that the Young's module and hardness for thin film materials normally are significantly higher than the values in the bulk form and increase with the decrease of the thickness of the films.

The correlated values of the threshold force (F_t) based on the AFM scratching experiments discussed earlier for six different materials are also reported in Table 1.3. Some of the tip radii were not reported in the cited references and the values of the pitch and depth of the grooves scratched were used to estimate the tip radius. As indicated in Table 1.3, the calculated values (F_{tc}) are all lower than the correlated or extrapolated values (F_t) and the differences based on F_t are from 20 to 80%. Such big differences between F_t and F_{tc} may be due to that the surface and adhesive forces between the tip and substrate are higher than those considered in Eq. (7) and that the fully elastic assumption may be highly underestimate the contact force in the nanoscale domain.

Normally, to have an effective scratch, the applied tip normal force, F_n, should be at least double of the threshold force, F_t. Consequently, to perform an AFM scratching, it will be a good recommendation to load a tip normal force, F_n, one order magnitude larger than the calculated value (F_{tc}). For the sake of convenience, Fig. 1.16 shows the calculated threshold forces (F_{tc}) versus the normalized elastic modulus (E_s/E_t) at a range of the tip radius, R, and of the hardness, H, where a diamond tip is used and the Poisson's ratios are assumed to be constant equal to 0.3 for all the scratched samples considered. The hardnesses of the substrates considered are varying from 1 to 10 GPa. For samples with hardness less than 1 GPa, a Si or Si_3N_4 tip is appropriate to be used in scratching.

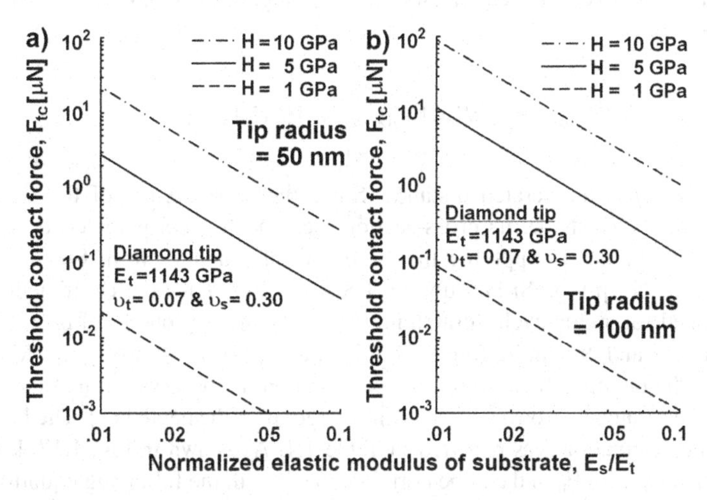

Fig. 1.16 Calculated threshold contact forces (F_{tc}) versus normalized substrate Young's modulus (E_s/E_t) using a diamond tip scratching substrates having different hardness (H): (**a**) tip radius R = 50 nm, (**b**) tip radius R = 100 nm

1.5.6 Wear Coefficient and AFM Scratchability

Wear can be considered as damage to a solid surface as a result of relative motion between it and another surface. The damage usually results in the progressive loss of material. No matter in using single or multiple scratching, one of the most popular parameters used to quantify the wear resistance or wear rate is the wear coefficient defined in Archard's wear equation, or simply Archard's law [112]. Srivastava, Grips, and Rajam [113] applied Archard's wear coefficient to quantify the wear resistances of seven different coating materials using a pin-on-disc wear tester with the goal to develop better coating for the friction parts used in the automobile industry. Chung et al. [114] used an AFM equipped with flattened Si and diamond tips to perform the sliding test to study the wear behavior of ZnO nanowires under applied tip loads varying from 50 to 150 nN, while Archard's law was applied to quantify the associated wear coefficients. Recently, Ogino et al. [74] employed a diamond-coated AFM tip to groove Si(100) substrates and found that the corresponding wear coefficient based on Archard's equation is highly dependent on the load applied. In this section, the wear coefficient will be evaluated for different AFM scratching conditions.

In Archard's wear equation, the loss of material by wear can be expressed in terms of volume [112]:

$$V = kLF_n/H \tag{8}$$

where V is the volume loss caused by wear or scratch, k is the wear coefficient, L is the sliding distance. The wear coefficient k is a dimensionless parameter: the higher the k value, the lower the wear resistance. Rearranging Eq. (8), k can be expressed as:

$$k = VH/(LF_n) = dw_fH/(N_oF_n) \tag{9}$$

where L becomes the scratch distance. Since the cross-section of the scratched groove is in a "V" shape, its cross-section area can be approximated by $d \times w_f$ and $V = dw_fL/N_o$. By applying the NiFe data of d and w_f from Fig. 1.12, and the d data of Ni, p-type Si and undoped Si from Fig. 1.14, the coefficient k can be calculated for single cycle scratching ($N_o = 1$) and is plotted in Fig. 1.17. The w_f data for Si and Ni can be found from [71] and [34], respectively, while the w_f values for the undoped Si were calculated based on the observation that the mouth width of the scratched grooves is four times larger than the depth [13]. The hardness used in the calculation was reported in Table 1.1. As shown in Fig. 1.17, k grows logarithmically with F_n and can be correlated well with the following equation:

$$k = \beta_w \mathrm{Ln}\,(F_n/F_{tw}) \tag{10}$$

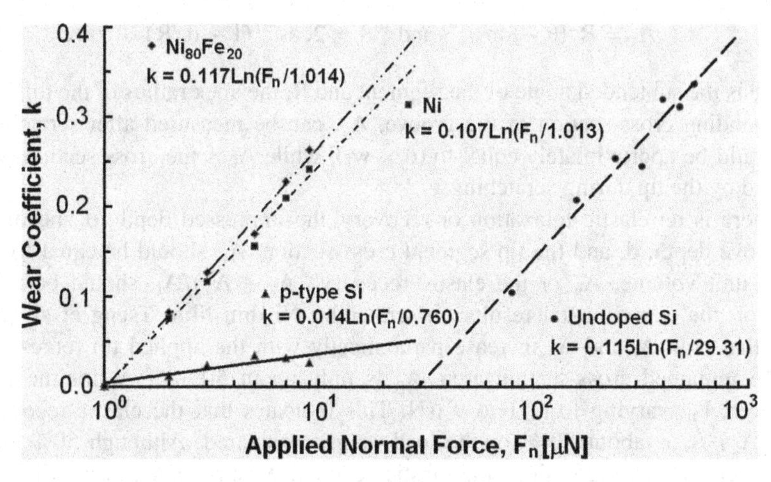

Fig. 1.17 Correlation of wear coefficient (k) with applied normal force (F_n) for various materials (Reproduced with permission from [5] by Elsevier B.V.)

where β_w is a dimensionless parameter for measuring the volumetric or material loss per unit scratching and F_{tw} is the threshold force based on wear coefficient correlation. The parameter β_w in Eq. (10) plays a similar role to α_1 in Eq. (2a) and is equal to 0.117, 0.107, 0.014, and 0.115 for scratching NiFe, Ni, p-type Si, and undoped Si, respectively, with F_{tw} equal to 1.014, 1.013, 0.760, and 29.31 μN. The F_{tw} values are somewhat smaller than the threshold values (F_t), as reported in Table 1.1, which are based on the groove depth correlation. The associated R^2 coefficients are 0.99, 0.99, 0.81, and 0.95, respectively, which implies that the accuracy of the correlation for Si is worse than that based on the groove depth correlation. As shown in Fig. 1.17, the wear coefficients of the two cases of Si scratching are quite different. This is especially the case because the β_f of the undoped Si is one order of magnitude larger than that of the p-type Si, whereas the ratio of β_w/F_{tw} of the undoped Si is almost five times smaller than the p-type Si value. These contradictory differences in the correlation parameters for the same type of materials may imply that the wear coefficient, k, is not suitable to be used to quantify the scratch property or manufacturability for AFM scratching. For the sake of comparison, the correlated β_w and F_{tw} values of the four cases considered are summarized in Table 1.2.

1.5.7 Recovery After Scratch

In Eq. (9), the value of the unit volume V/L is based on the scratched groove geometry, i.e., equal to $d \times w_f$. Since the unit volume is a measure of the amount of the material removed by scratching, it may be estimated by directly measuring the tip cross section impressed or indented into the sample substrate. If d_i is the impressed depth of the semispherical tip into the sample, the cross-section of the tip end segment (A_t) can be found as:

$$A_t = R^2(\theta - \sin\theta) \quad \text{and} \quad \theta = 2\cos^{-1}(1 - d_i/R) \tag{11}$$

where θ is the subtended angle of the segment and R, the apex radius of the tip. The corresponding cross-section of the groove, A_g, can be measured after scratching and should be approximately equal to ($d \times w_f$), while A_t is the cross-section area removed by the tip during scratching.

If there is no elastic relaxation or recovery, the impressed depth, d_i, becomes the groove depth, d, and the tip segment cross-section, A_t, should be equal to the groove unit volume, A_g or the elastic recovery, $(A_t - A_g)/A_t$, should be zero. Based on the experimental results on scratching Ni thin film, Tseng et al. [34] found that both A_t and A_s increase parabolically with the applied tip forces, F_n, and the remained cross-section area, A_g is only about 50% of A_t for the normal force, F_n, varying from 1 to 9 μN. This indicates that the elastic recovery, $(A_t - A_g)/A_t$ is about 50% for the load range considered. Although 50% elastic recovery seems high, since the elastic rebound or springback is usually less than 10% in conventional or macroscale machining or forming processes [115, 116], the results obtained by Tseng et al. in scratching Ni thin films are consistent with those reported by Fang, Chang, and Weng [67] and Xiang et al. [117]. Chang et al. [67] studied the nanoindentation behavior of Au and Pt using a diamond tip at a loading rate of 500 μN/s and found that the elastic recovery was approximately 18% for Pt and 24% for Au, where the elastic recovery was defined as the difference of the indentation depths measured during and after scratching. Xiang et al. [117] used a conventional scratch test to evaluate the elastic recovery of a range of semi-crystalline polymers, including nylon, polypropylene, low-density polyethylene (LDPE), and high-density polyethylene. They discovered that a relatively high visco-elastic recovery varying from 70% for LDPE to 85% for Nylon for the load range up to 30 N based on measuring the scratch depth during and after scratching [117]. The material parameters that influence the recovery behavior appear to be complex and should be subjected to further studies.

1.5.8 Effects of Scratch Speeds

Experiments have been conducted to study the effect of changing the scratching speed on the shapes of scratched grooves. Tseng et al. [71] performed groove scratching on a p-type Si(100) substrate with the scratching speed at 10^1, 10^2, 10^3, and 10^4 nm/s and the tip normal force, F_n, at 9 μN and reported that the groove depths and widths remain almost unchanged as the speed is varied from 10^1 to 10^4 nm/s with only slight perturbation at each speed. No deterioration of the diamond tip used was observed after scratching a total length of 150 μm at 10,000 nm/s. The associated standard deviations of the depth and width for the ten sets of the profile measurements are 4% and 3% of their respective means. Since the maximum difference among the mean depth or mean width is also less than 4%, which is not larger than the standard deviation of the data, the effects of changing speed can be considered to be negligible for the speed range considered. Similar conclusions of the effect of the scratching speed on the machined geometry were also made by Ogino, Nishimura, and Shirakashi [74] and Kawasegi et al. [13] for scratching Si using a similar diamond-coated tip and by Fang, Weng, and Chang [69] for scribing an Al thin film. Furthermore, based on limited quantitative data, Tsuchitani, Kaneko, and Hirono [86] observed that the groove dimensions can fluctuate unpredictably by ±5% of their mean values for a diamond tip scratching of an amorphous carbon film at different environments with the speed changing from 1 to 100 μm/s. It is suggested that scratching at higher speeds should be studied further so that high-speed scratching can be achieved to increase productivity.

1.6 Scratching with Dual or Combined Sources

As indicated in the preceding sections, the cantilevered tip has been mainly loaded with mechanical force or energy to perform AFM scratching. However, the tip–sample interaction can also be imposed or influenced by other energy sources, including thermal, electric, and magnetic energies or by the chemicals or contents of the environments, which can create different effects on the nanostructures created by AFM scratching. In this section, the effects of the mechanical energy combined with other types of energies on AFM scratching will be examined. Frequently, by combining additional physical or chemical source, the major limitations on AFM scratching including the shallowness of scratch depth and tool wear, could be alleviated.

1.6.1 Scratching with Thermal Energy

An AFM tip can be heated to a relatively high temperature by loading thermal energy. Heated tips have been used for scanning thermal microscopy [118] for measuring thermal-related information, such as temperature profiles and thermal conductivity. Recently, a heated tip has been used not only to locally melt or soften the scratched substrates but also to provide thermal energy to activate certain chemical reactions to break the intermolecular bonds or to modify the material structures of many organic substrates. More recently, the technique using heated tips for creating nanostructures is also known as thermomechanical or thermochemical nanolithography, abbreviated as TMNL or TCNL [119, 120].

A sizable amount of heat can be generated internally by the friction between the tip and scratched surface or externally by electrical or optical energy. Although the optical energy, such as laser pulse, has been used as the heat source in earlier work [38, 121], the electrical energy, especially a resistive heating element embedded in the cantilever, has become popular for heating the tip [122–124]. These elements are normally made of Si with different shapes and resistivities because they can be easily integrated into a Si cantilever by using a SOI wafer for fabrication [122, 123].

Because of its chemical stability and uniform temperature coefficient of resistivity, Pt has also been used as the heating element. Chiou et al. [120] demonstrated that a Si tip can be heated by an integrated Pt heater up to 120°C to scratch an 800-nm wide groove on a PMMA substrate. It is noteworthy that in SNOM lithography, the laser through the internal aperture of the tip can heat the tip up to a few hundred or even thousand degrees, eventually to melt and destroy the tip, if the diameter of the aperture is too small (much less than 100 nm) or the aperture tip is not appropriately designed [8].

Scratching behavior by a heated tip is more complex than the one with mechanical loading and is no longer dominated only by the magnitude of the force and the number of scratches applied. In fact, the speed, temperature, and duration of force applied also have major impacts on or influences to the scratched geometries. Okabe et al. [124] studied the atomic interaction of a diamond tip to a Si substrate under room and higher temperatures to examine the influence of the substrate temperature on the scratching process. Their experimental results indicated that the average depth of the scratched grooves at the same scratching conditions increased from less than 2 nm to larger than 4 nm as the substrate temperature rises from 290 to 470 K. Also the formation of the amorphous phase around the scratched groove under the higher temperature becomes a little bit larger than that under room temperature from the simulations.

Mamin [38] used an input power of 35 mW to heat the AFM tip to temperatures of up to 170°C for 4 ms to drill an array of sub–100 nm pits on PC substrates. The heat from the tip softens the PC in the contact region, at which point a very low tip force on the order of 100 nN can create a pit. Similarly, Vettiger et al. [22] successfully dug 15–20 nm holes with a similar pitch size in a 50-nm thick PMMA layer by a single heated tip. Initially, the heat transfer from the tip to polymer film, which is through a small contact area (10–40 nm^2), is very low and the tip needs to be heated to about 400 °C to initiate softening. After softening has started, the contact area increases and the loaded thermal energy can be more effectively to indent or write the polymer layer.

Heated tips can also be used to induce local chemical reactions for tearing covalently bound polymers. However, breaking a primary chemical bond by a heated tip at very fast time scales is not easy because of the large energy barriers of covalent bonds. Instead, Pires et al. [46] chose a special resist material, called phenolic compound, in which organic molecules are bound by secondary hydrogen-bonds. These H-bonds can still provide sufficient stability to the material for lithographic processing, but are weak enough to be efficiently thermally activated by the hot tip. Instead of using continuous heating tips for scratching, Pires et al. demonstrated that a 3D pattern can be written by simultaneously applying a force and temperature pulse for several μs. The force pulse pulls the tip into contact while the heat pulse heats the tip and triggers the breaking or patterning process, in which the phenolic compound is used as resist. Uniform depth of nanogrooves of the pixels can be created in the resist upon single exposure events, in which the groove depth can be controlled as a function of applied temperature and force. Patterning at a half pitch down to 15 nm without proximity corrections was demonstrated [46].

Research has also been focused on the development of different organic thin films, in which their thermally activated chemical reactions can be speeded up by a heated tip. Since chemical reaction rates generally increase exponentially with temperature, the thermal gradient, in principle, can be employed to change the substrate material's subsequent reactivity, surface energy, solubility, conductivity, or other property of interest as desired. Szoszkiewicz et al. [119] used a heated Si tip to write a sub-15 nm hydrophilic features onto the hydrophobic surface of the copolymer: p(THP-MA)$_{80}$ p(PMC-MA)$_{20}$ at the rate of 1.4 mm/s. The thermally activated chemical reactions and topography changes depend on the chemical composition of the copolymer, the scan speed, the temperature at the tip/sample interface, and the normal load. This capability can be useful in data storage application and complex nanofluidic devices. Hua et al. [125] utilized a heated cantilevered tip to thermally decompose a PMMA film. This technique can provide the pattern definition to act as a masking layer for the subsequent etching or deposition processes. Coulembier et al. [126] extended this concept by using a heated tip to scratch self-amplified depolymerization (SAD) polymers, in which the breaking of a single bond induces the spontaneous depolymerization of the entire polymer chain. As a result, 2D nanoscale patterns can be written with high efficiency.

Wang et al. [127] used a resistively heated tip to convert a sulfonium salt precursor polymer film into poly(*p*-phenylene vinylene) (PPV) in ambient conditions with the nanogrooves scratched shown in Fig. 1.18. The thickness of this precursor film was 100 nm. Grooves were scratched at a writing speed of 20 μm/s under a normal force of 30 nN with cantilever temperature ranging between 240 and 360°C. Figure 1.18c, d show the fluorescence and AFM images, respectively, of the PPV grooves created at temperature varying from 240 to 360 C. As shown, the nanogrooves started to show a visible fluorescent contrast at 240°C, while the contrast became clearer as the heating temperature was raised to 360°C. The typical cross-sectional profile of the grooves (outlined in d) scratched at 240 °C is shown in Fig. 1.18b indicating the FWHM width equal to approximately 70 nm.

Knoll et al. [7] also demonstrated that the decomposition reaction of phthalaldehyde SAD polymers trigged by hot tip scratching can be as fast as the tip scratch speed and the nanostructures can be written with less than 2 μs exposure time per pixel. The patterning capability has been extended to 3D by one-layer scratching, in which the scratch depth can vary from pixel to pixel by changing the tip normal force. Figure 1.19 shows that a 3D world nano-map was scratched into the SAD polymer using a heated tip with approximately 40 nm lateral and 1 nm vertical resolution. A Si tip heated to a temperature of 700°C with a force pulse duration of 14 μs was used in scratching. The nano map created is composed of 5×10^5 pixels with a pitch of 20 nm. The full map pattern was written within one layer at a pixel rate of 60 μs or a writing speed of 0.3 nm/s; the total patterning time amounted to 143 s. The digital geological-elevation profile of the world image of Fig. 1.19a was obtained from U.S. Geological Survey and was transformed into a digital 3D force profile by providing the depth-force scratching relationship of the phthalaldehyde polymer, which is similar to the one shown in Eq. (2), if the scratch temperature

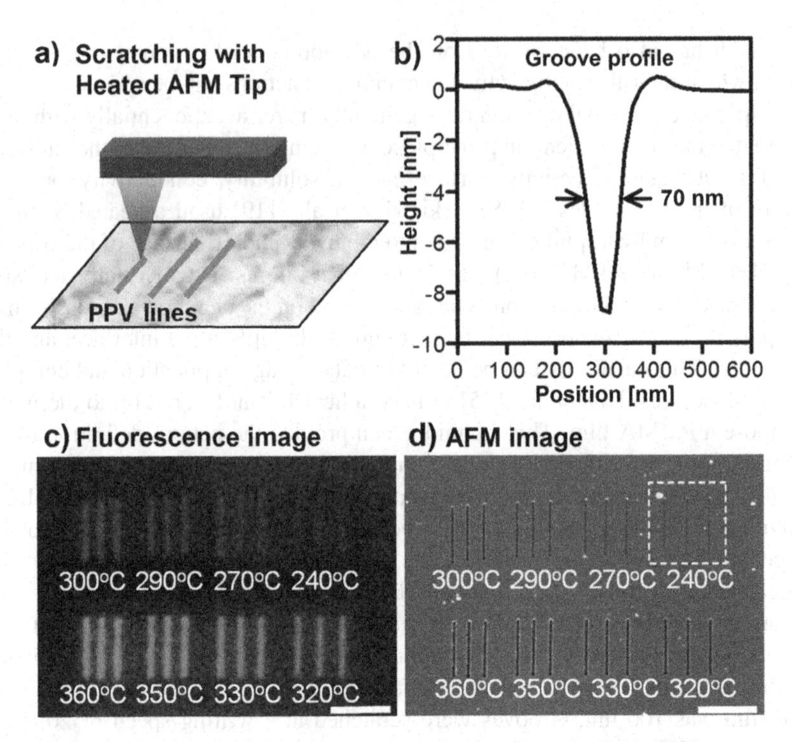

Fig. 1.18 Scratching with heated tip to create poly(p-phenylene vinylene) (PPV) nanogrooves: (**a**) scheme of thermal scratching of PPV, (**b**) average cross-section profile of grooves outlined in (**d**) showing FWHM width equal to 70 nm, (**c**) fluorescence and (**d**) AFM images of PPV nanogrooves created at temperatures from 240 to 360°C. Scale bars: 5 μm in (**c**) and (**d**) (Reprinted with permission from [127] by American Institute of Physics)

is kept constant. In scratching or writing the 3D nanomap, a depth of 8 nm corresponds to 1 km of real-world elevation. Figures 1.19b, c show a comparison of a sub-area in the programmed bitmap and the imaged relief, respectively. The white arrows indicate positions with features of 1 (in Fig. 1.19b) and 2 (in Fig. 1.19c) pixel width in the original bitmap. The 2 pixel wide features were correctly reproduced, corresponding to a resolution of approximately 40 nm for the patterning process. Figure 1.19d shows the cross-section profiles along the dotted line shown in Fig. 1.19a, for the blue line and the red line showing the original data and the relief reproduction, respectively, in the cited paper [7]. The cross-section cuts, from left to right, through the Alps, the Black Sea, the Caucasian mountains, and the Himalayas. Interestingly, the material also exhibits good RIE etch resistance, enabling the direct pattern-transfer into silicon substrates with a vertical amplification of six.

Furthermore, Lantz, Wiesmann, and Gotsmann [128] found that, in scratching these polymers, their chemical reactions can be thermally activated with a Si tip and the associated tip wear was dramatically reduced; because the tip shape is virtually not changed for a reasonable time period. In addition, the high efficiency associated

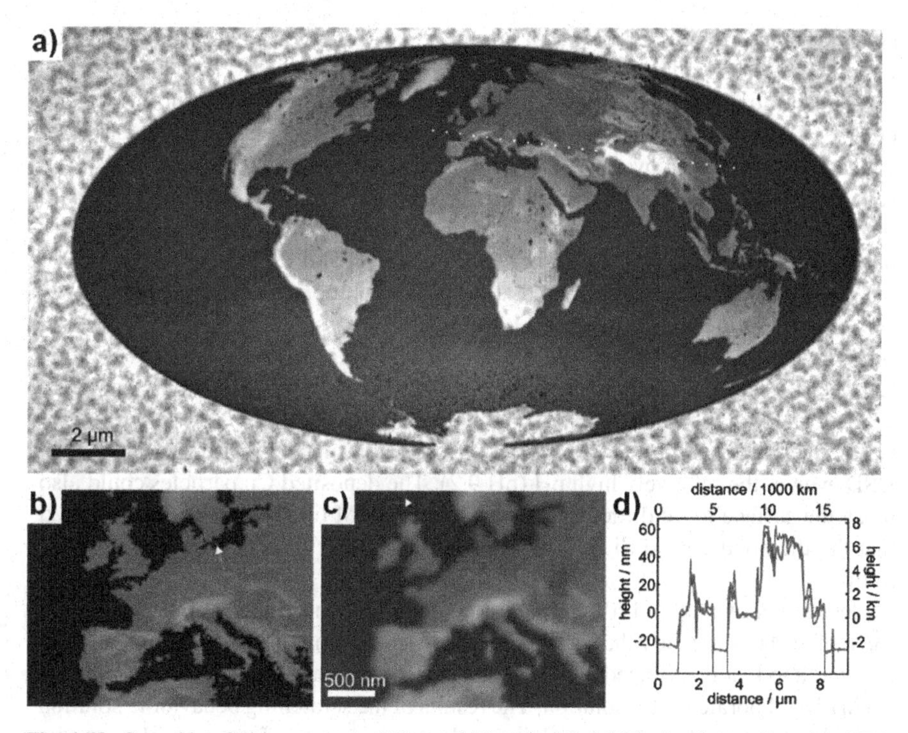

Fig. 1.19 Scratching 3D nanomap on 250-nm thick polyphthalaldehyd polymer film: (**a**) AFM image based on digital data of 3D world map provided by U.S. Geological Survey. A Si tip heated to a temperature of 700°C with a force pulse duration of 14 μs was used in scratching. The relief is composed of 5×10^5 pixels with a pitch of 20 nm. (**b**) the programmed bitmap of a sub-area, (**c**) the imaged relief the sub-area in b. The *white arrows* indicate positions with features of 1 (in b) and 2 (in c) pixel width in the original bitmap. The 2 pixel wide features were correctly reproduced, corresponding to a resolution of approximately 40 nm for the patterning process. (**d**) Cross-section profiles along the *dotted line* shown in (**a**), for the original data and the relief reproduction. The cross-section cuts, from left to right, through the Alps, the Black Sea, the Caucasian mountains, and the Himalayas (Reprinted with permission from [7] by Wiley-VCH Verlag GmbH & Co. KGaA)

with these hot tip scratching techniques can allow the removal of large volumes of material without altering the writing properties of the tip. As a result, not only can high-resolution nanostructures with more consistent dimensions be created but also can relatively large patterns be effectively achieved.

1.6.2 Chemical or Environmental Effects on Scratching

To prevent the substrate damage, an etching-aided scratching technique has been developed. Michler, et al. [129] have used an AFM equipped with a diamond tip to scratch a 5–7 nm deep nanopattern in a 10-nm thick SiO_2 film to reduce the possibility of damage beneath the Si substrate. Then, etching in HF was used to uniformly remove 4–5 nm of SiO_2 in both the scratched and non-scratched areas, thus exposing the scratched pattern on the Si substrate, while the non-scratched area was still covered with an oxide layer of 5–6 nm thick. Pd was selectively electrodeposited into the scratched pattern.

The solution etching process can also occur with scratching concurrently. Kato et al. [97] applied both Si and diamond tips to study the scratching behavior of Al, which originally was covered by 15 nm oxide at four different environments. The results indicated that the unit scratch depth (scratch depth per/load or USD) is in the order diluted NaOH solutions > $CuSO_4$ solutions > pure water > Cu-electroless plating solutions. The high USD for solutions NaOH and $CuSO_4$ may be owing to the relatively thin oxide film formed at pH 9.2 of solution NaOH and the relatively low pH of solution $CuSO_4$, where Al oxide tends to dissolve as $Al(OH)_4^-$ or Al^{3+}. In pure water, a stable, thick oxide–hydroxyl film is formed. The USD in solution Cu-electroless is the smallest in all the solutions. The small USD may be owing to Cu deposited on the Al surface covered with the thin oxide film. The deposition of Cu as fine particles may also inhibit the dissolution of Al, resulting in the lowest USD even at the relatively high pH (pH 9.2). The deposited Cu particles could also make the surface more difficult to scratch for an AFM tip. The results also showed that the diamond-coated Si tip has a much higher wear resistance than that of the Si tip.

Seo and Kawamata [130] adopted a conical diamond tip at a constant normal force of 300 μN to study the scratching behavior of a single-crystal Ta(100) surface that was anodically oxidized at 5.0 V at a constant current density of 50 $\mu A/cm^2$ for 1 h in pH 8.4 borate buffer solution. The results of the scratching behavior in solution were also compared with that of the scratching in air on the Ta surface after anodic oxidation. With the same normal tip force, the authors found that the average horizontal force along the scratch direction obtained with scratching in a solution was significantly larger than that obtained with scratching in air and increased linearly with the logarithm of time required for scratching. In contrast, the horizontal force obtained with scratching in air was independent of the time required for scratching. A stick–slip was always observed in horizontal force vs. horizontal displacement curves during scratching in solutions. The degree of stick–slip, however, decreased with increasing scratching rate. The protrusion at the scratching end of the groove produced by scratching in solution was observed from the AFM image, while no protrusion was observed for scratching in air. It was believed by the authors that the protrusion consists of anodic oxide that is accumulated on the moving front of the indenter tip during scratching in solution, and contributes to the increase in the horizontal force, that is, lateral force or friction coefficient.

Water contents can also affect the scratching behavior. Pendergas et al. [131] compared the scratching behavior of gold in water to the experiments performed in the ambient environment (~60% humidity). Both Si_3N_4 and diamond tips were used to perform both single and multiple scratching. Tests performed in the ambient environments resulted in slightly shallower scratch trenches than that of water wet scratches where the difference increases with the normal load. The different components of the scratch or horizontal force were investigated to explore the main contributors to the scratching friction created in gold scratching.

1.6.3 Scratching with Electric Bias

It is well-known that an AFM conducting tip is charged with a negative electric bias with respect to a substrate to induce surface oxide patterns, which is an electrochemical process, known as local anodic oxidation (LAO) [132]. By moving the tip close to the substrate surface, the negatively biased tip induces a high electric field that ionizes the water molecules from the ambient humidity between the tip (cathode) and the substrate (anode) and the OH^- ions produced provide the oxidant for the electrochemical reaction to form a localized oxide beneath the tip. It is one of the most popular nanolithographic processes and, sub–10 nm features can be fabricated on silicon, metallic films such as titanium, and even organic layers [2].

Miyake et al. [133] used an electrically conductive diamond tip by applying a load and an electrical bias to investigate a process that involves both scratching and LAO. The size and morphology of the nanostructures induced by a 45-nm radius tip on the p-type Si(100) substrate which initially was covered by a 2 nm thick native oxide layer was studied. The features of these nanostructures were affected by both the load and voltage. For pure scratching at relatively low loading, as the tip load decreases from 120 to 80 nN and from 80 to 40 nN, a protrusion formed and grew from 0.20 to 0.26 nm and from 0.26 to 0.40 nm, respectively. The protrusion was caused by a pressure-induced phase transformation [78, 79] as mentioned earlier. By applying bias voltages of 1, 2, and 3 V, a tip load of 2 nN is applied to keep the tip in contact with the substrate and a humidity range from 50 to 80%, LAOs with thickness of 0.44, 0.50, and 0.71 nm were formed, respectively. If the tip load is increased to 120 nN, the height of the LAO at biases of 1, 2, and 3 V, become 0.44, 0.56, and 0.64 nm, respectively. The height of the LAO becomes 0.44 nm at bias of 1 V and a load of 40 nN, 0.56 nm at bias of 2 V and a load of 80 nN and 0.64 nm at bias of 3 V and a load of 120 nN. As expected, the feature size is determined by the two competitive processes, which is enhanced by the electrical LAO but is depressed by the mechanical scratching at a relatively high load.

1.6.4 Scratching with Electric Bias in Solution Environments

The scratching process can also be loaded with an electric bias and performed in a chemical solution environment. Thus, the electrochemical process can occur with scratching concurrently, which is also known as electrochemical nanomachining (ECM). Koinuma and Uosaki [134] applied a Si_3N_4 tip with a radius of 40 nm immersed in a solution to induce electrochemical etching at an ambient temperature. By applying a tip force of 10 nN and a voltage bias of –0.05 V immersed in an etchant of 10 mM H_2SO_4 solution for 30 min, a groove with an 80 nm wide mouth and an 8 nm depth was chemical-mechanically formed on a single-crystalline Zn-doped p-GaAs(1 0 0) surface, which acted as an electrode. If the potential is increased to +0.05 V (0.1 V more positive potential), the groove depth is enhanced to ~18 nm. It has been found that the depths of the groove created were dependent not only on the applied tip force and the number of repeated cycles but also on the applied voltage bias. In addition to the groove pattern formed in the doped GaAs surface, the ECM has also been utilized to create different patterns in a range of other materials, which include nanosized bimolecular patterns [135], nano-scale polymer patterns [136, 137], nano-holes and lines on Au and Ti [138] and nanogrooves on Cu thin films [139].

Dependent on the solution and bias involved, in ECM, the basic mechanism for material removal can be mainly due to the electrochemical sources, which is similar in concept to the reverse electroplating process, that a current is passed between an electrode (the tip) and the workpiece (substrate), through an electrolytic material removal process having a negatively charged tip (cathode), a conductive fluid (electrolyte), and a conductive substrate (anode). The only difference is that the ECM is performed in a nanoscale cell. Sometime the material removing in ECM can be mainly driven through non-mechanical means, such as the field-emission current to cut CNT [91]. Since this type of ECM is not associated with mechanical scratching, no further discussion on this type of techniques will be included here.

1.7 AFM Scratching Applications

AFM scratching has been used to fabricate a fairly wide range of nanoscale devices and systems. In the preceding sections, by reviewing many subjects of AFM scratching, many applications related to these devices and systems have already been introduced. Therefore, only the representative applications, which have not been appropriately elaborated, are presented in this section.

1.7.1 Nanodevices Scratching

Rosa et al. [140] employed an electron beam deposited (EBD) tip to mechanically drill an array of holes in an InAs-AlSb heterostructure containing a two-dimension electron gas (2DEG) system. The array holes were directly perforated into the top 8-nm-thick InAs layer with a contact force of a few μN. The EBD tip is basically a needle-shape carbon tip with a tip curvature of approximately 30-nm, in which the carbon material was deposited to reinforce and sharpen the original Si tip by e-beams through the dissociation of background gas in the vacuum chamber. With this tip, holes with diameter of few nanometers have been drilled with a period of 55-nm. By comparison, an array of holes with a period of 85 nm has been obtained by chemical etching using an AFM-drilled photomask by pattern transferring. As a result, being a maskless process, the AFM direct drilling which can avoid the inaccuracy introduced in pattern transfer, can provide patterns with a much higher resolution.

Shumacher et al. [141] used an AFM tip with a contact force on the order of 50 μN at a scan speed of 100 μm/s to repeatedly scratch the 5-nm GaAs cap layer of a GaAs/AlGaAs heterostructure leading to a local depletion of a 2DEG system. The resistance of the scratched barriers was measured in situ. It was found that the more the heterostructure surface is removed, the higher the conduction band in the GaAs quantum well is lifted and thus, the stronger the local depopulation of the 2DEG. Eventually, after 120 scratches or scans, the depth of the machined line reached 12 nm and the 2DEG was completely depleted, achieving a background resistance of 55 MΩ. A new type of heterostructures with a compensating p-type doped cap layer has shown an electron enhancement if the cap layer is selectively removed. Also, a machined groove in these structures enables one to induce a one-dimensional electron system (1DES).

Versen et al. [142] used a Si tip for direct scratching of a GaAs surface resulting in a groove of approximately 3–4 nm deep and 30 nm wide. Beneath such a groove, a barrier arises in the electron channel of a GaAs/AlGaAs modulation-doped field effect transistor (FET). Using appropriate sub-100 nm line patterns, quantum point contacts and single electron devices were also created by Kunze [10]. At $T = 4.2$ K, the transconductance characteristics of this FET exhibit structures with signatures of either the quantized conductance or Coulomb-blockade effects.

Tseng et al. [71] used a diamond coated tip to scratch two grooves on a 20-nm thick planar NiFe nanowire to create nanoconstrictions in a magnetic structure as shown in Fig. 1.20. Because of their technological importance for magnetoelectronics, NiFe-based nanoconstrictions formed in nanowire type structures have attracted much scientific interest [143]. The nanowire was fabricated on a SiO_2/Si substrate through e-beam lithography using the standard lift-off technique with a source and a drain. AFM scratching was then performed across the nanowire in a selected location until the total thickness of the NiFe film was removed. The cross-sections of the nanowires before and after performing the scratches are depicted in Fig. 1.20 and indicate that the cross-sectional area of the NiFe nanowire left is less than 4% of the original cross-section, which is on the same order of magnitude of the standard deviation in the scratch experiment results shown in Figs. 1.9 and 1.12. The typical current-voltage characteristics measured before and immediately after ($\sim$ 1 h) performing the scratching is shown in Fig. 1.20c. As shown, the drain current after the scratching was clearly suppressed due to the removal of a nano-section of the nanowire at the constriction site and decreased from the order of μA to fA. This result suggests that the ferromagnetic nanostructures with the required properties can be fabricated by AFM scratching with control over current levels as low as few fA. Moreover, since the current flow can be achieved down to such low values, the proper real time control of resistivity can be made during scratching. As a result, the device can be on-line monitored and be conveniently fabricated to the required geometry with the required performance.

1.7.2 Polymeric Brushes and Gratings

Hirtz et al. [144] used a Si tip to perform AFM scratching on three polymer brushes made by polystyrene (PS), poly(n-butyl acrylate) (PNBA), and Poly(N-isopropylacrylamide) (PNIPAM), which can be prepared by spin-coating or self-assembly. The experiments were equipped with a standard Si probe with a spring

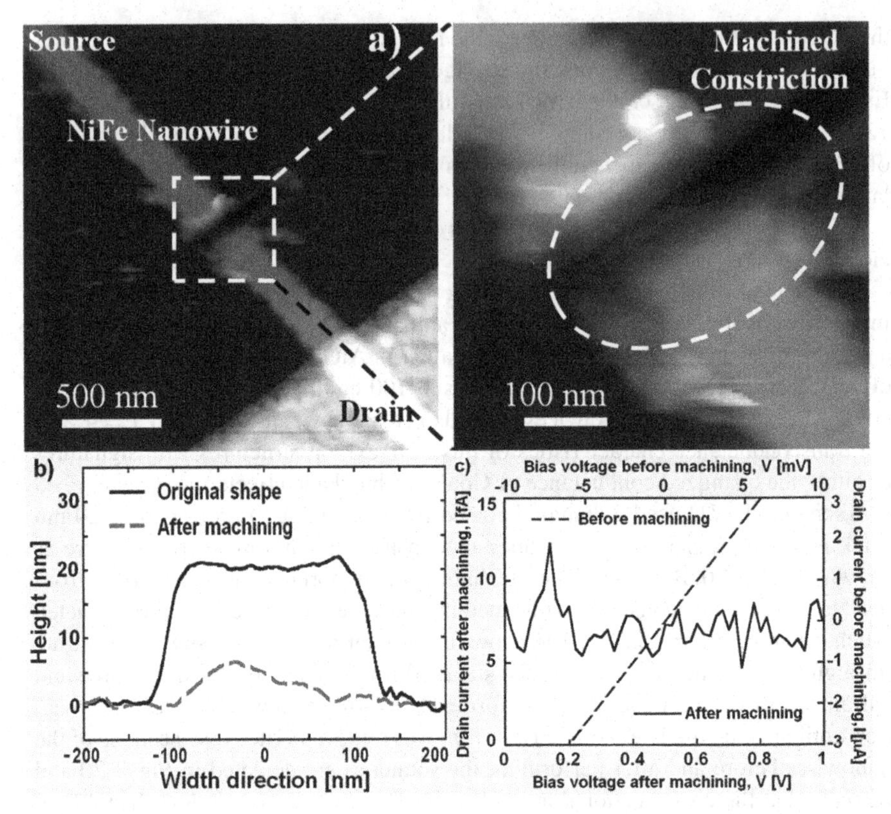

Fig. 1.20 NiFe Nanoconstriction made by AFM scratching: (**a**) AFM images of NiFe planar nanowire after scratching, (**b**) cross-sectional profiles before and after scratching, (**c**) I-V characteristics of nanoconstriction measured before and after scratching (Reprinted with permission from [71] by American Institute of Physics)

constant of 42 N/m operating in contact mode under a loading force of 22 μN. A polymer brush is a layer of polymers attached with one end to a surface. By structuring or patterning polymer brushes, functional surfaces with defined properties for the study of cell adhesion [145] and of cell alignment [146] can be obtained. Moreover, patterning of polymer brushes by AFM scratching allows the alteration and control of their wetting properties [147].

Choi et al. [54] made grating structures on the surface of a PC substrate using AFM scratching by a diamond coated Si tip. In order to obtain relatively deep patterns, a scanner movement method, in which the sample scanner moves along the z-axis was used. A grating of 100 μm × 150 μm was fabricated by the step and repeat method wherein the sample stage was moved in the direction of the xy-axis. The period and the depth of the grating are 500 and 50 nm, respectively, where light of 632.8 nm wavelength was diffracted on the surface of the PC substrate.

1.7.3 Nanostructured Masters or Molds

Zhao et al. [149] applied AFM scratching to create cavities in Si substrates and found that these cavity features on the Si substrates can be accurately transferred to polydimethylsiloxane (PDMS) stamps verified by an AFM morphologic imaging study. The authors claimed that AFM scratching for making these elastomeric PDMS stamps is particularly straightforward and economical, and can serve as masters to make the required elastomeric stamps for soft lithography.

Brousseau et al. [75] used Si tips to generate recessed features on Si and brass surfaces for stamp and mold making for microinjection molding of polymers. They studied the mold life and reported that the brass mold had longer mold life and was more appropriate for micromolding applications, as compared with the Si mold.

1.7.4 Scratching and Repairing Masks and Others

AFM scratching using a sharp tip is extremely effective to remove a small amount of materials and is an ideal tool for mask repairing or direct making a small mask, because of its precise position location to within angstroms combined with the *in-situ* metrology. Plecenik [150] used a W_2C coated Si tip with an apex radius of 30-nm in contact mode to repeatedly scratch a 1.4-μm thick positive photoresist, which was used as a mask for the subsequent optical lithographic process to fabricate variable-width and thickness bridges for a superconducting quantum interference device (SQUID) made from a MgB_2 superconducting thin film. The MgB_2 thin films with a roughness below 10 nm, a critical temperature of 31 K, and a critical current density of 5×10^6 A/cm^2 were prepared by magnetron sputtering and ex-situ annealing in vacuum. Since MgB_2 is a relatively hard material, it may cause problems to directly scratch MgB_2. Since its discovery in 2001, MgB_2 has attracted a great attention from researchers and is a very promising material to be used in cryoelectronic applications [151].

Based on electrophoretic migration, Iwata et al. [152] developed a method to effectively remove the scratched debris produced in an AFM photomask repair process. A small amount of aqueous solution was dropped on a cantilever, and then two electrodes placed on both sides of the cantilever form an electric field gradient around the cantilever. Under this condition, the particles of scratched debris were dispersed and became spontaneously charged in the liquid solution. Consequently, the scratched debris can be removed by electrophoresis migration. They demonstrated that the electrophoretic migration-based scratching technique was a very effective for photomask repairing. Robinson et al. [153] developed photolithographic simulation software to guide AFM to perform nanoscratching for mask repairing and for 2D shape reconstruction. Repair results were shown for various processes to highlight the relative strengths and weaknesses of the system developed. The advances in repair dimensional precision and overall imaging performance were also demonstrated.

As mentioned earlier, a heated tip can be used as a source to decompose polymeric materials, such as polymeric photoresists, for lithographic applications. In nanopatterning, a tip can be heated to a certain temperature to decompose sacrificial polymer and the decomposed polymer can be selectively removed. Henderson et al. [154] mentioned the use of sacrificial polymer which thermally decomposes after patterning in conjunction with non-decomposing permanent structural materials to create required masks or components used in microsystems. Jang et al. [155] used an AFM tip coated with an octadecanethiol solution to remove or repair a defective edge of a pattern made of 16-mercaptohexadecanoic acids without deactivating or damaging the underlying surface. Since the technique is considered by the authors as an application of DPN, the details of this technique will not be elaborated any further.

1.7.5 BioMedical Applications

DNA is not only the most fundamental and important building block for biological systems but also a kind of promising molecule as a nanolink to build or connect nano-devices due to its stable linear structure and certain conductivity. With supercoiled plasmid DNA deposited on a mica surface, Henderson [156] applied an AFM to tip to precisely cut the plasmid at a pre-determined location guided by the AFM on-line imaging. Small pieces of DNA (100–150 nm in length) were excised and deposited adjacent to the dissected plasmid, demonstrating that it was possible to remove and manipulate genomic DNA fragments from defined chromosomal locations by AFM. Hu et al. [157] aligned DNA strands on a solid substrate to form a matrix of 2D networks and used an AFM tip to cut the DNA network in order to fabricate fairly complex artificial patterns. The AFM tip was also used to push or manipulate the sliced DNA strands to form spherical nanoparticles and nanorods by folding up the DNA molecules into ordered structures in air. Since DNA can be easily amplified, it is needed to modify only one molecule in principle and is extremely suitable for the series nature of AFM operation. Consequently, the AFM-based manipulation, including cutting or scratching, pushing, folding, kneading, picking up and, dipping, which is also known as molecular surgery, should become an effective tool for modifying the DNA structure and have a wide impact on DNA technology.

Furthermore, the inability to discern a particular molecule in a solution or a particular position on a molecule becomes most conspicuous when handling DNA. Since genetic information of DNA is recorded as a linear sequence of bases, the position of the base has an essential meaning. The AFM based technique, which has the capability to control the position and confirmation of individual molecule, can overcome these difficulties [158]. A recent study indicated that with a proper surface treatment of the substrate, the AFM tip can pick a single DNA molecule with reasonable success rate at 75% [159]. As an extension, a new single DNA was allowed to hybridize with the picked DNA, and conjugated with the picker DNA by use of a ligase. Various applications of the tip for manipulating single molecules and preparing new nanomaterials are envisaged. The AFM tip can also act as a surgery instrument. Firtel et al. [59] applied an AFM tip to perform nanosurgery by mechanically removing large patches of outer cell walls after appropriate medical treatment.

1.7.6 Scratching Test

The advances in AFM scratching have prompted the adoption of AFM for the scratch test for the tribology study, especially in nanoscales. In certain aspects, AFM scratching is similar to the scratch test, which is used to find the threshold forces or critical loads for different types of material failures [160, 161]. Usually in a scratch test, a microscale diamond tip or indenter is moving over a sample under a normal load, which is increased either continuously or incrementally until a specifically defined failure is observed, at which point the corresponding load, called the critical load, is obtained. Randall, Favaro, and Frankel [160] used semi-spherical diamond tips with radii of 20–500 μm to study the adhesive failures of five different coatings on two different substrates. Li and Beres [161] gave a good review on the coating failures evaluated by conventional microscratch tests. Recently, AFM has been used to perform nanoscale scratch tests. Yasui et al. [85] applied a commercially-available AFM equipped with a diamond tip having a radius of 270 nm and a spring constant of 278 N/m to study the critical loads of diamond-like carbon overcoats used for the head and disk of 1 Tb/in^2 magnetic recording. As a result, the better understanding of AFM scratching should directly and indirectly lead to the better development of a scratch test.

1.7.7 Other Applications

In other applications, Lin et al. [162] fabricated metal nanostructures with sizes down to 20 nm and nanowires with widths ranging between 40 and 100 nm by a combination of AFM scratching and lift-off process. The localized surface plasmon resonance properties of the fabricated nanostructures for solar cell applications and the chemical sensing capability of a single nanowire based on resistance increase were demonstrated.

As mentioned earlier, IBM has applied multiple-tip array, called Millipede, for data-storage applications in which an array of heated tips is used to scratch (write), to image (read), and to melt (erase) nanoscale holes (data) on very thin polymer films. In imaging or reading, the cantilever, which is imposed with a heater and is originally used for writing, is provided an additional function of a thermal read-back sensor by monitoring the temperature-dependent resistance of the imposed heater [23]. The Millipede has exploited the parallel operation of a very large array (64 × 64) of AFM-type cantilevered tips for write/read/erase functionality and achieved with a space of 18 nm between tracks and 9 nm within a track, and depth of 1 nm leading to a storage density of more than 1 Tb/in^2 [22, 23].

The heated tip has also been used to provide thermal energy to activate certain chemical reactions to modify polymeric substrates for the applications of data storage and maskless nanolithography. Gotsmann et al. [163] demonstrated that with the sufficient heat energy provided by hot tips, the chemical bonds of a Diels–Alder (DA) material can be broken and subsequently the DA polymer can be decomposed into volatile monomer units. Since the DA polymers possess thermally reversible crosslinking, which allows switching between two different states: a rigid, highly crosslinked, low-temperature state, and a deformable, fragmented, high-temperature state. Consequently, this ability to cycle between two sets of properties in these materials opens up new perspectives in lithography and data storage. Examples of data storage with densities up to 1 Tb/in^2 and maskless lithography with resolution below 20 nm were demonstrated at writing times of 10 μs per bit/pixel [163].

Finally, graphene and CNT have been widely studied in recently year, because their unique electronic and mechanical properties and their potential applications in the field of electronics and sensing [164]. For example, by using the graphene oxide (GO) patterns as templates, the hybrid GO-Ag nanoparticle patterns were obtained, which could be useful for graphene-based device applications [89].

1.8 Concluding Remarks

Considering the advantages of an ultrahigh resolution capability and other unique features, including inexpensive equipment and relatively easy operation and control, AFM scratching has emerged as one of the essential technologies in the manufacturing of nanoscale structures. In this chapter, the recent developments of the AFM scratching have been reviewed with an emphasis on its ability to fabricate various nanostructures. The present review indicates that AFM scratch technique is extremely favorable for working with various materials for fabricating a wide rage of nanoscale components, devices and systems.

The major components of the AFM instrument and their advances for AFM scratching were first introduced. The techniques used for scratching different types of materials including polymers, self-assembled monolayers, metals, semiconductors, ceramic, and nanocarbonic materials were evaluated to illustrate the respective feasibilities and potentials of the specific technique applied. The effects of major scratch parameters, including the scratch direction, applied tip force, threshold force, scratch (scan) speed, and number of scratch, on the scratched geometry were presented with the specific correlations developed for providing guidelines for processing different materials at various scratching conditions. The material properties that govern the scratchability or machinability were also assessed and selected by comparing a wide range of materials and scratching conditions.

Moreover, the developments of the mechanical scratch enhanced by other energy sources, such as thermal, electrical or chemical, were presented with the emphasis on the combined effects caused by all the loaded energies. Basically, imposing both mechanical and non-mechanical sources concurrently on the tip can alleviate some of the limitations of a pure mechanical scratch process, including high tip wear, scratched geometries with low aspect ratio, and low scratch speed involved in scratching. On the other hand, with the tip loaded with dual energy sources, the material removing process becomes more complicated and more parameters are needed to be correctly controlled. In many cases, slight changes of some parameters, which could normally be insensitive or insignificant to the pure mechanical scratching process, could lead major impacts on the quality and properties of the nanostructures created. As a result, relevant phenomena for the dual-source scratching are needed to be properly understood and related research should be encouraged. Also, the basic elements in AFM instrumentation may be required to be redesigned and new processing strategy may need to be developed to satisfy the new challenges caused by the AFM charged with dual sources.

Although the AFM scratching has been extremely favorable for applications in making prototypes, critical to the ultimate success of the current AFM technology is the transfer of laboratory-scale successes to the creation of commercial products with the required throughput. For example, the simple process correlations presented in this chapter should be one of the major steps to develop a practical computer-aided manufacturing and control system for AFM automation. Furthermore, because of the serial nature of the tip movement, the low throughput is a common challenge pertaining to all AFM techniques. In addition to loading with dual sources and automation, the approaches used to improve the throughput by operating multiple-tip in parallel and by increasing the tip writing speed (rate), should be continuously emphasized. However, to operate multiple tips, the ability to control each tip independently still limits the number of tips in a system. A more versatile control strategy to handle a great number of tips should be developed. The approach to increase the writing speed has also been exploited, which makes the AFM comparable to the fast e-beam lithography systems. Furthermore, carbon nanotube tips have the potential to possess extraordinary mechanical, magnetic, and electric characteristics at reasonable scanning speeds and with an extremely low tip-wear rate. Certainly, further improvement of writing speed and tip hardness (or increasing tip life) is helpful, but a new approach would be more attractive. Furthermore, many efforts have also been dedicated to develop a microtip or macro-stamp to replace the nanotip to increase the working area or throughput [148, 165].

To achieve major improvement, it is believed that a multiresolution system should be developed to provide different accuracy or tolerance requirements in patterning different scales of nanostructures. For example, the tip diameter should be continuously changed in situ. This type of system has been available for many macroscale fabrication processes. To be a vital nanofabrication tool, it is imperative to be able to produce a functional device, which may contain both nanoscale and macroscale components. These components may come about as a combination of nanoscale and traditional fabrication processes. Therefore, the ability of the integration of nanofabricated components with subsequent manufacturing steps and the consolidation of nanostructures into micro/macroscopic objects becomes extremely critical.

In the coming decade, AFM scratching, especially loaded with multiple sources, is expected to quickly become an essential fabrication tool to many areas of science and technology and will have a revolutionary impact on every aspect of the nanofabrication industry. It is worth noting that the motivation of this chapter was to help the reader to appreciate the essential potential and trends in AFM scratching. A selection of topics and papers presented in this chapter was made without the intention of excluding valuable ones that provided important contribution to the development of the nanoscale material removing process.

Acknowledgements The author would like to acknowledge the support of Pacific Technology, LLC of Phoenix (USA) and the National Science Council (ROC) under Grant No. NSC99-2811-E-007-014 in funding a University Chair professorship at National Tsing Hua University (NTHU) in Hsinchu, Taiwan, where the author spent a semester in preparation of this manuscript in 2010. The author is grateful to Professors Wen-Hwa Chen, Hung Hocheng, and Chien-Chung Fu of NTHU for their hospitality and encouragement during the author's stay in Hsinchu. Special thanks are to Professor Jun-ichi Shirakashi of Tokyo University of Agriculture and Technology (Japan), Professor Noritaka Kawasegi of University of Toyama (Japan), Professor Zhuang Li of Chinese Academy of Sciences (China), Professor Patrik Schmuki of University of Erlangen-Nuremberg (Germany), Dr. Eugeniu Balaur of La Trobe University (Australia), Dr. Andrea Notargiacomo of CNR-IFN (Italy), and Dr. Luca Pellegrino of CNR-SPIN (Italy) for their fruitful discussions and useful digital data for illustration and presentation. The author is thankful for the assistance provided by Maggie S. Tseng and Parag S. Pathak of Arizona State University and by Ms Yichih Liu of NTHU in preparing this manuscript.

References

1. G. Binnig, C. F. Quate, C. Gerber, Atomic force microscope, *Phys. Rev. Lett.*, 56, 930 (1986).
2. A. A. Tseng, A. Notargiacomo, T. P. Chen, Nanofabrication by scanning probe microscope lithography: a review, *J. Vac. Sci. Technol. B*, 23, 877–894 (2005).
3. A. A. Tseng, S. Jou, A. Notargiacomo, T. P. Chen, Recent developments in tip-based nanofabrication and its roadmap, *Nanosci. Nanotechnol.*, 8, 2167 (2008).
4. Y. Kim, C. M. Lieber, Machining oxide thin films with an atomic force microscope: pattern and object formation on the nanometer scale, *Science*, 257, 375–377 (1992).
5. A. A. Tseng, A comparison study of scratch and wear properties using atomic force microscopy, *Appl. Surf. Sci.*, 256, 4246–4252 (2010).
6. Y. Yan, T. Sun, Y. Liang, S. Dong, Investigation on AFM-based micro/nano-CNC machining system, *Int. J. Mach. Tools Manuf.*, 47, 1651–1659 (2007).
7. A. W. Knoll, D. Pires, O. Coulembier, P. Dubois, J. L. Hedrick, J. Frommer, U. Duerig, Probe-based 3-D nanolithography using self-amplified depolymerization polymers, *Adv. Mater.*, *22*, 3361–3365 (2010).
8. A. A. Tseng, Recent developments in nanofabrication using scanning near-field optical microscope lithography, *Opt. Laser Technol.*, 39, 514 (2007)
9. C. Santschi, J. Polesel-Maris, J. Brugger, H. Heinzelmann, Scanning probe arrays for nanoscale imaging, sensing and modification, *in Nanofabrication: Fundamentals and Applications*, A. A. Tseng (Ed.), 65–126, World Scientific, Singapore, 2008.
10. U. Kunze, Nanoscale devices fabricated by dynamic ploughing with an atomic force microscope, *Superlatt. Microstruct.*, 31, 3–17 (2002).
11. C. K. Hyon, S. C. Choi, S. W. Hwang, D. Ahn, Y. Kim, E. K. Kim, Nano-structure fabrication and manipulation by the cantilever oscillation of an atomic force microscope, *Jpn. J. Appl. Phys.*, 38, 7257–7259 (1999).

12. K. Ashida, N. Morita, Y. Yoshida, Study on nano-machining process using mechanism of a friction force microscope, *JSME Int. J. Ser. C*, 44, 244–253 (2001).
13. N. Kawasegi, N. Takano, D. Oka, N. Morita, S. Yamada, K. Kanda, S. Takano, T. Obata, K. Ashida, Nanomachining of silicon surface using atomic force microscope with diamond tip, *ASME J. Manuf. Sci. Eng.*, 128, 723–729 (2006).
14. N. Kodera, M. Sakashita, T. Ando, A dynamic PID controller for high-speed atomic force microscopy, *Rev. Sci. Instrum.*, 77, 083704 (2006).
15. A. D. L. Humphris, M. J. Miles, J. K. Hobbs, A mechanical microscope: high-speed atomic force microscopy, *Appl. Phys. Lett.*, 86, 034106 (2005).
16. L. M. Picco, L. Bozec, A. Ulcinas, D. J. Engledew, M. Antognozzi, M. A. Horton, M. J. Miles, Breaking the speed limit with atomic force microscopy, *Nanotechnology*, 18, 044030 (2007).
17. J. H. Hafner, C. L. Cheung, T. H. Oosterkamp, C. M. Lieber, High-yield assembly of individual single-walled carbon nanotube tips for scanning probe microscopies, *J. Phys. Chem. B*, 105, 743–746 (2001).
18. C. L. Cheung, J. H. Hafner, T. W. Odom, K. Kim, C. M. Lieber, Growth and fabrication with single-walled carbon nanotube probe microscopy tips, *Appl. Phys. Lett.*, 76, 3136–3138 (2000).
19. K. Takami, M. Akai-Kasaya, A. Saito, M. Aono, Y. Kuwahara, Construction of independently driven double-tip scanning tunneling microscope, *Jpn. J. Appl. Phys. Part 2*, 44, L120 (2005).
20. X. F. Wang, C. Liu, Multifunctional probe array for nano patterning and imaging, *Nano Lett.*, 5, 1867–1872 (2005).
21. S. C. Minne, J. D. Adams, G. Yaralioglu, S. R. Manalis, A. Atalar, C. F. Quate, Centimeter scale atomic force microscope imaging and lithography, *Appl. Phys. Lett.*, 73, 1742 (1998).
22. P. Vettiger, G. Cross, M. Despont, U. Drechsler, U. Durig, B. Gotsmann, W. Haberle, M. A. Lantz, H. E. Rothuizen, R. Stutz, G. K. Binnig, The "millipede"-nanotechnology entering data storage, *IEEE Trans. Nanotechnol.*, 1, 39–54 (2002).
23. H. Pozidis, W. Haberle, D. Wiesmann, U. Drechsler, M. Despont, T. R. Albrecht, E. Eleftheriou, Demonstration of thermomechanical recording at 641 Gbit/in^2, *IEEE Trans. Magn.*, 40, 2531–2536 (2004).
24. B. W. Ahn, S. H. Lee, Characterization and acoustic emission monitoring of AFM nanomachining, *J. Micromech. Microeng.*, 19, 045028 (2009).
25. H. Chiriac, M. Pletea, E. Hristoforou, Magnetoelastic characterization of thin films dedicated to magnetomechanical microsensor applications, *Sens. Actuators A-Phys.*, 68 (1–3), 414–418 (1998).
26. T. Takezaki, D. Yagisawa, K. Sueoka, Magnetic field measurement using scanning magneto resistance microscope with spin-valve sensor, *Jpn. J. Appl. Phys. Part 1-Regular Papers Brief Communications & Review Papers*, 45, 2251–2254 (2006).
27. E. Gmelin, R. Fischer, R. Stitzinger, Sub-micrometer thermal physics – An overview on SThM techniques, *Thermochim. Acta*, 310, 1–17 (1998).
28. A. J. Bard, F. R. F. Fan, J. Kwak, O. Lev, Scanning electrochemical microscopy – introduction and principles, *Anal. Chem.*, 61, 132–138 (1989).
29. G. Boero, I. Utke, T. Bret, N. Quack, M. Todorova, S. Mouaziz, P. Kejik, J. Brugger, R. S. Popovic, P. Hoffmann, Submicrometer Hall devices fabricated by focused electron-beam-induced deposition, *Appl. Phys. Lett.*, 86, 042503 (2005).
30. A. J. Brook, S. J. Bending, J. Pinto, A. Oral, D. Ritchie, H. Beere, M. Henini, A. Springthorpe, Integrated piezoresistive sensors for atomic force-guided scanning Hall probe microscopy, *Appl. Phys. Lett.*, 82, 3538–3540 (2003).
31. S. Koshimizu, J. Otsuka, Detection of ductile to brittle transition in micro indentation and micro scratching of single crystal silicon using acoustic emission, *J. Mach. Sci. Technol.*, 5, 101–114 (2001).

32. D. E. Lee, I. Hwang, C. M. Valente, J. F. Oliveira, D. A. Dornfeld, Precision manufacturing process monitoring with acoustic emission, *Int. J. Mach. Tools Manuf.*, 46, 176–188 (2006).
33. L. Zhang, H. Tanaka, Towards a deeper understanding of wear and friction on the atomic scale-a molecular dynamics analysis, *Wear,* 211, 44–53 (1997).
34. A. A. Tseng, S. Jou, J. C. Huang, J. Shirakashi, T. P. Chen, Scratch properties of nickel thin films using atomic force microscopy, *J. Vac. Sci. Technol. B*, 28, 202–210 (2010).
35. M. Chandross, C. D. Lorenz, M. J. Stevens, G. S. Grest, Probe-tip induced damage in compliant substrates, *ASME J. Manuf. Sci Eng.*, 132, 0309161-0309164 (2010).
36. S. Kassavetis, K. Mitsakakis, S. Logothetidis, Nanoscale patterning and deformation of soft matter by scanning probe microscopy, *Mater. Sci. Eng. C*, 27, 1456–1460 (2007).
37. A. A. Tseng, J. Shirakashi, S. Nishimura, K. Miyashita, Z. Li, Nanomachining of permalloy for fabricating nanoscale ferromagnetic structures using atomic force microscopy, *J. Nanosci. Nanotechnol.*, 10, 456–466 (2010).
38. H. J. Mamin, Thermal writing using a heated atomic force microscope tip, *Appl. Phys. Lett.*, 69, 433 (1996).
39. S. McNamara, A. S. Basu, J. Lee, Y. B. Gianchandani, Ultracompliant thermal probe array for scanning non-planar surfaces without force feedback, *J. Micromech. Microeng.*, 15, 237–243 (2005).
40. H. J. Mamin, B. A. Gurney, D. R. Wilhoit, V. S. Speriosu, High sensitivity spin-valve strain sensor, *Appl. Phys. Lett.*, 72, 3220–3222 (1998).
41. B. A. Gozen, O. B. Ozdoganlar, A rotating-tip-based mechanical nano-manufacturing process: nanomilling, *Nanoscale Res. Lett.*, 5, 1403–1407 (2010).
42. I. A. Mahmood, S. O. R. Moheimani, Fast spiral-scan atomic force microscopy, *Nanotechnology,* 20, 365503 (2009).
43. S. K. Hung, Spiral scanning method for atomic force microscopy, *J. Nanosci. Nanotechnol.*, 10, 4511–4516 (2010).
44. Y. W. Kim, S. C. Choi, J. W. Park, D. W. Lee, The characteristics of machined surface controlled by multi tip arrayed tool and high speed spindle, *J. Nanosci. Nanotechnol.*, 10, 4417–4422 (2010).
45. A.A. Tseng, M. Tanaka, Advanced deposition techniques for freeform fabrication of metal and ceramic parts, *Rapid Prototyping J.*, 7, 6–17 (2001).
46. D. Pires, J. L. Hedrick, A. De Silva, J. Frommer, B. Gotsmann, H. Wolf, M. Despont, U. Duerig, A. W. Knoll, Nanoscale three-dimensional patterning of molecular resists by scanning probes, *Science*, 328, 732–735 (2010).
47. K. Bourne, S. G. Kapoor, R. E. DeVor, Study of a high performance AFM probe-based microscribing process, *ASME J. Manuf. Sci. Eng.*, 132, 030906 (2010).
48. Y. T. Mao, K. C. Kuo, C. E. Tseng, J. Y. Huang, Y. C. Lai, J. Y. Yen, C. K. Lee, W. L. Chuang, Research on three dimensional machining effects using atomic force microscope, *Rev. Sci. Instrum.*, 80, 0651051 (2009).
49. T. A. Jung, A. Moser, H. J. Hug, D. Brodbeck, R. Hofer, H. R. Hidber, U. D. Schwarz, The atomic force microscope used as a powerful tool for machining surfaces, *Ultramicroscopy*, 42–44 (Part 2), 1446–1451 (1992).
50. X. Jin, W. N. Unertl, Submicrometer modification of polymer surfaces with a surface force microscope, *Appl. Phys. Lett.*, 61, 657–659 (1992).
51. S. I. Yamamoto, H. Yamada, H. Tokumoto, Nanometer modifications of non-conductive materials using resist-films by atomic force microscopy, *Jpn. J. Appl. Phys.*, 34, 3396–3399 (1995).
52. S. F. Y. Li, H. T. Ng, P. C. Zhang, P. K. H. Ho, L. Zhou, G. W. Bao, S. L. H. Chan, Submicrometer lithography of a silicon substrate by machining of photoresist using atomic force microscopy followed by wet chemical etching, *Nanotechnology*, 8, 76–81 (1997).

53. U. Kunze, B. Klehn, Plowing on the sub-50 nm scale: nanolithography using scanning force microscopy, *Adv. Mater.*, 11, 473 (1999).
54. C. H. Choi, D. J. Lee, J. Sung, M. W. Lee, S. Lee, E. Lee, B. O, A study of AFM-based scratch process on polycarbonate surface and grating application, *Appl. Surf. Sci.*, 256, 7668–7671 (2010).
55. H. Sugihara, A. Takahara, T. Kajiyama, Mechanical nanofabrication of lignoceric acid monolayer with atomic force microscopy, *J. Vac. Sci. Technol. B*, 19, 593–595 (2001).
56. Y. Zhang, E. Balaur, P. Schmuki, Nanopatterning of an organic monolayer covered Si (1 1 1) surfaces by atomic force microscope scratching, *Electrochim. Acta*, 51, 3674–3679 (2006).
57. M. Müler, T. Fiedler, R. Gröger, T. Koch, S. Walheim, C. Obermair, T. Schimmel, Controlled structuring of mica surfaces with the tip of an atomic force microscope by mechanically induced local etching, *Surf. Interface Anal.*, 36, 189–192 (2004).
58. S. Kopta, M. Salmeron, The atomic scale origin of wear on mica and its contribution to friction, *J. Chem. Phys.*, 113, 8249–8252 (2000).
59. M. Firtel, G. Henderson, I. Sokolov, Nanosurgery: observation of peptidoglycan strands in Lactobacillus helveticus cell walls, *Ultramicroscopy*, 101, 105–109 (2004).
60. B. W. Muir, A. Fairbrother, T. R. Gengenbach, F. Rovere, M. A. Abdo, K. M. McLean, P. G. Hartley, Scanning probe nanolithography and protein patterning of low fouling plasma polymer multilayer films, *Adv. Mater.* 18, 3079–3082 (2006).
61. J. G. Park, C. Zhang, R. Liang, B. Wang, Nano-machining of highly oriented pyrolytic graphite using conductive atomic force microscope tips and carbon nanotubes, *Nanotechnology*, 18, 405306 (2007).
62. E. Gnecco, R. Bennewitz, E. Meyer, Abrasive wear on the atomic scale, *Phys. Rev. Lett.*, 88, 2155011 (2002).
63. R. Rank, H. Brueckl, J. Kretz, I. Moench, G. Reiss, Nanoscale modification of conducting lines with a scanning force microscope, *Vacuum*, 48, 467–472 (1997).
64. T. Sumomogi, T. Endo, K. Kuwahara, R. Kaneko, T. Miyamoto, Micromachining of metal surfaces by scanning probe microscope, *J. Vac. Sci. Technol. B*, 12, 1876–1880 (1994).
65. M. Watanabe, H. Minoda, K. Takayanagi, Fabrication of gold nanowires using contact mode atomic force microscope, *Jpn. J. Appl. Phys.*, 43, 6347–6349 (2004).
66. X. Li, P. Nardi, C. W. Baek, J. M. Kim, Y. K. Kim, Direct nanomechanical machining of gold nanowires using a nanoindenter and an atomic force microscope, *J. Micromech. Microeng.*, 15, 551–556 (2005).
67. T. H. Fang, J. G. Chang, C. I. Weng, Nanoindentation and nanomachining characteristics of gold and platinum thin films, *Mater. Sci. Eng. A*, 430, 332–340 (2006).
68. H. D. F. Filho, M. H. P. Mauricio, C. R. Ponciano, R. Prioli, Metal layer mask patterning by force microscopy lithography, *Mater. Sci. Eng. B*, 112, 194–199 (2004).
69. T. H. Fang, C. I. Weng, J. G. Chang, Machining characterization of the nano-lithography process using atomic force microscopy, *Nanotechnology*, 11, 181–187 (2000).
70. A. Notargiacomo, V. Foglietti, E. Cianci, G. Capellini, M. Adami, P. Faraci, F. Evangelisti, C. Nicolini, Atomic force microscopy lithography as a nanodevice development technique, *Nanotechnology*, 10, 458–463 (1999).
71. A. A. Tseng, J. Shirakashi, S. Nishimura, K. Miyashita, A. Notargiacomo, Scratching properties of nickel-iron thin film and silicon using atomic force microscopy, *J. Appl. Phys.*, 106, 044314 (2009).
72. N. Kawasegi, N. Morita, S. Yamada, N. Takano, T. Oyama, K. Ashida, H. Ofune, Nanomachining Zr-based metallic glass surfaces using an atomic force microscope, *Int. J. Mach. Mach. Mater.*, 2, 3–16 (2007).
73. L. Santinacci, T. Djenizian, H. Hildebrand, S. Ecoffey, H. Mokdad, T. Campanella, P. Schmuki, Selective palladium electrochemical deposition onto AFM-scratched silicon surfaces, *Electrochim. Acta*, 48, 3123–3130 (2003).
74. T. Ogino, S. Nishimura, J. Shirakashi, Scratch nanolithography on Si surface using scanning probe microscopy: influence of scanning parameters on groove size, *Jpn. J. Appl. Phys.*, 47, 712–714 (2008).

75. E. B. Brousseau, F. Krohs, E. Caillaud, S. Dimov, O. Gibaru, S. Fatikow, Development of a novel process chain based on atomic force microscopy scratching for small and medium series production of polymer nanostructured components, *J. Manuf. Sci. Eng.*, 132, 0309011 (2010).
76. A. A. Tseng, A. Notargiacomo, Nanoscale fabrication by nonconventional approaches, *J. Nanosci. Nanotechnol.*, 5, 683–702 (2005).
77. S. Miyake, J. Kim, Nanoprocessing of silicon by mechanochemical reaction using atomic force microscopy and additional potassium hydroxide solution etching, *Nanotechnology*, 16, 149–157 (2005).
78. J. W. Park, N. Kawasegi, N. Morita, D. W. Lee, Tribonanolithography of silicon in aqueous solution based on atomic force microscopy, *Appl. Phys. Lett.*, 85, 1766–1768 (2004).
79. N. Kawasegi, N. Morita, S. Yamada, N. Takano, T. Oyama, K. Ashida, Etch stop of silicon surface induced by tribo-nanolithography, *Nanotechnology*, 16, 1411–1414 (2005).
80. Y. D. Yan, T. Sun, X. S. Zhao, Z. J. Hu, S. Dong, Fabrication of microstructures on the surface of a micro/hollow target ball by AFM, *J. Micromech. Microeng.*, 18, 035002 (2008).
81. T. G. Bifano, T. A. Dow, R. O. Scattergood, Ductile regime grinding: a new technology for machining brittle materials, *J. Eng. Ind., Trans. ASME*, 113, 184–189 (1991).
82. H. T. Young, H. T. Liao, H. Y. Huang, Novel method to investigate the critical depth of cut of ground silicon wafer, *J. Mater. Process. Technol.*, 182, 157–162 (2007).
83. J. W. Park, C. M. Lee, S. C. Choi, Y. W. Kim, D. W. Lee, Surface patterning for brittle amorphous material using nanoindenter-based mechanochemical nanofabrication, *Nanotechnology*, 19, 085301 (2008).
84. R. Prioli, M. Chhowalla, E. L. Freire, Friction and wear at nanometer scale: a comparative study of hard carbon films, *Diam. Relat. Mater.*, 12, 2195–2202 (2003).
85. N. Yasui, H. Inaba, K. Furusawa, M. Saito, N. Ohtake, Characterization of head overcoat for 1 Tb/in^2 magnetic recording, *IEEE Trans. Magn.*, 45, 805–809 (2009).
86. S. Tsuchitani, R. Kaneko, S. Hirono, Effects of humidity on nanometer scale wear of a carbon film, *Trib. Int.*, 40, 306–312 (2007).
87. M. Bai, K. Koji, U. Noritsugu, M. Yoshihiko, X. Junguo, T. Hiromitsu, Scratch-wear resistance of nanoscale super thin carbon nitride overcoat evaluated by AFM with a diamond tip, *Surf. Coat. Technol.*, 126, 181–194 (2000).
88. F. Z. Fang, G. X. Zhang, An experimental study of optical glass machining, *Int. J. Adv. Manuf. Technol.*, 23, 155–160 (2004).
89. G. Lu, X. Zhou, H. Li, Z. Yin, B. Li, L. Huang, F. Boey, H. Zhang, Nanolithography of single-layer graphene oxide films by atomic force microscopy, *Langmuir*, 26, 6164–6166 (2010).
90. D. Li, M. B. Müller, S. Gilje, R. B. Kaner, G. G. Wallace, Processable aqueous dispersion of graphene nanosheets, *Nat. Nanotechnol.*, 3, 101–105 (2008).
91. D. H. Kim, J. Y. Koo, J. J. Kim, Cutting of multiwalled carbon nanotubes by a negative voltage tip of an atomic force microscope: a possible mechanism, *Phys. Rev. B Cond. Matt.*, 68, 113406 (2003).
92. Z. Lu, Y. Zhou, Y. Du, R. Moate, D. Wilton, G. Pan, Y. Chen, Z. Cui, Current-assisted magnetization switching in a mesoscopic NiFe ring with nanoconstrictions of a wire, *Appl. Phys. Lett.*, 88, 142507 (2006).
93. K. Miyake, S. Fujisawa, A. Korenaga, T. Ishida, S. Sasaki, The effect of pile-up and contact area on hardness test by nanoindentation, *Jpn. J. Appl. Phys. Part 1*, 43, 4602–4605 (2004).
94. Y. D. Yan, T. Sun, Y. C. Liang, S. Dong, Effects of scratching directions on AFM-based abrasive abrasion process, *Tribol. Int.*, 42, 66–70 (2009).
95. B. Mills, A. H. Redford, *Machinability of Engineering Materials*, Applied Science, London, 1983.
96. H. Kuramochi, K. Ando, T. Tokizaki, M. Yasutake, F. Perez-Murano, J. A. Dagata, H. Yokoyama, Large scale high precision nano-oxidation using an atomic force microscope, *Surf. Sci.*, 566–568, 343–348 (2004).

97. Z. Kato, M. Sakairi, H. Takahashi, Nanopatterning on aluminum surfaces with AFM probe, *Surf. Coat. Technol.*, 169–170, 195–198 (2003).
98. S. P. Timoshenko, J. N. Goodier, *Theory of Elasticity, 3rd ed.*, McGraw-Hill, New York, NY, 1970 (Name after H. Hertz, whose work was originally published in 1881).
99. R. W. Carpick, M. Salmeron, Scratching the surface: Fundamental investigations of tribology with atomic force microscopy, *Chem. Rev.*, 97, 1163–1194 (1997).
100. D. Tabor, *The Hardness of Metals*, Clarendon Press, Oxford, UK, 1951.
101. C. Klein, Anisotropy of Young's modulus and Poisson's ratio in diamond, *Mater. Res. Bull.*, 27, 1407–1414 (1992).
102. T. Namazu, S. Inoue, Characterization of single crystal silicon and electroplated nickel films by uniaxial tensile test with in situ X-ray diffraction measurement, *Fatigue Fract. Eng. Mater. Struct.*, 30, 13–20 (2007).
103. L. Zhou, Y. Yao, Single crystal bulk material micro/nano indentation hardness testing by nanoindentation instrument and AFM, *Mater. Sci. Eng. A*, 460–461, 95–100 (2007).
104. V. Mulloni, R. Bartali, S. Colpo, F. Giacomozzi, N. Laidani, B. Margesin, Electrical and mechanical properties of layered gold–chromium thin films for ohmic contacts in RF-MEMS switches, *Mater. Sci. Eng. B*, 163, 199–203 (2009).
105. A. A. Volinsky, J. Vella, I. S. Adhihetty, V. Sarihan, L. Mercado, B. H. Yeung, W. W. Gerberich, Microstructure and mechanical properties of electroplated Cu thin films, *Mat. Res. Soc. Symp.*, 649, Q5.3.1–Q5.3.6 (2001).
106. T. W. Wu, C. Hwang, J. Lo, P. Alexopoulos, Microhardness and microstructure of ion-beamsputtered, nitrogen-doped NiFe films, *Thin Solid Films,* 166, 299–308 (1988).
107. M. J. Madou, *Fundamentals of Microfabrication: The Science of Miniizturization, 2nd ed.*, CRC, Boca Raton, FL, 2002.
108. M. Kuruvilla, T. S. Srivatsan, M. Petraroli, L. Park, An investigation of microstructure, hardness, tensile behaviour of a titanium alloy: role of orientation, *Sadhana*, 33, 235–250 (2008).
109. Y. I. Golovin, Nanoindentation and mechanical properties of solids in submicrovolumes, thin near-surface layers and films: A review, *Phys. Solid State*, 50, 2205–2236 (2008).
110. L. Calabri, N. Pugno, A. Rota, D. Marchetto, S. Valeri, Nanoindentation shape effect: experiments, simulations and modelling, *J. Phys. Condens. Matter*, 19, 395002 (2007).
111. Z. Zhou, Y. Zhou, M. Wang, C. Yang, J. Chen, W. Ding, X. Gao, T. Zhang, Evaluation of Young's modulus and residual stress of NiFe film by microbridge testing, *J. Mater. Sci. Technol.*, 22, 345–348 (2006).
112. J. F. Archard, Contact and rubbing of flat surfaces, *J. Appl. Phys.*, 24, 981(1953).
113. M. Srivastava, V. K. W. Grips, K. S. Rajam, Electrochemical deposition and tribological behaviour of Ni and Ni-Co metal matrix composites with SiC nano-particles, *Appl. Surf. Sci.*, 253, 3814–3824 (2007).
114. K. H. Chung, H. J. Kim, L. Y. Lin, D. E. Kim, Tribological characteristics of ZnO nanowires investigated by atomic force microscope, *Appl. Phys. A*, 92, 267–274 (2008).
115. M. C. Kong, W. B. Lee, C. F. Cheung, S. To, A study of materials swelling and recovery in single-point diamond turning of ductile materials, *J. Mater. Process. Technol.*, 180, 210–215 (2006).
116. A. A. Tseng, K. P. Jen, T. C. Chen, R. Kondetimmamhalli, Y. V. Murty, Forming properties and springback evaluation of copper beryllium sheets, *Metall. Mater. Trans. A,* 26, 2111–2121 (1995).
117. C. Xiang, H. J. Sue, J. Chu, B. Coleman, Scratch behavior and material property relationship in polymers, *J. Polym. Sci. B*, 39, 47–59 (2001).
118. A. Majumdar, Scanning thermal microscopy, *Annu. Rev. Mater. Sci.*, 29, 505–585 (1999).
119. R. Szoszkiewicz, T. Okada, S. C. Jones, T. D. Li, W. P. King, S. R. Marder, E. Riedo, High-speed, sub-15 nm feature size thermochemical nanolithography, *Nano Lett.*, 7, 1064–1069 (2007).

120. C. H. Chiou, S. J.Chang, G. B. Lee, H. H. Lee, New fabrication process for monolithic probes with integrated heaters for nanothermal machining, *Jpn. J. Appl. Phys., Part 1*, 45, 208–214 (2006).
121. A. Boisen, O. Hanseny, S. Bouwstray, AFM probes with directly fabricated tips, *J. Micromech. Microeng.*, 6, 58–62 (1996).
122. B. W. Chui, T. D. Stowe, Y. S. Ju, K. E. Goodson, T. W. Kenny, H. J. Mamin, B. D. Terris, R. P. Ried, D. Rugar, Low-stiffness silicon cantilevers with integrated heaters and piezoresistive sensors for high-density AFM thermomechanical data storage, *IEEE J. Microelectromech. Syst.*, 7, 69–78 (1998).
123. D. W. Lee, T. Ono, M. Esashi, Electrical and thermal recording techniques using a heater integrated microprobe, *J. Micromech. Microeng.*, 12, 841–848 (2002).
124. H. Okabe, T. Tsumura, J. Shimizu, L. Zhou, H. Eda, Experimental and simulation research on influence of temperature on nano-scratching process of silicon wafer, *Key Eng. Mater.*, 329, 379 (2007).
125. Y. Hua, W. P. King, C. L. Henderson, Nanopatterning materials using area selective atomic layer deposition in conjunction with thermochemical surface modification via heated AFM cantilever probe lithography, *Microelectron. Eng.*, 85, 934–936 (2008).
126. O. Coulembier, A. Knoll, D. Pires, B. Gotsmann, U. Duerig, J. Frommer, R. D. Miller, P. Dubois, J. L. Hedrick, Probe-based nanolithography: self-amplified depolymerization media for dry lithography, *Macromolecules*, *43*, 572 (2010).
127. D. Wang, S. Kim, W. D. Underwood, A. J. Giordano, C. L. Henderson, Z. T. Dai, W. P. King, S. R. Marder, E. Riedo, Direct writing and characterization of poly(p-phenylene vinylene) nanostructures, *Appl. Phys. Lett.*, 95, 233108 (2009).
128. M. Lantz, D. Wiesmann, B. Gotsmann, Dynamic superlubricity and the elimination of wear on the nanoscale, *Nat. Nanotechnol.*, 4, 586 (2009).
129. J. Michler, R. Gassilloud, P. Gasser, L. Santinacci, P. Schmuki, Defect-free AFM scratching at the Si/SiO_2 interface used for selective electrodeposition of nanowires, *Electrochem Solid-State Lett.*, 7, A41 (2004).
130. M. Seo, D. Kawamata, Nano-scratching in solution to the single-crystal Ta(100) subjected to anodic oxidation, *J. Phys. D*, 39, 3150 (2006).
131. M. Pendergast, A. A. Volinsky, X. Pang, R. Shields, *in Nanoscale Tribology - Impact for Materials and Devices, Materials Research Society Symposium Proceedings*, Vol. 1085, 19–24 (2008).
132. A. A. Tseng, T. W. Lee, A. Notargiacomo, T. P. Chen, Formation of uniform nanoscale oxide layers assembled by overlapping oxide lines using atomic force microscopy, *J. Micro/Nanolith. MEMS & MOEMS*, 8, 043050 (2009).
133. S. Miyake, H. Zheng, J. Kim, M. Wang, Nanofabrication by mechanical and electrical processes using electrically conductive diamond tip, *J. Vac. Sci. Technol. B*, 26, 1660 (2008).
134. M. Koinuma, K. Uosaki, AFM tip induced selective electrochemical etching of and metal deposition on p-GaAs(100) surface, *Surf. Sci.*, 357–358, 565–570 (1996).
135. R. D. Piner, J. Zhu, F. Xu, S. Hong, C. A. Mirkin, "Dip-pen" nanolithography, *Science*, 283, 661–663 (1999).
136. L. Guangming, W. B. Larry, Controlled patterning of polymer films using an AFM tip as a nano-hammer, *Nanotechnology*, 18, 245302 (2007).
137. F. Yang, E. Wornyo, K. Gall, W. P. King, Nanoscale indent formation in shape memory polymers using a heated probe tip, *Nanotechnology*, 18, 285302 (2007).
138. X. Hu, J. Yu, J. Chen, X. Hu, Analysis of electric field-induced fabrication on Au and Ti with an STM in air, *Appl. Surf. Sci.*, 187, 173–178 (2002).
139. G. Lee, H. Jung, J. Son, K. Nam, T. Kwon, G. Lim, Y. H. Kim, J. Seo, S. W. Lee, D. S. Yoon, Experimental and numerical study of electrochemical nanomachining using an AFM cantilever tip, *Nanotechnology*, 21, 185301 (2010).
140. J. C. Rosa, M. Wendel, H. Lorenz, J. P. Kotthaus, M. Thomas, H. Kroemer, Direct patterning of surface quantum wells with an atomic force microscope, *Appl. Phys. Lett.*, 73, 2684–2686 (1998).

141. H. W. Schumacher, U. F. Keyser, U. Zeitler, R. J. Haug, K. Eberl, Nanomachining of mesoscopic electronic devices using an atomic force microscope, *Appl. Phys. Lett.*, 75, 1107 (1999).
142. M. Versen, B. Klehn, U. Kunze, D. Reuter, A. D. Wieck, Nanoscale devices fabricated by direct machining of GaAs with an atomic force microscope, *Ultramicroscopy*, 82, 159–163 (2000).
143. C. C. Faulkner, D. A. Allwood, R. P. Cowburn, Tuning of biased domain wall depinning fields at Permalloy nanoconstrictions, *J. Appl. Phys.*, 103, 073914 (2008).
144. M. Hirtz, M. K. Brinks, S. Miele, A. Studer, H. Fuchs, L. Chi, Structured polymer brushes by AFM lithography, *Small*, 5, 919–923 (2009).
145. R. R. Bhat, B. N. Chaney, J. Rowley, A. Liebmann-Vinson, J. Genzer, Tailoring cell adhesion using surface-grafted polymer gradient assemblies, *Adv. Mater.*, 17, 2802–2807 (2005).
146. S. Lenhert, A. Sesma, M. Hirtz, L. F. Chi, H. Fuchs, H. P. Wiesmann, A. E. Osbourn, B. M. Moerschbacher, Capillary-induced contact guidance, *Langmuir*, 23, 10216–10223 (2007).
147. M. Motornov, S. Minko, K. J. Eichhorn, M. Nitschke, F. Simon, M., Stamm, Reversible tuning of wetting behavior of polymer surface with responsive polymer brushes, *Langmuir*, 19, 8077–8085 (2003).
148. J. I. Shirakashi, Scanning probe microscope lithography at the micro- and nano-scales, *J. Nanosci. Nanotechnol.*, 10, 4486–4494 (2010).
149. X. L. Zhao, S. Dong, Y. C. Liang, T. Sun, Y. D. Yan, AFM for preparing Si masters in soft lithography, *Key Eng. Mater.*, 315–316, 762–765 (2006).
150. T. Plecenik, M. Gregor, M. Prascak, R. Micunek, M. Grajcar, A. Lugstein, E. Bertagnolli, M. Zahoran, T. Roch, P. Kúš, A. Plecenik, Superconducting MgB_2 weak links and superconducting quantum interference devices prepared by AFM nanolithography, *Physica C*, 468, 789–792 (2008).
151. J. Nagamatsu, N. Nakagawa, T. Muranaka, Y. Zenitani, J. Akimitsu, Superconductivity at 39 K in magnesium diboride, *Nature*, 410, 63–64 (2001).
152. F. Iwata, K. Saigo, T. Asao, M. Yasutake, O. Takaoka, T. Nakaue, S. Kikuchi, Removal method of nano-cut debris for photomask repair using an atomic force microscopy system, *Jpn. J. Appl. Phys.*, 48, 08JB2 (2009).
153. T. Robinson, A. Dinsdale, M. Archuletta, R. Bozak, R. White, Nanomachining photomask repair of complex patterns, in *Photomask Technology 2008, Proceedings of SPIE*, Vol. 7122, The International Society for Optical Engineering, 2008.
154. C. L. Henderson, W. P. King, C. E. White, H. R. Rowland, Microsystem manufacturing via embossing of thermally sacrificial polymers, *in Materials Research Society Symposium Procceedings,* EXS, 17-20, Materials Research Society, 2004.
155. J. W. Jang, D. Maspoch, T. Fujigaya, C. A. Mirkin, A 'molecular eraser' for dip-pen nanolithography, *Small*, 3, 600–605 (2007).
156. E. Henderson, Imaging and nanodissection of individual supercoiled plasmids by atomic force microscopy, *Nucleic Acids Res.*, 20, 445–447, 1992.
157. J. Hu, Y. Zhang, H. B. Gao, M. Q. Li, U. Hartmann, Artificial DNA patterns by mechanical nanomanipulation, *Nano Lett.*, 2, 55–57, 2002.
158. M. Washizu, Microsystems for single-molecule handling and modification, *in Micromachines as Tools for Nanotechnology*, H. Fujita (Ed.), Chapter 2, 22–43, Springer, Berlin, 2003.
159. D. Kim, N. K. Chung, J. S. Kim, J. W. Park, Immobilizing a single DNA molecule at the apex of AFM tips through picking and ligation, *Soft Matter*, 6, 3979–3984, 2010.
160. N. X. Randall, G. Favaro, C. H. Frankel, The effect of intrinsic parameters on the critical load as measured with the scratch test method, *Surf. Coat. Technol.*, 137, 146–151 (2001).
161. J. Li, W. Beres, Scratch test for coating/substrate systems – A literature review, *Can. Metall. Quart.*, 46, 155–174 (2007).

162. H. Y. Lin, H. A. Chen, Y. J. Wu, J. H. Huang, H. N. Lin, Fabrication of metal nanostructures by atomic force microscopy nanomachining and related applications, *J. Nanosci. Nanotechnol.*, 10, 4482–4485 (2010).
163. B. Gotsmann, U. Duerig, J. Frommer, C. J. Hawker, Exploiting chemical switching in a diels-alder polymer for nanoscale probe lithography and data storage, *Adv. Funct. Mater.*, 16, 1499–1505 (2006).
164. A. K. Geim, Graphene: status and prospects, *Science*, 324, 1530–1534 (2009).
165. N. Farkas, R. D. Ramsier, J. A. Dagata, High-voltage nanoimprint lithography of refractory metal films, *J. Nanosci. Nanotechnol.*, 10, 4423–4433 (2010).

Chapter 2
Local Oxidation Using Dynamic Force Mode: Toward Higher Reliability and Efficiency

Hiromi Kuramochi and John A. Dagata

Abstract Local oxidation by scanning probe microscopy (SPM) is used extensively for patterning nanostructures on metallic, insulating, and semiconducting thin films and substrates. Numerous possibilities for refining the process by controlling charge density within the oxide and shaping the water meniscus formed at the junction of the probe tip and substrate have been explored by a large number of researchers under both contact mode (CM) and dynamic-force mode (DFM) conditions. This article addresses the question of whether or not the oxide growth rate and feature size obtainable by each method arise from distinctly different kinetic processes or arise simply because charge buildup and dissipation evolve over different time scales for these two cases. We report simultaneous oxide-volume and current-flow measurements for exposures performed by CM and DFM and then go on to discuss the practical realization of enhanced reliability and energy efficiency made possible by a better understanding of the relation between oxidation time and ionic diffusion using DFM.

Keywords Meniscus formation · Oxidation time · Probe speed

2.1 Introduction

Local oxidation by scanning probe microscopy (SPM) [1] is a method for generating deliberate nanoscale patterns on a wide variety of substrates [See References 2–4 for reviews]. The fundamental principle of local oxidation is quite straightforward: A voltage applied between a conductive SPM tip and (positively biased) substrate, results in the formation of a highly non-uniform electric field, *E*, in the range of 10^8 V/m to 10^9 V/m. The *E* field attracts a stable water meniscus to the tip-sample junction, creates oxyanions from water molecules, and transports these oxyanions through the growing oxide film. At these field strengths almost all materials undergo reaction with oxygen anions. This simple scheme leads to oxidation of the substrate

H. Kuramochi (✉)
MANA, NIMS, Tsukuba 305-0044, Japan; NRI, AIST, Tsukuba 305-8568, Japan
e-mail: kuramochi.hiromi@nims.go.jp

A.A. Tseng (ed.), *Tip-Based Nanofabrication*, DOI 10.1007/978-1-4419-9899-6_2,

on a scale determined by the dimensions of the *nanocell* defined by the water meniscus [5, 6]. Size, shape, and stability of the nanocell, in addition to the properties of the E field, are therefore of crucial importance for successful nanolithography using local oxidation.

This leads to a classification of oxidation conditions into four basic categories: In the simplest implementation, the properties of the nanocell are determined by the spontaneous occurrence of a water meniscus when by the probe tip is in direct contact with a substrate, i.e., the microscope is operated in contact mode (CM); conversely, if the microscope is operated in dynamic-force mode (DFM), variation of the oscillation amplitude and tip-sample separation offer alternative strategies for forming and maintaining the water meniscus during tip positioning and lithographic exposure. In addition, the E field may be constant for the duration of the voltage pulse, i.e., the applied voltage and tip-sample separation are both constant (DC), or may be varied by employing a voltage modulated waveform (VM). In contact mode, the E field is then variable if a voltage waveform rather than a constant pulse is imposed. On the other hand, modulation of the E field by both voltage waveform and the oscillation of the tip becomes possible with DFM, enabling both the shaping of the nanocell and ion current flow to be controlled either independently or in a synchronized manner over a wide range of time scales.

In the present work we demonstrate that the relationship between current density and oxide formation is essentially identical for CM and DFM exposure. We do so by re-examining space-charge and water-bridge formation concepts – notions which have been largely treated separately in the past – in a coherent and consistent fashion. A significant implication of this result is that all of the most critical aspects of the oxidation process can be optimized by controlling meniscus lifetime in the DFM operating mode using principles established from the extensive kinetics studies done more conveniently in contact mode. This more inclusive view of the four basic modes of local oxidation allows us to better understand some of the parameter choices for DFM operation reported previously in the literature.

In striking a proper balance between oxidation time and ionic diffusion, this understanding points the way toward achieving *maximum reliability*, defined as the success rate of writing deliberate, continuous features, and *energy efficiency*, defined as operating as close as possible to the transition point between transient and steady-state growth. From the standpoint of practical application, these are the primary metrics for the implementation of DFM approaches to local oxidation.

2.2 Kinetics of Space Charge Buildup and Water Bridge Formation

2.2.1 Space Charge Buildup

Local oxidation by SPM is an example of a reaction-diffusion system [7, 8]. The simplest kinetic description of such a system is a pair of first-order, coupled reaction rates. As Alberty & Miller have shown [9], this simple system can be solved analytically to obtain expressions for the overall rate constants that describe the initial distribution of oxyanion binding sites in silicon and on the internal rearrangement of trapping defects as the oxide grows. Since the reaction occurs at low, i.e., room, temperature in a concentrated solid phase, reactants and products are not thermally equilibrated and time evolution of oxide is sub-diffusional. The extensive theoretical work of Plonka on dispersive kinetics for low-temperature, condensed phase systems addresses the time-dependent reactivities of a system not in thermal equilibrium with its environment [7]. The notion of anomalous diffusion [8] refers to the non-Gaussian propagation (in statistical physics terms) of the oxidation wave front through the substrate and depends on the initial distribution of oxyanion binding sites in silicon and on the internal rearrangement of trapping defects as the oxide grows. In particular, we have shown how Plonka's approach to time-dependent rate coefficients provides an excellent fit to experimental data for silicon over seven decades of time in air [10].

This approach is also consistent with the concepts of Uhlig [11] regarding the necessity of nonuniform charge distributions in the oxide film and the distinctions drawn by Fehlner & Mott about low-temperature oxidation [12]. These fundamental ideas form the basis for the subsequent description of silicon oxidation developed Wolters and Zegers-van Duynhoven [13] and Dubois & Bubbendorff [14] and us [10, 15–18]. Recent work by Kinser et al. [19] has extended this kinetic model to describe SPM oxidation of silicon in inert, organic solvents as well.

The oxide growth model for silicon is based on two assumptions, coupled kinetic pathways and time-dependent rate constants. Coupled kinetic pathways recognize that more than one reaction channel contributes to the overall growth rate. The first reaction channel represents the direct conversion of silicon and oxygen ions into oxide. The second channel corresponds to the build up of space charge defects in the oxide [15–17]. We imagine that local oxidation of silicon consists of the simplest possible example of coupled rate equations by considering a "direct" reaction represented by $A \rightarrow C$, with a first order rate constant k_4, and an "indirect" one, $A \rightarrow B \rightarrow C$, with successive rate constants k_1 and k_3 [10]. (Systems of coupled, first-order reaction rate equations were first solved by Alberty & Miller [9]. Numbering of rate constants follows their scheme.) Initial conditions are set such that $A(t = 0) = [A_o]$, $B(t = 0) = C(t = 0) = 0$, where A_o represents the initial reactant concentration of single-crystal silicon and oxygen; $B(t)$ represents an intermediate, defected SiO_x slows the transport of oxyanions; and $C(t)$ represents the local oxide in its final state, close to the density of SiO_2. However, simply applying this scheme on its own does not provide a quantitative fit to experimental data. This problem was encountered by Orians et al. [5] in their recent computational modelling effort. The source of this discrepancy has been treated in great detail by Plonka [7]: Low-temperature, condensed phase systems that are unable to achieve equilibrium with their environment exhibit time-dependent reactivity, i.e., the rate "constants" are not constant at all, but are strongly time-dependent. This means that local oxidation cannot be treated as a series of static steps since the system does not return to equilibrium, or "re-set", on a time scale comparable to the integrated exposure time. In the case of first-order rate processes, time dependence adds a factor $t^{\alpha-1}$ such that the propagation of the oxidation wave front through the substrate depends on both the initial distribution of binding sites in the substrate and the increasingly disordered internal rearrangements of trapping defects as the oxide grows. This is referred to as *anomalous diffusion* [8] since its effect is to reduce wave front propagation from the $t^{1/2}$ diffusive limit to $t^{1/2-\alpha}$.

2.2.2 Water Bridge Monitoring

Nanoscale control of water bridge formation is of fundamental interest, as well as practical concern, for nanocell control in nanolithography. Initial reports of DFM oxidation by Wang et al., in 1994, were performed using feedback ON conditions. Perez-Murano et al., also in 1994, suggested that switching the feedback OFF provided an alternative to voltage switching [20]. Then the conditions required to monitor the formation and maintenance of the nanocell – the water bridge between the tip and substrate – have been explored extensively in work carried out by Perez-Murano and Garcia and their co-workers [20–25].

Garcia and co-workers [21, 22, 24] began investigating voltage-induced meniscus formation in 1998 by monitoring the oscillation amplitude during feedback ON/OFF conditions as DC pulse voltages were applied in noncontact mode. The transitions of oscillation amplitude and tip-sample separation under successive switching of the voltage and feedback are illustrated in Fig. 2.1. Step I illustrates the cantilever oscillation amplitude (points) and tip-substrate separation (dashed line) for normal SPM operating conditions, i.e., voltage OFF, feedback ON. Step II demonstrates the reduction in both parameters under the voltage ON, feedback OFF state. This was interpreted as representing the impact on the cantilever oscillation due to the addition of the electrostatic force between the tip and substrate. Step III indicates that a reduced amplitude condition persists in the voltage OFF, feedback OFF state, although the tip-sample separation increases to an intermediate position between the Step I and Step II positions. This indicates that a water meniscus clearly forms in the

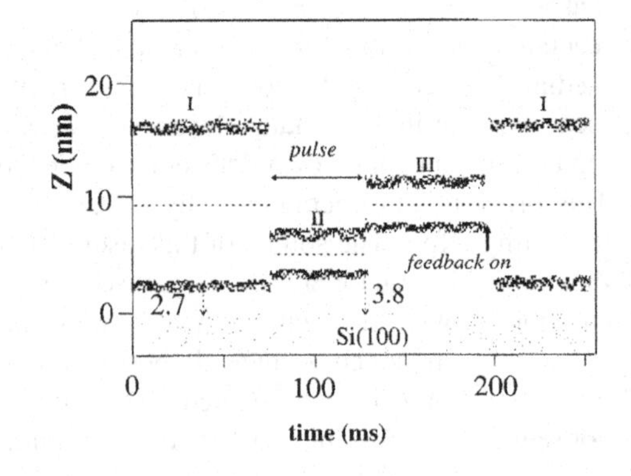

Fig. 2.1 Observation of electrostatic force and water bridge formation during dynamic force oxidation: Cantilever oscillation amplitude as a function of time before, during, and after the formation of an oxide dot. [Source: R. Garcia et al. [21], figure 1. Copyright © American Institute of Physics]

course of Step II, as the original authors conclude. The final step returns the system to the Step I state, increasing the larger oscillation amplitude corresponding to its feedback ON set point value, rupturing the meniscus and returning the tip-substrate separation to its pre-determined value.

The work by Calleja et al. [24] also investigated the threshold voltage required for the formation of the water bridge. The voltage pulse required to form a water bridge was found to be greater than that for the subsequent oxidation reaction for a given tip-substrate separation. The capacitance between water layers on surfaces of the energized tip and substrate is strongly non-uniform due to the curvature of the tip and this induces a surface tension gradient of these water films. More recent work has included computational efforts under continuum assumptions, i.e., without consideration of ionic concentration and its reorganization and mobility within these films. We examine this question in the following section since it is a critical element for understanding local oxidation in the short-pulse/fast-scan speed limit.

2.2.3 Combining the Two Concepts

Although many details of space charge buildup and water bridge formation have been established independently in recent years, there has not been a concerted effort to unify the basic principles articulated above into a coherent understanding of the nanocell, especially under dynamic conditions. This makes it particularly difficult to evaluate various approaches to DFM oxidation and reach a consensus on an entirely optimized methodology for local oxidation. This chapter certainly doesn't provide a final answer. However, in a pair of articles published in 2004 [26, 27] we discussed experiments examining current, capacitance, and charge under contact and noncontact conditions. Here we mention a few conclusions from that work which provides guidance for interpreting already published results, Section 2.3, as well as performing new DFM oxidation experiments, Section 2.4, based on these efforts to intentionally link the space charge and meniscus formation concepts.

Now, the experiment in Fig. 2.1 [21], on its own, provides no information about charge deposited and current flow that actually occurs as voltage is applied through the meniscus. Figure 2.2 presents results reported by us and our colleagues in 2003 [28]. In this work we monitor simultaneously the cantilever deflection and current measurements for a force-distance curve. Note that the voltage is 12 V, 40%*RH*, 40 N/m. This simple measurement reveals that electrostatic bending of the cantilever occurs until the force overcomes the resistance of the cantilever. At this point in the force curve, a current of about 15 pA flows momentarily through the tip-sample junction and drops off as oxidation takes place. The oxidation reaction persists until the mechanical pull-off force exceeds the attractive surface tension of the nanocell. Prior to pull-off, the attractive force associated with the nanocell reaches a point at which it is no longer in contact with the substrate. What is interesting here is that if we plot the current vs. the actual tip-surface separation (rather than z displacement); we see that current increases from 1 to 2 pA before the meniscus is ruptured. This apparently paradoxical current increase can be explained: By moving the tip out

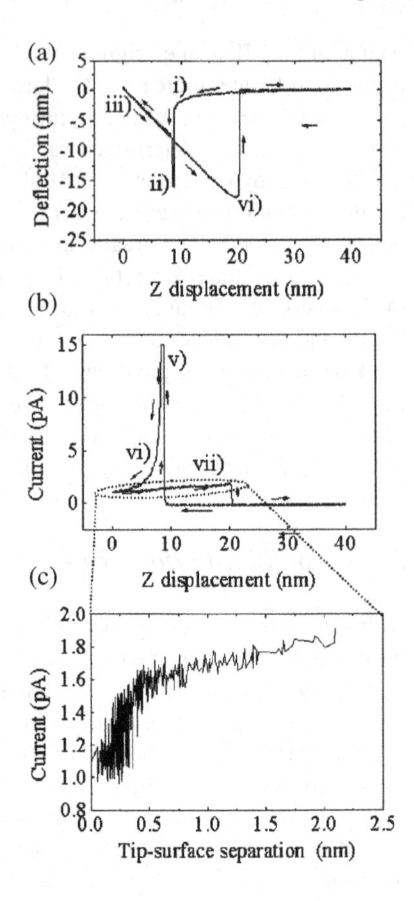

Fig. 2.2 Observation of simultaneous electrostatic cantilever bending and current flow during a force curve (voltage ON, feedback OFF): (**a**) Cantilever deflection and (**b**) current as a function of *z*-displacement. (**c**) Current as a function of tip-surface separation as the tip retracts from the substrate. Tip-surface separation is the *z*-displacement plus cantilever deflection. [Source: F. Perez-Murano et al. [28], figure 1. Copyright © American Institute of Physics]

of direct contact with the surface, ionic charge can be reorganized, and thereby reduced, allowing current across the junction to increase.

The measurement illustrated in Fig. 2.2 thus provides a complementary view to the one established in Fig. 2.1 in the sense that the connection between current flow and tip-substrate separation are correlated directly during the approach and retract cycles. Furthermore, the relationship between oxide volume and current can be investigated in detail. Figure 2.3 [27] again demonstrates the necessity of the water bridge but the experiment incorporates an important additional feature: Here oxide growth for identical pulse conditions (13 and 18 V, 1 s) performed under contact and noncontact conditions are compared. Note that the peak current obtained when the tip is in repulsive contact is on the order 10 nA whereas it is on the order of 1 nA if the current flows through a water meniscus. What is indeed interesting about this measurement is that the height, diameter, and volume of the oxide feature produced is nearly identical for the same exposure conditions regardless of whether or not the tip is in direct contact with the substrate. We interpreted this measurement to mean that considerable excess, i.e., non-faradaic, current occurs if the tip is in direct contact with the substrate; in contrast, essentially only faradaic current occurs

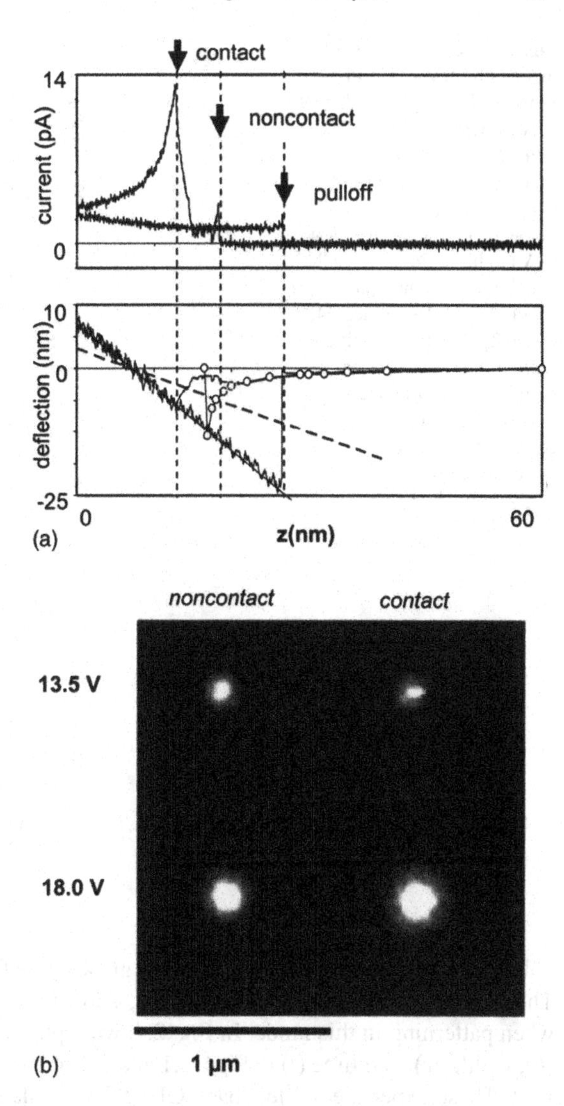

Fig. 2.3 (**a**) Observation of simultaneous electrostatic cantilever bending and current flow during a force curve (voltage ON, feedback OFF) for contact and noncontact force curves. (**b**) Oxide features patterned in contact and noncontact conditions using identical exposure conditions. [Source: J. A. Dagata et al. [27], figure 5. Copyright © American Institute of Physics]

if the tip is connected only through the water meniscus. Since non-faradaic current is either lost or contributes to the production of defects or charge traps within the growing oxide film, the preference for noncontact oxidation becomes obvious.

2.2.4 Water-Bridge Formation at High Voltage

The observation that a threshold voltage exists for the formation of a water bridge between the tip and substrate mentioned earlier was further examined by Calleja et al. [24], in particular, in a set of measurements that examined the effect of high

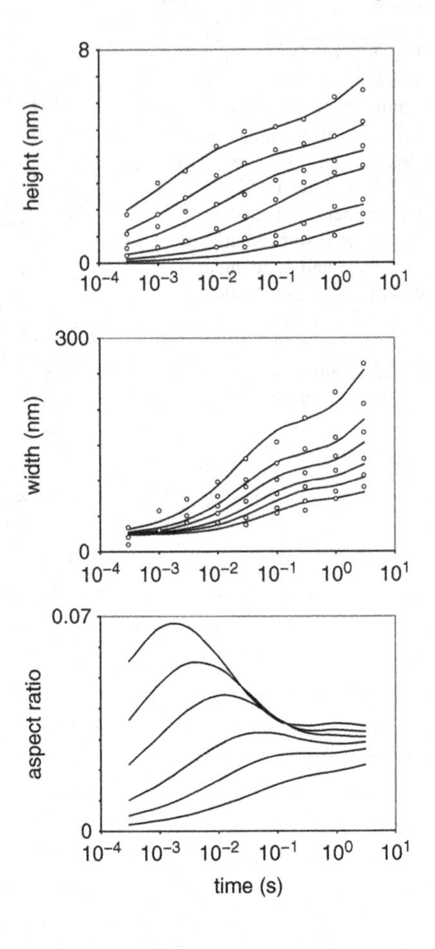

Fig. 2.4 Semi-log plots of oxide height, width, and aspect ratio (height: width) of *dots* formed during feedback OFF DFM oxidation as a function of exposure time. *Curves* increase in regular voltage increments from low (8 V) to high (24 V) in each panel. Experimental data (circles) from Reference 24, figure 2]. Calculated values (*continuous lines*) are based on previously published Alberty & Miller rate constants for silicon published in Reference 10

voltage and short exposure time on oxide growth rate for feedback OFF DFM mode. Their conclusion was that high-voltage, short-time pulses provide an advantage when patterning in this mode. In Fig. 2.4, we replot their data – presented as height (h), width (w), and time (t) in Figs. 2.1 and 2.2 of their article – as aspect ratio (h/w) vs. t. These experimental feedback OFF DFM oxidation data are compared to calculated values obtained from the Alberty & Miller kinetic model and rate constant parameters reported in Reference 10. What is interesting is that the applied voltage in this experiment ranged from 8 to 24 V, much higher than the range of voltage usually studied. Our manner of plotting the data reveals the appearance of a distinct maximum in the aspect ratio, at increasingly shorter times, as the applied voltage is increased. Figure 2.4 confirms their conclusion, but offers additional insight into a unique aspect of DFM lithography: The possibility of extremely fast oxidation, albeit if one is willing to accept limited oxide thickness of about 3 nm or so.

The benefits of high-voltage oxidation in DFM do not end with short pulse times. Simultaneous cantilever deflection and current measurements obtained during a force–distance curve using applied voltages in the range of 20 V to 30 V reveal

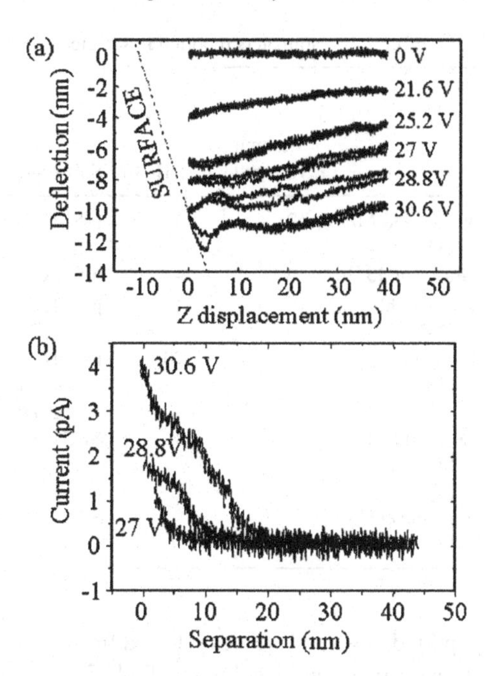

Fig. 2.5 (**a**) Cantilever deflection as a function of z-displacement during high voltage noncontact force curves. (**b**) Current as a function of tip-surface separation measured during a high-voltage, noncontact force curve. [Source: F. Perez-Murano et al. [28], figure 3. Copyright © American Institute of Physics]

that the onset of measurable current – and, hence, spontaneous meniscus formation – occurs even at a tip-sample separation of 20 nm, as indicated in Fig. 2.5. [See the text related to Reference 27, figure 7, and Reference 28, figure 2, for a more detailed discussion.] These results are consistent with the Lippmann equation, which describes the change in surface tension due to a change in surface potential arising from the capacitance of and mobile charges present in the water layers covering the tip and substrate. Extrapolating the results calculated by Garcia-Martin & Garcia to high voltage [29], we find that spontaneous meniscus formation at 20 nm tip-substrate separation is predicted for an applied voltage of 30 V. However, it should be noted that the existence of ions in the water layers cannot be neglected when we account for cyclic oxide formation during large-amplitude feedback ON DFM oxidation.

2.3 Key Concepts in Dynamic-Force Mode Methods

In addition to minimizing space charge effects during oxidation, repeatable and precise control of meniscus shape formation is a crucial element for SPM-based nanolithography. The first DFM oxidation reports performed in feedback ON conditions relied on extremely low set points to avoid instability of feedback operation due to the instantaneous switching of voltage [30]. For arbitrary pattern generation, it is desirable for the SPM system to switch smoothly from imaging mode to lithography mode as the SPM tip is brought into position. This section reviews work

Table 2.1 Parameters reported in selected DFM oxidation approaches (feedback ON)

	Maximum voltage (V)	RH(%)	Substrate	Oxidation time or scan speed	Tip coating / spring constant (N/m)	Oscillation amplitude (nm)
Vicary and Miles [34]	12	60	p-Si:H	2 cm/s	PtIr 42	5
Graf et al. [38]	16	42	n-GaAs	2 μm/s	TiPt 40	–
Kuramochi et al. [39]	10	45	p-Si:H	1 μm/s	Rh 20	258
Clement et al. [33]	30	70	n-Si:H w/oxide regrowth	2 μm/s	– 40	20
Legrand and Stievenard [32]	12	–	Si:H	–	–	12
Fontaine et al. [40]	12		Si:H	10 μm/s	PtTi 40	2

reported in several articles selected from the feedback ON DFM oxidation literature. These articles are listed in the Table 2.1 and were chosen because they specifically address control factors affecting DFM oxidation performance.

2.3.1 Synchronized Pulse, Part I

Voltage modulation (VM), i.e., using a series of unipolar or bipolar pulses, rather than a single continuous one, originated in our initial investigations of space charge buildup during local oxidation. In 1998 we were interested in validating the *interfacial* nature of the space charge as revealed by electrical force microscopy (EFM) measurements of oxide features produced under CM conditions as a function of the doping type and level of silicon substrates [15, 16]. The frequency dependence of the enhancement and the oxide density variation as a function of dopant form a necessary logical connection to the classic literature of low-temperature oxidation [11, 12].

The implication that voltage modulation enhances oxidation by reducing the build up of space charge tested a certain hypothesis. In a more practical sense, it was of interest for producing thicker oxide features and avoiding the slow-growth, steady-state regime and, especially, lateral diffusion that reduced the aspect ratio of oxide features. VM techniques provide an important benefit specific to DFM oxidation, as first described by Legrand & Stievenard in 1999 [31, 32]. Their principal concern was to overcome the old problem of avoiding the discontinuities produced by voltage switching during feedback ON operations. They studied the effect of synchronizing the voltage pulses to coincide with the phase of the cantilever oscillation. The electrostatic force, if applied under instantaneous feedback-on conditions,

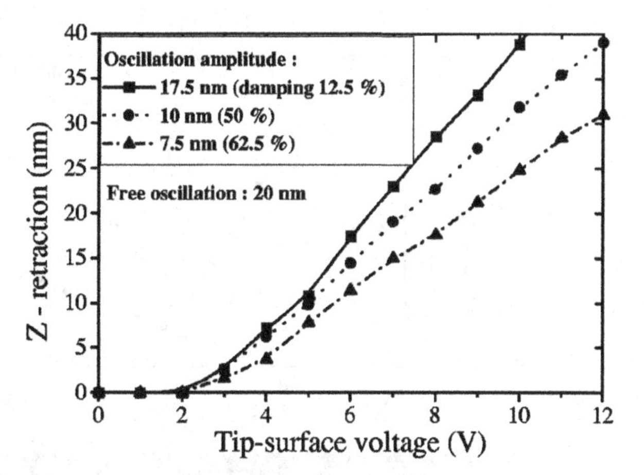

Fig. 2.6 Retraction of the z-piezo, or z-displacement, as a function of the bias voltage applied between an SPM tip and substrate. [Source: B. Legrand and D. Stievenard [31], figure 1. Copyright © American Institute of Physics]

induces instability in the mechanical oscillatory behavior of the cantilever. An example of the retraction arising from the application of the electrostatic force is shown in Fig. 2.6 [31]. Adjustment of the phase led to much reduced responses of the amplitude and tip-sample separation as a consequence of the time-averaged electrostatic effects due to modulated voltage waveforms. Figure 2.7 illustrates the control of the oxide height and width on phase and oxide lines produced at alternating phase angles. By controlling the damping of oscillation amplitude and modulation of phase between the pulsed voltages and the mechanical excitation, they showed that it was possible to minimize the electrostatic interaction and heighten the meniscus stability during large-scale patterning. To our knowledge, this work has not been followed up on. Our reasonably extensive exploration of this approach indicated that it produced excellent aspect ratio features once the system parameters were properly tuned; however, we abandoned the technique because of the propensity of a tip to stop oxidizing abruptly and once this happened could not be revived. [This behavior was confirmed by B. Legrand, private communication.]

On the other hand, the insights gained from the approach of Legrand & Stievenard also contributed to our fundamental understanding of balance between electrostatic and meniscus forces during cantilever oscillation, toward a more complete appreciation of the *dynamical* nanocell. This takes us beyond the static picture that feedback OFF characterization of meniscus properties offers.

2.3.2 High Humidity and Fast Scan Speed

In 2003 Clement et al. [33] proposed a novel set of DFM operating parameters, high voltage high humidity and demonstrated high oxidation rates for voltage pulses and relatively fast scan speeds, 2 μm/s. They concluded that oxidation rates – height of 1 nm with scan speeds of 1 cm/s could be achieved with DFM oxidation, Fig. 2.8. More recently, Vicary and Miles [34] reported oxidation heights for scan speeds as

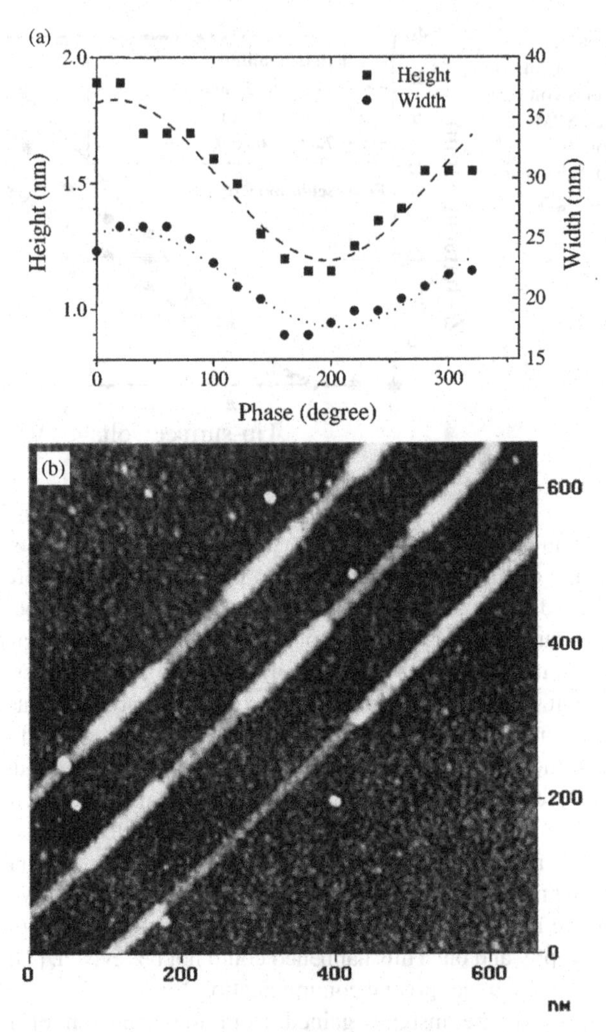

Fig. 2.7 Synchronized pulse DFM oxidation. (**a**) Variation of the oxide height and width as a function of the phase angle between voltage pulse and cantilever oscillation. (**b**) Effect of phase modulation on oxide height and width. [Source: B. Legrand and D. Stievenard [32], figures 2 and 3. Copyright © American Institute of Physics]

high as 2 cm/s and single pulse oxidation for times as short as 500 ns, Fig. 2.9. Note that the width or diameter of the oxide features is not an important consideration since these experiments were performed well within the transient region. Parameters for both are listed in the Table 2.1. Note the specific combination of (*V, RH, A*) – voltage, *RH*, and cantilever oscillation amplitude values used by Clement et al. (30 V, 70%*RH*, 20 nm) and Vicary & Miles (12 V, 60%*RH*, 5 nm) to obtain 1–2 nm of oxide height. Both authors attribute the persistence of the water meniscus as the key explanation for the success of their respective techniques.

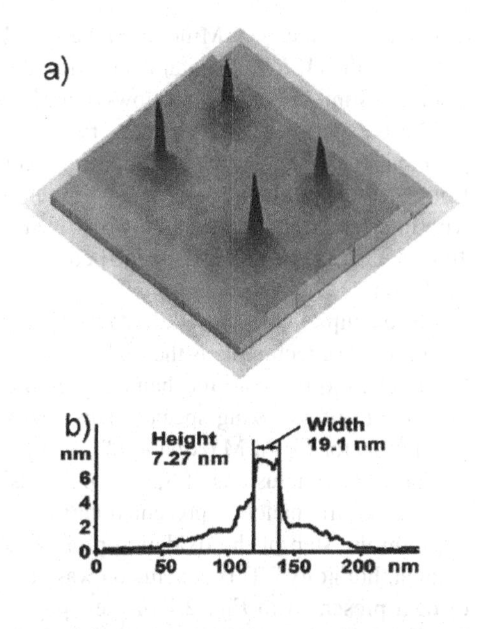

Fig. 2.8 SPM topographic image (**a**) and line profile (**b**) of oxide dots produced using high-voltage, high humidity DFM oxidation. [Source: N. Clement et al. [33], figure 1. Copyright © 2010 American Vacuum Society]

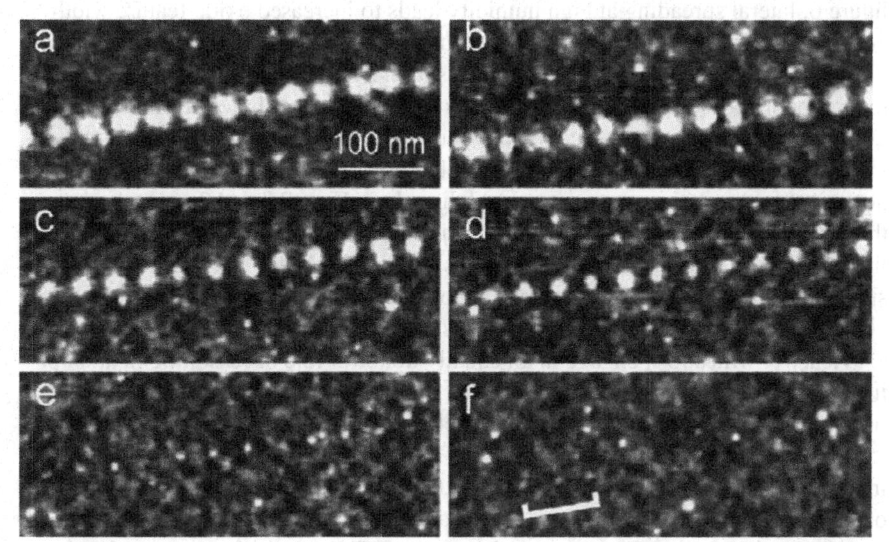

Fig. 2.9 SPM topographic image of *oxide dots* produced using high probe speed DFM oxidation. Exposure time varies from 100 μs in panel (**a**) to 500 ns in panel (**e**). [Source: J. A. Vicary and M. J. Miles [34], figure 2. Copyright © 2010 Elsevier B.V.]

On the basis of our unified concept of the nanocell, we say that all probe-based oxidation follows from the same fundamental process, i.e., that a single process operates from the transient, fast-growth regime, at the transition point, and throughout the steady-state, slow-growth regime. After all, this is what the coupled kinetic equations of Alberty & Miller are all about. If the fundamentals are shared, it must mean that the (V, RH, A) parameter space allows a reduced RH if A is increased, for instance. Similarly, higher V allows a larger A, and so forth.

These experimental results have been interpreted by the authors by framing the principal question in terms of a "limiting oxidation rate" and "mechanical pressure". Both Clement et al. and Vicary & Miles chose to pick up the discussion beginning with interpretations given in the work of Snow et al. [35–37] on CM oxidation and Tello & Garcia and co-workers [25] concerning differences between CM and DFM oxidation.

For example, Clement et al. state [33] "Therefore the main difference with the pulsed voltage technique is the higher tip oscillation amplitude combined with high humidity ratio to avoid mechanical pressure during oxide growth and to increase the amount of oxidizing species. This also suggests that oxidizing species are the limiting factor for AFM oxidation."

The early conclusions of Snow are inconsistent with the space charge model and the view of the nanocell presented earlier. Snow et al. [37] concluded that "...the rate-limiting step of the oxidation process is the production of O anions from the ambient humidity." This conclusion was based on a clearly erroneous interpretation of data presented in Fig. 2.1 of their paper. As we have shown in Reference 28, figure 6, lateral spreading at high humidity leads to increased oxide feature width – and at the near saturation levels, 95%RH, used in the Snow work – accounts quantitatively for their observations, not condensation from the vapor. Keep in mind that their vapour-to-liquid transition hypothesis was never actually tested. In Reference 29, figure 7 we demonstrate that ionic mobility within the water layers, not condensation from the vapour, explains their results. Current flow through the nanocell is necessary for oxidation, as demonstrated directly in Figs. 2.3 and 2.4. Furthermore, the long-range formation of a water bridge in Fig. 2.5 agrees qualitatively and quantitatively with the capacitance calculations of Garcia-Martin & Garcia [31]. Quite simply, water layers are always present on the tip and substrate surfaces above 20%RH and, if they are not, oxidation does not occur.

Vicary & Miles [34] engage in an equally doubtful line of reasoning in attempting to explain results shown in Fig. 2.2 of their paper: In order to account for the observation that oxide growth ceases entirely for pulse durations longer than 20 μs, they invoke a model proposed by Tello & Garcia [25] in which the growth rate during CM oxidation is suppressed because of the additional work done by the growing oxide feature to bend the cantilever. [Note, again, that this hypothesis has never been tested experimentally.] The idea applied to the DFM case is that, due to the short duration of the cantilever oscillation and oxidation pulse in their experiments, the feedback loop does not raise the tip in response to the growth of oxide. The difficulty with this reasoning is that it ignores the degrees of freedom that oxyanions possess in the nanocell – and here we specifically include lateral diffusion within both the water meniscus and the oxide. In the nanocell, anions are not constrained by a fixed boundary, rather, they respond by diffusing laterally, a dissipative mechanism at the heart of the space-charge concept.

2.3.3 Synchronized Pulse, Part II

Graf et al. [38] tried optimizing the ratio of oxidation time to total pulse cycle in their approach to voltage modulation during feedback ON DFM oxidation of Ga[Al]As and n-Si. To achieve the most stable and efficient lithographic conditions, these authors suggested that it is necessary to determine an effective oxidation time ratio for the system. They found that the feedback ON DFM requirement of operating with a low set point, i.e., 8% < DC set < 16% of the free oscillation amplitude, as others had reported, could be substantially increased to a range of 10% < VM set < 40%, because the time averaged electrostatic force experienced by the feedback loop is reduced, similar to the reasoning of Legrand & Stievenard [31], however, without the additional step of exactly synchronizing the voltage pulse and cantilever oscillation.

Another interesting result that they report is one bearing directly on the question of comparative reliability of VM versus DC exposure for feedback ON DFM oxidation. The series of lines shown in the SPM image of Fig. 2.10 are representative

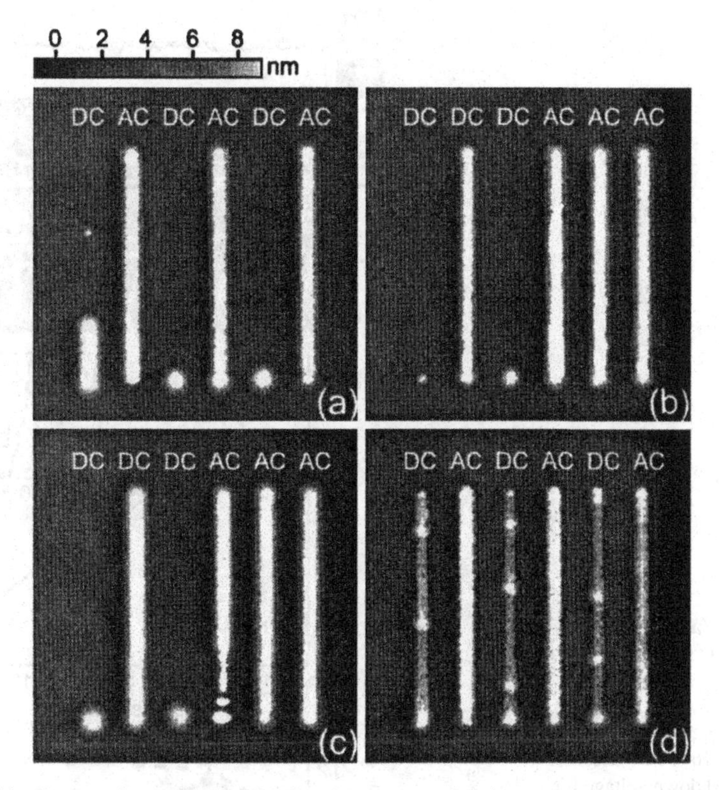

Fig. 2.10 Comparative reliability of DC and voltage modulation, referred to as VM in the text and labeled AC in the figure. SPM topographic image presents alternating series of oxide lines produced by DFM oxidation on undoped GaAs. [Source: D. Graf et al. [38], figure 5. Copyright © American Institute of Physics]

of a much larger set of exposures. As Graf et al. [38] note "In ac oxidation we have the chance to pick up the water film in each cycle. . .in the *dc* mode, however, once it failed at the beginning, the strong additional electrostatic damping will keep the tip too far from the sample surface for a water bridge to form." By operating in a pulsed mode situation, even an inevitable disruption of the nanocell can be repaired on the following voltage cycle. We gradually become aware of the possibility that a persistent meniscus is not actually necessary for reliable DFM oxidation. In other words, it may be essential for a stable feedback loop, but not for lithography. As we demonstrate shortly, the persistence of the meniscus is not even needed for that.

These authors also explore another aspect of practical concern to DFM oxidation in their comparative study of the breakdown voltage of DC and VM oxide lines. The density and electrical properties of oxide features are crucial to their performance as prototype etch masks and device structures. To study the voltage breakdown of DC and VM lines, they defined the elements of a Ga[Al]As two-dimensional electron gas (2DEG) device is shown in Fig. 2.11a, with DC isolation patterned in the

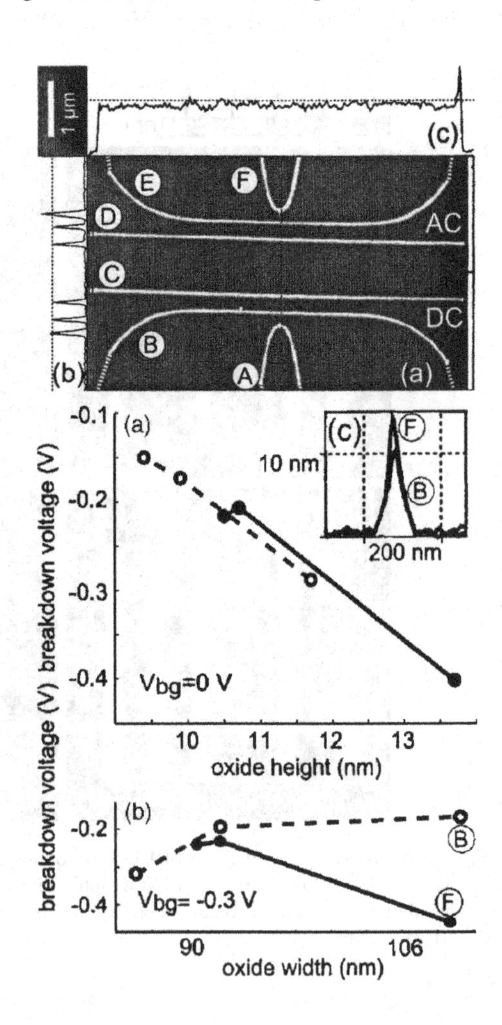

Fig. 2.11 (**a**) SPM topographic image of a GaAs 2DEG test device with lines fabricated by DC (lower half) and VM (upper half) DFM oxidation. (**b**) Comparison of oxide breakdown voltage for DC and VM DFM oxide lines. [Source: D. Graf et al. [38], figures 6 and 7. Copyright © American Institute of Physics]

lower half and VM isolation patterned in the upper half of the overall structure. A breakdown threshold corresponding to 10 pA was determined for each oxide line in the series and plotted in Fig. 2.11b. They attribute the substantial increase of 186% in breakdown voltage to the more uniformly continuous nature of the VM-patterned oxide. While the authors conclude that the final confinement potential of each region of the test device is comparable for DC and VM methods – in doing so they attach more significance to the magnetoresitive results than to oxide quality directly – it is useful to generalize from the special case of remote 2DEG patterning to other device applications.

Voltage breakdown and oxide homogeneity may be far more important metrics for silicon-oxide tunnel barriers or anisotropic etch resistance, for example. It is worthwhile to mention that some years ago we also examined the density variation for local silicon-oxide device structures patterned by DC and VM techniques [18]. Furthermore, annealing effects on the oxide density and interfacial defect density of the oxide features were pointed out and the reduction of such defects produced during the patterning of device structures by local oxidation is necessary in order to achieve acceptable device performance in these systems.

2.4 New Aspects of Reaction Control by Probe Speed

This section sheds new light on achieving the highest possible reliability, writing speed, and large-scale precision through simultaneous analysis of oxide-volume and current-flow measurements during DFM oxidation. We have already reviewed the formation of the nanocell on the basis of investigations in the feedback OFF condition in the foregoing section. While certainly adequate for gaining deeper understanding of the underlying kinetic process, operating in this mode limits patterning to single point-like features which is unsuitable for the practical applications of interest here.

It is for this reason that we focus our attention on feedback ON DFM in the rest of this chapter. Furthermore, we evaluate CM and feedback ON DFM oxidation as an explicit function of probe speed, v, during fabrication of continuous oxide lines and area features [39–45]. Since probe speed is more complicated than a simple inversion of time in the case of VM feedback ON DFM oxidation, when "time" is discussed in the following, a careful distinction between voltage-pulse duration and meniscus-formation cycles [39] must be made. Such a distinction does not arise for CM oxidation of course.

2.4.1 *Small and Large Amplitude*

Most of the authors of articles listed in the Table 2.1, Legrand & Stievenard, Fontaine et al., Vicary & Miles, and Graf et al., have emphasized how the stability of the cantilever oscillation depends on selecting a small value of the oscillation amplitude and a low amplitude set point during feedback ON DFM oxidation. By contrast, we and our colleagues have shown that there is another way to look at the

DFM control problem – by using DC voltage but working at a very *large amplitude*. Cyclic meniscus formation was also found to occur when oxidation was performed with the cantilever driven at high oscillation amplitudes and constant DC voltage was applied. This is comparable to performing oxidation using pulsed voltages. We found that during that part of the cycle when the tip-sample separation is smallest, the meniscus length is essentially constant for fixed voltage and *RH* from cycle to cycle. Thus a meniscus could be formed periodically at large amplitude settings (>200 nm). In this case, since maintaining feedback in its active state is possible under these conditions, we can avoid electrostatic damping, and that as the probe tip makes its closest approach to the substrate, a water meniscus is regenerated as it would be in the case of a voltage pulse. Meniscus lifetime could be controlled to sub-microsecond order by adjusting the cantilever amplitude setting or by modulating its oscillation amplitude by altering the probe-sample distance as illustrated in Fig. 2.12 [23]. This regeneration is both spontaneous and more reliable as revealed in the foregoing sections.

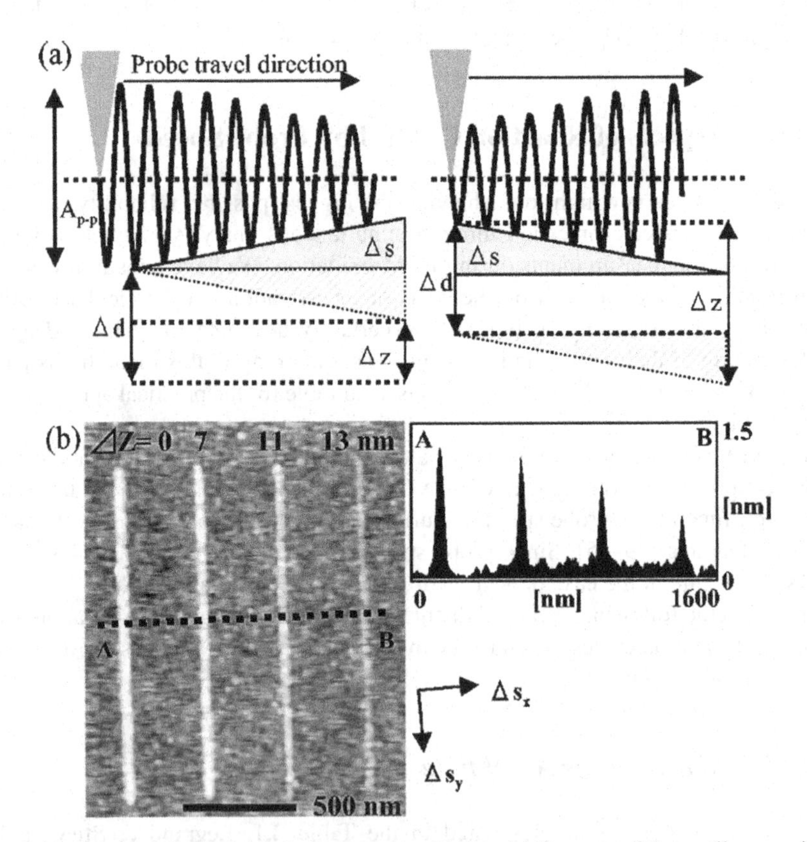

Fig. 2.12 Large-amplitude dynamic force mode oxidation. (**a**) Decrease of the cantilever oscillation amplitude, $\Delta z = \Delta d - \Delta s$, calculated from the z-piezo displacement, Δd, and the slope of the substrate, Δs. When used in the feedback OFF condition, the minimum oscillation amplitude necessary for stable meniscus rupture may be obtained. [Source: H. Kuramochi et al. [39], figure 1. Copyright © American Institute of Physics]

2.4.2 Comparison of Contact and Dynamic Force Mode Oxidation

Efforts to employ feedback ON DFM oxidation have proceeded in parallel with feedback OFF investigations. Figure 2.13 reproduces early results from Fontaine et al. comparing oxide growth rate for CM and feedback ON DFM oxidation as a function of probe speed [40]. The general trends are that the oxide height, or thickness, and the maximum probe speed that can be used for DFM are somewhat less than for CM, although the voltage-dependent slopes are similar. Note also that the oscillation of the cantilever is set to a remarkably low value of 2 nm in these experiments.

What we now wish to examine is how current and volume calculated for CM and DFM vary in such measurements. Moreover, we want to determine if rupturing and reforming the nanocell during each oscillation cycle of the probe tip – *large-amplitude* DFM lithography – produces a qualitatively similar oxide as low-amplitude DFM lithography. For this purpose, comparative CM and DFM experiments were carried out. Lithographic patterning was performed using an environmental control SPM unit at a constant relative humidity of 50% at room temperature. Probe speed was varied over a range of 0.1–20 μm/s. A hydrogen passivated p-type Si(001) sample (1 – 10 Ω · cm) was used as the sample. Rh-coated Si cantilevers for CM (~0.1 N/m), for DFM (~20 N/m) and modified cantilevers

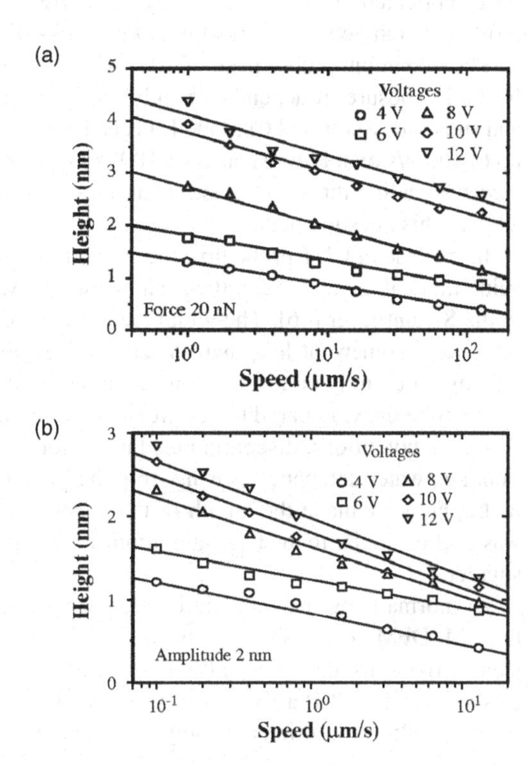

Fig. 2.13 Comparison of contact mode and dynamic force mode oxidation as a function of probe speed: (**a**) Oxide line height by CM. (**b**) Oxide line height by DFM. [Source: P. A. Fontaine et al. [40], figures 10 and 17. Copyright © American Institute of Physics]

(~3 N/m) with a conductive multi-walled CNT (MWNT) are employed. CNT probes were also used in DFM operation; hereafter experiments with CNT probes were called CNT. The oscillation amplitude set point was maintained at ~250 nm (DFM) and less than 10 nm (CNT) during nano-oxidation by maintaining active feedback control. Oxide production was decided by faradaic current detection and visual evaluation of SPM topographic and/or current images.

The effect of probe speed on the line width, volume, and current are summarized in Fig. 2.14. The decrease in average value of the width (FWHM), height and volume of the fabricated oxide lines with increasing probe speed are plotted in Fig. 2.14a–c. The volume of oxide features is standardized as all oxide lines have the same length of 1 μm. Lines in the graphs are to guide the eye. A large decrease in width and volume occurred with increasing probe speed for $v < 1$ μm/s. In the region of $v > 1$ μm/s, decreases in width and volume of oxide lines were slight. The amount of detected current was found to be in good agreement with the value of ionic charge calculated from the oxide volume for all experimental comparisons, indicating that the detected current was indeed the faradaic current. The faradaic current per unit time increased proportionally with probe speed at low probe speed and became virtually constant at $v = 10$ μm/s, as shown in Fig. 2.14d.

Similar experiments employing CM operating mode, Fig. 2.14e–h, and CNT probe tips, Fig. 2.14i–l, were performed under similar exposure conditions. The quantitative properties are basically similar to that in the case of DFM, thicker lines were fabricated at the slower speed and current per unit time increased at higher probe speed in good agreement with the results of Fontaine et al.

The maximum probe speed of 20 μm/s employed here is not the threshold speed for CM exposure. It depends strongly on the *RH* and material properties of the tip and substrate. Snow and Campbell succeeded fabricating oxide lines at probe speeds up to 1 mm/s on a H-passivated Si(100) surface [35]. Does DFM oxidation possess a corresponding threshold – and what does it mean physically? We will attempt to address this question next.

In the case of CNT probe tips, the applied sample voltage of 8 V is the maximum value that can be applied without damaging the welded attachment of the MWNT to the Si cantilever [46]. The width and normalized volume of the fabricated oxide lines were somewhat less than in the case of DFM fabrication. It is notable that making an entirely continuous line became difficult at $v > 5$ μm/s in the case of a CNT probe tip. Compared to our previous experience with *RH* dependence of DFM oxidation conditions, discontinuous fabrication is a result of poor meniscus formation, i.e., water deficiency, arising from the hydrophobicity of CNT probe tips. The hydrophobicity made the length of the water short, therefore, meniscus formation was cyclic, similar to the large-amplitude case, although the amplitude set point was smaller.

By normalizing the faradaic current density in terms of the oxide volume for the CM, DFM, and CNT datasets presented in Fig. 2.15a, we find that the progression of oxide formation for all three operating conditions follows a common path as shown in Fig. 2.15a. However, there is a key difference that distinguishes these three operating conditions, namely an effective time factor for oxidation – the time

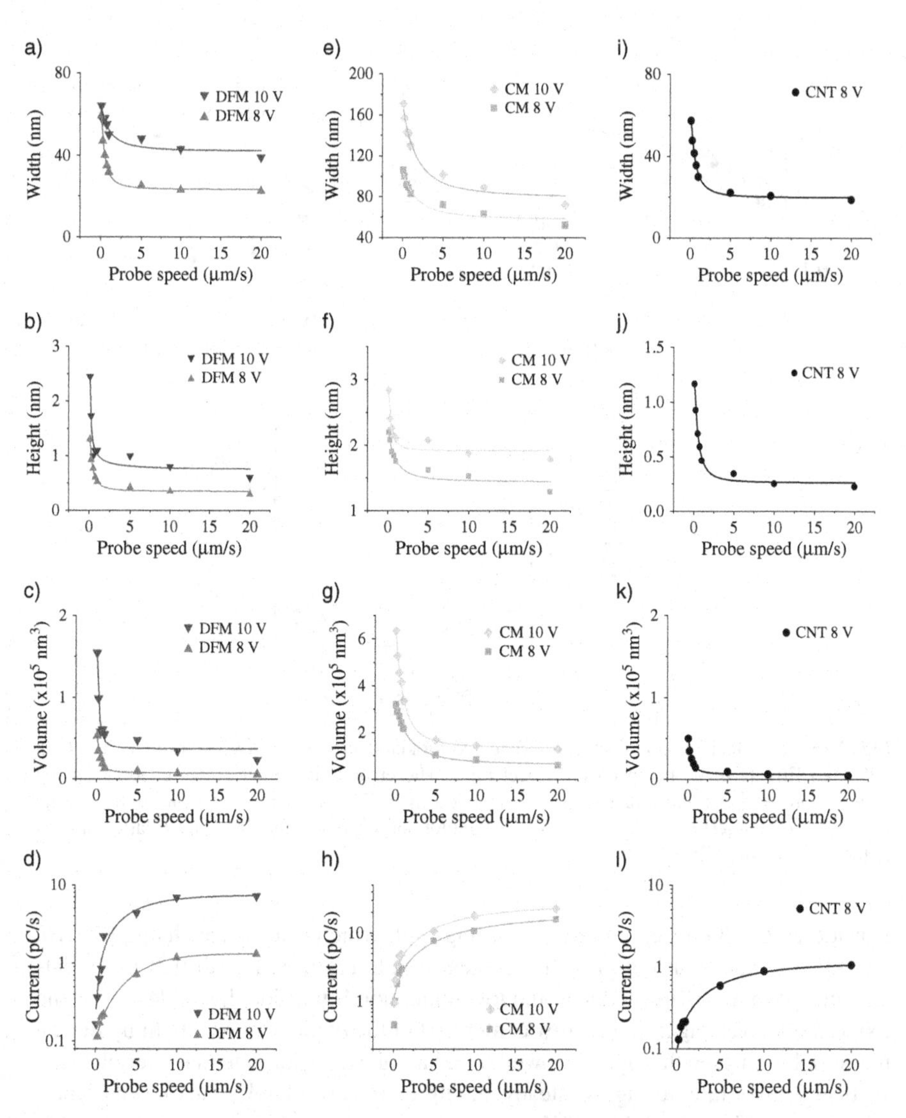

Fig. 2.14 Effects of probe speed on oxide volume and detected faradaic current in DFM (**a**)–(**d**), CM (**e**)–(**h**) and CNT (**i**)–(**l**) nano-oxidation.Oxide width with an accuracy of ±2 nm (**a**), (**e**), (**i**), and the volume (**c**), (**g**), (**k**) decreased when the probe speed increased under all conditions. The measurement errors in the width and height made the error in volume less than 10%. By contrast, detected current (**d**), (**h**), (**l**) increased with the increment of probe speed. The original detection limit of the current amplifier is ~50 fA

factor directly reflects the interplay of oxide growth kinetics and the cyclic nature of the water meniscus for large-amplitude DFM oxidation. Since the voltage pulse duration and humidity effect are strongly coupled to the rates of ionic diffusion and meniscus life time [26, 27, 39, 41, 43, 44], the growth mechanism must be

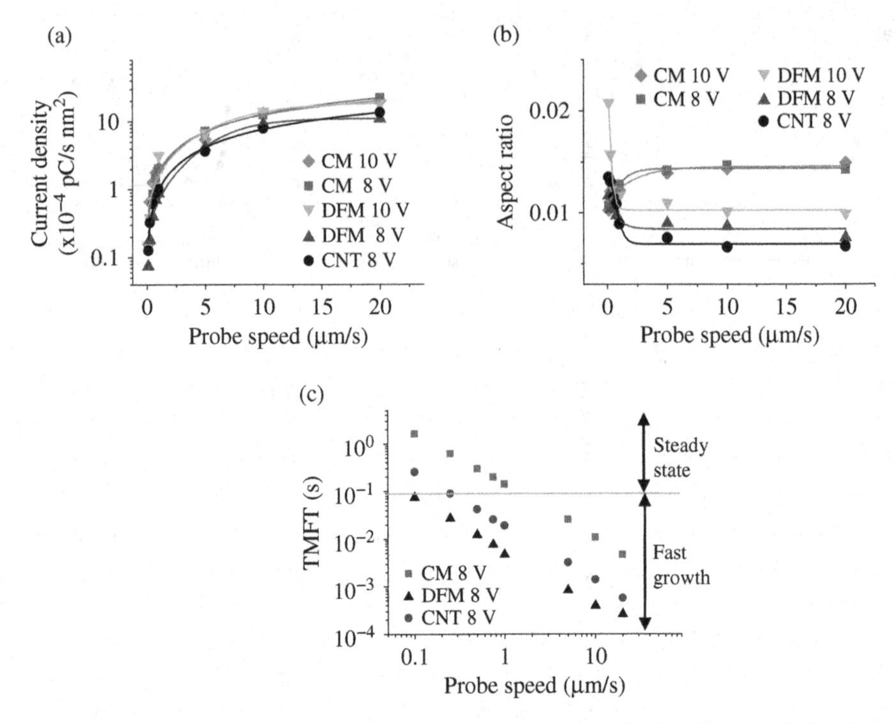

Fig. 2.15 (**a**) Effects of probe speed on detected faradaic current density. The error is less than 10%. (**b**) Effects of probe speed on aspect ratio. The error is less than 10%. (**c**) Relationship between probe speed and total meniscus formation time (TMFT). The *border line* indicates the transition point at $RH = 50\%$ and $Vs = 8$ V. If the plot point is below the line, the oxidation state is in the fast growth regime

considered as a two step process according to the role of the space charge [15, 16, 42, 43], i.e., the transient-growth and steady-state regimes. Recall that in the latter, surface ionic diffusion due to the low ionic mobility inside the oxide causes the oxide shape to change: The narrow center part of the oxide feature is built up during the initial high growth stage, followed by a lateral expansion rate that overtakes the vertical growth rate during the steady state growth stage. Under the present experimental conditions for which $RH = 50\%$, the humidity effect does not yet appear obvious [44], but surface ionic diffusion definitely occurs as space charge persists for longer and longer exposure times.

The role of cyclic meniscus formation becomes apparent by replotting these data in Fig. 2.15b, *aspect ratio vs. probe speed*, as we did earlier for DFM voltage pulses in Fig. 2.4, *aspect ratio vs. time*. Figure 2.15b also indicates a crossover from steady state growth conditions at low probe speed to transient growth conditions at high probe speed. In the case of CM oxidation, the aspect ratio *increases* somewhat with increasing v to become almost constant at high probe speed for probe speed above 5 μm/s; conversely, the contribution of lateral oxide growth to total oxide volume production is significant at low probe speed. By contrast, the aspect ratio *decreases* with the increase in probe speed in the case of DFM and CNT oxidation because vertical oxide growth continues during repeated oxidation of the same surface area – the system remains in the transient regime. The explanation is that meniscus formation is cyclical, the effective oxidation time is very short in DFM, and the oxidation reaction terminates for a portion of every oscillation cycle before lateral ionic diffusion occurs [43]. Moreover, as the probe tip is displaced in response to the voltage pulse, water flows onto the fabricated oxide, which is considerably more hydrophilic than the surrounding H-passivated Si surface. Water condensation is thus confined to a narrow region of the meniscus directly beneath the probe tip and oxidation at the base of the meniscus is suppressed.

2.4.3 Concept of the Total Meniscus Formation Time

As noted earlier, the first detailed investigations of DFM oxidation were performed under greatly reduced oscillation amplitude of the cantilever. Careful monitoring of the amplitude is necessary to ensure stable meniscus formation during voltage switching. Both the work of Garcia et al., Fig. 2.1, and Fontaine et al., Fig. 2.13, employed small oscillation amplitude, i.e., under feedback OFF and feedback ON DFM oxidation, respectively. The results shown in Figs. 2.12, 2.14 and 2.15 indicate that operating with such small oscillation amplitudes is not necessary.

However, operating successfully in the large-amplitude regime requires a separate analysis of cyclic meniscus formation since the physical situation is qualitatively different from the small-amplitude case. In this sense, feedback OFF DFM oxidation is more akin to CM exposure since the meniscus remains intact for the entire length of the voltage pulse. Thus we can define a *total meniscus formation time* (TMFT) at a given location on the substrate to be d ms. (Consider that if the probe speed is set at 1 μm/s, a meniscus of diameter d nm exposes a certain minimum area at a single location on the substrate for a duration of d ms.) But the TMFT for large-amplitude DFM exposure depends also on the oscillation frequency of the cantilever and the amplitude set point – the cyclical existence of a meniscus and therefore the propensity for oxidation through the nanocell and so may be quite different [17, 18].

To make an estimate of such difference, we may assume that the oscillation of the cantilever is harmonic. For example, using resonant frequencies of 110 and 150 kHz, setting amplitudes of 250 and 5 nm, and a meniscus length of 8 and 2 nm for DFM and CNT cases [41, 46], respectively, then the meniscus formation time may be estimated to be about 0.83 and 1.97 μs, respectively. TMFTs calculated for these exposure conditions follow a power-law relationship as shown in Fig. 2.15c. Now if oxidation exposure occurs continuously at a given substrate location, the growth rate decreases due to the buildup of space charge. The transition point is about 0.09 s for typical values of $RH = 50\%$ $Vs = 8$ V for CM oxidation [42].

On the other hand, if the TMFT is less than the transition point, the reaction state remains in the fast growth regime as confirmed by the single-exponential fit to the data in Fig. 2.15c, exposure conditions for DFM mode correspond to the fast growth regime. In the CNT case, $v = 0.25$ μm/s is just on the boundary between fast and slow growth conditions. If we want to operate CM oxidation in the fast growth regime, the probe speed must be set to 5 μm/s or higher. This trend is similar to that of the humidity effect in CM, DFM and CNT cases [43]. Operation in the fast growth regime improves reproducibility in the size and shape of oxide features since variability due to ionic diffusion is reduced. Now it is clear that we can control the reaction state by the probe speed.

2.5 Conclusion: Toward Higher Reliability and Efficiency

From a practical standpoint, quality of the oxide feature and performance of the process have primary importance. Not only the final accuracy of the fabrication, i.e., pattern placement and aspect ratio, but also the reproducibility, stability, and energy efficiency of the underlying process must be continuously refined. The role of the most obvious control factors – (V, RH, t) – have become standard parameters reported by all authors in the field. Through this reporting we gain a sense of the "typical" range of the values employed in the published literature and thus are in a better position to assess potential benefits of applying unusually large voltages or very high RH. Then, on the other hand, there are examples that point to an effective combination both large V and RH, as shown in the work of Fontaine et al. [40]. However, not all of these factors can at once be established in every experimental situation, leading to confusion over an optimized choice of parameters to employ in practical processing conditions. And this is why we need to develop an adequate understanding of the full range of possibilities available with DFM oxidation. Then the question rests on how to incorporate the insight from demonstrations of fast scan/small amplitude by Vicary & Miles for instance with our novel approach which shows that it is not essential for cantilever stability if the meniscus doesn't really remain intact.

Our own view emphasizes that monitoring faradaic current and understanding the existence of charged ions in the water layers, prior to and during the spontaneous formation of the water bridge are essential facts that account for the large-amplitude oxidation results during feedback ON DFM. As we have shown, this ultimately reduces uncertainty in feature size and increases reliability for large-scale patterning of fine features. It should be noted that measuring faradaic current in the case of high conductivity samples may not always be possible. In this case, relying on effective exposure "time" estimates calculated from the TMFT, although not exact, nevertheless provides a guide to suppressing lateral ionic diffusion and achieving higher aspect ratio and enhanced reproducibility. An important feature of TMFT is that it can be controlled externally by setting the *dynamic* operating parameters of the SPM instrument, namely, the resonance frequency, amplitude of the cantilever, RH, probe-sample distance and probe speed.

Energy efficiency is another important practical aspect of exposure control. The highest energy efficiency is achieved when the oxidation rate is maintained just below point at which the transition to lateral ionic diffusion becomes significant and oxidation remains in the fast growth regime as long as possible. Direct faradaic current detection during CM oxidation at a fixed point under actual process condition is the best way to determine this transition point. It is also possible to estimate it by a calculation of the growth rate from the oxide volume production. The DFM oscillation and probe speed can then be adjusted so that TMFT occurs just before the transition point. This is possible because very rapid, cyclic meniscus formation in high-amplitude DFM operation mode is more precisely controllable than voltage pulses defined solely in terms of time.

This chapter has attempted to bring together the more familiar aspects of the electrochemical nanocell with a critical discussion of feedback ON DFM oxidation. Future progress of local oxidation demands that novel refinements of these methods be investigated and the full implication for faster, more reliable operation uncovered and reported. Finally, let us express our hopes that such gains in processing efficiency will eventually find their way into truly high-throughput extensions of local oxidation, e.g., high-voltage nanoimprint lithography [46, 47] or parallel cantilevers [48].

References

1. J. A. Dagata, J. Schneir, H. H. Harary, C. J. Evans, M. T. Postek and J. Bennett, Appl. Phys. Lett. **56**, 2001 (1990).
2. A. A. Tseng, A. Notargiacomo and T. P. Chen, JVST B **23**, 877 (2005).
3. D. Stievenard and B. Legrand, Prog. Surf. Sci. **81**, 112 (2006).
4. J. A. Dagata, in Scanning Probe Microscopy: Electrical and electromechanical phenomena at the nanoscale, edited S. Kalinin and A. Gruverman, Springer Press, New York (2007), Vol. II, p. 858.
5. A. Orians, C. B. Clemons, D. Golovaty and G. W. Young, Surf. Sci. **600**, 3297 (2006).
6. S. Djurkovic, C. B. Clemons, D. Golovaty and G. W. Young, Surf. Sci. **601**, 5340 (2007).
7. A. Plonka, Prog. React. Kinet. **16**, 157 (1991).
8. R. Meltzer and J. Klafter, Phys. Rep. **339** 1 (2000) and references therein.
9. R. A. Alberty and W. G. Miller, J. Chem. Phys. **26**, 1231 (1957).
10. J. A. Dagata, F. Perez-Murano, G. Abadal, K. Morimoto, T. Inoue, J. Itoh, K. Matsumoto and H. Yokoyama, Appl. Phys. Lett. **76**, 2710 (2000).
11. H. H. Uhlig, Acta Metall. **4**, 541 (1956).
12. F. P. Fehlner and N. F. Mott, Oxid. Met. **2**, 59 (1970).
13. D. R. Wolters and A. T. A. Zegers-van Duynhoven, J. Appl. Phys. **65**, 5126 (1989).
14. E. Dubois and J-L. Bubendorff, J. Appl. Phys. **87**, 8148 (2000).
15. J. A. Dagata, T. Inoue, J. Itoh and H. Yokoyama, Appl. Phys. Lett. **73**, 271 (1998).
16. J. A. Dagata, T. Inoue, J. Itoh, K. Matsumoto and H. Yokoyama, J. Appl. Phys. **84**, 6891 (1998).
17. F. Perez-Murano, K. Birkelund, K. Morimoto and J. A. Dagata, Appl. Phys. Lett. **75**, 199 (1999).
18. K. Morimoto, F. Perez-Murano, and J. A. Dagata, Appl. Surf. Sci. **158**, 205 (2000).
19. C. R. Kinser, M. J. Schmitz, and M. C. Hersam, Adv. Mater. **18**, 1377 (2006).
20. F. Perez-Murano, G. Abadal, N. Barniol, X. Aymerich, J. Servat, P. Gorostiza and F. Sanz, J. Appl. Phys. **78**, 6797 (1995).
21. R. Garcia, M. Calleja and F. Perez-Murano, Appl. Phys. Lett. **72**, 2295 (1998).
22. M. Calleja, J. Anguita, R. Garcia, K. Birkelund, F. Perez-Murano and J. A. Dagata, Nanotechnology **10**, 34 (1999).

23. R. Garcia, M. Calleja and H. Rohrer, J. Appl. Phys. **86**, 1898 (1999).
24. M. Calleja and R. Garcia, Appl. Phys. Lett. **76**, 3427 (2000). We thank RG for providing the data in Figure 4 in numerical form.
25. M. Tello and R. Garcia, Appl. Phys. Lett. **79**, 424 (2001).
26. J. A. Dagata, F. Perez-Murano, C. Martin, H. Kuramochi and H. Yokoyama, J. Appl. Phys. **96**, 2386 (2004).
27. J. A. Dagata, F. Perez-Murano, C. Martin, H. Kuramochi and H. Yokoyama, J. Appl. Phys. **96**, 2393 (2004).
28. F. Perez-Murano, C. Martin, N. Barniol, H. Kuramochi, H. Yokoyama and J. A. Dagata, Appl. Phys. Lett. **82**, 3086 (2003).
29. A. Garcia-Martin and R. Garcia, Appl. Phys. Lett. **88**, 123115 (2006).
30. D. Wang, L. Tsau and K. L. Wang, Appl. Phys. Lett. **65**, 1415 (1994).
31. B. Legrand and D. Stievenard, Appl. Phys. Lett. **76**, 1018 (2000).
32. B. Legrand and D. Stievenard, Appl. Phys. Lett. **74**, 4049 (1999).
33. N. Clement, D. Tonneau, B. Gely, H. Dallaporta, V. Safarov, and J. Gautier, JVST B **21**, 2348 (2003).
34. J. A. Vicary and M. J. Miles, Ultramicroscopy **108**, 1120 (2008).
35. E. S. Snow and P. M. Campbell, Appl. Phys. Lett. **64**, 1932 (1994).
36. E. S. Snow, P. M. Campbell, and F. K. Perkins, Appl. Phys. Lett. **75**, 1476 (1999).
37. E. S. Snow, G. G. Jernigan, and P. M. Campbell, Appl. Phys. Lett. **76**, 1782 (2000).
38. D. Graf, M. Frommenwiler, P. Studerus, T. Ihn, K. Ensslin, D. C. Driscoll and A. C. Gossard, J. Appl. Phys. **99**, 053707 (2006).
39. H. Kuramochi, K. Ando, T. Tokizaki and H. Yokoyama, Appl. Phys. Lett. **88**, 093109 (2006).
40. P. A. Fontaine, E. Dunois and D. Stievenard, J. Appl. Phys. **84**, 1776 (1998).
41. H. Kuramochi, K. Ando, T. Tokizaki and H. Yokoyama, Jpn. J. Appl. Phys. **45**, 2018 (2006).
42. H. Kuramochi, F. Perez-Murano, J. A. Dagata and H. Yokoyama, Nanotechnology **15**, 297 (2004).
43. H. Kuramochi, T. Tokizaki, H. Yokoyama and J. A. Dagata, Nanotechnology **18**, 135703 (2007).
44. H. Kuramochi, K. Ando and H. Yokoyama, Surf. Sci. **542**, 56 (2003).
45. H. Kuramochi, K. Ando, T. Tokizaki and H. Yokoyama, Appl. Phys. Lett. **84**, 4005 (2004).
46. A. Yokoo, JVST B **21**, 2966 (2003).
47. M. Cavallini, P. Mei, F. Biscarini and R. Garcia, Appl. Phys. Lett. **83**, 5286
48. D. Wouters and U. S. Schubert, Nanotechnology **18**, 485306 (2007).

Chapter 3
Double Layer Local Anodic Oxidation Using Atomic Force Microscopy

Urszula Gasser, Martin Sigrist, Simon Gustavsson, Klaus Ensslin, and Thomas Ihn

Abstract Double-layer local anodic oxidation is a powerful method for fabricating complex semiconductor nanostructures. Here we review the application of this technique to Ga[Al]As heterostructures with titanium top gate electrodes. After short historical remarks, the details of the experimental oxidation setup, and the most relevant physical aspects of the involved materials are described. The experimental procedures and the influence of the key parameters are discussed for both the direct oxidation of Ga[Al]As and the oxidation of titanium. The power of the technique is corroborated by an overview over fabricated devices ranging from a single few-electron quantum dot to a complex quantum circuit comprising an Aharonov-Bohm ring with two embedded mutually coupled quantum dots, and an integrated charge read-out coupled capacitively.

Keywords Atomic force microscope · Scanning probe microscope · Scanning force microscope · AFM lithography · SFM lithography · Local anodic oxidation · GaAs/AlGaAs heterostructures · Electronic properties · Semiconductor nanostructures · Semiconductor patterning · Quantum point contacts · Quantum dots · Double quantum dots · Coupled nanostructures

Abbreviations

2DEG	Two-dimensional electron gas
AC	Alternating current
AFM	Atomic force microscope
DC	Direct current
DQD	Double quantum dot
FWHM	Full width at half maximum
HEMT	High-electron-mobility transistor
LAO	Local anodic oxidation
MBE	Molecular beam epitaxy
MOSFET	Metal-oxide-semiconductor field-effect transistor

T. Ihn (✉)
Solid State Physics Laboratory, ETH Zurich, 8093 Zurich, Switzerland
e-mail: ihn@phys.ethz.ch

A.A. Tseng (ed.), *Tip-Based Nanofabrication*, DOI 10.1007/978-1-4419-9899-6_3,

PMMA	Poly(methyl methacrylate)
QD	Quantum dot
QPC	Quantum point contact
SFM	Scanning force microscope
SPM	Scanning probe microscope
STM	Scanning tunneling microscope

3.1 Introduction

Lithography using a local anodic oxidation process is a powerful tool to define nanoscale patterns on surfaces. The method is based on applying an electric voltage between the tip of a scanning probe microscope (SPM) and a grounded surface. The pioneering work in this field was done in 1990 by Dagtata *et al.* [1] and by Becker *et al.* [2], where atomic hydrogen was selectively removed from a H-passivated Si surface using the biased tip of a scanning tunneling microscope (STM). However, the major disadvantage of the STM patterning method is the coupling between the feedback loop controlling the tip-sample distance and the actual lithography mechanism. For the STM the tunneling current depends exponentially on the size of the vacuum gap between the sample and the tip. Moreover, the tip voltage required for lithography cannot be chosen independently. This problem can be avoided if, instead of a STM, an atomic force microscope (AFM) with a metallic tip is used [3]. Another advantage of the AFM over the STM is that the AFM does not necessarily need a conducting sample surface. Metals, semiconductors and insulators can be scanned with similar feedback-loop parameters, whereas scanning insulating areas with the STM is basically impossible. In the first experiment of this kind, the AFM tip acted as an electron source for the local exposure of PMMA photoresist creating nanoscale patterns [3]. The metal-coated AFM tips were further used to oxidize silicon surfaces locally [4–9], and the first AFM-defined MOSFET transistor was reported in 1995 by Minne *et al.* [10]. In parallel, the STM oxidation technique was extended to thin metallic films, namely to titanium [11–14]. Shortly afterwards, the AFM-based local oxidation of a thin chromium film was demonstrated [15]. This work was followed by *in-situ* measurements of the electronic properties of AFM-defined narrow channels on Al [16, 17], Ti [18–21] and Nb [22].

The developments of this field are fruitfully continued [23, 24] but this review is focused on the subfield of the AFM lithography on two-dimensional electron gas confined in GaAs/AlGaAs heterostructures combined with the AFM lithography on thin Ti films. The next milestone in this subfield was achieved in 1995, when the local depletion of a two-dimensional electron gas was successfully demonstrated [25]. In such a way, charge carriers could be laterally confined by local oxidation. The following work showed that shallow two-dimensional electron gases can be fully depleted during oxidation [26–28], opening the doors for the fabrication of quantum devices such as quantum point contacts and quantum dots. To overcome the problem of limited tunability and to utilize the third dimension for patterning, the double-layer AFM lithography process was developed in 2004 [29]. In this method, a thin metallic top-gate layer is deposited above the GaAs structure. Then, the top-gate is laterally patterned into individual segments that control particular parts of the device. This method allowed for the fabrication of high quality complex nanostructures.

AFM nanooxidation is a very comprehensive research area, and many techniques and applications have been reported and reviewed [23, 30]. Here, we focus on the AFM local oxidation of GaAs/AlGaAs heterostructures. First, we introduce the principle of the method for semiconducting and metallic surfaces. Then we discuss the most important properties of GaAs/AlGaAs heterostructures. In Section 3.4, we present the details of single-layer lithography on GaAs, and a selection of devices fabricated by this method. Next, we move to a detailed description of the second-layer lithography, followed by examples of state-of-the-art nanodevices.

3.2 Local Anodic Oxidation of Semiconducting and Metallic Surfaces

In this section we describe the principles of local anodic oxidation. First, we present the typical setup used for lithography. Then we focus on electrochemical reactions that lead to the formation of the oxide under an applied tip voltage. After that we discuss the influence of key parameters for the oxidation including the type of conductive tips, writing speed, applied voltage and relative humidity.

3.2.1 Experimental Setup

The most common setup designed for AFM lithography consist of a high quality AFM placed on a vibration isolation table, and enclosed in a hood necessary for establishing acoustic isolation, as well as constant temperature and humidity during the oxidation process (see Fig. 3.1). The relative humidity is controlled by a constant, non-turbulent flow of humidified nitrogen gas as explained in Section 3.2.7. The conductive tip is connected to an external voltage source and the sample is grounded. Figure 3.2 shows a close-up on the AFM head indicating the most important elements of the setup.

Local oxidation can either be achieved in standard contact mode or in tapping mode. The first experiments were done in contact mode while keeping constant tip-substrate force. This repulsive force was typically set to 10–20 nN [3, 9]. However, successful anodization was even demonstrated at 100 nN [6] or 1 nN [5, 7]. In 1999 Irmer *et al.* reported local oxidation in the so-called tapping mode, which minimizes tip degradation by simultaneously enhancing resolution and reliability of the oxidation process [18]. In tapping mode, the cantilever oscillates near its resonance frequency (around 200–300 kHz) "tapping" on the surface. The amplitude of the free oscillation is typically between 100 and 200 nm. As the tip approaches the surface, the oscillation amplitude decreases. The feedback loop withdraws the

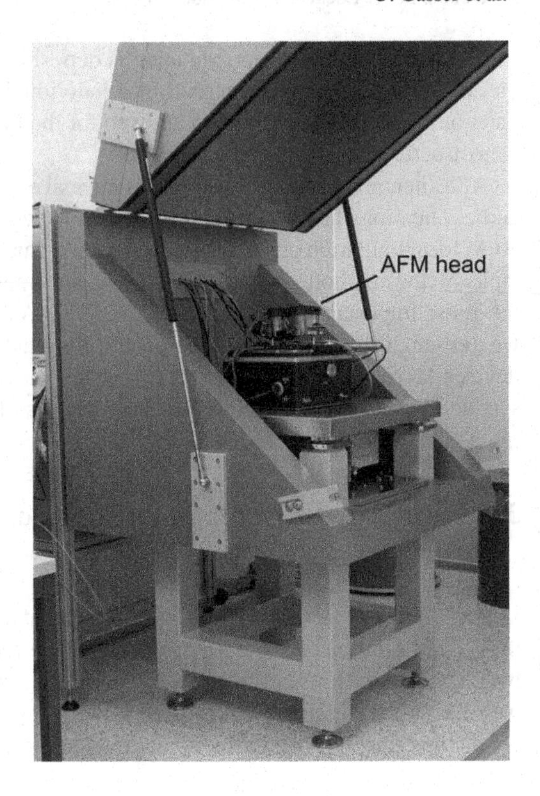

Fig. 3.1 The MFP Asylum Research AFM setup used for anodic oxidation in the nanophysics group at ETH Zurich. The AFM head is placed inside a hood providing acoustic isolation. The specially designed table prevents from mechanical vibrations

tip to keep the oscillation amplitude constant providing the height profile of the scanned surface.

During the oxidation, a negative voltage is applied to the tip. The electric field attracts the cantilever towards the surface. The related force gradient $\partial F/\partial z$ changes the cantilever force constant k to $k^* = k - \partial F/\partial z$ (in a simplifying linear approximation) and therefore shifts the resonance frequency from the initial ω_0 to $\omega_0^* = \omega_0\sqrt{k^*/k}$. In tapping mode, the cantilever is excited at the fixed frequency $\omega < \omega_0^* < \omega_0$ and by applying negative voltage the oscillation amplitude increases. By measuring the amplitude-distance curve, the optimal working parameters can be set. If the feedback loop is turned on, the overall loading force remains unchanged even for large applied voltages [19]. The tapping (dynamic) mode significantly enhances the lifetime of the tips and prevents undesired tip-crashes.

3.2.2 The Oxidation Mechanism

Topography scans and *in-situ* electron transport measurements show that a negative bias applied to the tip of the AFM can modify the metallic or semiconducting surface. A detailed analysis (e.g. by Auger electron spectroscopy) proves that the

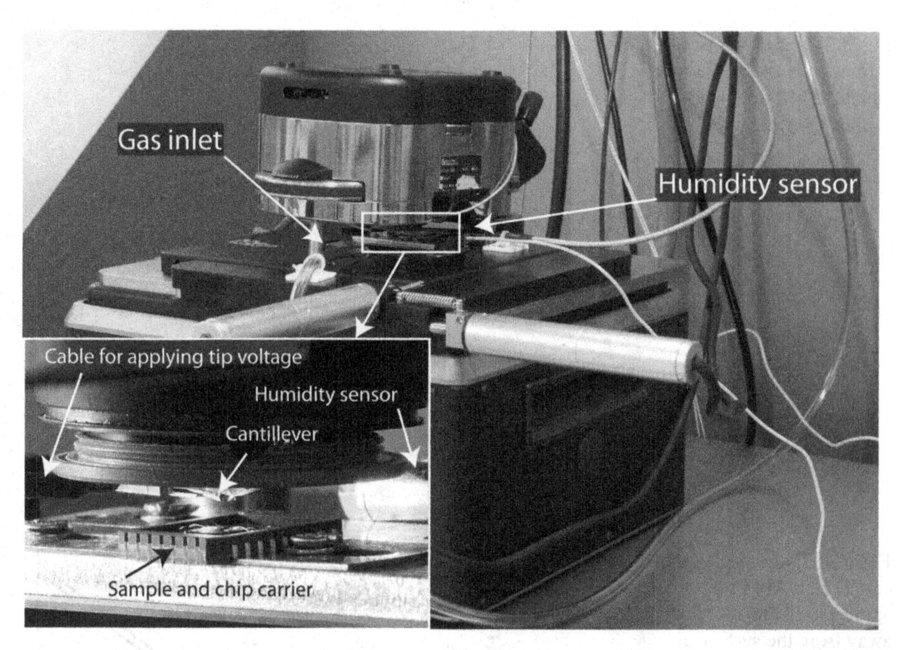

Fig. 3.2 Close-up of the AFM setup from Fig. 3.1. The humidity sensor controls the humidity near the sample. The flow of dry nitrogen reduces the humidity and the flow of water saturated nitrogen gas increases humidity. The inset shows a zoom in the sample area. The sample is glued on a chip carrier and the golden wires provide metallic contact. The chip carrier is fixed to a grounded plate. The voltage is applied to a cantilever

structures created in this way on the silicon or titanium surfaces consist of silicon oxide and titanium oxide, respectively [5, 14].

Under ambient conditions with finite humidity, water adsorbs on the sample surface and on the AFM tip. Due to capillary forces, a water meniscus forms between the tip and the sample, as illustrated in Fig. 3.3. As the tip approaches the sample surface, it makes a sudden jump towards the surface ("jump-in") as soon as the force-gradient exceeds the spring constant of the cantilever as shown in Fig. 3.4. On the way back, it exhibits hysteretic behavior and jumps away from the surface at larger distance, because capillary forces tend to pull the tip towards the surface [15]. For lithography purposes, it is important to approach the tip so close to the surface that the water meniscus forms. The electric field produced by a biased tip may exceed 1 V/nm. Such strength can alter the distribution of electrons in the water molecules between bonding and antibonding orbitals [10]. An electric field hydrolyzes the water film in the meniscus creating oxygen anions OH^- and O^{2-}. The anions travel through the growing oxide layer and further oxidize the surface [6]. Such a process is called anodic oxidation or electric field enhanced oxidation and can be explained using a space charge model [31], in which the growth of the oxide is limited by the diffusion of oxidizing species [7, 8].

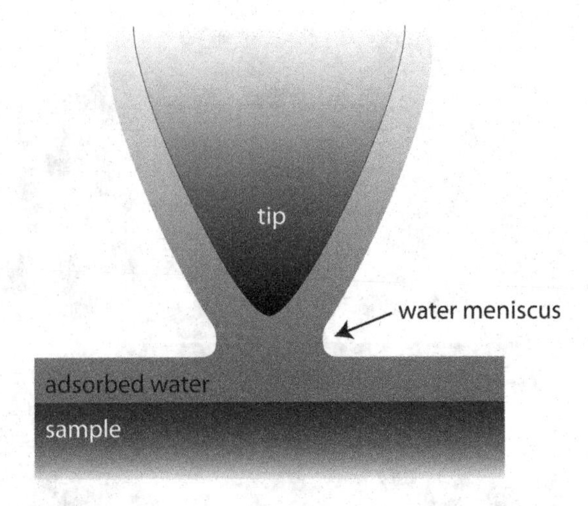

Fig. 3.3 Water meniscus formation between the sample and the AFM tip due to capillary forces

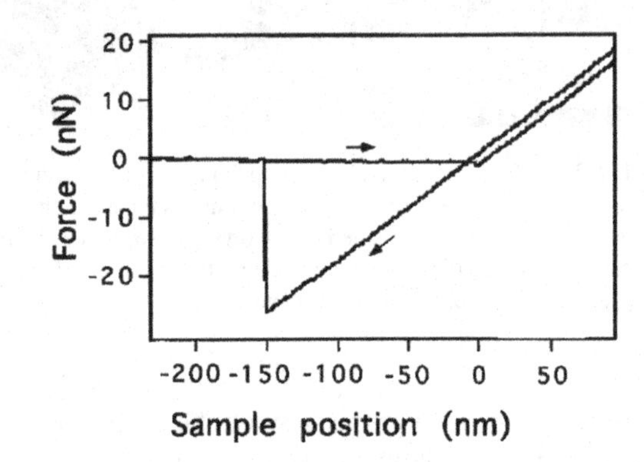

Fig. 3.4 Force-distance curve of an AFM tip approaching and moving away from the surface as denoted by *arrows*. On the left side, the scanner tube is fully retracted and the net force on the cantilever is zero. Reprinted with permission from [15]. Copyright 1995, American Institute of Physics

3.2.3 Cantilever Tip

The quality of the tip is a key parameter for AFM lithography. The oxidation process is very sensitive to the exact shape and properties of the tip and cantilever. The tip has to be conductive, sharp and robust. It may happen that among nominally identical tips taken from a particular batch, some are good for lithography, while some others do not oxidize the surface at all. The only way to test a particular tip is to try to oxidize the surface. During oxidation, the tip is exposed to extreme conditions such as high humidity, large voltages, and small tip-sample distances, which significantly increase the probability of electrostatic discharge and tip-crashes. Even a tip working well degrades slowly during writing and scanning. This degradation leads to worse oxide quality, blunt or double tips, discharges, tip-crashes or to the removal of the conductive tip coating. Therefore, choosing a superb tip is one of the most important prerequisites for successful AFM lithography.

In the first experiment on electric field enhanced AFM lithography, a negatively biased tip was used as a highly localized electron source to pattern the PMMA resist.

Table 3.1 Typical parameters of TiN coated n-doped silicon tips successfully used for AFM lithography on GaAs and Ti surfaces

Resonant frequency	200–350 kHz
Force constant	5–40 N/m
Tip height	10–15 μm
Full tip cone angle	≤30°
Tip curvature radius	≤40 nm
Tip coating thickness	10–15 nm

The tip used in this work was a standard, commercially available silicon nitride (Si_3N_4) tip. In order to make it conductive, it was coated with 50 nm of gold. Later on, a large variety of tips has been used for AFM lithography. Most common are tips coated with 10–30 nm Cr [4], Au [5, 7], Ti [6, 8], or Pt [22]. Another kind of tips successfully used for AFM lithography is degenerately n- or p-type doped Si tips with radii down to 5 nm [9, 15, 18–20]. One type of tips that proved to be highly reliable and well performing is n-doped silicon tips coated with 10–25 nm of Ti, W_2C or TiN [32]. The typical parameters of this kind of tips are presented in Table 3.1.

3.2.4 Writing Speed

Another important parameter controlling the oxidation is the writing speed. In general, too low writing speeds result in broad lines [8, 14], whereas writing too fast may cause discontinuities and decrease the height of the oxide lines. For silicon, the oxide line height was found to be proportional to the square root of the writing speed [7] and the optimal writing speed is about 0.5 μm/s [8, 10]. In case of titanium, the height of an oxide line increases with each writing cycle (repetition of the oxidation along the same path) and this effect is the more pronounced the lower the humidity. However, after a certain number of cycles, the height of the oxide saturates at a value independent of the relative humidity, but dependent on the bias voltage applied during the oxidation as shown in Fig. 3.5. This effect demonstrates that the rate of local anodization is limited by the thickness of the water film at the beginning of the anodization process. Figure 3.6 shows that the spatial resolution decreases with the anodization time, especially at high humidity. In general, the lines have to be thick enough to create working devices and at the same time as narrow as possible to optimize the lateral resolution. Depending on the lithographic purposes the optimal writing speed varies in the range 0.05–0.5 μm/s.

3.2.5 Applied Voltage

The applied voltage is a driving force for the AFM lithography process and has a strong influence on the height and the width of the oxide lines. For the first lines written by an AFM tip on a p-type silicon surface 25 V were applied to the sample and the tip was kept at zero potential [4]. The resulting lines were 530 nm wide and

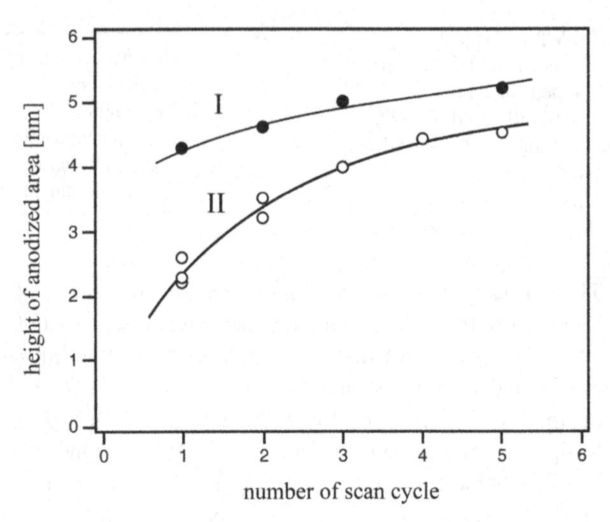

Fig. 3.5 The height of the oxide pattern on a Ti film increases with the number of writing cycles. Filled circles (I) represent data at high relative humidity (95%), empty circles (II) correspond to low humidity (below 25%). Adapted with permission from [12]. Copyright 1993, American Institute of Physics

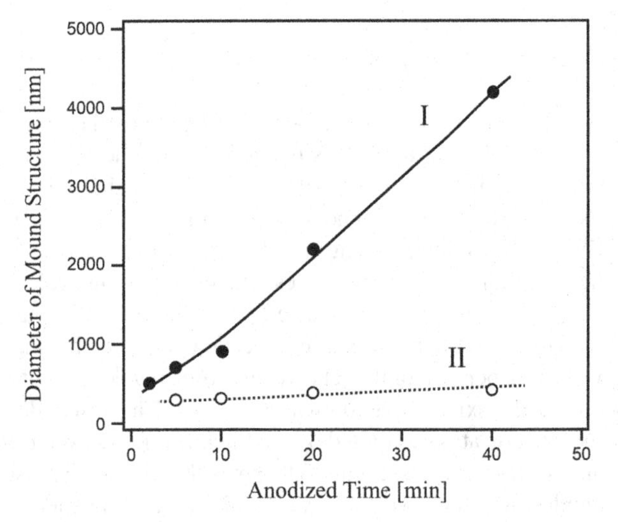

Fig. 3.6 Diameter of the spot created by the STM microscope over a Ti film at a fixed position as a function of anodization time. The applied voltage was 5 V and the STM current 0.5 nA. Filled circles (I) represent data at high relative humidity (95%), empty circles (II) correspond to low humidity (below 25%). Adapted with permission from [12]. Copyright 1993, American Institute of Physics

about 2.8 nm high as shown in Fig. 3.7. For silicon, the height of the oxide lines varies linearly with the applied voltage and saturates at a value equivalent to that found in native oxidation [7–9]. In the first successful attempt to oxidize titanium locally, bias voltages in the range between 3 and 5 V were applied to the sample while the tip was grounded [11]. The maximum achievable height of the lines was 5 nm at a width of 300 nm, while the minimum width of the oxide line obtained in this experiment was 70 nm. There exists a sample and tip dependent low-voltage threshold, below which the oxidation process is not possible [9]. Applying voltages

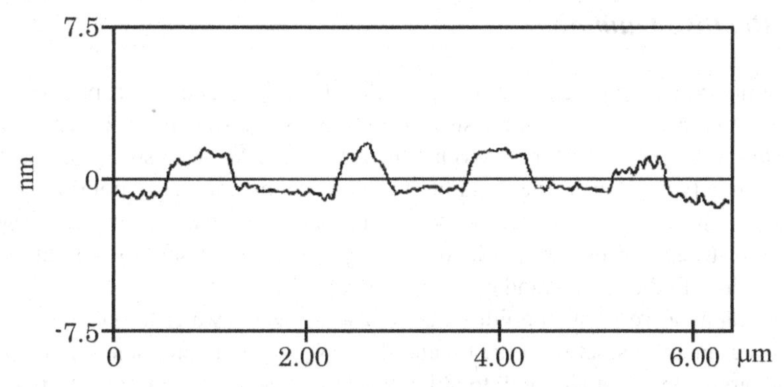

Fig. 3.7 Cross section of four oxide lines written on p-type silicon. The voltage applied to the substrate was 25 V and the writing speed was 0.1 μm/s. The lines are approximately 530 nm wide and 2.8 nm high. Adapted with permission from [4]. Copyright 1993, American Institute of Physics

larger than the high-voltage threshold will not increase the oxide height. Instead, the lines will become inhomogeneous and the risk of a tip crash will increase.

Applying an AC voltage on top of the DC signal greatly improves the reliability and reproducibility of AFM lithography. On GaAs it can enhance the so-called aspect ratio (height-to-width ratio) of the oxide line by 15% [32]. The aspect ratio is an important parameter, because high and narrow lines are desired for the design of complex nanostructures. The modulation of the AC voltage should be around 1 kHz, which is low compared to the resonant frequency of the tip. Using even lower frequencies may cause fragmentation of the oxide lines. This can be avoided by decreasing the writing speed. The optimal magnitude of the AC voltage depends strongly on the DC voltage set-point. The choice of an optimal combination of these two voltages leads to high-quality oxide lines.

3.2.6 Working Distance

In tapping mode, the (static) distance between the surface and the tip is determined by the oscillation amplitude. Here, it is useful to define the set-point amplitude as the operating amplitude normalized to the vacuum amplitude of the cantilever (without tip-sample interactions). For standard scanning, the safest set-point amplitude is the maximum amplitude at which the topography of the surface can be mapped. This corresponds typically to a set-point amplitude of 60–90%. However, for local oxidation the oscillation amplitude has to be decreased to move the tip closer to the surface and form the water meniscus. It is convenient to try the oxidation at a set-point amplitude of about 20%, and decrease it slowly to the point where the line height is close to the desired one [33]. Of course, there is a strong correlation between the magnitude of the voltage and the set-point amplitude. For small set-points (below 5%) the voltage should be decreased in order to prevent electrostatic discharges on the surface [32].

3.2.7 Relative Humidity

Relative humidity is a crucial parameter for local anodic oxidation. It provides the thin water film covering the sample surface and serves as a source of oxygen anions. To create stable conditions for the oxidation process, the AFM is usually placed in a chamber. In a typical setup, the humidity is controlled by constantly pouring water-saturated or dry nitrogen gas. It is important to make sure that the nitrogen flow is neither too strong nor too direct. The optimum position of the humidity-monitoring sensor is close to the sample and opposite the nitrogen outlet.

Lithography at very low humidity may not work, as the water film cannot form, or the water meniscus between the tip and the sample may break, resulting in electric discharge. On the other hand, too high humidity leads to broad lines decreasing the spatial resolution of the designed structure. For example, oxide lines on Ti are broadened at a relative humidity of 90% as compared to a relative humidity below 25% [12], while the height of the lines is only slightly affected. Under ambient conditions, the Ti film is covered with at least 2–3 nm of native oxide layer. This layer is strongly hydrophilic and absorbs water even if humidity is very low. Therefore, lithography on Ti is possible even under low humidity conditions [13], which is not necessarily true for other materials.

3.3 Two-Dimensional Electron Gases in GaAs/AlGaAs Heterostructures

Two-dimensional electron gases confined in modulation doped semiconductor heterostructures serve as the base material for laterally defined nanoscale devices. Lateral patterning of high quality GaAs/AlGaAs heterostructures is a common technique for the fabrication of semiconductor nanostructures. This chapter provides a brief overview over the most important properties of two-dimensional electron gases confined in GaAs/AlGaAs heterostructures.

3.3.1 Two-Dimensional Electron Gas

A two-dimensional electron gas (2DEG) is a gas of electrons free to move in two dimensions (x, y), but tightly confined in the third dimension (z). In the simplified description based on the effective mass approximation, the dispersion relation for a 2DEG is parabolic and the eigenenergies of the electrons with the momentum $\vec{k}$ are

$$E = \varepsilon_n + \frac{\hbar^2}{2m^*}\left(k_x^2 + k_y^2\right), \quad (1)$$

where the index n numbers successive subbands, ε_n is the lowest energy of electrons in the n-th subband, and m^* is the effective mass of an electron. In most cases it is desired that only the lowest subband is occupied at sufficiently low temperatures (two-dimensional electron gas in the quantum limit).

In such a system, electron states can be described as freely propagating plane waves and their in-plane wavefunctions are

$$\Psi(x, y) = \frac{1}{\sqrt{S}} \exp(ik_x x) \exp(ik_y y), \tag{2}$$

where the factor $1/\sqrt{S}$ is a normalization to the area S [34].

The resulting spin-degenerate density of states $D(E)$ for a 2DEG is constant and determined by the effective mass:

$$D(E) = \frac{m^*}{\pi \hbar^2} = \text{const.} \tag{3}$$

At low temperature, electrons with energies close to the Fermi energy E_F contribute to the electrical current. An important length scale is therefore the Fermi wavelength λ_F, which for a given 2D sheet density n_{2D} is

$$\lambda_F = \frac{2\pi}{k_F} = \sqrt{\frac{2\pi}{n_{2D}}}. \tag{4}$$

The related Fermi velocity υ_F is expressed as a function of Fermi wave number k_F

$$\upsilon_F = \frac{\hbar k_F}{m^*}. \tag{5}$$

In high-quality semiconductors, the Fermi wavelength can be of the order of several ten nanometers. Therefore, by using modern processing techniques (e.g. AFM local oxidation), it is possible to create devices with sizes comparable to λ_F. In this regime quantum effects start to play an important role at sufficiently low temperatures.

The quality of the semiconductor wafer is mainly determined by its mobility $\mu = \tau_m e/m$. Here, τ_m is the momentum relaxation time, the average time between two back-scattering events of an electron. Usually, density and mobility are known from measurements of the longitudinal resistivity, and the Hall effect [35]. Another characteristic length associated with a 2DEG is the mean free path $l_e = \tau_m \upsilon_F$. The mean free path is the average distance that an electron travels without being back-scattered. In a typical 2DEG based on GaAs/AlGaAs heterostructure, the mean free path can be of the order of micrometers at a temperature of 4.2 Kelvin. If the size of the nanostructure is smaller than l_e, the regime of ballistic transport is reached.

3.3.2 GaAs/AlGaAs Heterostructures

Due to the high mobility of the 2DEG, modulation doped GaAs/$Al_xGa_{1-x}As$ is often chosen as the starting material for the fabrication of semiconductor nanostructures. The index x denotes the percentage of Ga atoms substituted with Al atoms (typically $x = 0.3$ and the index is omitted). The highest quality wafers are grown with atomic resolution in growth direction, atomic layer by atomic layer, using molecular beam epitaxy (MBE). The schematic layer sequence in a typical GaAs/AlGaAs heterostructure is shown in Fig. 3.8. The electron affinities are different for pure GaAs and for AlGaAs. Furthermore, the band gap energy depends strongly on the aluminum concentration. Careful engineering of the growth allows the creation of a triangular potential, confining electrons at the GaAs/AlGaAs interface. The width of the potential is of the order of the Fermi wavelength λ_F. The energy of the electron is quantized along the growth direction. When only the lowest subband is occupied, electrons can move freely only in the plane parallel to the surface. Typical electron densities are of the order of $5 \cdot 10^{11}$ cm^{-2} and typical mobilities are around 500 000 cm$^2 \cdot$ V$^{-1} \cdot$ s^{-1}.

3.4 First Layer – Depletion of Two-Dimensional Electron Gas

Local AFM oxidation of heterostructures containing a two-dimensional electron gas is a powerful method for fabricating nanoscale devices. In this chapter we explain the principles of the 2DEG depletion and characterize the electronic properties of the oxide lines. Then we present a variety of devices fabricated with this technique and discuss the limitations of the method.

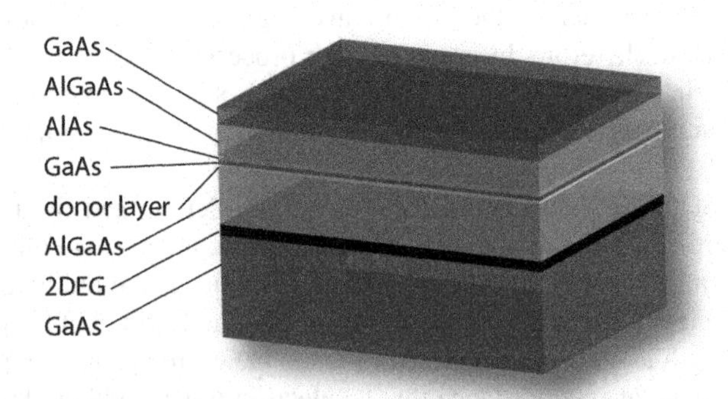

Fig. 3.8 Schematic diagram showing the different layers of the heterostructure containing a two-dimensional gas. The electron gas is formed 34 nm below the surface. The successive layers and their typical thicknesses: GaAs cap (5 nm), $Al_{0.3}Ga_{0.7}As$ (8 nm), AlAs (2 nm), GaAs (2 nm), δ-doped donor layer, $Al_{0.3}Ga_{0.7}As$ (17 nm), 2DEG at the $Al_{0.3}Ga_{0.7}As$-GaAs interface, GaAs wafer

3.4.1 Sample Preparation and Surface Treatment

The first step in preparing the wafer containing a 2DEG for AFM lithography is to define a mesa. Using optical lithography, a large region of the 2DEG is removed by chemical wet etching, leaving individual electrically conducting "fingers". The fingers connect a small square mesa to ohmic contacts at the edge of the chip as illustrated in Fig. 3.9. The ohmic contacts are made by evaporating and annealing a eutectic mixture of gold and germanium. In this particular sample, the metal composition of the ohmic contact is Ge (18 nm)/Au (50 nm)/Ge (18 nm)/Au (50 nm)/Ni (40 nm) /Au (100 nm). During annealing at about 440°C the first four layers create a Ge-Au alloy that diffuses tens of nanometers below the surface contacting the 2DEG layer. Ni acts as a diffusion barrier and the topmost thick layer of gold serves as the contact pad to which a bonding wire will be attached. All these steps have to be performed in a clean-room environment. For AFM lithography the surface roughness of the sample should be below 1 nm. It may therefore be essential to expose the sample prior to oxidation to an oxygen plasma [10] or even dip the sample in hydrochloric acid. These steps remove the residuals of photoresist, some dirt particles and etch away the native oxide. A clean and contacted mesa is the starting point for AFM lithography.

3.4.2 Oxidation and Depletion of the 2DEG

In 1995 Ishii *et al.* wrote oxide lines across a high-electron-mobility transistor (HEMT) using an AFM [25]. The selective oxidation produced a depletion layer

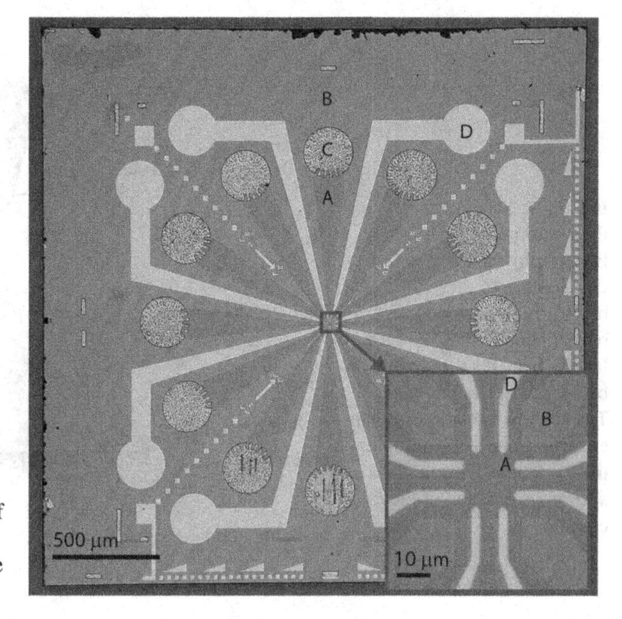

Fig. 3.9 GaAs/AlGaAs sample prepared for AFM lithography. Darker regions (**a**) contain the 2DEG below the surface, whereas in the brighter regions (**b**) the 2DEG was removed by wet etching. The 2DEG is contacted to metallic pads (**c**) later used to apply voltages. The golden fingers deposited on the surface and electrically isolated from the 2DEG (**d**) are necessary for the second layer lithography. The inset shows the middle of the sample with the 20 × 20 μm mesa. The AFM structure will be written on the mesa

in the 2DEG channel, leading to an increased resistance of the HEMT device. The increase of the resistance was correlated with the height of the oxide lines. This showed that the height of the oxide line allows one to control the depth of the depletion.

Figure 3.10 shows a cross section through a GaAs/AlGaAs heterostructure. The sample is grounded and a thin layer of water covers its surface. A negatively biased AFM tip locally oxidizes the GaAs surface. The electrochemical oxidation reaction is given by:

$$2\text{GaAs} + 12\text{h}^+ + 10\text{H}_2\text{O} \rightarrow \text{Ga}_2\text{O}_3 + \text{As}_2\text{O}_3 + 4\text{H}_2\text{O} + 12\text{H}^+, \quad (6)$$

where h^+ denotes a hole provided by the anode, while the hydrogen ions combine to molecular hydrogen [26, 36, 37]. Chemical analysis of the AFM generated oxide showed that the main constituents of the GaAs anodic oxide are indeed Ga_2O_3 and As_2O_3 [38].

After the oxidation, the crystalline semiconductor surface is closer to the 2DEG layer than before. This results in an increased surface area, and simultaneously, increased number of surface states. In the 2DEG used for AFM lithography, about 90% of the donor electrons are trapped by surface states that lead to mid-gap Fermi-level pinning at the surface, and only the remaining 10% contribute to electron transport in the 2DEG layer. If the surface of the sample gets closer to the 2DEG, more electrons will be trapped by surface states and the electron density below the oxide line will be depleted. Figure 3.11 presents several oxide lines of different heights. After taking this image, the oxide was selectively removed and the

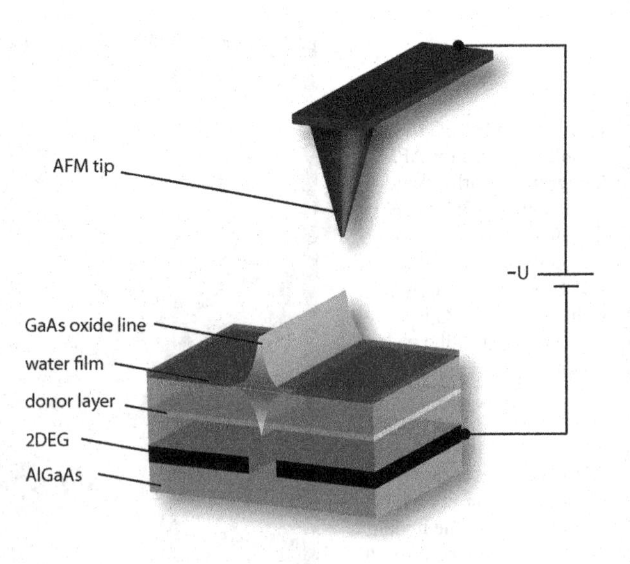

Fig. 3.10 Schematic illustration of AFM lithography on an GaAs/AlGaAs heterostructure. A negative bias applied to the AFM tip locally oxidizes the sample below and above the surface. The shallow two-dimensional electron gas is depleted below the oxide line. This method allows the creation of electric circuits

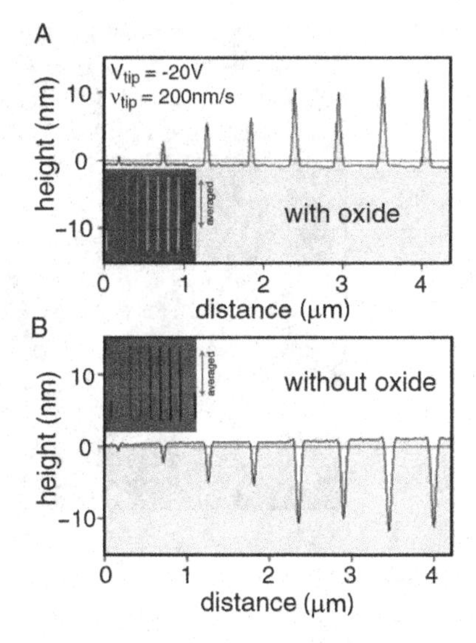

Fig. 3.11 (**a**) Profile of the oxide lines. The set-point of the tapping amplitude was reduced from left to right, effectively bringing the tip closer to the surface and increasing the height of the oxide lines. The inset shows a full two-dimensional scan of the structure. (**b**) The sample in (**a**) was immersed in a developer that selectively etches the oxide. As the oxide was removed, protrusions form in the GaAs. This proves that the oxide height is a good measure for the depth of the oxidation. Adapted from [33]. Copyright 2002, with permission from Elsevier

remaining material is pure GaAs. The lines changed to trenches of a depth similar to the previous height of the oxide. This indicated that the oxide grows not only on top of the surface, but it also penetrates into the sample towards the 2DEG. The trenches with increased GaAs surface leading to depletion are clearly visible [33].

Total depletion of the 2DEG at low temperatures can be achieved, if the 2DEG is not too deep and if the oxide lines are high enough. The oxidation is possible only for 2DEGs lying less than 50 nm below the surface [26–28, 33, 39, 40]. On the other hand, the 2DEG should not be too close to the surface because of the strong scattering of electrons at the doping atoms situated between the 2DEG and the surface.

Figure 3.12 shows a typical oxide line written on a GaAs/AlGaAs heterostructure. The 2DEG is placed 34 nm below the surface. The line is about 18 nm high and 80 nm wide. At low temperatures, the 2DEG is depleted below the line. Here, the 2DEG on one side of the line is grounded and a voltage is applied to the other side. There is no measurable current between the upper and lower voltage thresholds, beyond which leakage currents start to flow. The voltage threshold depends on the line height, the depth of the 2DEG and it may be wafer dependent. There exists a minimum height of the line required to deplete the 2DEG.

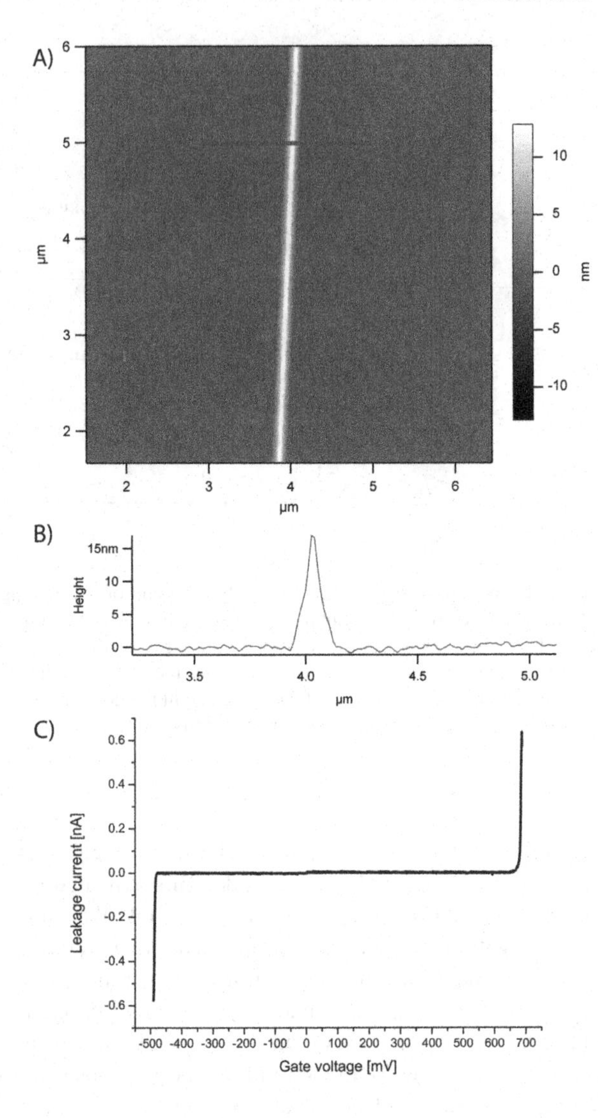

Fig. 3.12 (**a**) AFM scan of an oxide line written on GaAs. The 2DEG is 34 nm below the surface. (**b**) Cross section of the line corresponding to the *red line* marked in (**a**). The line is 18 nm high and 80 nm wide. (**c**) Leakage current across the oxide line measured at 20 mK. The 2DEG on one side of the line was grounded and a gate voltage was applied to the other side of the line. No current is measurable in the gate range of –500 to +700 mV, proving the insulating properties of the oxide line. Above these thresholds, leakage current starts to flow

3.4.3 Optimal Oxidation Parameters

There are no universal parameters that guarantee successful device fabrication. The interplay of all parameters discussed in Section 3.2 and the specific wafer determine under which conditions the oxidation is optimal. However, there exists a parameter range, for which the chances to fabricate a good working device are best. The goal is to define a complicated electronic structure by depleting the 2DEG below the surface. Depending on the complexity of the sample, the total length of the oxide lines can exceed hundreds of microns. All these lines must be high enough at each point to ensure electrical insulation. This task is not easy, but possible if we choose proper parameters.

First, the sample's surface must be free of dirt particles and the average surface roughness should not exceed 1–2 nm. Second, the tip has to be sharp, conductive and the driving amplitude of the voltage applied to the cantilever operating in the tapping mode should be below 300 mV. Note that it may happen that despite a seemingly perfect tip the oxidation will not work. The humidity necessary for writing good insulating lines varies between 36% and 39% in our setup. Smaller values will lead to electrostatic discharge, whereas larger values will lead to very broad lines. The starting set-point (damping amplitude of the tip) is around 15% and it may be necessary to decrease it even to 2%. There is a tendency of getting higher lines at smaller set-points, but one has to be careful as small set-points can trigger electrostatic discharge or simply cause a tip-crash. At the beginning of the oxidation, the DC and the AC voltages should be relatively low. For example, a good working recipe for the lithography is the following: we want to write a 0.5 μm long test line. The line should be at least 15 nm high and its FWHM should be around 80 nm. We start with a large set-point around 15% and relatively low DC voltage of –10 V. Additionally, an AC voltage can be applied (peak-to-peak amplitude somewhere in-between 1 and 12 V at a frequency of 1 kHz). The writing speed is 0.2 μm/s and the relative humidity is 38%. After writing, we scan over the region, to check if the surface has been oxidized. If not, then we slowly increase the voltage or lower the set-point. The effect of the set-point on the line height is demonstrated in Fig. 3.11. The maximum negative voltage is about –25 V. Above this value electrostatic discharge is likely to happen. If we see at some point a small oxide line, then we need to fine-tune its height by changing the applied voltage. A small decrease in relative humidity will help to narrow the oxide lines. If the parameters are good and we can successively write 2–3 reproducible test lines, we immediately move to the desired spot on the mesa and write the final structure. It is important to avoid scanning as much as possible as it may alter the tip. After writing the structure, the so-called connection lines have to be written to cut the 2DEG outwards to the edges of the mesa. An example of a sample where the first-layer lithography is finished is presented in Fig. 3.13. The structure contains two quantum dots connected in series marked with black circles. The dots are connected to the source lead on one side and to the drain lead on the other side. Two quantum point contacts (QPC1 and QPC2) are defined near the quantum dots. The oxide lines extend to the edges of the mesa depleting the 2DEG and creating three independent electrical circuits. Each lead is connected to an ohmic contact, via which voltages will be applied.

3.4.4 Barrier Height and Tunability

To determine the oxide barrier height it is necessary to measure the current-voltage characteristics as a function of the top-gate or back-gate voltages. In the experiment presented in Fig. 3.14, the entire oxide line defined on a 34 nm deep 2DEG is covered by a metallic gate. The voltage drop across the line vs source-drain bias voltage

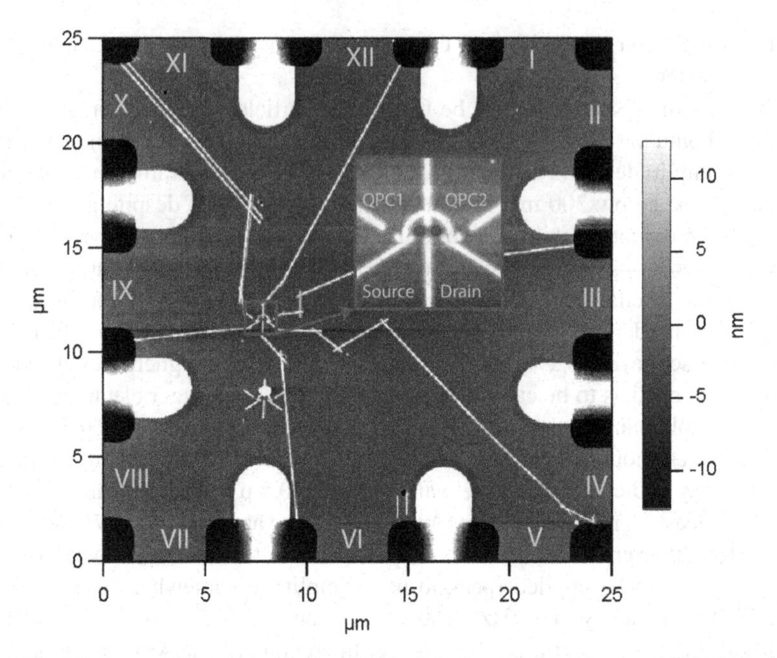

Fig. 3.13 AFM scan of the mesa depicted in Fig. 3.9. Here, the first layer of AFM lithography is finished. The oxide lines deplete the 2DEG and define connections to the ohmic contacts. For example, the source lead is connected to contacts VII and VIII and the drain lead is connected to contacts V and VI. The structure in the inset consist of a double quantum dot (*circles*) and two adjacent quantum point contacts

Fig. 3.14 Current-voltage characteristic across the oxide line for different top-gate voltages. At negative voltages below –250 mV, the 2DEG is depleted by the top-gate. At top-gate voltages above 50 mV the barrier created by the oxide line is lifted. Reprinted from [33]. Copyright 2002, with permission from Elsevier

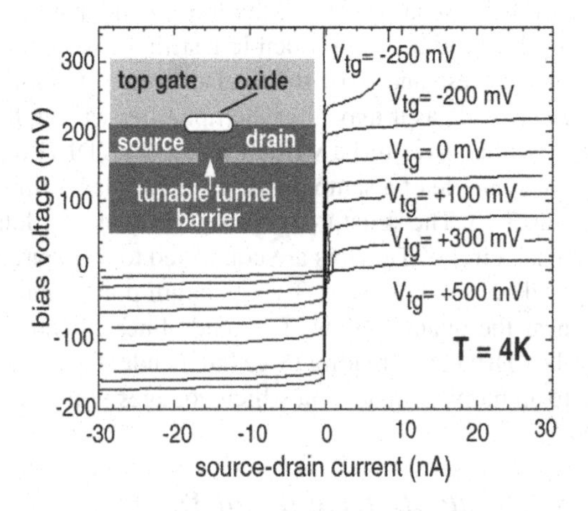

is measured for different top-gate voltages V_{tg}. At zero top gate voltage, the breakdown threshold is about ± 150 mV. For top-gate voltages below –220 mV the 2DEG is completely depleted as expected from the plate capacitor model. The capacitance C of the plate capacitor with accumulated charge ΔQ is given by

$$C = \frac{\Delta Q}{\Delta V} = \frac{\varepsilon\varepsilon_0}{d} A, \tag{7}$$

where ε_0 is the vacuum permittivity, ε=13 is the dielectric constant of GaAs, d = 34 nm is the distance between the 2DEG and the top-gate and A is the area of the capacitor. The change of charge ΔQ can be expressed as the charge of the electron density Δn:

$$\Delta Q = -e\Delta n A. \tag{8}$$

Combining Eq. (7) and (8) leads to:

$$\Delta V = e\Delta n \frac{d}{\varepsilon\varepsilon_0}. \tag{9}$$

The typical electron density of a 2DEG is about $5 \cdot 10^{11} \mathrm{cm}^{-2}$. Therefore, to deplete the 2DEG completely we need to apply a negative voltage of the order of –230 mV. This value is in good agreement with the data presented in Fig. 3.14. In this experiment, the lever arm of the top-gate was $dE/dV_{tg} = 0.025$ eV/V, where E denotes the energy of the conduction band bottom. The oxide barrier is lifted for top-gate voltages exceeding 500 mV and the estimated barrier height is 12.5 meV. These are typical parameters for a shallow 2DEG.

The nanostructures exclusively defined by AFM lithography can be tuned by so-called in-plane gates. The in-plane gate (also called side-gate) consists of the 2DEG adjacent to the oxide line and isolated from the other electrical circuits. By applying a voltage between the in-plane gate and the circuit on the other side of the oxide line, the electron density changes close to the oxide line on both sides. This effect allows us to control the depletion length in AFM-defined nanodevices. For example, the resistance of a narrow wire defined by two parallel oxide lines can be changed by a few orders of magnitude using in-plane gates.

3.4.5 Devices Fabricated Using LAO on GaAs/AlGaAs

A large variety of devices defined by local anodic oxidation on a shallow 2DEG exists. The full discussion of their properties is beyond the scope of this review. Here, we will focus on a selection of structures that contain rich and still not completely understood physics.

3.4.5.1 Quantum Point Contacts

A quantum point contact (QPC) is a narrow constriction connected to two conductive leads. The width of the constriction is of the order of the Fermi wavelength λ_F. For a typical 2DEG, λ_F is about 35 nm. The conductance of a quantum point contact is quantized in units of $2e^2/h$ due to the existence of discrete quantum confined

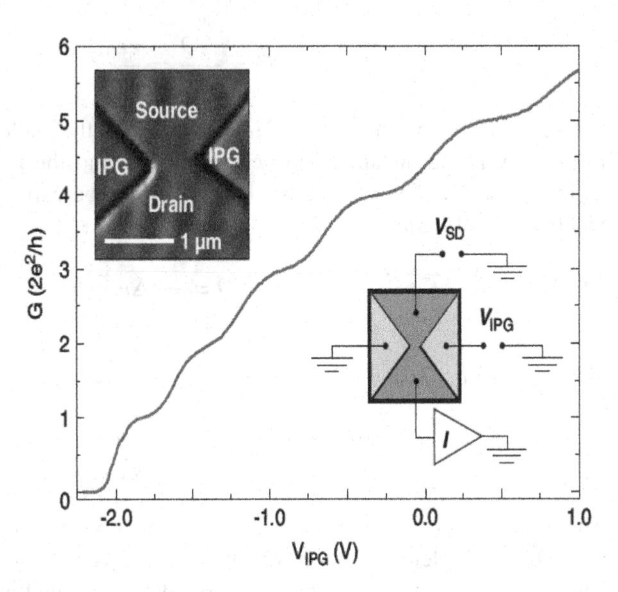

Fig. 3.15 Quantized conductance through a QPC as a function of the in-plane gate voltage. The upper inset shows the AFM micrograph of the surface. The lower inset shows the electrical scheme applied during the measurement. Reprinted from [42]. Copyright 2006, with permission from Elsevier

modes propagating through the constriction [41]. An example of a QPC fabricated on a two-dimensional electron gas is presented in Fig. 3.15. The QPC is defined at the smallest separation of two oxide lines having a kink. The voltage applied between the in-plane gates and the QPC channel effectively changes the width of the constriction [42]. At voltages below –2 V the QPC is pinched-off and no current can flow. As the in-plane gate voltage is increased, the constriction gets wider and the conductance increases reaching a plateau at $2e^2/h$. The next plateaus occur at $4e^2/h$, $6e^2/h$ etc. The quantum point contact is the simplest nanodevice exhibiting quantum confinement effects.

3.4.5.2 Quantum Dots

An example of a so-called zero-dimensional structure, where electrons are confined in all three spatial dimensions is a quantum dot (QD). A quantum dot can be imagined as an island filled with electrons, and weakly connected via tunneling barriers to source and drain leads. It is a building block for more complex quantum circuits. AFM lithography is a common tool for the definition of quantum dots [28, 43]. Figure 3.16 shows an AFM image of a typical quantum dot. If all in-plane gate voltages are fixed, the little electron island (marked with a black circle) usually contains a fixed number of electrons. The energy levels in a quantum dot are discrete like those of an atom. If there is no energy level in the bias window given by the source-drain voltage, the transport is blocked and no current flows through the quantum dot. However, by changing the voltage on the center gate we can shift the energy levels of the QD up or down. If at least one of the levels lies between the Fermi energies of the source and drain contact, an electron can hop onto the dot and leave it on the other side, a phenomenon called single-electron tunneling. Such an event

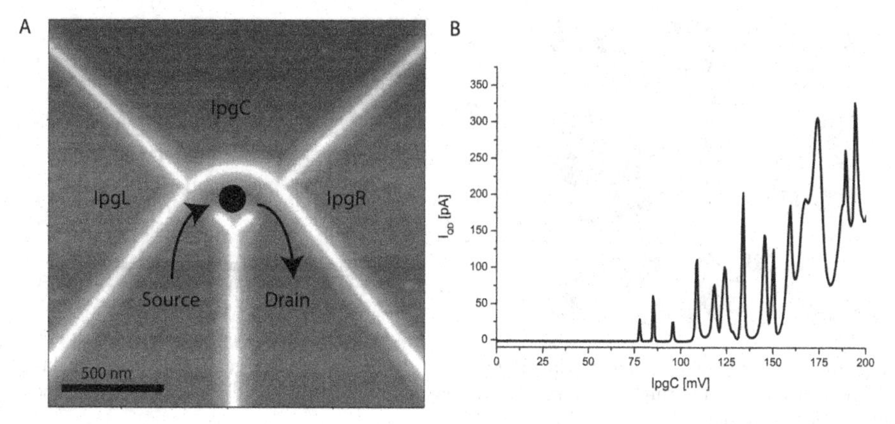

Fig. 3.16 (**a**) A quantum dot defined on the 2DEG and is denoted with a *black circle*. For Coulomb resonances, the electrons can hop on the dot from the source lead and leave the dot via the drain lead. The gates IpgL and IpgR can tune the width of the source-dot and drain-dot constrictions. The center gate IpgC is used to shift the quantized energy levels in a QD. (**b**) Coulomb oscillations of the quantum dot. In the regions between the peaks the transport is blocked and no electrons can travel through the QD

will contribute to the average current and it will lead to the conductance resonances shown in Fig. 3.16a.

3.4.5.3 Quantum Rings

A quantum ring geometry allows us to investigate the wave nature of electrons by studying interference effects. In this type of experiments, a coherent beam of electrons is split into two partial waves that enclose a magnetic flux Φ. At the exit of the ring, the partials waves interfere and their interference pattern can be detected by measuring the electric current. This type of experiment was proposed by Aharonov and Bohm in order to verify the physical significance of electromagnetic potentials in quantum theory [44]. The Aharonov-Bohm effect is seen as h/e periodic oscillations of the current as a function of the magnetic flux. Quantum rings can be successfully fabricated using AFM lithography [45, 46]. Figure 3.17a presents a micrograph of a quantum ring. The electrons are injected on the source side, have some probability amplitude to travel along the two paths of the interferometer and leave the system on the drain side. The ring current measured as a function of the perpendicular magnetic field reveals oscillations with periodicity h/e as presented in Fig. 3.17b.

3.4.6 Limitations of the Single-Layer Method

As shown in the previous section, AFM lithography is a powerful tool to create semiconductor nanostructures. However, to go beyond these simple structures, additional

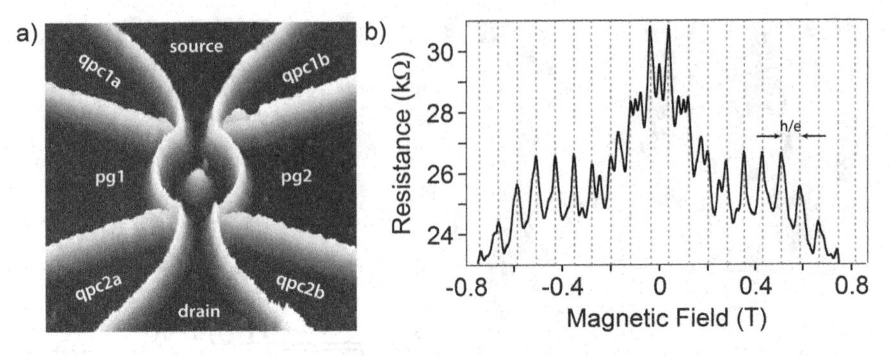

Fig. 3.17 (**a**) AFM micrograph of a ring. Black letters denote the in-plane gates. (**b**) Measurements of the resistance of the ring in a perpendicular magnetic field B at liquid He temperatures. The period of the oscillations is h/e

tunability is required. Every nanostructured element (i.e., each QPC, each QD) of the structure needs to have a gate. For easy manipulation it is desired that one gate tunes preferentially one element and, in a perfect world, does not influence another. For example, a quantum dot consist of three elements: two quantum point contacts (source and drain constrictions) and the quantum dot itself. Therefore, we need three gates to efficiently tune each of them. We want to minimize cross-talk between gates, since if we tune the dot levels we do not wish to open the tunneling contact to the source lead at the same time. The cross-talk effect is visible in Fig. 3.16b, where at large positive voltage the middle gate opens the source and drain contacts. Because of the enhanced dot-lead coupling, the Coulomb peaks are broader. This is the side-effect of the large capacitive coupling between the 2DEG parts. Another challenge for single-layer AFM lithography is that the structure must be perfectly fabricated. If the alignment accuracy for neighboring lines is less than 10 nm, the device will not work. Many of these critical issues arising in single-layer lithography can be significantly improved by using two-layer AFM lithography.

3.5 Second Layer – Local Oxidation of Titanium

One of the possibilities to overcome the challenges mentioned in the previous chapter is to define a second, metallic layer on top of the GaAs surface, and to pattern isolated top-gates into it, just above the circuit elements that need to be tuned. This minimizes the capacitive cross-talk between the gates. A metallic top-gate has typically a five times stronger effect on the quantum dot or the quantum point contact than an in-plane gate. Therefore, by using designated top-gates, we gain substantially in tunability. Furthermore, the top-gate patterning technique makes use of the third spatial dimension [29], which overcomes geometrical limitations for placing a large number of gates in a single plane.

In this chapter, we discuss the lithography of the second layer including sample preparation and the oxidation process. Then we present the most important properties of thin Ti top-gates.

3.5.1 Preparation of the Surface and Ti Film Evaporation

After having completed the first lithography layer as described before, the sample surface is clean and ready for evaporation of a Ti film. In the pioneering work on STM local anodization of Ti [11], the metal was deposited by an electron beam evaporator in ultrahigh vacuum at a very low rate of 0.1 nm/s. Such a slow rate leads to a smooth Ti surface with a roughness below 1.5 nm required for subsequent local oxidation. The thickness of the evaporated film should be around 3.8 nm. After exposure to ambient conditions, the surface of the film immediately oxidizes forming a 2–3 nm thick native oxide. However, below this oxide, the remaining thin Ti layer remains metallic and conductive. The thickness of the evaporated film is crucial for the lithography. A too thin film will not be conductive after the native oxide has formed, and if the film is too thick it will be impossible to locally oxidize through the remaining Ti layer. The thickness of the film measured with the AFM after formation of the native oxide should be between 4 and 6 nm. The two-terminal resistance of the top-gate is a good measure of the quality of the film. At room temperature it should be between 5 and 25 kΩ [29].

3.5.2 Oxidation Process

The oxidation of Ti is analogous to the oxidation of GaAs described in detail in Section 3.4.2. Under an anodic bias applied to Ti, the electrochemical reactions take place at the tip-water-sample interfaces [11]. The reactions lead to the oxidation of Ti,

$$\mathrm{Ti} + 4\mathrm{H}^+ + 2\mathrm{H_2O} \rightarrow \mathrm{TiO_2} + 4\mathrm{H}^+ \tag{10}$$

with the simultaneous reduction of water

$$4\mathrm{H_2O} + 4e^- \rightarrow 2\mathrm{H_2} + 4\mathrm{OH}^-. \tag{11}$$

As a result, the Ti film under the native oxide is oxidized locally beneath the tip. The maximum height of the oxide lines is around 3–4 nm, in agreement with the expected change in density and molecular weight [19].

The oxidation scheme is shown in Fig. 3.18. Here, the Ti and water films cover the entire structure. With appropriate parameters, oxidation across oxide lines defined below the Ti film on the GaAs surface is possible.

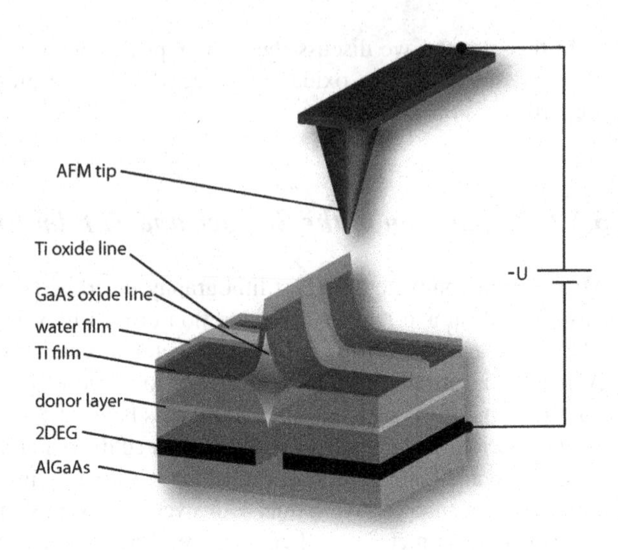

Fig. 3.18 A thin titanium film is deposited on the AFM patterned GaAs surface. A water layer covers the Ti film. The voltage applied to the AFM tip locally oxidizes the Ti film, which results in the creation of mutually insulating electric barriers

3.5.3 Barrier Height

In 1995 Matsumoto *et al.* demonstrated the first device patterned on a thin Ti layer [14]. First, they deposited a 4 nm titanium layer on a silicon substrate. Then they wrote an 18 nm wide and 3 nm high insulating oxide line across the metallic surface creating a metal-insulator-metal diode. The current-voltage characteristics at 77 K showed a suppressed current below a threshold voltage of 4 V. The Ti oxide line acted as a potential barrier for the electrons. At higher bias voltages, a strongly nonlinear tunneling current flowed across the line. The electronic properties of a thin metallic wire were studied in more detail by Irmin *et al.* [18]. Figure 3.19 shows the AFM picture of a metallic Ti wire surrounded by an oxidized region. In the middle of the channel a 21 nm wide oxide line was drawn. Before the oxidation, the current-voltage characteristics of the metallic channel were perfectly ohmic. As shown in Fig. 3.20, upon the oxidation, the barrier was created and the conductance of the device dropped by four orders of magnitude showing a highly non-linear behavior. Figure 3.21 shows the conductance through such barriers for different widths of the oxide lines. In the simplest picture, the width of the oxide line w is related to the width of the tunneling barrier. The current through the device j depends exponentially on the w:

$$j \propto \exp\left(-w\sqrt{\Phi}\right), \tag{12}$$

which indicates that the tunneling is the dominant conduction mechanism. The typical barrier height obtained for TiO_x oxide lines is around 200 meV [14, 18, 20].

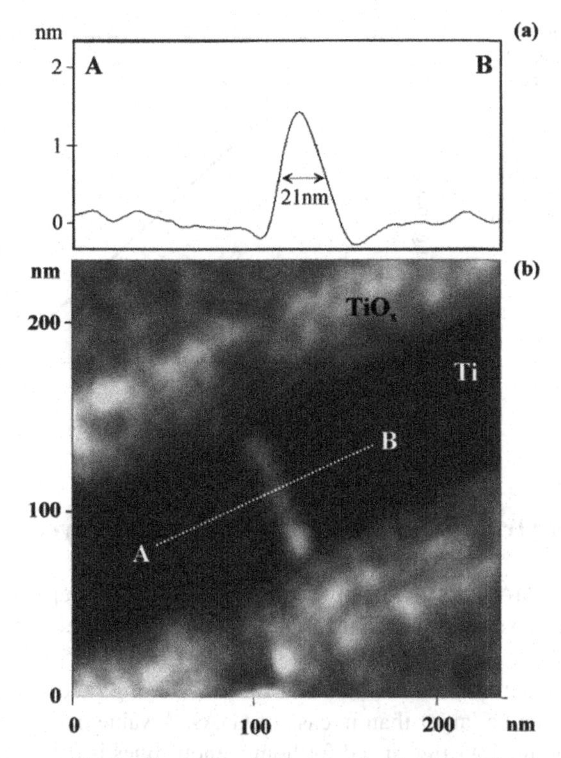

Fig. 3.19 (**a**) The cross section along the metallic wire shown in (**b**). The titanium oxide line is approximately 1.5 nm high. (**b**) The AFM picture of the structure. Titanium wire (dark) and the titanium oxide (bright). Reprinted with permission from [18]. Copyright 1997, American Institute of Physics

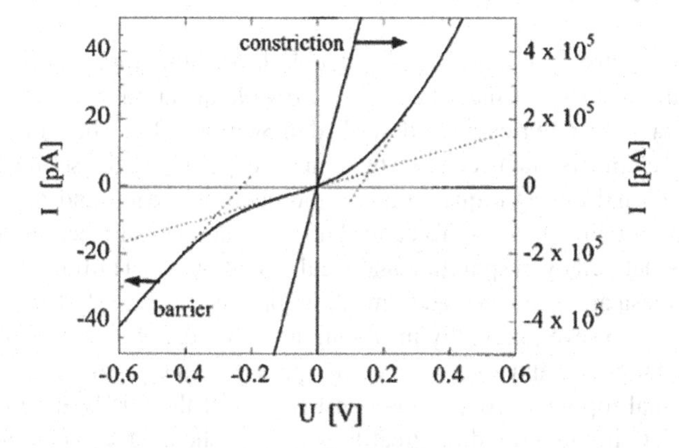

Fig. 3.20 Current-voltage characteristic of the Ti metallic wire before the formation of the barrier (right axis) and with the 21 nm wide oxide barrier across the wire (left axis). *Dotted lines* are tangents to the zero bias and high bias regions [18]. Reprinted with permission from [18]. Copyright 1997, American Institute of Physics

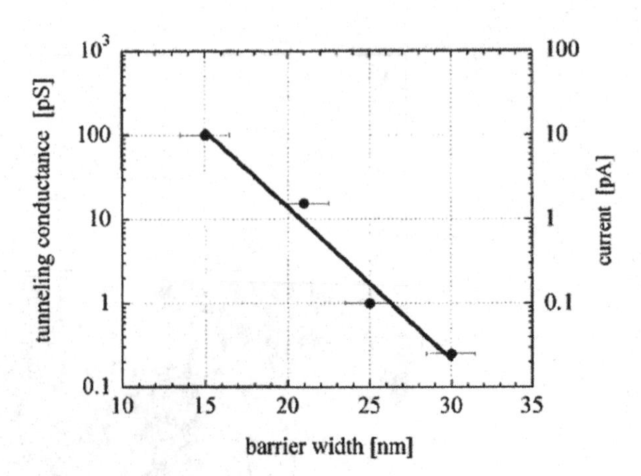

Fig. 3.21 Tunneling conductance for different TiO_x barrier widths. The conductance depends exponentially on the barrier thickness. Reprinted with permission from [18]. Copyright 1997, American Institute of Physics

3.5.4 Optimal Oxidation Parameters for Second Layer Lithography

As compared to GaAs, the oxidation of titanium is less tip-dependent and more reliable. In order to oxidize through the thin Ti film for creating electrically isolating top-gates, the voltage should be in the range between –10 and –15 V. Larger negative voltages will cause unwanted oxidation of the GaAs lying below the Ti. The set-point amplitude can be larger than in case of GaAs. A value of 20–25% is usually sufficient. The optimal writing speed for homogenous lines is 0.1 μm/s, a factor of 2 slower than for GaAs. The relative humidity for obtaining 100–120 nm wide lines should be adjusted to 41–43%.

3.5.5 Design and Tunability

An example of a structure fabricated by double-layer lithography is presented in Fig. 3.22. In the first step of the lithography, a double quantum dot with two quantum point contacts was fabricated as described in Section 3.4.3. The entire structure contains seven elements that need to be tuned in an experiment: the source and drain barriers, the left and the right quantum dot, the interdot barrier, and the two independent quantum point contacts. To control all these degrees of freedom, patterned top-gates were defined by evaporation and local oxidation of a Ti film. The film covers the entire mesa and is connected to metallic top-gate fingers. By local oxidation the film is split into seven mutually insulating metallic regions. Each region serves as a designated top-gate and a voltage can be applied via a top-gate finger.

The individual top-gates control the constrictions of the QPCs and of the double quantum dot circuits providing great tunability of the device. The effect of the top-gates on the structure at liquid helium temperature is presented in Fig. 3.23. Plot A shows the current through the double quantum dot circuit as a function of the voltages applied to top-gate 1 and top-gate 2. A small DC bias voltage was

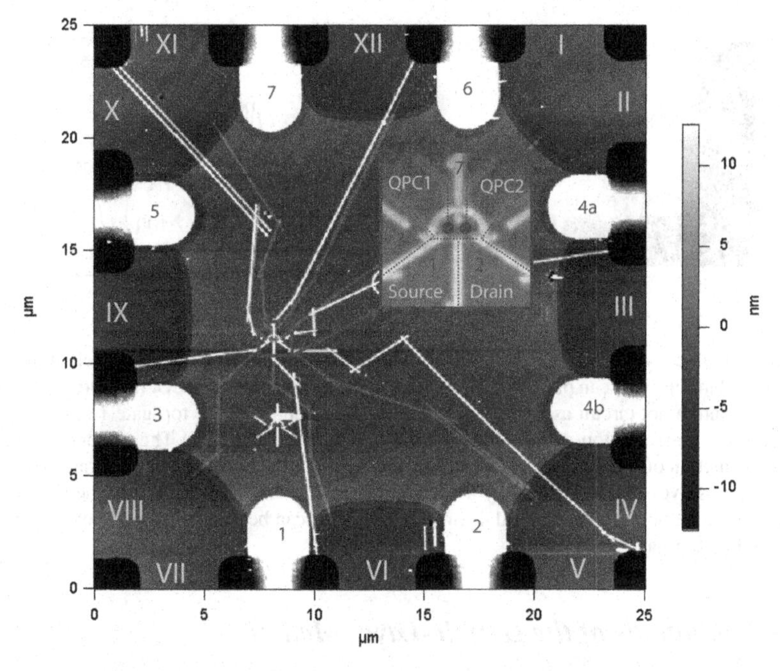

Fig. 3.22 The picture of the mesa presented in Fig. 3.13. The 4 nm thick Ti film was evaporated in the middle of the mesa. It is contacted to the golden top-gate fingers labeled with arabic numbers. The inset shows the top-gate pattern written over the double quantum dot. The *dashed lines* indicate the oxide lines on Ti. They split the metallic film into seven independent top-gates. Each top-gate is connected to a metallic top-gate finger at the edge of the mesa. Top-gates 1 and 2 are designed to tune the width of the source and drain constrictions. Analogously, top-gates 3 and 4 control the width of QPC1 and QPC2. The size of the quantum dots is changed by top-gates 5 and 6. The coupling between the dots can be modified using the middle top-gate 7

applied in the 2DEG plane between source and drain. At positive top-gate voltages, the current flows through the DQD. As the voltages are decreased, the constrictions become more resistive and the current is eventually blocked. The pinch-off voltage for top-gate 1 is around –50 mV and for top-gate 2 it occurs at +50 mV. The pinch-off voltage of top-gate 1 is independent of the voltage applied to top-gate 2 and vice-versa. The reduced cross-talk between the gates is one of the advantageous properties of double layer lithography. The characteristics of the remaining three gates defined in the DQD circuit are plotted in Fig. 3.23b. For this measurement, the source and drain top-gates were kept fixed in the open regime indicated by the black circle in A. The negative voltage was applied separately to either top-gate 5, 6 or 7. Each of the top-gates has a lever-arm strong enough to pinch the circuit off, and the pinch-off voltages are around –200 mV. Similarly, the QPC1 and QPC2 constrictions can be opened or closed by tuning top-gates 3 and 4 as shown in Fig. 3.23c. In general, the pinch-off voltages may be cool-down dependent, but the large tunability provided by individual top-gates helps to overcome this problem.

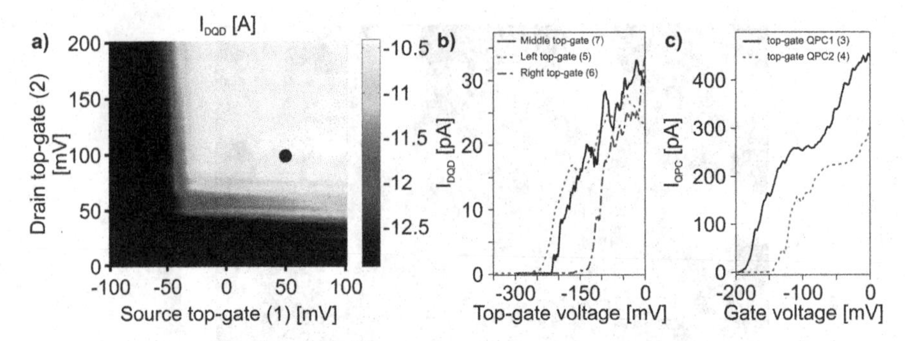

Fig. 3.23 Double layer lithography provides full tunability of the nanostructures. Each of the top-gates defined in Fig. 3.22 can pinch-off the constrictions defined on the 2DEG. (**a**) Current through a double quantum dot circuit as a function of the voltages applied to the top-gate 1 (covering the source constriction) and top-gate 2 (covering the drain constriction). (**b**) The current through a DQD as a function of top-gates 5, 6 or 7. During this measurement, the source and drain top-gates were set to positive values (denoted by a black circle in A). The functionality of the QPC1 and QPC2 circuit tuned by top-gates 3 and 4. Both of the QPCs can be easily opened or closed by the corresponding top-gate

3.5.6 Limitations of the Double-Layer Method

It is found experimentally, that top-gate sweeps induce on average more charge rearrangements than sweeps of in-plane gates. This effect is most likely due to charge traps residing at the GaAs surface or in the doping plane. The way around this problem is to find a suitable working regime by using the top-gates and then sweep only the in-plane gates. Another side-effect of double layer lithography is that the presence of metallic top-gates effectively screens electrostatic interactions. This provides sample designers with a means for tailoring the interaction strength, but it may also reduce the coupling between a charge detector and a quantum dot [47].

3.6 Semiconductor Nanostructures Fabricated Using Double Layer AFM-Lithography

Quantum point contacts, quantum dots, double quantum dots and rings can be coupled to each other and form complex quantum circuits. In this chapter, we present three devices fabricated by double layer lithography.

3.6.1 Few-Electron Quantum Dot

Few-electrons quantum dots are considered to be model systems for quantum computing [48]. A quantum dot with one electron can be considered as an artificial hydrogen atom. If the second electron is added onto a dot, it forms a spin singlet ground state, like in the helium atom. Few-electron quantum dots have allowed experimentalists to perform many interesting experiments, such as the investigation of energy spectra [49, 50] or spin effects [51, 52] in artificial atoms. Moreover, they are used to study the physics of the Kondo [53] and Fano [54] effects as well as higher-order tunneling processes [55].

It is difficult to create few-electron quantum dots exclusively by single-layer AFM lithography. The biggest problem is that the plunger gate used to change the number of electrons in a QD has a strong influence on the source and drain constrictions. As a result, the barriers become opaque for electrons and block the current before the dot is empty. This can be avoided by applying the double layer local oxidation technique [56]. The top-gate used for controlling the electron number in the QD has little impact on the coupling to source and drain leads. The design of a few-electron QD is presented in Fig. 3.24a. The bright lines are the oxide lines defining the quantum dot connected to source and drain leads. Below the dot, there is a quantum point contact. The oxide lines written on Ti defining four top-gates are marked with black lines. Applying a negative voltage to the top-gate *tpg* reduces the number of electrons in the quantum dot as shown in Fig. 3.24b. The last Coulomb peak is visible around –74 mV. Below this value the dot is empty. To prove that

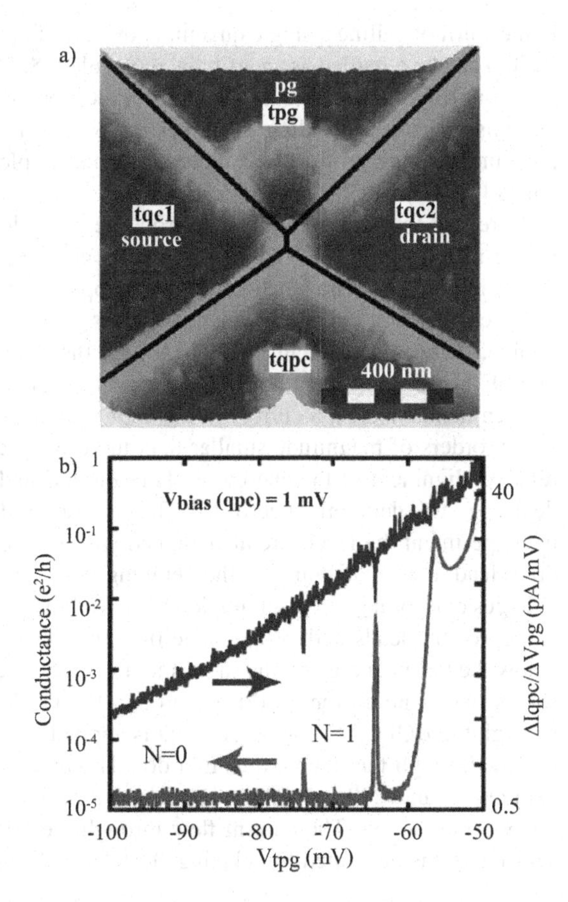

Fig. 3.24 Conductance of a quantum dot as a function of the plunger top-gate *tpg* (left axis). The corresponding transconductance signal of the quantum point contact shows dips at the positions of each Coulomb peak. N indicates the number of electrons in a dot. Adapted from [56]. Copyright 2006, with permission from Elsevier

indeed the last electron has been removed, one can use a so-called charge read-out. A QPC placed in the vicinity of the QD and tuned to the regime between the pinch-off and the first quantized conductance plateau serves as good charge detector [57]. The current flowing through the QPC at constant applied bias voltage is sensitive to electrostatic potential changes that occur in the QD. Whenever an electron hops onto the dot, the QPC current is reduced as a result of the repulsive Coulomb interaction between QD and QPC electrons. When the transconductance through the QPC is measured, a resonant dip arises exactly at the position of a Coulomb peak in the QD current. The charge read-out can detect changes in the charge state of the dot even if the direct current through the QD is too small to be measured directly. In Fig. 3.24b, the last dip in the QPC transconductance coincides with the last observable Coulomb peak in the direct QD current. This, together with finite bias measurements [56] proves that this is a few-electron quantum dot that can be filled with single electrons starting from zero.

3.6.2 Double Quantum Dot with Integrated Charge Read-Out

In the spirit of calling a single quantum dot an artificial atom, a double quantum dot (DQD) can be regarded as an artificial molecule [58, 59]. It consists of two quantum dots coupled via electrostatic interaction. Additionally, tunneling coupling between the dots may occur, leading to further splitting of resonant energy levels. Double quantum dots are promising candidates for the implementation of solid state spin qubits [48, 60].

Here, we focus on the transport properties of a double quantum dot influenced by the action of neighboring quantum point contacts. The design of the sample is shown in Fig. 3.22. The functionality of the top-gates was discussed in Section 3.5.5. In the experiment, the double quantum dot is kept at zero source-drain bias. In this regime, there is no current flowing through the double quantum dot. However, if one of the adjacent quantum point contacts is driven with a current exceeding a few ten nanoamperes, a current in the DQD circuit is observed. The DQD current is 4–5 orders of magnitude smaller than the driving current and it depends on the relative alignment of the energy levels in the left and the right dot, defined as the detuning. The detuning is zero if the levels in both dots have the same energy. In the experiment the levels are also aligned with the Fermi energy of the source and drain lead at zero detuning. If the detuning is increased (by applying the top-gate voltages controlling the quantum dots) the level in one dot is raised above the Fermi energy of the leads and, at the same time, the level in the second dot is lowered below the Fermi energy by the same amount. The trace with the squares in Fig. 3.25 shows that if no current is driven through the QPCs, there is no current flowing through the DQD. As soon as a current is applied to one of the QPCs, current starts to flow through the double quantum dot. The maximum current is observed for a detuning around 200 μeV. This effect can be explained in the framework of electron-phonon interaction. The current flow through the QPC causes energy dissipation. This energy is generated by relaxing electrons causing the emission of phonons in

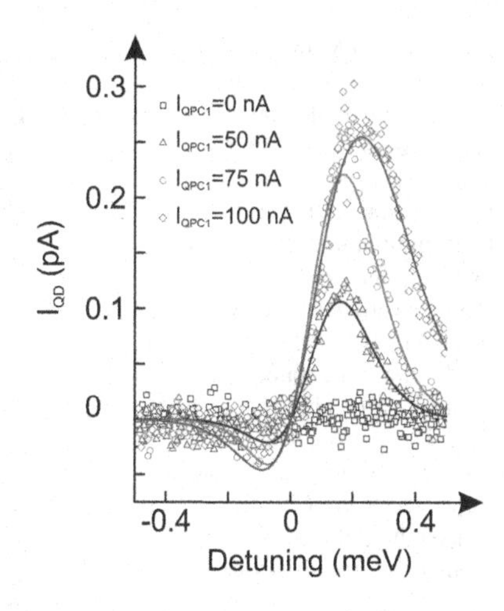

Fig. 3.25 Measurement of the current through a double quantum dot at zero source-drain bias voltage. Different measurement curves correspond to different currents applied to the quantum point contact. *Solid lines* are fits to a theoretical model [61]

the GaAs crystal. As a result the temperature of the phonon bath increases above the equilibrium temperature of the electron gas. The lack of thermodynamic equilibrium between the electronic system and the phonon bath is the driving force for the DQD current. The electrons in the DQD absorb energy from the phonon bath via electron-phonon interactions, and are excited into a state higher in energy. From this state they can leave the dot and contribute to the current. The transport cycle closes if the next electron jumps into the ground state of the DQD from the other lead. The solid lines in Fig. 3.25 are fits to a theoretical model based on electron-phonon interaction [61]. The exact shape of the energy-dependent DQD current is mainly determined by the phonons' energy distribution and by the geometry of the double quantum dot.

3.6.3 Double Quantum Dot in a Ring with Integrated Charge Read-Out

First-order tunneling processes that contribute to the conductance dominate the electron transport through quantum dots for small bias voltages. However, if the dot is tuned off-resonance, second-order tunneling processes start to play a dominant role at sufficiently large tunneling coupling, leading to the flow of a small co-tunneling current. If, after an individual co-tunneling process, the dot is left in the same energetic state, then the process is called elastic. Otherwise, if the dot is left in an excited state, the process is called inelastic. To determine if co-tunneling processes conserve the phase of the tunneling electrons, it is necessary to place the dot in an Aharonov-Bohm ring [62].

The structure containing a double quantum dot in a ring is presented in Fig. 3.26. The bright lines are oxide lines depleting the 2DEG below. The sophisticated top-gate pattern covers the entire structure. Top-gates tqc1, tqc2, tqc3 and tqc4 are used to control the width of the constrictions used to form the double quantum dot. The

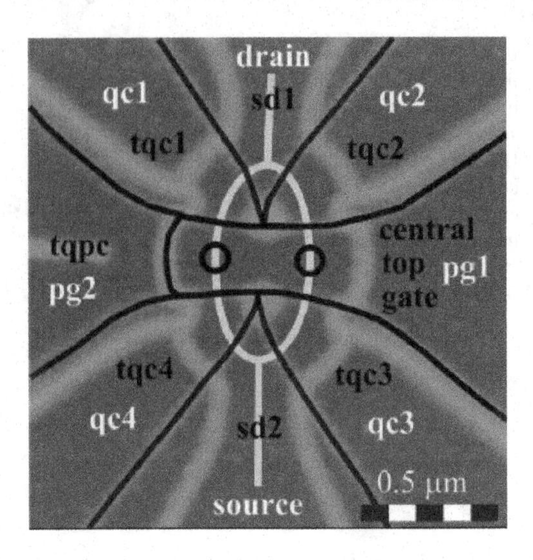

Fig. 3.26 AFM image of the structure. The wide *bright lines* are the oxide lines patterned on GaAs. White letters correspond to the in-plane gates. The oxide lines on titanium are marked by black lines and the top-gates are denoted with black letters. The *bright circle* shows the trajectories of the electrons in the ring. The black circles mark the positions of the quantum dots. Reprinted from [62]. Copyright 2006, with permission from Elsevier

top-gates sd1 and sd2 control the entrance and the exit of the ring. The central top-gate can change the electron number and the coupling within the double quantum dot system. Additionally, the structure contains also a quantum point contact charge detector, which can be tuned by the top-gate tqpc.

In the measurement, all the top-gates are set on suitable voltages in order to form a double quantum dot and to keep both arms of the ring open. Then the in-plane gate voltages *pg1* and *pg2* are used to control the electron population of both quantum dots. The so-called stability diagram of the double quantum dot is shown in Fig. 3.27a. The black horizontal line represents a Coulomb peak associated with dot 1 and the vertical line corresponds to a Coulomb peak of dot 2. The lines cross at the degeneracy point, at which the energy levels of both dots are aligned. In the white region of small conductance, the current is dominated by second-order elastic co-tunneling processes. In the next measurement, the gates *pg1* and *pg2* are swept simultaneously in 60 steps along the arrows drawn on the stability diagram. For each of the steps the magnetic field is swept and the conductance is measured. The result is shown in Fig. 3.27b. For all the gate configurations, Aharonov-Bohm oscillations are visible. At the positions of the Coulomb peaks, a sudden phase jump of π is observed. The experiment demonstrates the phase coherence of elastic co-tunneling processes in quantum dots.

3.7 Conclusions

Double layer lithography allows for the creation of complex coupled electronic nanoscale systems. It yields reliable and reproducible results and significantly enhances the tunability of the resulting devices compared to single-layer methods [63]. Additionally, it reduces the cross-talk between the gate electrodes, which

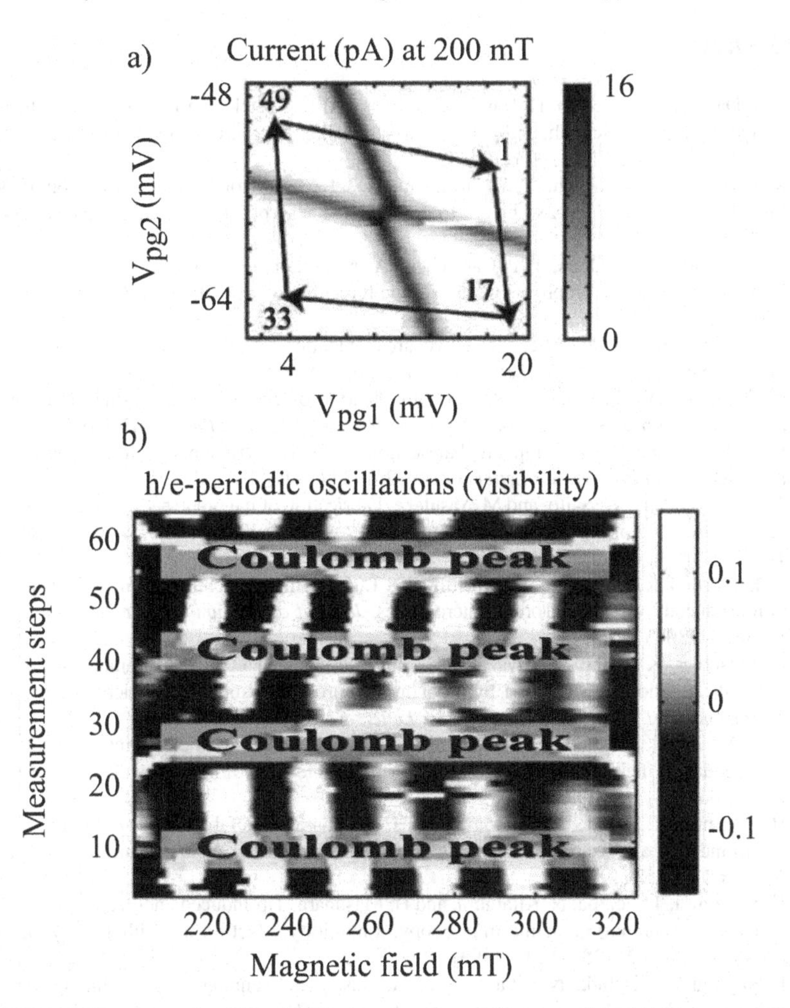

Fig. 3.27 (**a**) The current through a double quantum dot as a function of the in-plane gate voltages at finite magnetic field. The almost-horizontal line corresponds to the Coulomb resonance in dot 1 and the *vertical line* corresponds to the resonance in dot 1. At the crossing of the lines, the energy levels in dot 1 and dot 2 are aligned. The *arrows* denote the direction of the voltage changes in the following measurement. (**b**) For each step, the conductance of the double quantum dot is measured as a function of magnetic field. The experimental data are filtered. Clear Aharonov-Bohm oscillations are visible, except for the regions corresponding to the Coulomb peaks. Adapted from [62]. Copyright 2006, with permission from Elsevier

simplifies the process of tuning the structure into the regime desired. The alignment of the pattern in the Ti film relative to the oxide lines on the GaAs is straightforward and only limited by the resolution of the AFM. The experimental results presented in the previous section demonstrate the high quality of the fabricated devices and promise further innovative designs for highly tunable semiconductor nanostructures with tailored electronic properties.

References

1. J. A. Dagata, J. Schneir, H. H. Harary, C. J. Evans, M. T. Postek, and J. Bennett, Modification of hydrogen-passivated silicon by a scanning tunneling microscope operating in air, *Applied Physics Letters* **56**, 2001–2003 (1990).
2. R. S. Becker, G. S. Higashi, Y. J. Chabal, and A. J. Becker, Atomic-scale conversion of clean Si(111):H-1x1 to Si(111)-2x1 by electron-stimulated desorption, *Physical Review Letters* **65**, 1917 (1990).
3. A. Majumdar, P. I. Oden, J. P. Carrejo, L. A. Nagahara, J. J. Graham, and J. Alexander, Nanometer-scale lithography using the atomic force microscope, *Applied Physics Letters* **61**, 2293–2295 (1992).
4. H. C. Day, and D. R. Allee, Selective area oxidation of silicon with a scanning force microscope, *Applied Physics Letters* **62**, 2691–2693 (1993).
5. M. Yasutake, Y. Ejiri, and T. Hattori, Modification of silicon surface using atomic force microscope with conducting probe, *Japanese Journal of Applied Physics* **32**, L1021 (1994).
6. E. S. Snow, and P. M. Campbell, Fabrication of Si nanostructures with an atomic force microscope, *Applied Physics Letters* **64**, 1932–1934 (1994).
7. T. Hattori, Y. Ejiri, K. Saito, and M. Yasutake, Fabrication of nanometer-scale structures using atomic force microscope with conducting probe, *Journal of Vacuum Science & Technology A* **12**, 2586–2590 (1994).
8. A. E. Gordon, R. T. Fayfield, D. D. Litfin, and T. K. Higman, Mechanisms of surface anodization produced by scanning probe microscopes, *Journal of Vacuum Science & Technology B* **13**, 2805–2808 (1995).
9. T. Teuschler, K. Mahr, S. Miyazaki, M. Hundhausen, and L. Ley, Nanometer-scale field-induced oxidation of Si(111):H by a conducting-probe scanning force microscope: doping dependence and kinetics, *Applied Physics Letters* **67**, 3144–3146 (1995).
10. S. C. Minne, H. T. Soh, P. Flueckiger, and C. F. Quate, Fabrication of 0.1 mu m metal oxide semiconductor field-effect transistors with the atomic force microscope, *Applied Physics Letters* **66**, 703–705 (1995).
11. H. Sugimura, T. Uchida, N. Kitamura, and H. Masuhara, Nanofabrication of titanium surface by tip-induced anodization in scanning tunneling microscopy, *Japanese Journal of Applied Physics* **32**, L553 (1993).
12. H. Sugimura, T. Uchida, N. Kitamura, and H. Masuhara, Tip-induced anodization of titanium surfaces by scanning tunneling microscopy: a humidity effect on nanolithography, *Applied Physics Letters* **63**, 1288–1290 (1993).
13. H. Sugimura, T. Uchida, N. Kitamura, and H. Masuhara, Scanning tunneling microscope tip-induced anodization for nanofabrication of titanium, *The Journal of Physical Chemistry* **98**, 4352–4357 (1994).
14. K. Matsumoto, S. Takahashi, M. Ishii, M. Hoshi, A. Kurokawa, S. Ichimura, and A. Ando, Application of STM nanometer-size oxidation process to planar-type MIM diode, *Japanese Journal of Applied Physics* **34**, 1387 (1995).
15. D. Wang, L. Tsau, K. L. Wang, and P. Chow, Nanofabrication of thin chromium film deposited on Si(100) surfaces by tip induced anodization in atomic force microscopy, *Applied Physics Letters* **67**, 1295–1297 (1995).
16. E. S. Snow, and P. M. Campbell, AFM fabrication of sub-10-nanometer metal-oxide devices with in situ control of electrical properties, *Science* **270**, 1639–1641 (1995).
17. E. S. Snow, D. Park, and P. M. Campbell, Single-atom point contact devices fabricated with an atomic force microscope, *Applied Physics Letters* **69**, 269–271 (1996).
18. B. Irmer, M. Kehrle, H. Lorenz, and J. P. Kotthaus, Fabrication of Ti/TiO_x tunneling barriers by tapping mode atomic force microscopy induced local oxidation, *Applied Physics Letters* **71**, 1733–1735 (1997).
19. M. K. B. Irmer, H. Lorenz, and J. P. Kotthaus, Nanolithography by non-contact AFM-induced local oxidation: fabrication of tunnelling barriers suitable for single-electron devices, *Semiconductor Science and Technology* **13**, A79 (1998).

20. P. Avouris, R. Martel, T. Hertel, and R. Sandstrom, AFM-tip-induced and current-induced local oxidation of silicon and metals, *Applied Physics A: Materials Science & Processing* **66**, S659-S667 (1998).
21. R. Held, T. Heinzel, P. Studerus, K. Ensslin, and M. Holland, Semiconductor quantum point contact fabricated by lithography with an atomic force microscope, *Applied Physics Letters* **71**, 2689–2691 (1997).
22. J. Shirakashi, K. Matsumoto, N. Miura, and M. Konagai, Single-electron charging effects in Nb/Nb oxide-based single-electron transistors at room temperature, *Applied Physics Letters* **72**, 1893–1895 (1998).
23. X. N. Xie, H. J. Chung, C. H. Sow, and A. T. S. Wee, Nanoscale materials patterning and engineering by atomic force microscopy nanolithography, *Materials Science and Engineering: R: Reports* **54**, 1–48 (2006).
24. B. Bhushan, *Scanning Probe Microscopy in Nanoscience and Nanotechnology,* Springer, Berlin, 2010.
25. M. Ishii, and K. Matsumoto, Control of current in 2DEG channel by oxide wire formed using AFM, *Japanese Journal of Applied Physics* **34**, 1329 (1995).
26. R. Held, T. Vancura, T. Heinzel, K. Ensslin, M. Holland, and W. Wegscheider, In-plane gates and nanostructures fabricated by direct oxidation of semiconductor heterostructures with an atomic force microscope, *Applied Physics Letters* **73**, 262–264 (1998).
27. R. Held, S. Luscher, T. Heinzel, K. Ensslin, and W. Wegscheider, Fabricating tunable semiconductor devices with an atomic force microscope, *Applied Physics Letters* **75**, 1134–1136 (1999).
28. S. Luscher, A. Fuhrer, R. Held, T. Heinzel, K. Ensslin, and W. Wegscheider, In-plane gate single-electron transistor in Ga[Al]As fabricated by scanning probe lithography, *Applied Physics Letters* **75**, 2452–2454 (1999).
29. M. Sigrist, A. Fuhrer, T. Ihn, K. Ensslin, D. C. Driscoll, and A. C. Gossard, Multiple layer local oxidation for fabricating semiconductor nanostructures, *Applied Physics Letters* **85**, 3558–3560 (2004).
30. X. N. Xie, H. J. Chung, C. H. Sow, and A. T. S. Wee, Microdroplet and atomic force microscopy probe assisted formation of acidic thin layers for silicon nanostructuring, *Advanced Functional Materials* **17**, 919 (2007).
31. J. A. Dagata, T. Inoue, J. Itoh, K. Matsumoto, and H. Yokoyama, Role of space charge in scanned probe oxidation, *Journal of Applied Physics* **84**, 6891–6900 (1998).
32. D. Graf, M. Frommenwiler, P. Studerus, T. Ihn, K. Ensslin, D. C. Driscoll, and A. C. Gossard, Local oxidation of Ga[Al]As heterostructures with modulated tip-sample voltages, *Journal of Applied Physics* **99**, 053707–053707 (2006).
33. A. Fuhrer, A. Dorn, S. Luscher, T. Heinzel, K. Ensslin, W. Wegscheider, and M. Bichler, Electronic properties of nanostructures defined in Ga[Al]As heterostructures by local oxidation, *Superlattices and Microstructures* **31**, 19–42 (2002).
34. S. Datta, *Electronic Transport in Mesoscopic Systems,* Cambridge University Press, Cambridge, 1995.
35. T. Heinzel, *Mesoscopic Electronics in Solid State Nanostructures,* Wiley, Weinheim, 2003.
36. S. K. Ghandhi, *VLSI Fabrication Principles,* Wiley, New York, 1994.
37. S. R. Morrison, *The Chemical Physics of Surfaces,* Plenum Press, New York, 1977.
38. Y. Okada, Y. Iuchi, M. Kawabe, and J. J. S. Harris, Basic properties of GaAs oxide generated by scanning probe microscope tip-induced nano-oxidation process, *Journal of Applied Physics* **88**, 1136–1140 (2000).
39. U. F. Keyser, H. W. Schumacher, U. Zeitler, R. J. Haug, and K. Eberl, Fabrication of a single-electron transistor by current-controlled local oxidation of a two-dimensional electron system, *Applied Physics Letters* **76**, 457–459 (2000).
40. N. J. Curson, R. Nemutudi, N. J. Appleyard, M. Pepper, D. A. Ritchie, and G. A. C. Jones, Ballistic transport in a $GaAs/Al_xGa_{1-x}As$ one-dimensional channel fabricated using an atomic force microscope, *Applied Physics Letters* **78**, 3466–3468 (2001).

41. C. W. J. Beenakker, H. van Houten, E. Henry, and T. David, Quantum transport in semiconductor nanostructures, *Solid State Physics* **44**, 1–228 (1991).
42. C. Fricke, J. Regul, F. Hohls, D. Reuter, A. D. Wieck, and R. J. Haug, Transport spectroscopy of a quantum point contact created by an atomic force microscope, *Physica E: Low-dimensional Systems and Nanostructures* **34**, 519–521 (2006).
43. R. Schleser, E. Ruh, T. Ihn, K. Ensslin, D. C. Driscoll, and A. C. Gossard, Time-resolved detection of individual electrons in a quantum dot, *Applied Physics Letters* **85**, 2005–2007 (2004).
44. Y. Aharonov, and D. Bohm, Significance of electromagnetic potentials in the quantum theory, *Physical Review* **115**, 485 (1959).
45. A. Fuhrer, S. Luscher, T. Ihn, T. Heinzel, K. Ensslin, W. Wegscheider, and M. Bichler, Energy spectra of quantum rings, *Nature* **413**, 822–825 (2001).
46. U. F. Keyser, et al., Aharonov-Bohm oscillations of a tuneable quantum ring, *Semiconductor Science and Technology* **17**, L22 (2002).
47. T. Ihn, S. Gustavsson, U. Gasser, B. Küng, T. Müller, R. Schleser, M. Sigrist, I. Shorubalko, R. Leturcq, and K. Ensslin, Quantum dots investigated with charge detection techniques, *Solid State Communications* **149**, 1419–1426 (2009).
48. D. Loss, and D. P. DiVincenzo, Quantum computation with quantum dots, *Physical Review A* **57**, 120 (1998).
49. S. Tarucha, D. G. Austing, T. Honda, R. J. van der Hage, and L. P. Kouwenhoven, Shell filling and spin effects in a few electron quantum dot, *Physical Review Letters* **77**, 3613 (1996).
50. J. M. Elzerman, R. Hanson, J. S. Greidanus, L. H. Willems van Beveren, S. De Franceschi, L. M. K. Vandersypen, S. Tarucha, and L. P. Kouwenhoven, Few-electron quantum dot circuit with integrated charge read out, *Physical Review B* **67**, 161308 (2003).
51. R. Hanson, L. P. Kouwenhoven, J. R. Petta, S. Tarucha, and L. M. K. Vandersypen, Spins in few-electron quantum dots, *Reviews of Modern Physics* **79**, 1217–1249 (2007).
52. R. Hanson, B. Witkamp, L. M. K. Vandersypen, L. H. W. van Beveren, J. M. Elzerman, and L. P. Kouwenhoven, Zeeman energy and spin relaxation in a one-electron quantum dot, *Physical Review Letters* **91**, 196802 (2003).
53. S. Sasaki, S. De Franceschi, J. M. Elzerman, W. G. van der Wiel, M. Eto, S. Tarucha, and L. P. Kouwenhoven, Kondo effect in an integer-spin quantum dot, *Nature* **405**, 764–767 (2000).
54. T. Otsuka, E. Abe, S. Katsumoto, Y. Iye, G. L. Khym, and K. Kang, Fano effect in a few-electron quantum dot, *Journal of the Physical Society of Japan* **76**, 084706 (2007).
55. S. De Franceschi, S. Sasaki, J. M. Elzerman, W. G. van der Wiel, S. Tarucha, and L. P. Kouwenhoven, Electron cotunneling in a semiconductor quantum dot, *Physical Review Letters* **86**, 878 (2001).
56. M. Sigrist, S. Gustavsson, T. Ihn, K. Ensslin, D. Driscoll, A. Gossard, M. Reinwald, and W. Wegscheider, Few-electron quantum dot fabricated with layered scanning force microscope lithography, *Physica E: Low-dimensional Systems and Nanostructures* **32**, 5–8 (2006).
57. M. Field, C. G. Smith, M. Pepper, D. A. Ritchie, J. E. F. Frost, G. A. C. Jones, and D. G. Hasko, Measurements of Coulomb blockade with a noninvasive voltage probe, *Physical Review Letters* **70**, 1311 (1993).
58. C. Livermore, The Coulomb blockade in coupled quantum dots, *Science* **274**, 1332 (1996).
59. R. H. Blick, R. J. Haug, J. Weis, D. Pfannkuche, K. v. Klitzing, and K. Eberl, Single-electron tunneling through a double quantum dot: the artificial molecule, *Physical Review B* **53**, 7899 (1996).
60. W. G. van der Wiel, S. De Franceschi, J. M. Elzerman, T. Fujisawa, S. Tarucha, and L. P. Kouwenhoven, Electron transport through double quantum dots, *Reviews of Modern Physics* **75**, 1 (2002).
61. U. Gasser, S. Gustavsson, uuml, B. ng, K. Ensslin, T. Ihn, D. C. Driscoll, and A. C. Gossard, Statistical electron excitation in a double quantum dot induced by two independent quantum point contacts, *Physical Review B* **79**, 035303 (2009).

62. M. Sigrist, T. Ihn, K. Ensslin, M. Reinwald, and W. Wegscheider, Phase coherence in the cotunneling regime of a coupled double quantum dot, *Physica E: Low-dimensional Systems and Nanostructures* **34**, 497–499 (2006).
63. M. Sigrist, Phase-coherent transport in quantum circuits containing dots and rings, ETH Zurich, Zurich (2006).

Chapter 4
Nanomanipulation of Biological Macromolecules by AFM

Guoliang Yang

Abstract AFM-based nanomanipulation has been used to study every type of biological macromolecules and revealed important, previously unknown properties and functional mechanisms. The capacity of the AFM for the isolation, transfer, positioning and assembling of individual macromolecules with nanometer spatial resolutions has significantly advanced the field of bionanotechnology toward the fabrication of functional structures and devices with macromolecules as building blocks. This review focuses on the studies of proteins, nucleic acids and polysaccharides using AFM-based nanomanipulation technique. An introduction to the principles of the technique is followed by reviewing the types of problems investigated, with emphasis on the capacities of the technique to reveal novel structural and functional mechanisms of biological macromolecules.

Keywords Scanning probe microscopy · Atomic force microscopy · Force spectroscopy · Nanomanipulation · Single molecule manipulation · Macromolecules · Protein folding · Mechanical unfolding · Mechanical stability · Energy landscape · Membrane proteins · Protein ligand interactions

Abbreviations

AFM	Atomic force microscope
Dopa	3,4-dihydroxy-L-phenylalanine
dsDNA	Double stranded DNA
DTT	Dithiothreitol
FJC	Freely-joint chain
FRET	Förster resonance energy transfer
GFP	Green fluorescent protein
PCR	Polymerase chain reaction
SPM	Scanning probe microscope
ssDNA	Single stranded DNA
STM	Scanning tunneling microscope

G. Yang (✉)
Department of Physics, Drexel University, Philadelphia, PA 19104, USA
e-mail: gyang@drexel.edu

A.A. Tseng (ed.), *Tip-Based Nanofabrication*, DOI 10.1007/978-1-4419-9899-6_4,

TIRF	Total internal reflection fluorescence
TMV	Tobacco mosaic virus
WLC	Worm-like chain

4.1 Introduction

The invention and development of scanning probe microscopes (SPM) have not only revolutionized microscopic techniques, but also greatly expanded the scope and capacity of nanoscale manipulation methods. Nanomanipulation is the precise control of the relative positions and relations of individual atoms, molecules, and nanometer-scale objects or components, which can be realized by pulling, pushing, twisting, cutting, transporting, assembling, or other means. Through nanomanipulation, structural and dynamical properties of nanoscale objects can be characterized, mechanisms of nanoscale processes can be elucidated, nanoscale structured materials with novel properties can be fabricated, and miniature devices with nanoscale components can be made. These advances will have significant and long lasting impacts on all aspects of science and technology, though there are still many technological challenges before the potential of nanomanipulation can be fully exploited [1].

Different methods can be used to exert forces on nanoscale objects and to measure their displacements, such as electrical fields, magnetic fields, optical traps, hydrodynamic fields, flexible micropipettes, microneedles, and SPM probes [2–6]. Of these methods, the SPM has brought nanomanipulation to a new height. In particular, the ability of SPMs to assemble nanometer scale components into predefined structures has greatly enhanced the prospective of using the bottom-up approaches to fabricate functional microscopic devices and to develop novel nanostructured materials. The first demonstration of SPM's unparallel capability for nanomanipulation was reported by Eigler and Schweitzer in 1990 [7] using the inaugural member of the SPM family, *i.e.,* the scanning tunneling microscope (STM). They managed to spell out their institution's logo using individual Xenon atoms at low temperatures on a nickel surface. Several years later, nanomanipulation using the atomic force microscope (AFM) was demonstrated by Junno et al. [8] through the controlled arrangement of aerosol nanoparticles. The AFM has since become a powerful and popular tool for manipulating a wide variety of objects, from individual atoms [9], molecules to live cells [10]. The advantages of the AFM for nanomanipulation include the diversity of nanoscale objects that can be manipulated, as well as the capability to manipulate in various environments and to generate topographic images at the same time. The availability of user friendly commercial instruments has made it easy for research groups without extensive instrumentation expertise to carry out nanomanipulation experiments.

Manipulation and measurement on single macromolecules have attracted the interest of researchers in different fields because these experiments can provide previously unattainable information on the structural and dynamic properties of many biological systems. The three-dimensional structures and functions of biological macromolecules are originated from a large number of weak interactions, which are fundamentally different from small organic and inorganic molecules. Due to the complexity conferred by these weak interactions, individual inhomogeneities in conformations and dynamic trajectories are characteristic of a population of biological macromolecules. Bulk measurement techniques, such as NMR and protein engineering ϕ-value analysis [11], have been very powerful at answering many questions about the structures and functions of biological macromolecules. However, in these experiments only ensemble measurements can be made and the average properties of the molecules are obtained. To follow the dynamic changes at the molecular level using ensemble measurements, it would require the synchronization of all molecules not only at the beginning but over their time trajectories, which is not possible. In single-molecule measurements, the dynamic conformational changes of a macromolecule can be directly characterized, therefore, the molecular level insights into the structural and functional mechanisms of macromolecules can be obtained. Single molecule measurements can be carried out in very low concentrations, thus circumventing the difficulties caused by aggregation of many biological molecules at concentrations necessary for ensemble measurements. In addition, certain molecular species exists in very low copies (even one) in cells [12], single molecule characterization becomes necessary to understand the function of such molecules in living cells.

Over the past two decades, AFM has been used extensively for imaging and manipulating biological macromolecules as well as fabricating nanoscale structures with macromolecules [13, 14]. The readers are referred to many recent review papers on these topics that are cited in the following sections. In this chapter, recent applications of the AFM for manipulating biological macromolecules are reviewed. Instead of an exhaustive account of the literature in this area, I focus on the description and illustration of the technical capabilities of the AFM-based nanomanipulation and the significant new information that has been obtained from the investigation of biological macromolecules using this technique.

4.2 AFM Based Techniques for Macromolecular Manipulation

Among the currently available methods of manipulating biological macromolecules, AFM is unique in its capacities for simultaneous precise position control, sensitive force measurement and high-resolution topographic imaging. Using AFM, controlled forces can be applied to individual macromolecules in physiologically relevant solution conditions. A macromolecule, without the need of fixing or tagging, can be pushed, pulled, or cut by the AFM tip, resulting in the determination of its structural, mechanical and dynamic properties, or the placement in the desired location of a predefined structure.

4.2.1 Operation Principles of AFM

The atomic force microscope was invented by Binnig, Quate and Berber in 1986 [15] in an effort to develop a technology for obtaining atomic resolution images of an insulating surface after the stunning success of the scanning tunneling

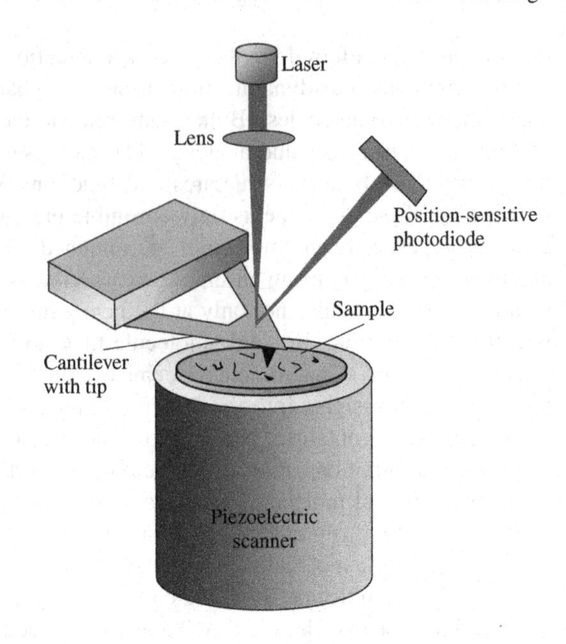

Fig. 4.1 Schematic of the operation principle of an atomic force microscope. The scanner can move the sample in three dimensions with respect to the tip. The optical lever system can detect the cantilever deflection to sub-nanometer precision

microscope on conducting surfaces [16, 17]. The AFM works by detecting the interaction force between the sample and a sharp tip mounted at the end of a flexible cantilever, as shown in Fig. 4.1. The detection of the force-induced cantilever bending is achieved by using a laser beam focused at the end of the cantilever. The movement of the reflected laser beam is measured by a position-sensitive quadrant photodiode. The sample (or the tip) can be moved in three dimensions (x, y and z) by a piezoelectric scanner. The lateral motion of the tip (or the sample) can be controlled to a precision of 0.1 nm and the cantilever deflection as small as 0.01 nm can be detected.

When the AFM is used for imaging purposes, two imaging modes are generally used. In the contact mode imaging, a soft cantilever with a spring constant $k \sim 0.1$ N/m is used. As the sample is raster scanned, the cantilever deflection (thus the interaction force) is kept constant by adjusting the z position via a feedback mechanism. Correlating each z value with the xy position of the tip yields a topographic image of the scanned area. In tapping mode imaging, a stiff cantilever ($k \sim 100$ N/m) is made to oscillate near its resonant frequency (typically ~ 300 kHz). The tip touches the sample transiently during each cycle (tapping), and this interaction reduces the amplitude of the cantilever. By keeping this amplitude constant during scanning via the feedback loop, an image is obtained from the z position of the sample and the xy location of the tip. The tapping mode is beneficial in imaging biological samples in that it minimizes the sample deformation and damage. Both contact and tapping mode imaging can be performed in vacuum, air and liquids. When the AFM is used for nanomanipulation purposes, the tip is controlled using the piezoelectric stage to make the prescribed motions relative to the macromolecule attached to the substrate. Manipulation is carried out by

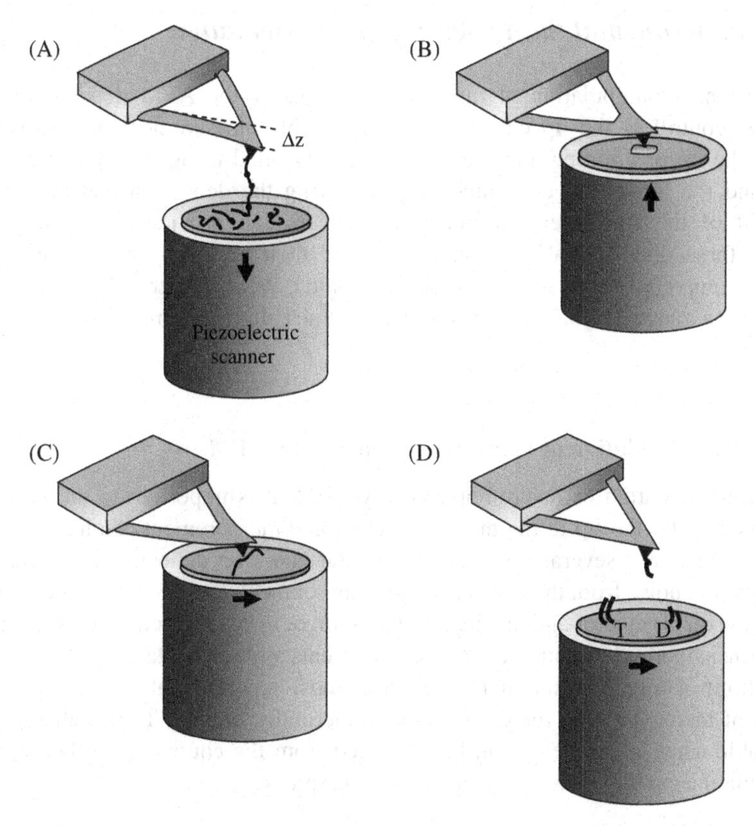

Fig. 4.2 Schematic of AFM-based nanomanipulation of macromolecules. (**a**) Pulling molecule tethered between the tip and the substrate surface by moving the surface away from the tip. The force exerted by the cantilever is determined by its deflection from the unstressed state, Δz. (**b**) Pushing a macromolecule by moving the surface toward the tip. (**c**) Cutting a macromolecule by moving surface laterally while the tip is pressed on the surface. (**d**) Transporting a molecule by picking it up at the depot site "D", moving it to target site "T", and depositing it there

pulling, pushing, transporting, or cutting the macromolecule of interest, as shown in Fig. 4.2. In a commonly used mode of manipulation, the macromolecule under study is positioned between the tip and the substrate surface, and the conformation of the macromolecule is changed continuously by moving the tip relative to the substrate. The deflection of the cantilever, which provides the force being exerted on the molecule, is simultaneously monitored. The correlation between the force and the conformational change provides the desired properties of the molecule. In another mode of manipulation, a feedback loop keeps the applied force at a constant value by adjusting the tip-surface distance, while the conformational of the macromolecule, such as its end-to-end distance, is recorded as a function of time [18]. Such "force-clamp" can be used to investigate the structural and dynamic properties under the quasi-equilibrium conditions.

4.2.2 Experimental Strategies and Instrumentation

Effective nanomanipulation of macromolecules using the AFM depends on the accurate control of the tip movement and a reliable measurement of the exerted force. The tip positioning and movement are controlled using piezoelectric scanners, and the forces are determined by measuring the deflection and the spring constant of the cantilever. Piezoelectric scanners can provide angstrom precision in three dimensional position control and their accuracy can be calibrated using commercially available calibration standards. Precise force calibration at the piconewton range is challenging, though several calibration methods have been developed.

4.2.2.1 The Resolution of Force Measurements by AFM

In nanomanipulation of macromolecules by AFM, it is important to measure the force exerted by the tip to the molecule. The force measurement resolution of the AFM is affected by several factors, including the cantilever deflection measurement accuracy, the noise from the instrument and environment, the data acquisition bandwidth, the quality factor and the thermal vibration of the cantilever. Of these factors, the intrinsic thermal vibration of cantilever presents a fundamental limitation on the lower limit of force measurement, and under most experimental conditions, is the dominant factor, *i.e.*, the force resolution is thermally limited. The smallest force detectable using a cantilever can be estimated from the equipartition theorem by modeling the cantilever as a one dimensional harmonic oscillator:

$$\frac{1}{2}\kappa\left\langle z^2\right\rangle = \frac{1}{2}k_B T \tag{1}$$

where κ is the bending spring constant of the cantilever, $<z^2>$ is mean-squared amplitude of the thermal oscillation of the cantilever, k_B is the Boltzmann constant and T is the temperature. A cantilever can be treated as a linear spring obeying Hooke's law $F = \kappa z$, therefore, the noise due to thermal vibration in force measurement, ΔF, can be expressed as

$$\sqrt{\left\langle \Delta F^2\right\rangle} = \kappa\sqrt{\left\langle z^2\right\rangle} = \sqrt{\kappa k_B T} \tag{2}$$

Although cantilevers with spring constants as low as 0.1 pN/nm have been fabricated and used to measure forces in the attonewton regime at cryogenic temperatures [19], for manipulation of biological macromolecules, cantilevers need to have spring constants higher than about 10 pN/nm, which set the lower limit of detectable forces in AFM to about 7 pN at physiologically relevant temperatures.

4.2.2.2 Determination of the Spring Constants of AFM Cantilevers

The force exerted on the tip of an AFM cantilever, F, can be expressed as

$$F = \kappa \cdot \Delta z \tag{3}$$

where κ is the spring constant and Δz is the deflection distance by the tipped end of the cantilever, see Fig. 4.2a. Δz can be determined to sub-angstrom precision through calibration using the piezoelectric scanner, then, the accuracy of force measurements using the AFM is directly proportional to the accuracy in the determination of the spring constant of the cantilever. In the sub-nanonewton force regime, there is no convenient and accurate method to measure the spring constant of a cantilever. The values given by the manufacturers can only be used as a rough guide, individual cantilevers in the same batch can have spring constants differing by a factor of two due to the heterogeneity in dimensions and material properties. For example, the spring constant depends on its thickness to the third power, and the uniformity of cantilever thickness is the most difficult parameter to control in the fabrication process. As a result, the spring constant must be determined individually for each cantilever when accurate force measurement is desired.

Currently, the most widely used method for spring constant determination is the "thermal vibration" method [20], in which the spring constant is obtained from a measurement of the cantilever vibration due to thermal agitations from the Brownian motion of the fluid molecules in the surrounding medium. In this method, the energy equipartition theorem is used by assuming the oscillational motion of cantilever to be the same as that of a one-dimensional simple harmonic oscillator. The cantilever's fluctuations in the absence of any externally applied forces are measured as a function of time, and the power spectrum is obtained from the data via Fast Fourier Transformation [21]. By fitting the power spectrum to a Lorentzian line shape, the rms value of the cantilever's thermal oscillation amplitude is obtained, from which the cantilever's spring constant is calculated using Eq. (1). This method has an acceptable accuracy with the advantages of being fast, simple, nondestructive, and in-situ. Calculations based on the method have been made automatic and imbedded in the control software in some newer models of commercial SPMs. Several other methods for cantilever spring constant determination have been reported, such as the added mass method [22], reference cantilever method [23]. These methods offer various degrees of accuracies, but are not in widespread use due to inconveniences.

4.2.2.3 Overcoming Mechanical Drifts

AFMs are designed and built for controlling and measuring displacements and forces at the molecular scale, therefore, any drifts in the tip position relative to the sample cause errors in the measurement results. All piezoelectric elements suffer from hysteresis, creep and non-linearity. The non-uniform thermal expansions of AFM parts also cause drifts. Various design features have been implemented in commercial instruments to circumvent these problems but they are not sufficient to maintain a long time scale stability required for certain nanomanipulation experiments. Most of the reported methods for overcoming the drift problems of AFM are for imaging purposes. There are a few instruments with drift-correction mechanisms that are compatible with nanomanipulation of biological macromolecules, such as the differential force microscope using two cantilevers [24], and the ultra-stable AFM that control the tip position in three dimensions to a precision of 40 pm by monitoring laser signals scattering off the AFM tip and a marker on the sample surface [25].

4.2.2.4 Theoretical Models of Biological Macromolecules

Many biological macromolecules are polymers that are flexible when the molecules are in certain states. Nanomanipulation of macromolecules often involves applying a stretching force between two points, and observing the induced structural or conformational changes in the macromolecule. In interpreting the data from such experiments, it is often required to model the elasticity of macromolecules using polymer theories [26]. Two models are commonly used to quantitatively describe the elasticity of flexible macromolecules, *i.e.*, the worm-like chain (WLC) and the freely jointed chain (FJC) models [27]. In the WLC model, the polymer is treated as a semi-flexible, non-extensible chain whose stiffness is characterized by a parameter called the persistence length. The force and extension relationship of a WLC is given by

$$F = \left(\frac{k_B T}{p}\right)\left[\frac{1}{4(1-x/L)^2} - \frac{1}{4} + \frac{x}{L}\right] \tag{4}$$

where F is the stretching force, x is the end-to-end distance, L is the contour length, p is the persistence length, k_B is the Boltzmann constant, and T is the temperature. A polymer with a smaller persistence length tends to form random coils, and a higher force is required to extend the polymer.

In the FJC model, the polymer is considered as a chain of rigid, orientationally independent statistical segments (the Kuhn segments). A stretching force is required to extend the chain because an extended chain is in a higher energy conformation due to the lowered entropy. The length of each segment, b, known as the Kuhn length, is a direct measure of the chain stiffness and is related to the contour length L by $L = nb$. The extension, *i.e.*, the end-to-end distance, of a FJC, x, is related to the stretching force F by:

$$x = L\left[\coth\left(\frac{Fb}{k_B T}\right) - \frac{k_B T}{Fb}\right], \tag{5}$$

At low forces, $Fb << k_B T$, the relation can be simplified to

$$F = \frac{3k_B T}{b}\left(\frac{x}{L}\right) \tag{6}$$

Within this force regime, the FJC behaves like a linear spring with the spring constant proportional to 1/b. In AFM studies, it has been shown that the elastic behaviors of unfolded proteins can be described well by the WLC model [28], whereas that of polysaccharides and single stranded DNA follow the FJC model more closely [29].

4.2.2.5 Viscosity Effects

Most macromolecular manipulation experiments using AFM are performed in buffer solutions, where the cantilever experiences a viscous drag force due to its motion in the solution. This viscous force superimposes onto the force generated by the macromolecule under study, causing ambiguity in the measured forces and difficulties in data interpretation. The viscous drag effect becomes significant when the speed of the cantilever is high [30] or the viscosity of the solution is high [31]. Janovjak et al. [30] and Alcaraz et al. [32] have used the scaled spherical model to treat the cantilever as a sphere-spring system and utilized the Stokes' Law to determine the viscous force. These methods have provided a reasonable estimate of the viscous drag on the AFM cantilevers, however, there is a deficiency in these approaches because the cantilever's mode of motion during the viscous force determination is different from that during a macromolecular manipulation experiment. Recently, Liu et al. [33] reported a different approach by modeling the cantilever as a rotating beam and treating the viscous drag on the cantilever in macromolecular manipulation experiments as a superposition of the viscous force on a static cantilever in a moving liquid and that on a bending cantilever in a static liquid. The viscous force on the cantilever was measured under both hydrodynamic conditions, yielding parameters that are amenable to the experimental conditions, and corrections of the viscous force induced errors were then carried out using these parameters.

4.3 Manipulation of Protein Molecules

The manipulation of individual protein molecules by AFM has generated a wealth of novel and important information on the mechanisms of the mechanical functions of proteins, the significance of the rare events during protein folding, the properties of the energy landscape of protein molecules, the details of the interactions between protein molecules, the influence of force on the enzymatic activities of proteins, and many other properties that are crucial to understand the various functions of protein molecules. This section is not an exhaustive review of the literature in the field, but to present representative experiments that addressed important and novel questions on proteins. The readers are referred to recent reviews for additional information [2, 34–49].

4.3.1 *Force-Induced Protein Unfolding and Refolding*

The mechanisms that determine how a folded protein molecule resists mechanical stresses are still not well understood. Conventional biophysical and biochemical techniques are unable to directly probe the problem due to the difficulties to exert a controlled force on individual proteins. However, this is an important problem to solve since all proteins are exposed to certain amount of mechanical stress, especially those involved in tissue strain. Disease-causing mutations in many proteins have signatures in their mechanical stability, *i.e.,* the maximum force a folded protein domain can withstand [50]. Furthermore, the characterization of the conformational changes of a protein induced by an externally applied force can be used to probe the folding energy landscape of the protein. The breakthrough in the direct characterization of the mechanical properties of proteins came in 1997 when three papers were published almost simultaneously in which individual molecules of the protein titin were stretched with AFM and laser tweezers [28, 51, 52]. The globular titin domains were observed to unfold when the applied force reached a certain level. Up to date, there are about five dozens of different proteins that have been studied this way [53].

4.3.1.1 Elucidating the Mechanism of the Mechanical Functions of Proteins

Proteins with mechanical functions often have modular structures composed of multiple individually folded globular domains. Of these proteins, titin is probably the best characterized using AFM and other single molecule techniques. Titin is a large modular protein, with a molecular weight of $\sim$4 MDa, containing hundreds of sequentially connected Ig and fibronectin type III (Fn3) domains. Titin functions as a molecular spring and ensures the return of the sarcomere to its initial dimensions after muscle relaxation [54, 55].

In studying the mechanical properties of titin, the protein is deposited on a substrate surface, and the tip is then pushed onto the sample surface with a force of a few nanonewtons for several seconds to allow the molecules to interact with and attach to the tip [28]. The tip is then retracted from the surface at a specified speed and the force is measured as a function of the tip-sample separation. If a single titin molecule is tethered between the tip and the substrate, the measured force displays a characteristic sawtooth pattern, as shown in Fig. 4.3, where each peak corresponds to the unfolding of a single folded domain. The ability to observe individual unfolding events comes from the fact that the unfolding of one domain causes a sudden lengthening of the molecule, which reduces the tension and prevents another unfolding event from occurring immediately.

AFM experiments have shown that titin globular domains unfold when stretched and refold when relaxed, thus making titin extremely extensible because of the large length gain upon the unfolding of each domain [28, 56]. The refolding of titin Ig domains can occur under relatively large forces, indicating a robust refolding mechanism that can operate over a large range of sarcomere lengths [57]. It has been suggested that the reversible unfolding of the titin globular domains serves as a

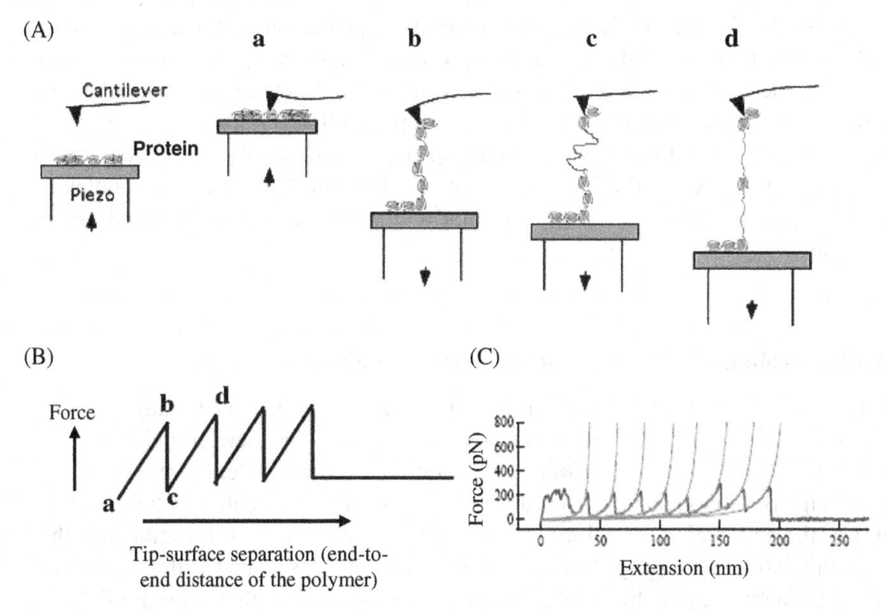

Fig. 4.3 Mechanical unfolding of single protein molecules using AFM. (**a**) Procedure of the experiment. Polymeric proteins are deposited on a surface, and picked up by the AFM tip via nonspecific adsorption. The individual protein domains are unfolded by force as the surface is moved away from the tip. The procedure is first reported in reference [28]. (**b**) Schematic of the data, a force versus extension curve, generated from the procedure shown in (**a**). Each peak corresponds to the unfolding of a single domain in the polymer. (**c**) A experimental curve from pulling a protein polymer. The features at the beginning of the curve are due to unspecific interactions between the tip and the surface when they are close to each other

safety mechanism that protects titin and the sarcomere from mechanical damage in case of extreme stretch during stress or pathological conditions [54, 58–61]. Many mechanical proteins have a modular construct, with a similar safety mechanisms built in, such as the cell adhesion protein tenascin [62].

Proteins involved in cytoskeleton are another family of proteins whose mechanical properties are closely related to their functions. Cytoskeleton is a protein network that supports cellular structure and also plays an important role in intercellular transport, cell division, and cell motility [63]. Spectrin is the major component of the membrane-associated cytoskeleton in erythrocytes [64], which are known for their high deformability [65]. The investigation of the force induced structure and conformation changes of spectrin by AFM nanomanipulation has provided a molecular basis for the elastic properties of erythrocytes. Each spectrin is made of two subunits (the α and β chains) that form a lateral dimer. Each of these chains has multiple domains that consist of bundles of three anti-parallel coiled-coil [66–68]. Rief et al. measured the force required to unfold spectrin repeats and obtained values in the range of 25–35 pN [69]. They found that individual spectrin repeats unfold following a two state process, and the unfolded polypeptide refolds efficiently when the strand is relaxed. The unfolding force of spectrin repeats is almost a order of magnitude smaller than that of the titin Ig domains [28], but the two protein domains have a similar melting temperature [70, 71], reflecting the structural and functional differences of the two proteins. Law et al. [72] reported that serially adjacent spectrin repeats unfold cooperatively when stretched, and Ortiz et al. [73] studied the forced unfolding pathways of the linkers between tandem spectrin repeats and obtained insights in understanding the cooperative unfolding behaviors of spectrin and similar proteins.

4.3.1.2 Probing the Protein Folding Energy Landscape

Understanding the protein folding problem is of great significance because a majority of proteins depend on the correct folded structures for their proper biological functions. The folding and unfolding transitions of certain proteins are also critical to a range of biological processes and systems, such as molecular trafficking, control of the cell cycle and cell movements [74, 75]. Also, it is well established that protein misfolding, aggregation and fibrillogenesis are associated with a number of diseases, such as cystic fibrosis, Alzheimer's disease, Huntington's disease [75].

The energy landscape theory for protein folding [76–81] depicts the folding process as following a path on a partially rough, funnel-shaped free energy surface, with the native state at the bottom of the funnel. From the energy landscape, all of the observable properties of the folding process can be inferred, including folding rates and transition-state structure [82]. However, it has been difficult to determine the characteristics of the energy landscape using experimental measurements [83]. The dynamical responses of a protein to mechanical force can be used to characterize its free energy landscape, therefore the AFM based nanomanipulation technique has provided a method to directly probe properties of the energy landscape of proteins [2, 84–89].

Model systems used for protein folding studies are mostly small globular proteins. Such proteins are not amenable to direct nanomanipulation experiments using the AFM because the tethering surfaces, *i.e.,* the AFM tip and the substrate, have much larger radii of curvatures than the dimensions of the proteins, as shown in Fig. 4.4. The nonspecific interactions between the AFM tip and the substrate may be stronger than the forces required to unfold the protein when these surfaces are a few nanometers apart. To circumvent these difficulties, globular protein molecules have been made into artificial polymers, which are then used in the nanomanipulation experiments [84, 86, 90]. A single protein polymer, or a polyprotein, can contain a number of identical or different proteins, or "domains", arranged in a defined manner. With such constructs, the unfolding and refolding of individual globular proteins in the polymer can be identified unambiguously from the sawtooth pattern in the force curves (see Fig. 4.4). By combining protein polymer synthesis with the capacity of AFM to apply and measure forces in the sub-nanonewton regime, and to measure distances in the nanometer scale, it has become possible to determine detailed features of a protein's energy landscape.

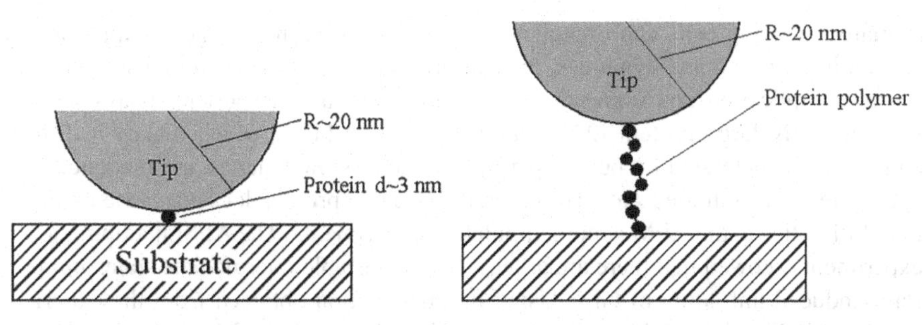

Fig. 4.4 Schematic showing the necessity to use protein polymers in carrying out mechanical unfolding experiments. When the AFM tip and surface are close to each other, unspecific interactions due to the surface roughness and surface adsorbed proteins interfere with the measurement of the unfolding forces. Using polymers circumvents the difficulties, while the individual unfolding events can still be observed (see Fig. 4.3)

Dietz and Rief [91] used heteropolymers of a GFP (green fluorescent protein) molecule flanked by smaller known protein domains on both sides to investigate the properties of the energy landscape of GFP. The unfolding of GFP in the mechanical unfolding experiments was indentified from its large size in the sawtooth pattern, see Fig. 4.5. GFP has a great significance in biotechnology [92] because of its applicability as a marker to monitor biomolecular processes, both temporarily and

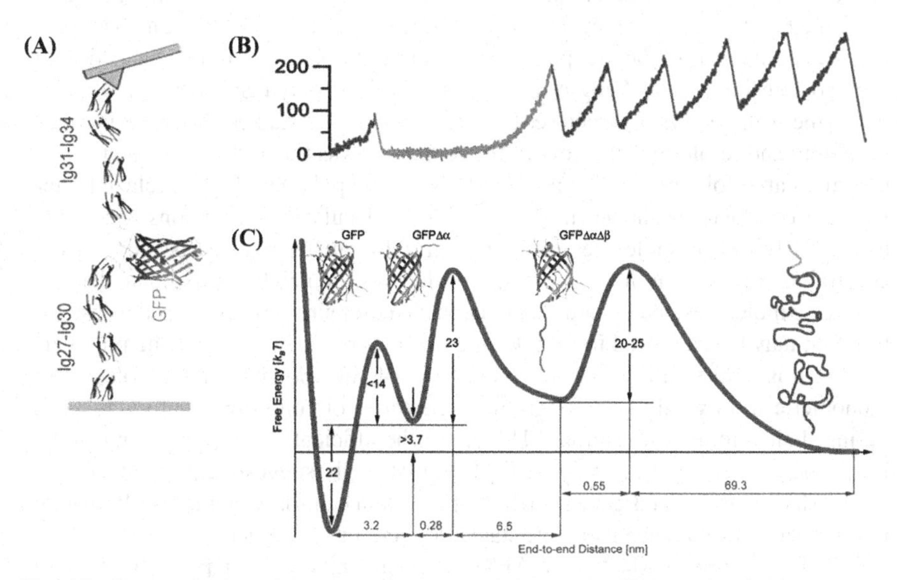

Fig. 4.5 Characterization of the energy landscape of the protein GFP. (**a**) GFP is inserted in a polymer of titin domains. (**b**) The unfolding event of the GFP is identified by the greater increase in polymer length. (**c**) The measured energy landscape of GFP plotted as a function of the end-to-end distance of the protein. Reprinted with permission from reference [91], Copyright (2004) National Academy of Sciences, U.S.A

spatially, in living cells and organisms. These processes include gene expression, protein localization and dynamics, protein-protein interactions, cell division, chromosome replication and organization, and intracellular transport pathways. GFP is a relatively large protein (238 amino acids, 26.9 kDa), which makes it difficult to be characterized by bulk experiments due to slow kinetics and tendency to aggregate in the unfolded state. However, the size of a protein does not cause particular difficulties in the AFM nanomanipulation experiments. The AFM manipulation experiments showed that there are at least two metastable intermediate states in the force induced unfolding of GFP. Using the information obtained from the experiments, including the unfolding force distribution, the position of the transition state, and the lifetimes of the intermediate states, the energy landscape of GFP was reconstructed along the direction of the applied mechanical force, see Fig. 4.5c. The rough energy landscape reflects a complex topology of GFP, and the knowledge can be useful for the development of GFP mutants for various applications, such as molecular force sensors.

With their force-clamp instrumentation [18], Berkovich et al. characterized the folding free energy landscapes of ubiquitin and titin domain I27 using the force-quench technique, where after unfolding a protein at a high force, the pulling force is abruptly reduced to a low value and the collapse process of the extended polypeptide chain is monitored [93]. Analysis of the collapse trajectories yielded the parameters describing the free energy landscape along a well defined coordinate (the end-to-end distance of the protein), such as the height of the energy barrier, the curvature of the free-energy minima, and the location of the transition states. Using a double-jump method [94] in their mechanical unfolding experiments, Schlierf and Rief [95] determined the energy landscape of the protein ddFLN4 to be funnel shaped with a broad transition state. Yang et al. [84] synthesized polymers of the protein T4 lysozyme using a crystal structure directed method, and studied the force-induced unfolding and refolding behaviors of the protein. It was found that the proteins in a polymer can refold efficiently and rapidly when the polymer chain is relaxed after they are unfolded, as shown in Fig. 4.6. The molecular configurations accessible to the T4 lysozyme molecules in the polymer are greatly reduced because of the constraints imposed on the two linking residues by the neighboring molecules. T4 lysozyme molecules do not naturally exist in the polymerized form, as is the case for titin that may have evolved to fold despite the constraints imposed by the polymerization. It is unlikely that a T4 lysozyme molecule in a polymer and a molecule in monomeric form would go through the same series of conformational microstates during their refolding transitions. Therefore, the efficient and rapid folding of T4 lysozyme molecules in the polymerized form strongly suggests the existence of a funnel shape energy landscape on which the protein can follow multiple alternative folding pathways to reach its native state at bottom of the funnel.

These experiments show that AFM nanomanipulation is a powerful tool to directly measure the properties of the energy landscape of proteins. In particular, the technique can explore regions of the energy landscape that are not accessible in bulk experiments in which it is difficult to study proteins in well extended conformation. The energy landscape of proteins is a high dimensional surface, the AFM

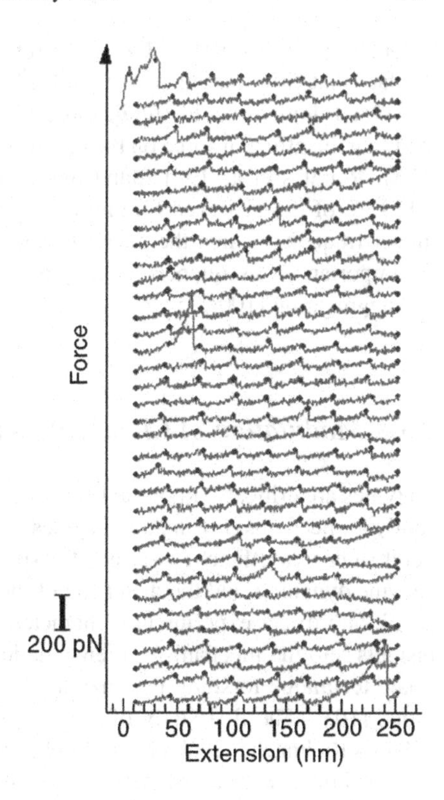

Fig. 4.6 Repeated unfolding-refolding of T4 lysozyme molecules in a polymer [84]. The curves were obtained by stretching and relaxing the same polymer multiple times (refolding parts are not plotted). The dots mark the positions where unfolding occurs. The force curves in these plots are shifted vertically for clarity, so they do not share a common zero force level

experiments can only probe the surface along a particular direction. As the technique advances, force can be applied along several different directions on the same protein molecule, and the energy landscape can be explored more thoroughly.

4.3.1.3 Observing Rare Misfolding Events

In biological systems, many states are transient, which may not be represented in macroscopic measurements of the population. The average properties may reflect the dominant population in a system, but not the biologically active individuals. The activity of even a small minority population may be the significant population for understanding the physiological activity of this molecule [96]. Therefore, identification and characterization of rare molecular events and infrequent subpopulations are crucial in understanding the functional mechanisms of many biological systems, and AFM nanomanipulation has been shown to have such a capacity [97–99]. Oberhauser et al. [97] observed that the sawtooth patterns obtained in repeated unfolding and refolding experiments on titin domain I27 occasionally had a missing force peak. Detailed analysis of the data with missing peaks showed that they represent the rare misfolding of the I27 domains, in which two neighboring domains fold into a single structure. Such misfolding was found to be very rare, accounting for fewer than 1% of the refolding events. The low frequency reflects the fact that titin domains have evolved to fold efficiently to avoid misfolding, a misfolded domain could significantly affect the function of the titin molecule. In studying the effect of force on the unfolding cooperativity of spectrin domains using AFM, Randles et al. [98] found that adjacent spectrin repeats occasionally misfold (3% of the observed folding events), and that such misfolding occurs with a higher frequency when the adjacent domains are identical in sequence, suggesting the importance of sequence diversity of modular proteins in minimizing misfolding and aggregation [100].

4.3.2 Modulation of Chemical and Enzymatic Reactions by Force

Enzymes are critically important for all living organisms, but our knowledge on the catalytic mechanism of many enzymes is still far from complete and it is still difficult to design efficient enzymes for specific purposes [101]. It was realized that enzyme dynamics and the distortion of the substrate could be essential for catalysis [102], and direct experimental characterization of these process can provide new insights into the mechanism of enzymatic catalysis. AFM has provided the appropriate technique to study the enzyme dynamics by using mechanical force as the probe [103]. The first study of enzyme dynamics with the AFM was reported in 1994 by Radmacher et al. [104]. By placing an AFM tip on the lysozyme molecules adsorbed on a surface, the conformational changes of the lysozyme molecules during catalysis were directly observed as the cantilever fluctuations increased when the substrate (oligoglycoside) was added. While in the presence of an inhibitor (chitobiose), the fluctuations decreased to the level without the substrate. In recent years, more sophisticated measurements using AFM nanomanipulation have been carried out to investigate the molecular level details of enzymatic activities.

Wiita et al. [105] studied the effect of force on the reduction cleavage of a disulfide bond, which was engineered in the protein titin domain I27, by dithiothreitol (DTT). The covalent disulfide bonds, formed between the thiol groups of two cysteine residues in a protein, play an important role in the folding and stability of certain proteins [106]. Many proteins that are exposed to mechanical stress contain disulfide bonds [107], therefore, the effect of force on disulfide bond reduction can reflect the functional mechanisms of these proteins. By using the force-clamp AFM, it was shown that the reaction of disulfide reduction by DTT is sped up by force, and the reaction rate depends on the applied force exponentially. The same laboratory also studied the catalysis mechanism of the enzyme *E. Coli* thioredoxin (Trx) for its activities in the reduction of disulfide bonds in proteins [108]. Trx exists in all organism, understanding how this enzyme modulates the disulfide bond cleavage reaction is of great importance. The most interesting and significant finding from the study is the observation that the reaction rate of disulfide bond reduction by Trs decreases fourfold when the applied force is increased from 25 to 250 pN, and then increases threefold when the force is increased from 250 to 600 pN, as shown in

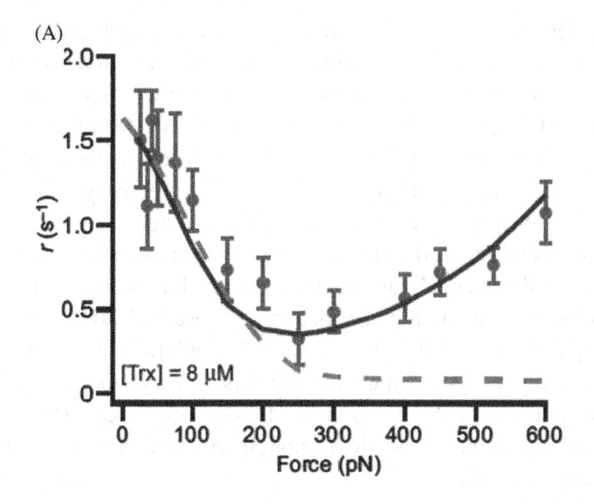

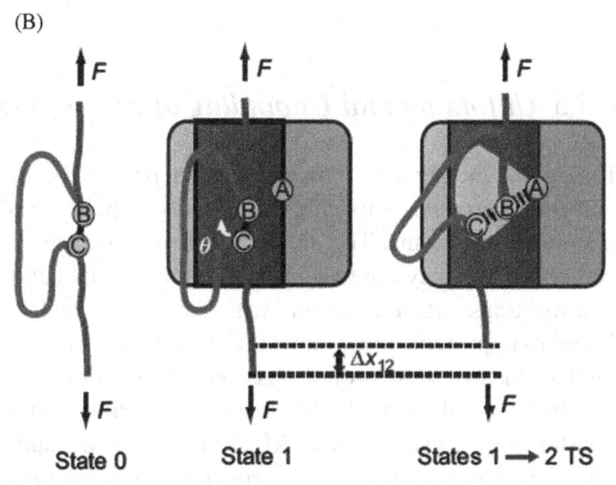

Fig. 4.7 The effect of force on Trx catalysis. (**a**) The dependence of the reaction rate on the applied force. The *dashed line* is a model fit. (**b**) Schematic of a disulphide bond reduction by Trx when the bond is subject to a stretching force. The bond (between *B* and *C*) has to rotate to reach the correct geometry at the transition state (TS), causing a shortening of the substrate polypeptide by an amount Δx_{12}. Reprinted by permission from Macmillan Publishers Ltd: Nature [108], copyright (2007)

Fig. 4.7. The result shows that Trx can catalyze the disulfide bond reduction in two distinct routes, without a common intermediate. The first route requires a reorientation of the substrate disulphide bond in order to achieve the correct geometry at the transition state. The substrate rotation leads to a shortening of the protein polymer chain tethered between the AFM tip and the surface, therefore, it is opposed by the stretching force. The second route of the catalysis reaction elongates the disulphide bond, therefore, it is favored at high forces. The finding that a force can alter the catalytic routes of Trx has important implications. The authors argued that mechanical stresses applied to tissues may completely change the enzymatic chemistry from that observed in solution biochemistry, and the effects may be particularly significant in tissues exposed to pathological force levels such as those experienced during a mechanical injury.

Gumpp et al. [109] used a combined AFM-TIRF instrument [110] to investigate the effect of a mechanical force on the activity of the enzyme lipase B from *Candida Antarctica* (CalB), which converts the fluorogenic substrate 5-(and-6)-carboxyfluorescein diacetate (CFDA) into the highly fluorescent product 5-(and-6)-carboxyfluorescein. The enzyme was tethered between a glass coverslip and the AFM cantilever, and its catalytic activity was monitored by fluorescence while it was periodically stretched and relaxed. It was observed that releasing the stretching force leads to a higher enzymatic activity. A theoretical analysis of the data reveals the detailed energy landscape of the enzyme that governs the active and inactive conformations. These and other experiments clearly prove the capacity of the AFM based nanomanipulation technique for studying the effect of forces on enzyme catalysis reactions. In order to carry out such studies effectively, the activity of the enzyme needs to yield a clear detectable signal, and the reaction rate needs to be slow enough to be compatible with the time scale of AFM measurements.

4.3.3 Unfolding and Unbinding of Membrane Proteins

Biological membranes, composed of a phospholipid double layer with various embedded proteins, surround cells (and organelles) and define their physical borders in all organisms. The amount of membrane proteins differ among species and cell types, and they are responsible for many vital cellular activities, such as solute and ion transport, interactions with other cells, and sensory stimuli transduction. Membrane proteins are encoded by about 30% of genes in most organisms [111], but the study of these proteins has been difficult by many conventional biochemical and biophysical methods due to their insolubility in aqueous solutions and their resistance to crystallization. AFM imaging and manipulation have proved to be effective in characterizing the properties of membrane proteins, in particular, the AFM based techniques can investigate membrane proteins in their native environment, *i.e.*, embedded in the lipid bilayer and immersed in a physiologically relevant solution [38, 39, 112–116]. These studies have revealed novel and important insights about how membrane proteins work, their interactions with lipids and other proteins, and their dynamics and structures.

The first study of membrane proteins using AFM nanomanipulation was reported by Muller et al. [117]. Utilizing both the imaging and force measurement capacities of AFM, these authors measured the unbinding force of the hexagonally packed intermediate (HPI) layer protein from the membrane and directly visualized the correlation of the structural changes with protein unbinding events. Oesterhelt et al. [118] used a similar approach to study the seven-helix transmembrane protein bacteriorhodopsin in purple membrane patches from *Halobacterium salinarum*. Individual protein molecules were first localized from AFM imaging and then extracted from the membrane with the AFM tip. It was found that the anchoring forces of the helices in bacteriorhodopsin to the membrane are between 100 and

200 piconewtons. The helices are unfolded upon extraction, with each helix following a different unfolding pathway, which reflects the local interactions with the membrane and within the protein.

Goncalves et al. [119] reported high-resolution imaging and force spectroscopy measurements on unsupported membranes using a two-chamber AFM setup that the authors had designed. The experimental setup has the advantage of combining structural and functional studies of membrane proteins since the two aqueous chambers separated by the membrane can mimic the environments of cytoplasm and extracellular space, respectively. The sample of nonsupported membrane for AFM studies was prepared by spanning the surface layers (S layers) of *Corynebacterium glutamicum* on a silicon surface with holes having diameters of 90–250 nm. The protein pore structures in the membrane were identified by imaging, the elastic properties and membrane rupture forces were measured by indenting and puncturing the membranes with the AFM tip. Functional studies of the membrane were demonstrated by monitoring pH changes induced by bacteriorhodopsin proton pumping. The experimental setup represents a significant technical advance for investigating membranes and membrane proteins under more physiological conditions.

Kessler et al. reported the study of the refolding of membrane proteins back into the bilayer using the AFM [120]. Experiments were carried out by first extracting and partially unfolding an individual bacteriorhodopsin molecule from the purple membrane using an AFM tip. Then, the protein was slowly lowered toward the membrane to allow reentry and refolding while the force was monitored by the cantilever deflection. It was found that certain helices can refold against a pulling force of several tens of pN. During the refolding process against the stretching force, several folding intermediate states with differing complexity were identified.

The AFM is currently the only technique that can perform high resolution measurements on the structure and dynamics of membrane proteins under physiological conditions and without the need of labeling. The AFM studies have not only revealed the functional mechanisms of the membrane proteins but also helped the development of new drugs for various diseases since the majority of drugs target membrane proteins.

4.3.4 Measurement of Protein Ligand Interactions

The interactions between proteins and between a protein and its ligand are important for many biological functions of protein molecules, such as in signal transduction and cell motility. The studies of these interactions by conventional methods on an ensemble of molecules in solution under equilibrium conditions provide average properties and a static viewpoint, while inside cells, such interactions generally occur on surfaces under nonequilibrium conditions. During the past decade, AFM based nanomanipulation methods have provided important complementary information on the dynamics of protein-protein and protein-ligand interactions

[36, 121]. AFM can probe conformational transitions of individual molecules, reveal the structural and functional heterogeneity in a population, and directly quantify the magnitudes and working distances of forces involved in the interactions.

For single molecule measurements of protein–ligand interactions using the AFM, the ligand is attached to the AFM tip and the protein is immobilized on a surface (or the other way around). The AFM tip is first brought into contact with the surface to form a protein-ligand complex, and then the tip is retracted from the surface while the force and distance are recorded. In such experiments, the unbinding force, the dissociation rate constants and bonding energy landscape can be determined. Benoit et al. [122] investigated the adhesion between two live cells of the eukaryote *D. discoideum* by attaching one cell on the AFM cantilever and the other on a surface. Previous experiments showed that the interactions of these cells are governed by the glycoprotein CsA, which reacts with each other to form non-covalent bonds between adjacent cells. AFM force measurements revealed that the de-adhesion process was continuous upon separating the cell, but the final detachment was found to occur mostly at a force of 23 pN, which was identified as the unbinding force between two CsA molecules. This relatively low interaction force in cell-adhesion is consistent with the ability of motile cells to glide against each other as they become integrated into a multicellular structure.

Recently, Lim et al. [123] reported the investigation of the adhesion strength and kinetic properties of the homophilic interactions between the two extracellular loops in the protein Claudin-2 (Cldn2) and the full-length Cldn2. Claudins are transmembrane proteins that regulate the paracellular transportation of solutes across epithelia. The two extracellular loops of Cldns of adjacent cells interact to form the paracellular tight junction strands. By linking one of the interaction partners (the two extracellular loops C2E1, C2E2 and the full-length protein Cldn2) to the AFM tip and immobilizing another on a glass coverslip, the interaction forces and kinetics between the partners were determined. It was found that the first extracellular loop of Cldn2 is the major determinant of trans-interactions involving Cldn2. Dissociation of hemophilic Cldn2/Cldn2 and C2E1/C2E1 complexes follow a two-energy-barrier model within the range of loading rates used (10^2–10^4 pN/s). These results provide an insight into the trans-interaction of Cldn2-mediated adhesion. In another recent report, Zhang et al. [124] measured the binding properties between the transmembrane protein cadherines using single-molecule FRET and AFM, in order to elucidate the functional organization of cadherin adhesive states. Cadherins are Ca^{2+}-dependent cell-cell adhesion proteins regulating organization of cells and maintaining the structural integrity of tissues. The measurements showed that cadherin binding involves two stages, which helped to clarify some controversies regarding the cadherin mediated cell-cell interactions. Lee et al. [125] reported an interesting study aimed at understanding the molecular mechanisms of mussel adhesion to various surfaces. Mussels are remarkable in that they can maintain strong and long-lasting adhesion in a wet environment, and they can adhere to all types of inorganic and organic surfaces. It was known that the adhesive proteins in mussels contain 3,4-dihydroxy-L-phenylalanine (dopa). The AFM experiments were carried out to directly measure the interaction forces between dopa and model surfaces. Dopa was tethered on the AFM tip via a linker, and its adhesion to an inorganic (Ti) surface and an organic (amine-modified Si) surface were characterized in aqueous solutions. The measured adhesion force, dissociation energy, and theoretical calculation provide an explanation of mussel adhesion. On inorganic surfaces the unoxidized dopa forms high-strength yet reversible coordination bonds, whereas on organic surfaces oxidized dopa is capable of adhering via covalent bond formation. The knowledge can be used for development of anchoring agents for biological macromolecules onto various surfaces.

4.4 Manipulation of Nucleic Acid Molecules

DNA and RNA play vital roles in all organisms, and their properties and functions have been investigated with various techniques. During the last 12 years, it has been increasingly realized that the mechanical properties of DNA and RNA are important for understanding the functional mechanisms of these nucleic acid molecules in many biomolecular processes, such as the wrapping of DNA in a nucleosome, the packing of DNA into a phage capsid, and DNA strand separation during transcription. AFM nanomanipulation technique provides a powerful tool for the direct measurement of the mechanical properties of nucleic acids and the determination of the effect of mechanical forces on the functions of these molecules [4, 126–130]. AFM nanomanipulation has also been used to perform controlled delivery of DNA [131] and to assemble molecular structures on surfaces with nucleic acid molecules [110, 126].

4.4.1 Elastic Property and Force-Induced Structural Transitions in DNA

The elastic properties of individual DNA molecules and the force-induced structure transition of DNA were first studied using the magnetic tweezers and laser tweezers [132, 133]. The investigation of the mechanical properties of nucleic acid molecules using AFM nanomanipulation has the advantages that shorter molecules can be studied and higher forces can be exerted, thus a wider range of properties can be probed.

Rief et al. used the AFM to stretch individual dsDNA molecules for the determination of their sequence-dependent mechanical properties [29]. By pressing the AFM tip against the DNA layer on a gold surface, individual molecules of the digested λ-DNA were picked up by the tip and subsequently stretched as the tip was moved away from the surface, yielding a force (tension) and extension relationship for the molecule. At a force of 65 pN, the previously known transition from B-form to S-form DNA [133] was observed. At a force of ~150 pN a new transition was found, which was identified as the split of the double helix into single strands. The elastic behaviors of the single stranded DNA could be well described by both WLC and FJC models at low forces (<100 pN), but these models failed at higher forces. It was found that the B-S transition (at 65 pN) was independent of pulling speeds, indicating an equilibrium process, while the melting transition (at 150 pN) did show a pulling speed dependence, suggesting that this transition occurs in nonequilibrium. The sequence dependence of the mechanical properties was investigated using synthetic constructs of double-stranded poly(dG-dC) and poly(dA-dT). For poly(dG-dC), the B-S transition was found to be at 65 pN, and the melting transition occurred at ~250 pN. For poly(dA-dT), the B-S transition was at 35 pN and the double strand was found to begin melting during this B-S transition. These sequence-dependent properties of DNA could have biological relevance in modulating the positioning of nucleosomes and the interactions with proteins. Morii et al. [134] measured the elastic properties of double stranded DNA using AFM in aqueous solutions, and found that two types of transitions from B-form to S-form. The first B-S transition occurred at a force of 67 pN, while the second one was observed at 116 pN. The structural transition at the higher force was attributed to the unnicked DNA molecules with both ends of each strand tethered to the surfaces, such that the untwisting of the double helix was blocked. The Young's modulus of the DNA molecules was measured to be around 200 MPa in water and became lower as the ionic strength of the solution was increased.

During transcription and replication, the relevant regions of the DNA become single stranded DNA (ssDNA). Ke et al. [135] investigated the elasticity of two types of single-stranded synthetic DNA, poly(dA) and poly(dT), to elucidate the effect of base stacking on the elastic and structural properties of the ssDNA. It was found from these AFM force measurements that poly(dT) exhibits the expected entropic elasticity behavior, while poly(dA) displays two overstretching transitions at 23 pN and 113 pN. The transitions are the mechanical signature of base-stacking interactions among adenines, which may have important implications in modulating sequence specific functions, such as binding to enzymes and drug molecules.

4.4.2 Inter-Strand Interaction Forces in Duplex DNA

During many important biomolecular processes, the two strands of a dsDNA molecule need to be separated, while in other situations, the DNA needs to have a stable double helical structure. Therefore, the inter-strand interactions in a DNA double helix are central to the understanding of its structure and various functions. The first direct measurement of the forces between the two strands in a dsDNA by AFM was reported by Lee et al. [136]. By immobilizing complementary oligonucleotides on the AFM tip and a substrate surface respectively, the rupture forces between a single pair of oligonucleotides involving 20, 16 and 12 base pairs were determined. Rief et al. [29] used the hairpins formed by self-complementary sequence to measure the forces required to unzip the DNA double helix by pulling the two strands apart with the AFM cantilever, as shown in Fig. 4.8. The experiment

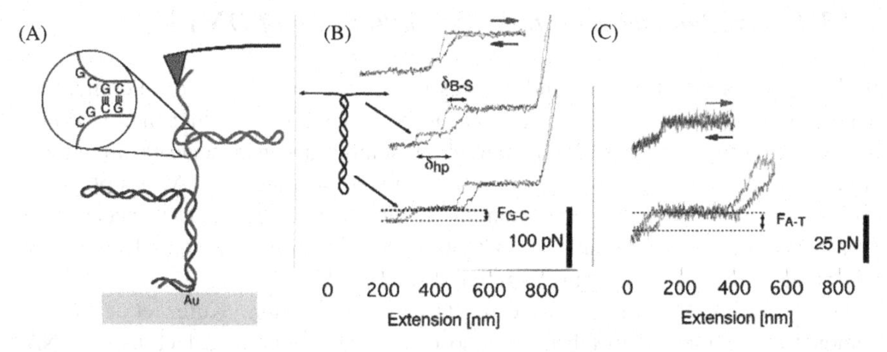

Fig. 4.8 AFM measurement of base-pairing forces in a double stranded DNA. (**a**) The self-complementary sequence forms hairpins, which can be unzipped to provide a direct measurement of the base-pairing forces. (**b**) Extension (*red*) and relaxation (*blue*) curves of one poly(dG-dC) molecule. (**c**) Extension and relaxation curves of a poly(dA-dT) DNA molecule. Reprinted by permission from Macmillan Publishers Ltd: Nature Structural Biology [29], Copyright (1999)

revealed that the average base pair-unbinding force for G-C is 20 pN and that for AT is 9 pN. The sequence dependent unzipping force measurement can serve as a first step toward sequencing individual DNA molecules using mechanical means. The information is also important in understanding the sequence specific stability and functions of DNA.

To get a better sensitivity of the sequence dependence of the unzipping force, Krautbauer et al. [137] carried out DNA unzipping experiments using shorter DNA molecules with both artificial and natural sequences. Complementary DNA oligonucleotides were chemically attached to the AFM tip and a glass surface respectively. As the tip was brought close to the surface, the complementary strands on the tip and the surface interacted and formed a double stranded helix, with the tethered ends of both strands on the same side. As the tip moved away from the surface, the double strand was opened in a zipper like fashion. The DNA sequence was discriminated from the unzipping force with a resolution of 10 base pairs. Sattin et al. [138] investigated the effects of base mismatches on the inter-strand interaction forces of DNA by using AFM and an oligonucleotide microarray. A short ssDNA was attached to the AFM tip, and ssDNA fragments with sequences complementary to tip-attached DNA, and fragments with one or two mismatched bases to the tip-attached DNA were immobilized on designated areas of the microarray surface. By measuring the unbinding forces between the tip-attached DNA strand and the surface-immobilized strands, a direct comparison was made between the interaction forces of the perfectly matched sequences and those containing a single or double mismatches. Since the measurements were made under exactly the same experimental conditions, experimental errors were reduced and the contribution of a single base pair to the inter-strand interaction force was determined. The measured results show that AT and GC base pairs have similar contributions to the rupture forces of the two strands in a double helix, suggesting that base stacking is more important than hydrogen bonding to the mechanical stability of DNA.

4.4.3 Controlled Delivery and Nanopatterning of DNA

Building nanoscale structures and devices is a central task in the development of nanotechnology. Nanoscale fabrications can be realized using either the top-down or the bottom-up approach. Molecular self assembly and nanoscale manipulation are the two technologies for the bottom-up fabrication strategy. AFM has played a critical role in the development of nanofabrication techniques due to its capacity to perform controlled manipulation, positioning and delivering atoms, molecules and other nanoscale entities with high spatial resolutions.

Using the AFM and the selective DNA hybridization, Kufer et al. [110] demonstrated a method to fabricate nanoscale structures on a surface using DNA molecules. In this method, ssDNA oligomers to be transferred and used for assembly (the cargo strand) are "stored" in the depot area by surface-tethered ssDNA oligomers (the support strand) that are based paired to a portion of the cargo strand. Another ssDNA oligomer (the tip strand) that is complementary to the non-base paired portion of the cargo strand is attached to the AFM tip. When the tip is positioned at the depot and lowered to the surface, a double strand helix is formed between the tip strand and the cargo strand. Sequences of the ssDNA strands are designed in such a way that when the AFM tip is pulled away, the cargo strand and support strand are unzipped, which requires a lower force than that to unbind the cargo strand from the tip strand, where the force is applied along the axis of the double helix, as shown in Fig. 4.9. As a result, the cargo strand is picked up by the AFM tip and then transferred to the designated spot in the target area, where the free end of the cargo strand binds to the surface-immobilized target strand by

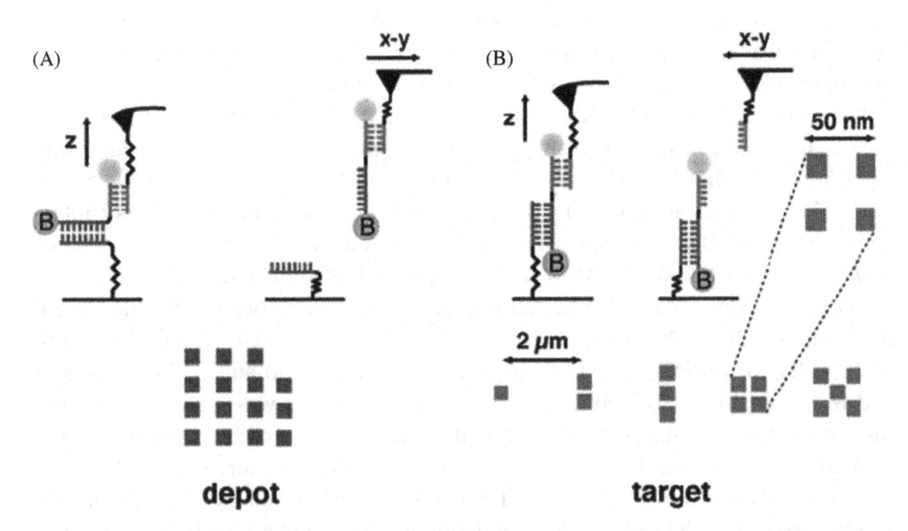

Fig. 4.9 Assembling patterns with DNA oligomers using AFM nanomanipulation. (**a**) Individual functional units (DNA oligomers) are picked up one at a time by the AFM tip and transferred to the target area. (**b**) The transferred DNA oligomer is deposited in the target area. From [110], reprinted with permission from AAAS

a double helix. The portion of the cargo strand bound to the target strand is larger than the portion bound to the tip strand, therefore, the cargo strand is separated from the tip strand when the AFM tip is moved away, leaving the cargo strand at the target area to form the programmed structure. The effectiveness of the method was demonstrated by spelling the letter "M" using 400 fluorescently labeled oligonulotides transferred from the depot to the target area with a spatial resolution of 10 nm [110]. It was also shown that an AFM tip could be used to transport and deliver more than 5,000 molecules over a period of 10 h. The authors suggested that, with the development of massively parallel operating AFM cantilevers [139], this molecular assembly technique could find many applications. Using a similar approach, Duwez et al. [131] used the AFM to deliver polymer molecules onto a Si surface, where the molecules were covalently immobilized and could be subsequently modified by further chemical reactions. DNA molecules deposited on a solid surface can also be manipulated with the AFM tip to form designated patterns. Hu et al. [126] demonstrated this method by spelling the letters D, N, and A on a mica surface using DNA molecules.

The efficient delivery of specific molecules to target cells is important in biotechnology and biomedicine. Efforts have been made to use AFM for such a purpose. Afrin et al. [140] reported the transfection of DNA containing the gene for a fluorescent protein into living fibroblast cells using the AFM. The AFM tip with the DNA molecules attached was inserted into the cell under a high loading force. After the insertion procedure, the cells were incubated and the level of the fluorescent protein expression was checked by fluorescence microscopy. The results showed a successful transfection of the DNA molecules into the cells. Han et al. [141] reported a low invasive method for delivering DNA into living cells using 200 nm diameter nanoneedles that were made by etching the Si AFM tips. DNA molecules to be delivered were immobilized on the nanoneedle, which was then inserted into a living cell by the force exerted from the AFM cantilever. The depth of insertion of the nanoneedle was monitored and controlled by the AFM. It was found that cells could still proliferate normally after repeated insertions of the nanoneedle, reflecting the low invasiveness of the method. The DNA molecules were found to remain on the nanoneedle during insertion and removal. The authors are attempting to develop a transient transcription technique that could utilize the immobilized DNA while the nanoneedle is inserted in the cell.

4.4.4 Dissection and Isolation of Fragments from DNA Molecules

The isolation of a specific section in a DNA molecule for analysis is often required in research, clinical analysis, and biotechnology. The operation is normally performed using restriction enzymes on a large number of molecules. In recent years, it has been shown that AFM tip can be used to manipulate DNA on solid surfaces [142–144] and to isolate a designated segment from a single DNA molecule [145]. Lu et al. [145] reported the cutting and isolation of a segment from a long

DNA molecule at the designated position by a precise AFM nanomanipulation. The isolated fragment was subsequently amplified using PCR. The DNA molecules (pBR322) were first deposited and stretched on a mica surface using a molecular combing technique [146]. After an AFM image was taken, a specific section of a DNA molecule was selected for dissection. The selected section was cut from the molecule by scanning the tip at the designated locations (ends of the selected segment) with high forces. The tip then scanned over the dissected fragments several times with moderately high forces to loosen and pick up the segment from the surface. The AFM tip containing the isolated DNA fragment was then transferred to a sterile tube, and a single-molecule PCR was carried out for amplification. It was found that 7 out of 20 tips used in such experiments picked up the dissected DNA fragment. The successful PCR amplification suggests no significant damage to the isolated DNA fragment, which makes it amenable for subsequent analysis. AFM has also been used to extract DNA from chromosomes following similar procedures [37, 147–150].

4.4.5 AFM Manipulation of RNA Molecules

In a recent report, Liu et al. [151] investigate the RNA- protein interactions within tobacco mosaic virus (TMV) in an aqueous solution by pulling the RNA genome out of a TMV particle using the AFM. TMV is a rod-shaped virus with its capsid containing 2,130 molecules of the coat protein. One molecule of genomic RNA ($\sim$6,390 bases) is wrapped inside the capsid. In the reported study, the TMV particle was immobilized on a gold surface. When the AFM tip was brought into contact with the particle, the viral RNA could stick to the AFM tip and be pulled out of the capsid. The measured force as a function of the pulling distance revealed the strength of the interactions between the RNA and the coat proteins. The effects of pulling speed and pH on RNA-protein interactions were also investigated. The authors argued that the results could help gain new insights into the mechanism of virus infection as well as the rational design of novel nanomaterials using TMV as a template.

Osada et al. [152] showed that messenger RNA (mRNA) molecules could be extracted from living cells using an AFM tip. A silicon nitride AFM tip was inserted into the cytosol of a cell (a rat fibroblast-like cell or a mouse osteoblast-like cell), the β-actin mRNA molecules attached to the tip and were subsequently extracted. The mRNA was then washed off the tip and quantified by reverse PCR. The method has the benefit of measuring the mRNA level without destroying the cell, thus can be used to monitoring time-dependent gene expression. Marsden et al. [153] studied the functional mechanisms of two RNA helicases to unwind localized structures in long RNA molecules. By tethering the RNA molecule between the AFM tip and a surface, a controlled amount of tension was applied to the RNA, thus mimicking the situation in cells where the RNA might experience a force from the translocating biomolecular complex. The results revealed qualitative and quantitative mechanistic differences between the two types of helicases.

4.5 Manipulation of Polysaccharide Molecules

Polysaccharides are polymeric carbohydrate molecules, either linear or branched, that are composed of mono- or di-saccharide units linked via glycosidic bonds. They exist in all types of organisms (plants, animals, fungus and bacteria) and perform various functions. Polysaccharides have also found important applications in biomedicine and other fields [154]. The properties of polysaccharides have been extensively investigated by various techniques, but the large sizes and structural complexity have made it difficult to characterize these molecules with many conventional techniques. In recent years, the AFM manipulation technique has been shown to have the capacity to characterize the structural and conformational properties of certain polysaccharides at the single molecule level and in aqueous solutions [26, 155–159]. Such studies are particular interesting in understanding the mechanical functions of polysaccharides, such as those in bacterial and plant cell walls.

4.5.1 Force-Induced Conformational Changes in Polysaccharides

The first polysaccharide studied using AFM nanomanipulation is dextran [160]. Dextran molecules functionalized with streptavidin were attached to a gold surface through an epoxy-alkanethiol bond. The biotin-functionalized AFM tip picked up the dextran via the strong avidin-biotin interaction and exerted a tensile force on the molecule. To perform experiment at higher forces, a hydrophobic tip was used to attach the free end of the dextran via non-specific adsorption. The dextran molecules were found to behave as an entropic chain that could be described by the FJC model. Under high tensil forces, the molecule exhibited conform changes governed by a twist of bond angles. The conformational change was found to be reversible and its structural basis was revealed by molecular dynamics calculations. Another significant single molecule study of polysaccharides was carried out by Marszalek et al. [161] who measured the force-induced conformation changes of amylose, dextran and pullulan in order to understand the structural units responsible for elastic properties of polysaccharides. The polysaccharide molecules were adsorbed onto a glass surface, and after the AFM tip was pressed on the surface for several seconds, individual molecules were picked up by the tip and subsequently stretched. A transition in the elasticity of the polysaccharide molecules was observed upon the cleavage of pyranose rings, suggesting that the pyranose ring is the structural unit governing the elastic behaviors of polysacchrides. The results showed that the applied force induced an elongation of the ring structure in polysaccharides, and caused the ring conformation to change from a chair-like structure to a boat-like structure. Such conformation changes can regulate polysaccharide functions, such as its binding to proteins. Therefore, the observed force structure relationship can be of biological significance as the polysaccharide molecules often experience mechanical stresses in their natural environments. The force-induced conformational changes can also serve as mechanical signatures of polysaccharides, based on which

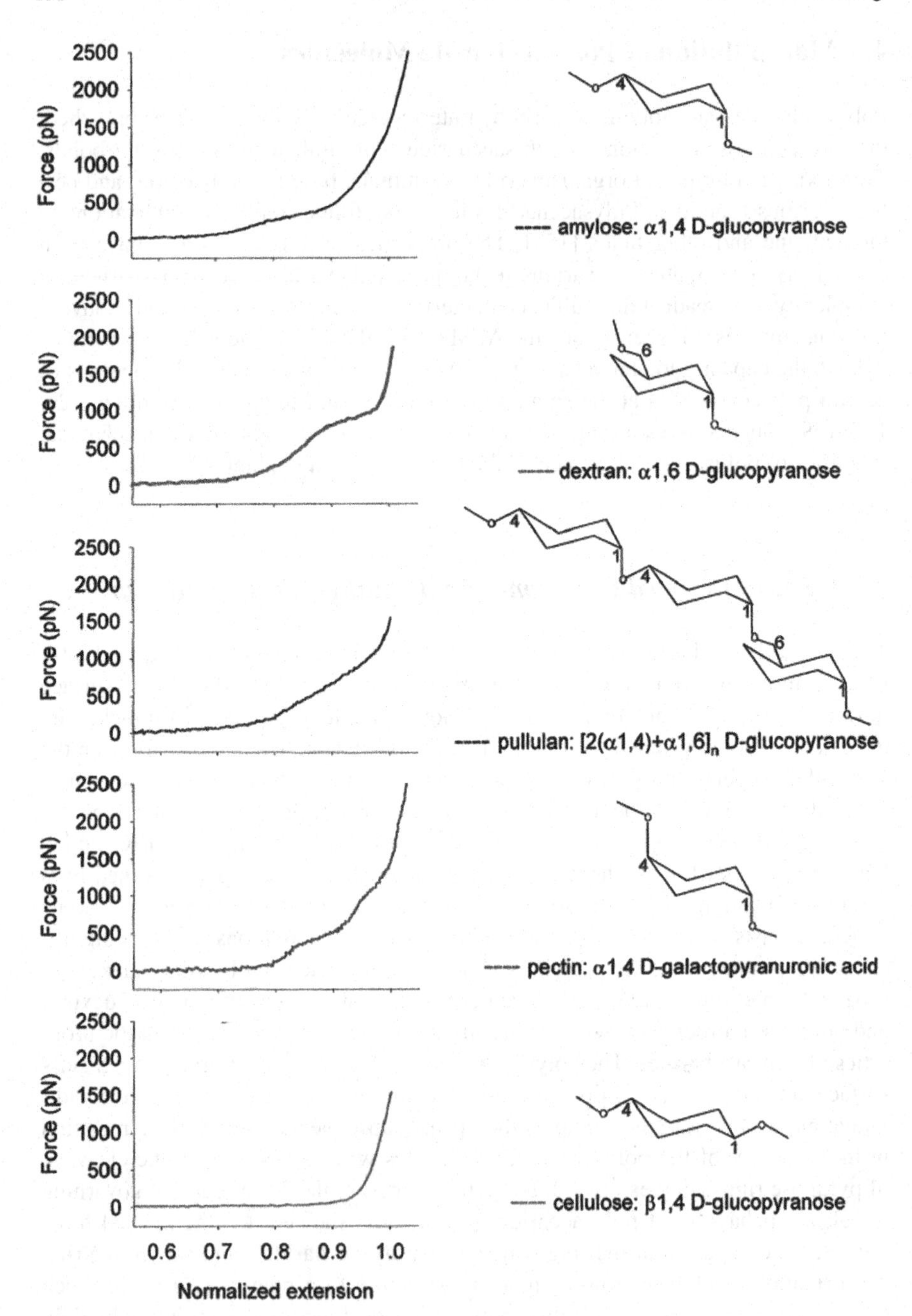

Fig. 4.10 Identification of polysaccharides based on their elastic behaviors. The left panel shows the force-extension curves of single polysaccharides obtained from vertically stretching the molecules with AFM. The right panel shows the monomer structures and the type of linkages in the polysaccharides. Reprinted by permission from Macmillan Publishers Ltd: Nature Biotechnology [162], Copyright (2001)

individual polysaccharide molecules can be identified in a solution containing a mixture of different types of polysaccharides [162], as shown in Fig. 4.10. The method constitutes an important advance in the analytical techniques of polysaccharides.

4.5.2 Properties of Polysaccharides on Living Cell Surfaces

In a recent report, Francius et al. studied the properties of polysaccharides on the surface of live bacteria [163]. Polysaccharides on bacterial surfaces serve many purposes, such as protection, cellular recognition, and bacterial adhesion. In this study, the bacterial cells were immobilized in a porous membrane, the cell surface morphology was obtained by AFM imaging and the cell wall elasticity was mapped by AFM tip indentation measurements. The surface polysaccharides were investigated with tips functionalized to recognize a specific type of polysaccharide. Pulling the surface polysaccharides using the functionalized tip revealed the distribution of different types of polysaccharides on the cell surface, and the differences in the molecular size and distribution between wide-type and mutant bacterial cells. Using a similar approach, Alsteens et al. [164] investigated the surface structure, cell wall elasticity and polymer properties of living cells from two yeast strains, and found that the two strains have similar surface ultrastructures, but different cell wall elasticitis and polysaccharide properties. The polysaccharides on one strain were observed to be more extended, which helps to explain the different aggregation properties of the two types of cells. The AFM studies have revealed previous unknown properties of polysaccharides and thus broadened our understanding of the mechanisms of their biological functions in various organisms, especially the effects of mechanical forces.

4.6 Conclusion

As one of the few single molecule techniques, AFM-based nanomanipulation has been used to study all types of biological macromolecules and revealed important, previously unknown properties and functional mechanisms. The capacity of the AFM for the isolation, transfer, positioning and assembling of individual macromolecules with nanometer spatial resolutions has significantly advanced the field of bionanotechnology toward the fabrication of functional structures and devices with macromolecules as building blocks.

Precise manipulation of macromolecules will certainly continue to be one of the important techniques in our effort to understand the intricacies of the macromolecular processes in biological cells. Many of these processes are mechanical in nature, and can only be properly described in terms of fundamental physical concepts such as force, torque, work, and energy. AFM has the capacity to directly measure these quantities on individual macromolecules, thus becoming a unique and indispensable tool that will be used by more researchers to investigate an increasing range of biological problems. The AFM nanomanipulation technique, together with ensemble measurement methods and molecular dynamics simulations, will make major contributions in advancing our knowledge on the mechanisms of important biological processes.

Despite all the recent progress, manipulation of biological macromolecules with the AFM is still at a beginning stage of its development. There are many technical limitations and challenges to be overcome before this technique can be employed to solve a wider range of problems in biology and biotechnology. The relative large size and high stiffness of the cantilevers limit the ability to measure weak forces and follow fast biological processes. In many manipulation experiments, the difficulties to control and identify the desired interactions between the AFM tip and the macromolecule of interest make the data interpretation ambiguous. The requirements for surface attachment and the polymerization restrict the type of macromolecules that can be investigated. Thermal drifts make slow manipulation experiments unreliable, and viscous drag on the cantilever complicates fast manipulation experiments.

Future technical developments in AFM nanomanipulation are aiming at circumventing these technical limitations, further improving the spatial and temporal resolutions, expanding the information content, and enhancing the throughput of the measurements. Fast AFM and small cantilevers are being developed to facilitate fast measurements, efforts are being made to combine force measurement with fluorescence detection, and techniques are advancing for parallel measurements using cantilever arrays. Manipulation of macromolecules in living cells and controlled delivery of functional molecules into the cells have been attempted. Optimization of these methods will find many significant biological and biomedical applications. Another exciting potential application of the AFM nanomanipulation technique is the ultra-fast sequencing of individual DNA molecules, which has been predicted based on molecular dynamics simulation [165]. Realization of such a technique will revolutionize many aspects of biomedicine and biological research.

Acknowledgements The author thanks the NIH for financial support.

References

1. J.P. Desai, A. Pillarisetti, A.D. Brooks, Engineering approaches to biomanipulation, *Annu Rev Biomed Eng, 9*, 35–53 (2007).
2. A. Borgia, P.M. Williams, J. Clarke, Single-molecule studies of protein folding, *Annu Rev Biochem, 77*, 101–125 (2008).
3. J. Voldman, Electrical forces for microscale cell manipulation, *Annu Rev Biomed Eng, 8*, 425–454 (2006).
4. U. Bockelmann, Single-molecule manipulation of nucleic acids, *Curr Opin Struct Biol, 14*, 368–373 (2004).
5. P.Y. Chiou, A.T. Ohta, M.C. Wu, Massively parallel manipulation of single cells and microparticles using optical images, *Nature, 436*, 370–372 (2005).
6. P.R. Selvin, T. Ha, *Single-molecule techniques: A laboratory manual.* Cold Spring Harbor Laboratory Press, New York, 2008.
7. D.M. Eigler, E.K. Schweizer, Positioning single atoms with a scanning tunneling microscope, *Nature, 344*, 524–526 (1990).

8. T. Junno, K. Deppert, L. Montelius, L. Samuelson, Controlled manipulation of nanoparticles with an atomic force microscope, *Appl Phys Lett, 66*, 3627–3629 (1995).
9. A.A. Tseng, Z. Li, Manipulations of atoms and molecules by scanning probe microscopy, *J Nanosci Nanotech, 7*, 1–14 (2007).
10. P. Dörig, P. Stiefel, P. Behr, E. Sarajlic, D. Bijl, M. Gabi, J. Vörös, J.A. Vorholt, T. Zambelli, Force-controlled spatial manipulation of viable mammalian cells and micro-organisms by means of fluidfm technology, *Appl Phys Lett, 97*, 023701 (2010).
11. A. Matouschek, J.T.J. Kellis, L. Serrano, A.R. Fersht, Mapping the transition state and pathway of protein folding by protein engineering, *Nature, 340*, 122–126 (1989).
12. P. Guptasarma, Does replication-induced transcription regulate synthesis of the myriad low copy number proteins of escherichia coli, *Bioessays, 17*, 987–997 (1995).
13. Y. Seo, W. Jhe, Atomic force microscopy and spectroscopy, *Rep Progr Phys, 71*, 016101 (2008).
14. C. Bustamante, S.B. Smith, J. Liphardt, D. Smith, Single-molecule studies of DNA mechanics, *Curr Opin Struct Bio, 10*, 279–285 (2000).
15. G. Binnig, C.F. Quate, C. Gerber, Atomic force microscope, *Phys Rev Lett, 56*, 930–933 (1986).
16. G. Binnig, H. Rohrer, Scanning tunneling microscopy, *Helv Phys Acta, 55*, 726–735 (1982).
17. G. Binnig, H. Rohrer, C. Gerber, E. Weibel, Surface studies by scanning tunneling microscopy, *Phys Rev Lett, 49*, 57–60 (1982).
18. J.M. Fernandez, H. Li, Force-clamp spectroscopy monitors the folding trajectory of a single protein, *Science, 303*, 1674–1678 (2004).
19. H.J. Mamin, D. Ruger, Sub-attonewton force detection at millikelvin temperatures, *Appl Phys Lett, 79*, 3358–3360 (2001).
20. J.L. Hutter, J. Bechhofer, Calibration of atomic force microscope tips, *Rev Sci Instrum, 64*, 1868–1873 (1993).
21. E.-L. Florin, M. Rief, H. Lehmann, M. Ludwig, C. Dornmair, V.T. Moy, H.E. Gaub, Sensing specific molecular interactions with the atomic force microscope, *Biosens Bioelectron, 10*, 895–901 (1995).
22. J.P. Cleveland, S. Manne, D. Bocek, P.K. Hansma, A nondestructive method for determining the spring constant of cantilevers for scanning force microscopy, *Rev Sci Instrum, 64*, 403–405 (1993).
23. A. Torii, M. Sasaki, K. Hane, S. Okuma, A method for detemining the spring constant of cantilevers for atomic force microscopy, *Meas Sci Technol, 7*, 179–184 (1996).
24. J.L. Choy, S.H. Parekh, O. Chaudhuri, A.P. Liu, C. Bustamante, M.J. Footer, J.A. Theriot, D.A. Fletcher, Differential force microscope for long time-scale biophysical measurements, *Rev Sci Instrum, 78*, 043711 (2007).
25. G.M. King, A.R. Carter, A.B. Churnside, L.S. Eberle, T.T. Perkins, Ultrastable atomic force microscopy: Atomic-scale stability and registration in ambient conditions, *Nano Lett, 9*, 1451–1456 (2009).
26. G. Francius, D. Alsteens, V. Dupres, S. Lebeer, S.D. Keersmaecker, J. Vanderleyden, H.J. Gruber, Y.F. Dufrene, Stretching polysaccharides on live cells using single molecule force spectroscopy, *Nat Protoc, 4*, 939–946 (2009).
27. A. Janshoff, M. Neitzert, Y. Oberdorfer, H. Fuchs, Force spectroscopy of molecular systems – single molecule spectroscopy of polymers and biomolecules, *Angew Chem Int Ed, 30*, 3212–3237 (2000).
28. M. Rief, M. Gautel, F. Oesterheld, J. Fernandez, H. Gaub, Reversible unfolding of individual titin immunoglobulin domains by AFM, *Science, 276*, 1109–1112 (1997).
29. M. Rief, H. Clausen-Schaumann, H.E. Gaub, Sequence-dependent mechanics of single DNA molecules, *Nat Struct Biol, 6*, 346–349 (1999).
30. H. Janovjak, J. Struckmeier, D.J. Muller, Hydrodynamic effects in fast AFM single-molecule force measurements, *Eur Biophys J, 34*, 91–96 (2005).

31. J.-M. Yuan, C.-L. Chyan, H.-X. Zhou, T.-Y. Chung, H. Peng, G. Ping, G. Yang, The effects of macromolecular crowding on the mechanical stability of protein molecules, *Protein Sci, 17*, 2156–2166 (2009).
32. J. Alcaraz, L. Buscemi, M. Puig-de-Morales, J. Colchero, A. Baró, D. Navajas, Correction of microrheological measurements of soft samples with atomic force microscopy for the hydrodynamic drag on the cantilever, *Langmuir, 18*, 716–721 (2002).
33. R. Liu, M. Roman, G. Yang, Correction of the viscous drag induced errors in macromolecular manipulation experiments using atomic force microscope, *Rev Sci Instrum, 81*, 063703 (2010).
34. E.M. Puchner, H.E. Gaub, Force and function: Probing proteins with AFM-based force spectroscopy, *Curr Opin Struct Biol, 19*, 605–614 (2009).
35. W.A. Linke, A. Grützner, Pulling single molecules of titin by AFM – recent advances and physiological implications, *Pflugers Arch - Eur J Physiol, 456*, 101–115 (2008).
36. J.W. Weisel, H. Shuman, R.I. Litvinov, Protein-protein unbinding induced by force: Single-molecule studies, *Curr Opin Struct Biol, 13*, 227–235 (2003).
37. D. Fotiadis, S. Scheuring, S.A. Müller, A. Engel, D.J. Müller, Imaging and manipulation of biological structures with the AFM, *Micron, 33*, 385–397 (2002).
38. A. Engel, H.E. Gaub, Structure and mechanics of membrane proteins, *Annu Rev Biochem 77*, 127–148 (2008).
39. D.J. Müller, A. Engel, Strategies to prepare and characterize native membrane proteins and protein membranes by AFM, *Curr Opin Colloid Interface Sci, 13*, 338–350 (2008).
40. Y.F. Dufrêne, Towards nanomicrobiology using atomic force microscopy, *Nat Rev Microbiol, 6*, 674–680 (2008).
41. F. Kienberger, G. Kada, H. Mueller, P. Hinterdorfer, Single molecule studies of antibody–antigen interaction strength versus intra-molecular antigen stability, *J Mol Biol, 347*, 597–606 (2005).
42. A.A. Deniz, S. Mukhopadhyay, E.A. Lemke, Single-molecule biophysics: At the interface of biology, physics and chemistry, *J R Soc Interface, 5*, 15–45 (2008).
43. A. Engel, D.J. Müller, Observing single biomolecules at work with the atomic force microscope, *Nat Struct Biol, 7*, 715–718 (2000).
44. A.N. Kapanidis, T. Strick, Biology, one molecule at a time, *Trend Biochem Sci, 34*, 234–243 (2009).
45. A.F. Oberhauser, M. Carrion-Vazquez, Mechanical biochemistry of proteins one molecule at a time, *J Biol Chem, 283*, 6617–6621 (2008).
46. C. Bustamante, In singulo biochemistry: When less is more, *Annu Rev Biochem, 77*, 45–50 (2008).
47. C. Bustamante, Y.R. Chemla, N.R. Forde, D. Izhaky, Mechanical processes in biochemistry, *Ann Rev Biochem, 73*, 705–748 (2004).
48. J. Castillo, M. Dimaki, W.E. Svendsen, Manipulation of biological samples using micro and nano techniques, *Integr Biol, 1*, 30–42 (2009).
49. A. Galera-Prat, A. Gomez-Sicilia, A.F. Oberhauser, M. Cieplak, M. Carrion-Vazquez, Understanding biology by stretching proteins: Recent progress, *Curr Opin Struct Biol, 20*, 63–69 (2010).
50. H. Li, M. Carrion-Vazquez, A.F. Oberhauser, P.E. Marszalek, J.M. Fernandez, Point mutations alter the mechanical stability of immunoglobulin modules, *Nat Struct Biol, 7*, 1117–1120 (2000).
51. L. Tskhovrebova, J. Trinick, J.A. Sleep, R.M. Simmons, Elasticity and unfolding of single molecules of the giant muscle protein titin, *Nature, 387*, 308–312 (1997).
52. M.S. Kellermayer, S.B. Smith, H.L. Granzier, C. Bustamante, Folding-unfolding transitions in single titin molecules characterized with laser tweezers, *Science, 276*, 1112–1116 (1997).
53. J.I. Sulkowska, M. Cieplak, Stretching to understand proteins – a survey of the protein data bank, *Biophys J, 94*, 6–13 (2008).

54. T.I. Garcia, A.F. Oberhauser, W. Braun, Mechanical stability and differentially conserved physical – chemical properties of titin Ig-domains, *Proteins, 75*, 706–718 (2009).
55. S. Labeit, B. Kolmerer, Titins: Giant proteins in charge of muscle ultrastructure and elasticity, *Science, 270*, 293–296 (1995).
56. H. Li, W.A. Linke, A.F. Oberhauser, M. Carrion-Vazquez, J.G. Kerkvliet, H. Lu, P.E. Marszalek, J.M. Fernandez, Reverse engineering of the giant muscle protein titin, *Nature, 418*, 998–1002 (2002).
57. B. Bullard, T. Garcia, V. Benes, M.C. Leake, W.A. Linke, A.F. Oberhauser, The molecular elasticity of the insect flight muscle proteins projectin and kettin, *Proc Natl Acad Sci USA, 103*, 4451–4456 (2006).
58. W.A. Linke, Sense and stretchability: The role of titin and titin-associated proteins in myocardial stress-sensing and mechanical dysfunction, *Cardiovasc Res, 77*, 637–648 (2008).
59. L. Tskhovrebova, J. Trinick, Properties of titin immunoglobulin and fibronectin-3 domains, *J Biol Chem, 279*, 46351–46354 (2004).
60. S. Labeit, B. Kolmerer, W.A. Linke, The giant protein titin. Emerging roles in physiology and pathophysiology, *Circ Res, 80*, 290–294 (1997).
61. W.A. Linke, Titin elasticity in the context of the sarcomere: Force and extensibility measurements on single myofibrils, *Adv Exp Med Biol, 481*, 179–202 (2000).
62. Y. Cao, H. Li, Single molecule force spectroscopy reveals a weakly populated microstate of the fniii domains of tenascin, *J Mol Biol, 361*, 372–381 (2006).
63. E. Frixione, Recurring views on the structure and function of the cytoskeleton: A 300-year epic, *Cell Motil Cytoskeleton, 46*, 73–94 (2000).
64. A. Elgsaeter, B.T. Stokke, A. Mikkelsen, D. Branton, The molecular basis of erythrocyte shape, *Science, 234*, 1217–1233 (1986).
65. D.R. Markle, E.A. Evans, R.M. Hochmuth, Force relaxation and permanent deformation of erythrocyte membrane, *Biophys J, 42*, 91–98 (1983).
66. D.M. Shotton, B.E. Burke, D. Branton, The molecular structure of human erythrocyte spectrin. Biophysical and electron microscopic studies, *J Mol Biol, 131*, 303–329 (1979).
67. J. Pascual, M. Pfuhl, D. Walther, M. Saraste, M. Nilges, Solution structure of the spectrin repeat: A left-handed antiparallel triple-helical coiled-coil, *J Mol Biol, 273*, 740–751 (1997).
68. D.W. Speicher, V.T. Marchesi, Erythrocyte spectrin is comprised of many homologous triple helical segments, *Nature, 311*, 177–180 (1984).
69. M. Rief, J. Pascual, M. Saraste, H.E. Gaub, Single molecule force spectroscopy of spectrin repeats: Low unfolding forces in helix bundles, *J Mol Biol, 286*, 553–561 (1999).
70. A.S. Politou, D.J. Thomas, A. Pastore, The folding and stability of titin immunoglobulin-like modules, with implications for the mechanism of elasticity, *Biophys J, 69*, 2601–2610 (1995).
71. T.M. DeSilva, S.L. Harper, L. Kotula, P. Hensley, P.J. Curtis, L. Otvos, Jr, D.W. Speicher, Physical properties of a single-motif erythrocyte spectrin peptide: A highly stable independently folding unit, *Biochemistry, 36*, 3991–3997 (1997).
72. R. Law, P. Carl, S. Harper, P. Dalhaimer, D.W. Speicher, D.E. Discher, Cooperativity in forced unfolding of tandem spectrin repeats, *Biophys J, 84*, 533–544 (2003).
73. V. Ortiz, S.O. Nielsen, M.L. Klein, D.E. Discher, Unfolding a linker between helical repeats, *J Mol Biol, 349*, 638–647 (2005).
74. R.J. Ellis, F.U. Hartl, Principles of protein folding in the cellular environment, *Curr Opin Struct Biol, 9*, (1999).
75. S.E. Radford, C.M. Dobson, From computer simulations to human disease: Emerging themes in protein folding, *Cell, 97*, 291–298 (1999).
76. J.D. Bryngelson, P.G. Wolynes, Spin glasses and the statistical mechanics of protein folding, *Proc Natl Acad Sci USA, 84*, 7524–7528 (1987).
77. J.D. Bryngelson, J.N. Onuchic, P.G. Wolynes, Funnels, pathways, and the energy landscape of protein folding: A synthesis, *Proteins, 21*, 167–195 (1995).

78. K.A. Dill, S. Chan, From levinthal to pathways to funnels, *Nat Struct Biol, 4*, 10–19 (1997).
79. K.A. Dill, S.B. Ozkan, M.S. Shell, T. R.Weikl, The protein folding problem, *Annu Rev Biophys, 37*, 289–316 (2008).
80. P.E. Leopold, M. Montal, J.N. Onuchic, Protein folding funnels: A kinetic approach to the sequence-structure relationship, *Proc Natl Acad Sci USA, 89*, 8712–8715 (1992).
81. S.S. Plotkin, J.N. Onuchic, Understanding protein folding with energy landscape theory part i: Basic concepts, *Quart Rev Biophys, 35*, 111–167 (2002).
82. J.E. Shea, J.N. Onuchic, C.L. Brooks, Energetic frustration and the nature of the transition state in protein folding, *J Chem Phys, 113*, 7663–7671 (2000).
83. M. Scalley-Kim, D. Baker, Characterization of the folding energy landscapes of computer generated proteins suggests high folding free energy barriers and cooperativity may be consequences of natural selection, *J Mol Biol, 338*, 573–583 (2004).
84. G. Yang, C. Cecconi, W.A. Baase, I.R. Vetter, W.A. Breyer, J.A. Haack, B.W. Matthews, F.W. Dahlquist, C. Bustamante, Solid-state synthesis and mechanical unfolding of polymers of t4 lysozyme, *Proc Natl Acad Sci USA, 97*, 139–144 (2000).
85. D. Craig, A. Krammer, K. Schulten, V. Vogel, Comparison of the early stages of forced unfolding for fibronectin type III modules, *Proc Natl Acad Sci USA, 98*, 5590–5595 (2001).
86. M. Carrion-Vazquez, A.F. Oberhauser, S.B. Fowler, P.E. Marszalek, S.E. Broedel, J. Clarke, J.M. Fernandez, Mechanical and chemical unfolding of a single protein: A comparison, *Proc Natl Acad Sci USA, 96*, 3694–3699 (1999).
87. K.A. Scott, A. Steward, S.B. Fowler, J. Clarke, Titin: A multidomain protein that behaves as the sum of its parts, *J Mol Biol, 315*, 819–829 (2002).
88. A. Krammer, H. Lu, B. Isralewitz, K. Schulten, V. Vogel, Forced unfolding of the fibronectin type III module reveals a tensile molecular recognition switch, *Proc Natl Acad Sci USA, 96*, 1351–1356 (1999).
89. P.M. Williams, S.B. Fowler, R.B. Best, J.L. Toca-Herrera, K.A. Scott, A. Steward, J. Clarke, Hidden complexity in the mechanical properties of titin, *Nature, 422*, 446–449 (2003).
90. C.-L. Chyan, F.-C. Lin, H. Peng, J.-M. Yuan, C.-H. Chang, S.-H. Lin, G. Yang, Reversible mechanical unfolding of single ubiquitin molecules, *Biophys J, 87*, 3995–4006 (2004).
91. H. Dietz, M. Rief, Exploring the energy landscape of GFP by single-molecule mechanical experiments, *Proc Natl Acad Sci USA, 101*, 16192–16197 (2004).
92. M. Zimmer, Green fluorescent protein (GFP): Applications, structure, and related photophysical behavior, *Chem Rev, 102*, 759–781 (2002).
93. R. Berkovich, S. Garcia-Manyes, M. Urbakh, J. Klafter, J.M. Fernandez, Collapse dynamics of single proteins extended by force, *Biophys J, 98*, 2692–2701 (2010).
94. I. Schwaiger, M. Schleicher, A.A. Noegel, M. Rief, The folding pathway of a fast-folding immunoglobulin domain revealed by single-molecule mechanical experiments, *EMBO reports, 6*, 46–51 (2005).
95. M. Schlierf, M. Rief, Single-molecule unfolding force distributions reveal a funnel-shaped energy landscape, *Biophys J, 90*, L33–35 (2006).
96. S. Wennmalm, S.M. Simon, Studying individual events in biology, *Annu Rev Biochem, 76*, 419–446 (2007).
97. A.F. Oberhauser, P.E. Marszalek, M. Carrion-Vazquez, J.M. Fernandez, Single protein misfolding events captured by atomic force microscopy, *Nat Struct Biol, 6*, 1025–1028 (1999).
98. L.G. Randles, R.W.S. Rounsevell, J. Clarke, Spectrin domains lose cooperativity in forced unfolding, *Biophys J, 92*, 571–577 (2007).
99. A.F. Oberhauser, P.E. Marszalek, H.P. Erickson, J.M. Fernandez, The molecular elasticity of the extracellular matrix protein tenascin, *Nature, 393*, 181–185 (1998).
100. C.F. Wright, S.A. Teichmann, J. Clarke, C.M. Dobson, The importance of sequence diversity in the aggregation and evolution of proteins, *Nature, 438*, 878–881 (2005.).
101. C. Jäckel, P. Kast, D. Hilvert, Protein design by directed evolution, *Annu Rev Biophys, 37*, 153–173 (2008).

102. A. Mesecar, B.L. Stoddard, D.E.J. Koshland, Orbital steering in the catalytic power of enzymes: Small structural changes with large catalytic consequences, *Science, 277*, 202–206 (1997).
103. J. Alegre-Cebollada, R. Perez-Jimenez, P. Kosuri, J.M. Fernandez, Single-molecule force spectroscopy approach to enzyme catalysis, *J Biol Chem, 285*, 18961–18966 (2010).
104. M. Radmacher, M. Fritz, H.G. Hansma, P.K. Hansma, Direct observation of enzyme activity with the atomic force microscope, *Science, 265*, 1577–1579 (1994).
105. A.P. Wiita, S.R.K. Ainavarapu, H.H. Huang, J.M. Fernandez, Force-dependent chemical kinetics of disulfide bond reduction observed with single-molecule techniques, *Proc Natl Acad Sci USA, 103*, 7222–7227 (2006).
106. C.S. Sevier, C.A. Kaiser, Formation and transfer of disulphide bonds in living cells, *Nat Rev Mol Cell Biol, 3*, 836–847 (2002).
107. B. Yan, J.W. Smith, Mechanism of integrin activation by disulfide bond reduction, *Biochemistry, 40*, 8861–8867 (2001).
108. A.P. Wiita, R. Perez-Jimenez, K.A. Walther, F. Grater, B.J. Berne, A. Holmgren, J.M. Sanchez-Ruiz, J.M. Fernandez, Probing the chemistry of thioredoxin catalysis with force, *Nature, 450*, 124–127 (2007).
109. H. Gumpp, E.M. Puchner, J.L. Zimmermann, U. Gerland, H.E. Gaub, K. Blank, Triggering enzymatic activity with force, *Nano Lett, 9*, 3290–3295 (2009).
110. S.K. Kufer, E.M. Puchner, H. Gumpp, T. Liedl, H.E. Gaub, Single-molecule cut-and-paste surface assembly, *Science, 319*, 594–596 (2008).
111. E. Wallin, G. von Heijne, Genome-wide analysis of integral membrane proteins from eubacterial, archaean, and eukaryotic organisms, *Protein Sci, 7*, 1029–1038 (1998).
112. D.J. Muller, AFM: A nanotool in membrane biology, *Biochemistry, 47*, 7986–7998 (2008).
113. D.J. Müller, J. Helenius, D. Alsteens, Y.F. Dufrêne, Force probing surfaces of living cells to molecular resolution, *Nat Chem Biol, 5*, (2009).
114. A.J. García-Sáez, P. Schwille, Single molecule techniques for the study of membrane proteins, *Appl Microbiol Biotechnol, 76*, 257–266 (2007).
115. P.L.T.M. Frederix, P.D. Bosshart, A. Engel, Atomic force microscopy of biological membranes, *Biophys J, 96*, 329–338 (2009).
116. R.P. Goncalves, S. Scheuring, Manipulating and imaging individual membrane proteins by AFM, *Surf Interface Anal, 38*, 1413–1418 (2006).
117. D.J. Muller, W. Baumeister, A. Engel, Controlled unzipping of a bacterial surface layer with atomic force microscopy, *Proc Natl Acad Sci USA, 96*, 13170–13174 (1999).
118. F. Oesterhelt, D. Oesterhelt, M. Pfeiffer, A.G. Engel, H. E., D.J. Muller, Unfolding pathways of individual bacteriorhodopsins, *Science, 288*, 143–146 (2000).
119. R.P. Goncalves, G. Agnus, P. Sens, C. Houssin, B. Bartenlian, S. Scheuring, Two-chamber AFM: Probing membrane proteins separating two aqueous compartments, *Nat Methods, 3*, 1007–1012 (2006).
120. M. Kessler, K.E. Gottschalk, H. Janovjak, D.J. Müller, H.E. Gaub, Bacteriorhodopsin folds into the membrane against an external force, *J Mol Biol, 357*, 644–654 (2006).
121. C.-K. Lee, Y.-M. Wang, L.-S. Huang, S. Lin, Atomic force microscopy: Determination of unbinding force, off rate and energy barrier for protein – ligand interaction, *Micron, 38*, 446–461 (2007).
122. M. Benoit, D. Gabriel, G. Gerisch, H.E. Gaub, Discrete interactions in cell adhesion measured by single-molecule force spectroscopy, *Nat Cell Biol, 2*, 313–317 (2000).
123. T.S. Lim, S.R.K. Vedula, W. Hunziker, C.T. Lim, Kinetics of adhesion mediated by extracellular loops of claudin-2 as revealed by single-molecule force spectroscopy, *J Mol Biol, 381*, 681–691 (2008).
124. Y. Zhang, S. Sivasankar, W.J. Nelson, S. Chu, Resolving cadherin interactions and binding cooperativity at the single-molecule level, *Proc Natl Acad Sci USA, 106*, 109–114 (2009).
125. H. Lee, N.F. Scherer, P.B. Messersmith, Single-molecule mechanics of mussel adhesion, *Proc Natl Acad Sci USA, 103*, 12999–13003 (2006).

126. J. Hu, Y. Zhang, B. Li, H.B. Gao, U. Hartmann, M.Q. Li, Nanomanipulation of single DNA molecules and its applications, *Surf Interface Anal, 36*, 124–126 (2004).
127. J.W. Efcavitch, J.F. Thompson, Single-molecule DNA analysis, *Annu Rev Anal Chem, 3*, 109–128 (2010).
128. H.G. Hansma, K. Kasuya, E. Oroudjev, Atomic force microscopy imaging and pulling of nucleic acids, *Curr Opin Struct Biol*, 380–385 (2004).
129. F. Ritort, Single-molecule experiments in biological physics: Methods and applications, *J Phys Condens Matter, 18*, R531–R583 (2006).
130. D. Anselmetti, J. Fritz, B. Smith, Fernandez-Busquets, Single molecule DNA biophysics with atomic force microscopy, *Single Mol, 1*, 53–58 (2000).
131. A.-S. Duwez, S. Cuenot, C. Jérôme, S. Gabriel, R. Jérôme, S. Rapino, F. Zerbetto, Mechanochemistry: Targeted delivery of single molecules, *Nat Nanotech, 1*, 122–125 (2006).
132. S.B. Smith, L. Finzi, C. Bustamante, Direct mechanical measurements of the elasticity of single DNA molecules by using magnetic beads, *Science, 258*, 1122–1126 (1992).
133. S.B. Smith, Y. Cui, C. Bustamante, Overstretching B-DNA: The elastic response of individual double-stranded and single-stranded DNA molecules, *Science, 271*, 795–799 (1996).
134. T. Morii, R. Mizuno, H. Haruta, T. Okada, An AFM study of the elasticity of DNA molecules, *Thin Solid Films, 464–465*, 456–458 (2004).
135. C. Ke, M. Humeniuk, H. S-Gracz, P.E. Marszalek, Direct measurements of base stacking interactions in DNA by single-molecule atomic-force spectroscopy, *Phys Rev Lett, 99*, 018302 (2007).
136. G.U. Lee, L.A. Chrisey, R.J. Colton, Direct measurement of the forces between complementary strands of DNA, *Science, 266*, 771–773 (1994).
137. R. Krautbauer, M. Rief, H.E. Gaub, Unzipping DNA oligomers, *Nano Lett, 3*, 493–496 (2003).
138. B.D. Sattin, A.E. Pelling, M.C. Goh, DNA base pair resolution by single molecule force spectroscopy, *Nucleic Acid Res, 32*, 4876–4883 (2004).
139. P. Vettiger, M. Despont, U. Drechsler, U. Durig, W. Haberle, M.I. Lutwyche, H.E. Rothuizen, R. Stutz, R. Widmer, G.K. Binnig, The "millipede"-more than one thousand tips for future AFM data storage, *IBM J Res Devel, 44*, 323–340 (2000).
140. R. Afrin, U.S. Zohora, H. Uehara, T. Watanabe-Nakayama, A. Ikai, Atomic force microscopy for cellular level manipulation: Imaging intracellular structures and DNA delivery through a membrane hole, *J Mol Recognit, 22*, 363–372 (2009).
141. S.W. Han, C. Nakamura, I. Obataya, N. Nakamura, J. Miyake, A molecular delivery system by using AFM and nanoneedle, *Biosen Bioelectron, 20*, 2120–2125 (2005).
142. J.H. Lu, Nanomanipulation of extended single-DNA molecules on modified mica surfaces using the atomic force microscopy, *Colloids Surf B Biointerfaces, 39*, 177–180 (2004).
143. J. Hu, Y. Zhang, H. Gao, M. Li, U. Hartmann, Artificial DNA patterns by mechanical nanomanipulation, *Nano Lett, 2*, 55–57 (2002).
144. Z. Liu, Z. Li, G. Wei, Y. Song, L. Wang, L. Sun, Manipulation, dissection, and lithography using modified tapping mode atomic force microscope, *Microsc Res Tech, 69*, 998–1004 (2006).
145. J.-h. Lü, H.-k. Li, H.-j. An, G.-h. Wang, Y. Wang, M.-q. Li, Y. Zhang, J. Hu, Positioning isolation and biochemical analysis of single DNA molecules based on nanomanipulation and single-molecule pcr, *J Am Chem Soc, 126*, 11136–11137 (2004).
146. A. Bensimon, A. Simon, A. Chiffaudel, V. Croquette, F. Heslot, D. Bensimon, Alignment and sensitive detection of DNA by a moving interface, *Science, 265*, 2096–2098 (1994).
147. K. Yamanaka, M. Saito, M. Shichiri, S. Sugiyama, Y. Takamura, G. Hashiguchi, E. Tamiya, AFM picking-up manipulation of the metaphase chromosome fragment by using the tweezers-type probe, *Ultramicroscopy, 108*, 847–854 (2008).
148. S. Thalhammer, R.W. Stark, S. Muller, J. Wienberg, W.M. Heckl, The atomic force microscope as a new microdissecting tool for the generation of genetic probes, *J Struct Biol, 119*, 232–237 (1997).

149. R.W. Stark, F.J. Rubio-Sierra, S. halhammer, W.M. Heckl, Combined nanomanipulation by atomic force microscopy and uv-laser ablation for chromosomal dissection, *Eur Biophys J, 32*, 33–39 (2003).
150. X.M. Xu, A. Ikai, Retrieval and amplification of single-copy genomic DNA from a nanometer region of chromosomes: A new and potential application of atomic force microscopy in genomic research, *Biochem Biophys Res Commun, 248*, 744–748 (1998).
151. N. Liu, B. Peng, Y. Lin, Z. Su, Z. Niu, Pulling genetic RNA out of tobacco mosaic virus using single-molecule force spectroscopy, *J Am Chem Soc, 132*, 11036–11038 (2010).
152. T. Osada, H. Uehara, H. Kim, A. Ikai, Mrna analysis of single living cells, *J Nanobiotechnology, 1*, 2 (2003).
153. S. Marsden, M. Nardelli, P. Linder, J.E. McCarthy, Unwinding single RNA molecules using helicases involved in eukaryotic translation initiation, *J Mol Biol, 361*, 327–335 (2006).
154. Z. Liu, Y. Jiao, Y. Wang, C. Zhou, Z. Zhang, Polysaccharides-based nanoparticles as drug delivery systems, *Adv Drug Deliv Rev, 60*, 1650–1662 (2008).
155. T.E. Fisher, P.E. Marszalek, J.M. Fernandez, Stretching single molecules into novel conformations using the atomic force microscope, *Nat Struct Biol, 7*, 719–724 (2000).
156. N.I. Abu-Lail, T.A. Camesano, Polysaccharide properties probed with atomic force microscopy, *J Microscopy, 212*, 217–237 (2003).
157. M.A.K. Williams, A. Marshall, R.G. Haverkamp, K.I. Draget, Stretching single polysaccharide molecules using AFM: A potential method for the investigation of the intermolecular uronate distribution of alginate? *Food Hydrocolloids, 22*, 18–23 (2008).
158. H. Li, M. Rief, F. Oesterhelt, H.E. Gaub, Single-molecule force spectroscopy on xanthan by AFM, *Adv Mater, 3*, 316–319 (1998).
159. H. Li, M. Rief, F. Oesterhelt, H.E. Gaub, X. Zhang, J. Shen, Single-molecule force spectroscopy on polysaccharides by AFM -nanomechanical fingerprint of α-(1,4)-linked polysaccharides, *Chem Phys Lett, 305*, 197–201 (1999).
160. M. Rief, F. Oesterhelt, B. Heymann, H.E. Gaub, Single molecule force spectroscopy on polysaccharides by atomic force microscopy, *Science, 275*, 1295–1297 (1997).
161. P.E. Marszalek, A.F. Oberhauser, Y.-P. Pang, J.M. Fernandez, Polysaccharide elasticity governed by chair±boat transitions of the glucopyranose ring, *Nature, 396*, 661–664 (1998).
162. P.E. Marszalek, H. Li, J.M. Fernandez, Fingerprinting polysaccharides with singlemolecule atomic force microscopy, *Nat Biotech, 19*, 258–262 (2001).
163. G. Francius, S. Lebeer, D. Alsteens, L. Wildling, H.J. Gruber, P. Hols, S.D. Keersmaecker, J. Vanderleyden, Y.F. Dufrene, Detection, localization and conformational analysis of single polysaccharide molecules on live bacteria, *ACS Nano, 2*, 1921–1929 (2008).
164. D. Alsteens, V. Dupres, K.M. Evoy, L. Wildling, H.J. Gruber, Y.F. Dufrene, Structure, cell wall elasticity and polysaccharide properties of living yeast cells, as probed by AFM, *Nanotech, 19*, 384005 (2008).
165. S. Qamar, P.M. Williams, S.M. Lindsay, Can an atomic force microscope sequence DNA using a nanopore? *Biophys J, 94*, 1233–1240 (2008).

Chapter 5
Nanografting: A Method for Bottom-up Fabrication of Designed Nanostructures

Tian Tian, Zorabel M. LeJeune, Wilson K. Serem, Jing-Jiang Yu, and Jayne C. Garno

Abstract Nanografting is a scanning probe-based technique which takes advantage of the localized tip-surface contact to rapidly and reproducibly inscribe arrays of nanopatterns of thiol self-assembled monolayers (SAMs) and other nanomaterials with nanometer-scale resolution. Scanning probe-based approaches for lithography such as nanografting with self-assembled monolayers extend beyond simple fabrication of nanostructures to enable nanoscale control of the surface composition and chemical reactivity from the bottom-up. Commercial scanning probe instruments typically provide software to control the length, direction, speed and applied force of the scanning motion of a tip, analogous to a pen-plotter. Nanografting is accomplished by force-induced displacement of molecules of a matrix SAM, followed immediately by the surface self-assembly of *n*-alkanethiol *ink* molecules from solution. Desired surface chemistries can be patterned by choosing SAMs of different lengths and terminal groups. By combining nanografting and designed spatial selectivity of *n*-alkanethiols, in situ studies provide new capabilities for nanoscale surface reactions with proteins, nanoparticles or chemical assembly. Methods to precisely arrange molecules on surfaces will contribute to development of molecular device architectures for future nanotechnologies.

Keywords Nanografting · Scanning probe lithography · Nanolithography · Self-assembled monolayer · Atomic force microscopy · Nanopatterning · Alkanethiols · Protein patterning · DNA · Porphyrins · Nanoparticles · Nanostructures

Abbreviations

AFM	Atomic force microscope
APDES	Aminopropyldiethoxysilane
bps	Base pairs
BSA	Bovine serum albumin
C8DMS	Octyldimethylmonochlorosilane

J.C. Garno (✉)
Department of Chemistry, Louisiana State University, Baton Rouge, LA 70803, USA
e-mail: jgarno@lsu.edu

A.A. Tseng (ed.), *Tip-Based Nanofabrication*, DOI 10.1007/978-1-4419-9899-6_5,

C10	Decanethiol
C12	Dodecanethiol
C18	Octadecanethiol
CAM	Computer-assisted manufacturing
DNA	Deoxyribonucleic acid
DPN	Dip-pen nanolithography
DPP	5,10-diphenyl-15,20-di-pyridin-4-yl-porphyrin
dsDNA	Double-stranded DNA
EDC	1-ethyl-3-(3-dimethylaminopropyl) carbodiimide hydrochloride
EG	Ethylene glycol
GIXD	Grazing incidence X-ray diffraction
IgG	Immunoglobulin G
MBP	Maltose binding protein
MCH	6-mercaptohexan-1-ol
16-MHA	16-mercaptohexadecanoic acid
MHP	*n*-(6-mercapto hexyl) pyridinium bromide
MPA	3-mercaptopropionic acid
11-MUA	11-mercaptoundecanoic acid
11-MUD	11-mercaptoundecanol
NEXAFS	Near-edge X-ray absorption fine structure spectroscopy
NHS	N-hydroxysuccinimide
NPRW	Nanopen reader and writer
ODT	Octadecanethiol
OTS	Octadecyltrichlorosilane ($CH_3(CH_2)_{17}SiCl_3$)
SAMs	Self-assembled monolayers
SpA	Staphylococcal protein A
SPL	Scanning probe lithography
ssDNA	Single-stranded DNA

5.1 Introduction

Scanning probe lithography (SPL) enables bottom-up fabrication of nanostructures on surfaces for producing features with nanoscale dimensions. Methods using the probe of an atomic force microscope (AFM) have been used to fabricate sophisticated architectures at the molecular level with high spatial precision. A number of AFM-based approaches for SPL have been developed such as nanoshaving [1–5], nanografting [6–9], dip-pen nanolithography (DPN) [10, 11], NanoPen Reader and Writer (NPRW) [12–14], catalytic probe lithography [15–17], and bias-induced nanolithography [18, 19]. This chapter will focus specifically on the capabilities of nanografting for inscribing patterns of diverse composition from the bottom-up, to produce complicated surface designs with well-defined chemistries. Nanografting provides a versatile tool for generating nanostructures of organic and biological molecules, as well as nanoparticles. Protocols of nanografting are accomplished in liquid media, providing a mechanism for introducing new reagents for successive in situ steps for 3-D fabrication of complex nanostructures.

Nanografting was first introduced in 1997 by Xu, et al. and is accomplished by applying mechanical force to an AFM probe to generate nanostructures within a matrix film [8]. The molecules to be patterned are dissolved in the imaging media, and the substrates are precoated with a protective layer to prevent nonspecific adsorption of molecules throughout areas of the surface. When the tip is operated in liquid media under low force (less than 1 nN), high resolution characterizations of surfaces can be acquired in situ. When the force applied to the probe is increased to a certain displacement threshold the tip becomes a tool for surface fabrication. The exquisite resolution achieved with nanografting is mainly attributable to liquid imaging. When AFM experiments are conducted in liquid media, very low force can be used to accomplish imaging or nanofabrication. The geometry of the apex of the probe is preserved by operating at low forces, because liquid media serves to minimize the strong capillary forces of attraction that cause adhesion between the tip and sample [20, 21].

5.1.1 General Procedure for Nanografting

The basic steps for nanografting are presented in Fig. 5.1. In the first step, the surface of a self-assembled monolayer (SAM) prepared on a Au(111) substrate is imaged using low force in liquid media that contains the molecule or nanomaterial to be patterned. When the tip is operated at low force the surface is not damaged or altered by the scanning probe (Fig. 5.1a). A suitable flat area can be selected for inscribing

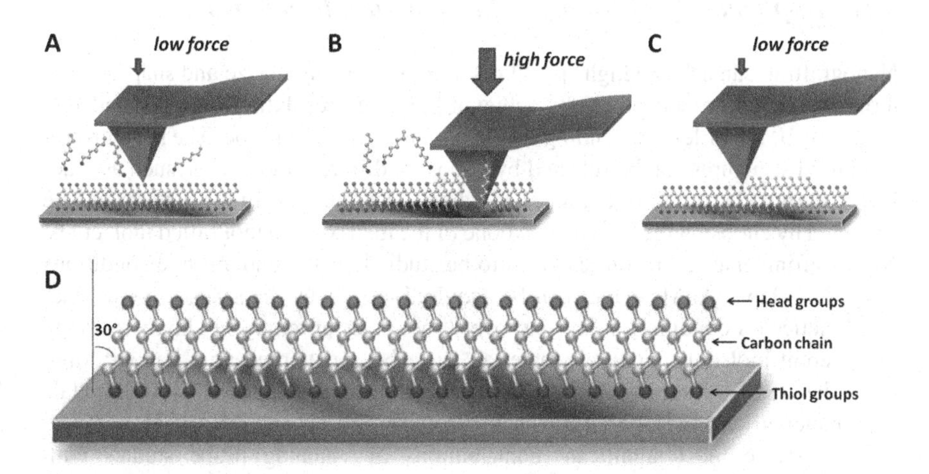

Fig. 5.1 Steps for producing patterns of *n*-alkanethiols with nanografting by changing the mechanical force applied to the AFM probe. The process is accomplished under liquid imaging media containing the molecules to be patterned. (**a**) Characterization is accomplished when the tip is operated at low force; (**b**) patterns are nanografted when the force is increased to a certain displacement threshold; (**c**) returning to low force, the patterns are characterized in situ. (**d**) Model of an *n*-alkanethiol self-assembled monolayer

patterns that has few defects or contaminants. Next, the tip is raster scanned across the surface using higher force to sweep away selected regions of the matrix SAM. During the fabrication step (Fig. 5.1b), fresh molecules from solution bind to the exposed areas of the substrate immediately following the pathway of the scanning probe to produce nanopatterns. Finally, the pattern that was grafted can be characterized in situ by returning to a low force for nondestructive imaging (Fig. 5.1c). Patterning and imaging are accomplished in situ with the same AFM tip, within a few minutes or less. The entire process can be automated to reproducibly write multiple patterns [22, 23].

A key requirement for nanografting is to determine the necessary amount of force for cleanly removing local areas of the matrix monolayer without damaging the tip. To find the appropriate force, one can monitor surface changes in situ while successively increasing the load applied to the tip. As the force is gradually increased at small increments, images will clearly show changes in surface morphology at a certain threshold. The optimum force must be derived for each experiment for several reasons. At the nanoscale, the actual geometry of tips is never identical and thus the sharpness will vary from probe to probe. Also, different amounts of force are necessary for matrix layers of different thicknesses or compositions. The requisite force needed for imaging in various liquid media will change according to dissolution parameters, for example the forces required for nanografting in aqueous media are not the same as for ethanolic media. For each system, the amount of force to be applied for fabrication must be determined experimentally.

5.1.2 Applicability of Nanografting for In Situ Studies

Nanografting can achieve high spatial resolution. The length, size and shape of patterns can be controlled precisely, achieving an edge resolution of 1 nm and line widths of 10 nm or less, depending on the dimensions of the probe. The head groups of grafted structures can be selected by choosing different molecules, such as aldehydes, carboxylates, thiols, amines, and others. The thickness of the patterns can be designed by choosing the carbon backbone of the matrix and nanografted molecules. Nanografting enables in situ reactions to be studied locally under dilute conditions [24]. Time-lapse AFM images can be acquired at selected intervals to view reaction kinetics for conditions that occur over time scales of minutes to hours. A range of different molecules and nanomaterials have been patterned with nanografting, examples will be described in this chapter for *n*-alkanethiol SAMs [14, 25], metals [26], nanoparticles [27], porphyrins [28], proteins [29–32] and DNA [33].

Among the most significant contributions of scanning probe studies with nanografting are the possibilities for studying step-wise surface reactions in real time with a molecular-level view. Imaging in liquid media provides a means for exchanging liquids to introduce new reagents in successive steps to build nanostructures from the bottom-up. To date, the primary examples that have been reported demonstrate nanografted patterns of *n*-alkanethiol SAMs, often as a foundation for attaching other molecules and nanomaterials. Further chemistries for nanografting experiments are likely to be extended to other types of surface binding motifs, such as phosphonic acids on metal substrates [34]; siloxane binding, pyridyl-[28] or thiol-[35] functionalized porphyrins, thiolated proteins [36, 37], thiolated DNA [33] or peptides and other types of surface linkers.

5.2 Patterning *n*-Alkanethiol Self-Assembled Monolayers (SAMs by Nanografting)

As a starting point, SAMs of *n*-alkanethiols prepared on gold substrates provide a model system for nanografting experiments. Thiol end groups furnish a functional handle for surface attachment, mediated by sulfur-gold chemisorption. The self-assembly process and surface structures of *n*-alkanethiols on Au(111) have been previously described [38, 39]. The carbon backbones of the molecules consist of tilted alkane chains (Fig. 5.1d), the lengths of which can be designed to define the thickness of the matrix areas and nanografted patterns. For *n*-alkanethiol SAMs, chain lengths ranging from 2 to 37 carbons have been nanografted successfully. The head groups of *n*-alkanethiols provide a way to attach other molecules and nanomaterials with spatial selectivity; for example, experiments can be designed to define patterned sites for specific adsorption of proteins, nanoparticles or DNA, within a matrix monolayer that resists binding of molecules or nanomaterials.

Nanopatterns of octadecanethiol (18 carbon backbone or C18) were nanografted side-by-side within a matrix SAM of decanethiol (10-carbon backbone or C10) as shown in Fig. 5.2a [8]. The square patterns measured 0.88 nm taller than the matrix.

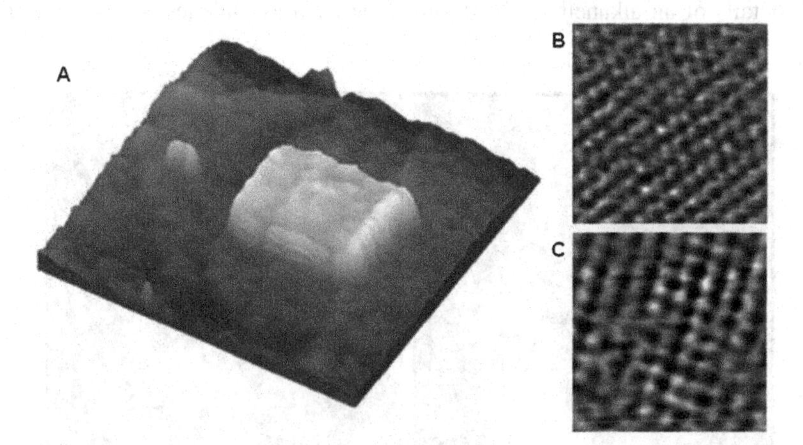

Fig. 5.2 Patterns of *n*-octadecanethiol were nanografted within a matrix monolayer of decanethiol. (**a**) AFM topography view (130 × 130 nm^2); (**b**) zoom-in view of the pattern surface (5 × 5 nm^2); (**c**) Zoom view from an area of the C10 matrix (5 × 5 nm^2). (Reprinted with permission from Ref. [8]. Copyright © American Chemical Society)

The dimensions of the smaller feature are 3 nm $\times$ 5 nm, in which approximately 60 thiol molecules were grafted. The size of the larger nanopattern is 50 $\times$ 50 nm^2. Zoom-in views of both the nanografted pattern of C18 and the C10 matrix are shown by in situ AFM topography images in Fig. 5.2b, c, respectively. The molecularly resolved images show that molecules within the nanopatterns display a periodic $(\sqrt{3} \times \sqrt{3})$ R30° lattice, thus the packing arrangement of thiols is preserved for alkanethiol nanostructures produced by nanografting.

Nanografted structures can be erased and rewritten in situ by exchanging the imaging media with different molecular adsorbates for patterning. Results for writing two parallel line patterns of octadecanethiol within a decanethiol matrix with a distance of 20 nm between patterns were shown by Xu and others [9]. One of the lines was erased by replacing the liquid imaging media with a solution of decanethiol and scanning at high force over one of the C18 patterns to replace the previous nanostructure with C10 molecules. After the line pattern was "erased" the imaging media was exchanged again to introduce a fresh solution of C18SH molecules to graft a line pattern spaced 65 nm from the previous pattern. Accomplishing this experiment required a scanning probe microscope with high stability, however this clearly demonstrates the flexibility for introducing and exchanging reagent solutions for multiple synthetic steps when imaging with AFM in liquids.

Different shapes and molecular components can be patterned by nanografting. Several letter patterns that spell the acronym "AFM" are shown in Fig. 5.3 that are terminated with carboxyl head groups. The line widths of the letter patterns are less than 10 nm, indicating that the very sharp AFM probe was not damaged by the physical process of scanning with the tip under high force. Although the AFM images of the patterns were captured after the writing process, we can still resolve the ultra-fine distinctive features of the matrix monolayer of decanethiol, resolving the characteristic details of an alkanethiol SAM landscape such as pinholes, scars, molecular

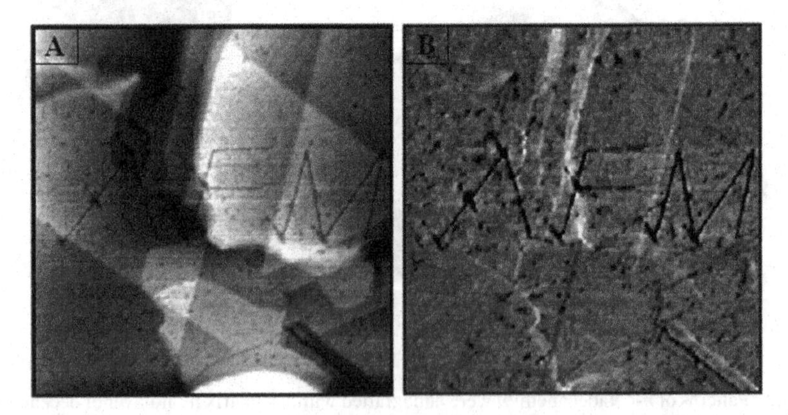

Fig. 5.3 Nanografted letters of 3-mercaptopropionic acid written within a decanethiol matrix SAM. (**a**) Topographic image (600 $\times$ 600 nm^2); (**b**) concurrent lateral force image of the same area

island vacancies [40] and overlapping gold terrace steps. The patterns are composed of 3-mercaptopropionic acid written within a decanethiol matrix. The difference in terminal chemistry is clearly distinguishable in the simultaneously acquired lateral force AFM image of Fig. 5.3b. Lateral force images do not show changes in height, instead the image contrast reveals nanoscopic differences in frictional and adhesive forces between the tip and surface. In this example, the tip-surface interactions are markedly different for the dark areas of the nanografted letters which are terminated with thiol head groups, as compared to the brighter areas of the surrounding methyl-terminated matrix SAM.

The simplicity of SAM preparation is another benefit of nanografting protocols. A matrix monolayer can be prepared by simply immersing a clean substrate into a dilute solution of *n*-alkanethiol in ethanol or sec-butanol for one or more hours. After a SAM film is formed on the metal substrate, the sample can be stored for several weeks in a solution of clean solvent, and often can be recycled and used for several experiments. Nanografted patterns can be engineered to incorporate diverse head group chemistries, such as methyl, alcohol, glycol, aldehyde, amide and carboxylate. Table 5.1 lists examples of thiol self-assembled monolayers which have been patterned using nanografting. Methyl-terminated SAMs of decanethiol or octadecanethiol have been commonly used as matrix monolayers for nanografting. Either ethanol or 2-butanol are most frequently used as solvents for liquid imaging. Patterns of diverse shapes, such as squares, rectangles and rings have been reported ranging up to 500 nm in size, with the dimensions of the smallest pattern measuring 3 nm × 5 nm.

5.2.1 Automated Nanografting

Beyond simple patterns of lines or rectangles, nanografting can be used to fabricate complicated designs with modern computer automation. The William Blake quotation "What is now proved was once only imagined" was nanografted with mercaptohexadecanoic acid by Cruchon-Dupeyrat, et al., using computer-assisted manufacturing (CAM) software [23]. The entire quotation was written in less than 20 s, inscribed within a $1.85 \times 0.9\ \mu m^2$ area. Arrays of circles, squares, lines and even mouse ear designs were produced by automated nanografting of different functionalized alkanethiols by Ngunjiri and others [22]. A sophisticated example was demonstrated by Maozi Liu, et al. for nanografting the design of the University of California at Davis' seal with a 10 nm line resolution using an aldehyde terminated alkanethiol within a decanethiol SAM [25]. The design was patterned inside an $8 \times 8\ \mu m^2$ area and was completed in 10 min.

The speed and ease of nanografting for AFM experiments has been greatly improved by advances in software for commercial instruments. Louisiana State University implemented nanografting experiments in physical chemistry laboratories starting in 2005 to teach and showcase the concepts of chemistry and nanoscience to undergraduate students [41]. Nanografted patterns can be produced

Table 5.1 Examples of thiol SAMs that have been successfully nanografted

Nanografted molecule	Pattern dimensions	Matrix film	Liquid media	References
1-hexanethiol	5.2 nm × 5.2 nm	Thiolated biotin SAMs	Ethanol	[32]
1-octadecanethiol	3 nm × 5 nm 50 nm × 60 nm	1-decanethiol	2-butanol	[8]
1-decanedithiol	100 nm × 100 nm	1-decanethiol	2-butanol	[61]
Dodecanethiol	300 nm × 300 nm	1,9-nonanedithiol	Ethanol	[86]
1-octadecanethiol	20 nm × 60 nm	1-decanethiol or 1-octadecanethiol	2-butanol	[9]
Docosanethiol	25 nm × 60 nm			
2-mercaptoethanol	75 nm × 100 nm			
16-mercapto-hexadecanoic acid	70 nm × 300 nm			
3-mercapto-1-propanoic acid	400 nm × 400 nm	$C_{11}(EG)_6$	Water	[73]
11-mercapto-1-undecanal	50 nm × 50 nm 100 nm × 100 nm	1-octadecanethiol	Decahydro-naphthalene	[62]
11-mercapto-undecanoic acid	Rings with diameter of 100 nm	1-octadecanethiol	Ethanol	[29]
1,8-octanedithiol	500 nm × 500 nm	Hexanethiol	Ethanol	[86]
6-mercaptohexan-1-ol	400 nm × 400 nm	$C_{11}(EG)_6$	Water	[73]
Biphenyl 4,4'-dithiol	100 nm × 100 nm	1-decanethiol	2-butanol	[61]
Mixed-*n*-alkanethiols	200 nm × 200 nm	1-decanethiol: 1-octadecanethiol =10:1	Ethanol or 2-butanol	[53]
10:1 ODT:decanethiol $CF_3(CF_2)_9(CH_2)_2SH$	200 nm × 200 nm 15 nm × 15 nm 300 nm × 300 nm	Hexanethiol Dodecanol Mixed SAM matrices	Ethanol, 2-butanol or poly-α-olefin oil	[53]
1-octadecanethiol	70 nm × 50 nm 175 nm × 225 nm 20, 50, 100, 200 nm	Decanethiol Mixed SAMs	Ethanol, 2-butanol or hexadecane	[54]

within a few minutes and thus are an excellent venue for providing hands-on training for students. At present, scanning probe-based lithography is primarily used for laboratory research rather than as a tool for industry. Knowledge and experience in modern methods of surface measurements and analysis will be pivotal to the eventual transfer of the technology gained with academic nanoscience research to benefit industry. The latest advances in automation of scanning probe instruments enable new possibilities for educational modules for engaging students with modern and compelling course activities, such as with nanografting studies.

5.2.2 Evaluating the Tip Geometry with Nanografting

For both imaging and nanofabrication with an AFM probe, the shape of the apex of the tip is critical for high resolution. Nanografting provides a way to evaluate the shape of an AFM tip, to help discern if images show artifacts or represent the true shape of surface structures [42]. Line patterns of alkanethiol SAMs are first fabricated using nanografting with a single scan, and then imaged using the same tip. The tip size and tip-surface contact area can be derived from cursor profiles in AFM topography views. The shape of the apex of the tip can be reconstructed by imaging small surface features of nanografted SAMs with known dimensions. When the tip is engaged for a sweeping a single line pattern, the width of the trench or pattern provides a reliable estimate of the tip-surface contact area. Tips with multiple asperities produce multiple nanopatterns. This approach is especially helpful for identifying tips with multiple asperities that are difficult to characterize by other techniques.

5.2.3 Nanografted Patterns of n-Alkanethiols Furnish a Molecular Ruler

Since the dimensions of methyl-terminated *n*-alkanethiols have been well-established, the height and orientation of other molecules can be evaluated by nanografting experiments, by referencing the thickness of *n*-alkanethiols as an in situ molecular ruler. Methyl-terminated *n*-alkanethiols can be prepared reproducibly with predictable, well-defined surface structures, thus nanografted patterns furnish a reliable height reference for nanoscale measurements of film thickness. Self–assembled monolayers of *n*-alkanethiols spontaneously form hexagonally-packed crystalline layers upon adsorption to metal surfaces, with an intermolecular spacing of ~0.5 nm [43]. The well-ordered packing of *n*-alkanethiol SAMs results from a strong affinity to the substrate through chemisorptive binding to produce a commensurate structure, and also from intermolecular chain-chain interactions of Van der Waals forces between the carbon backbones. Methyl-terminated *n*-alkanethiols form SAMs with a single thiol end group chemisorbed to Au(111) oriented in an upright configuration, with all-trans carbon chains. Studies conducted using IR, near-edge X-ray absorption fine structure (NEXAFS) spectroscopy, and grazing incidence X-ray diffraction (GIXD) indicate that the alkyl chains of SAMs are tilted $\sim 30^\circ$ with respect to surface normal [44–47]. The consistency for preparing reproducible molecular structures of *n*-alkanethiols provides predictable dimensions as a means to study structures of other patterned molecules using side-by-side local measurements of height differences with AFM-based nanografting protocols [12, 48, 49].

By labeling the DNA 3′ end with a fluorophore and immobilizing it onto a gold surface through thiol modification of the 5′ end, a pH-driven DNA nanoswitch can be reversibly actuated. By cycling the solution pH between 4.5 and 9, a conformational change is produced between a four-stranded and a double-stranded DNA structure which either elongates or shortens the separation distance between the 5′ and 3′ ends of the DNA. The nanoscale motion of the DNA produces mechanical work to lift up and bring down the fluorophore from the gold surface by at least 2.5 nm and transduces this motion into an optical "on-and-off" nanoswitch. Nanografting was used to measure the thickness of the monolayers of thiolated "motor" DNA under changing pH conditions by Dongsheng Liu, et al., [50]. Before nanografting, a DNA SAM prepared on template-stripped gold surface was first imaged under low force (0.2–0.5 nN) in phosphate buffered saline (pH 4.5) containing 1 mM of 2-mercaptoethanol. The area for nanografting was repeatedly scanned at 4–5 Hz under higher forces ($\sim$30 nN) to scratch away the DNA SAM, creating a freshly exposed gold surface that was immediately grafted with a SAM of 2-mercaptoethanol. After nanografting, a wider scan area was characterized under low force. Changes in the thickness of the DNA film measured at pH 4.5 and 9 were attributed to differences in the electrostatic interactions between the tip and the DNA layer.

5.2.4 Evaluating Properties Such as Friction, Elastic Compliance or Conductivity of Nanografted Patterns

Friction mapping can be accomplished with AFM to provide useful information about the composition and chemical properties of a surface with nanoscale sensitivity. A systematic study of differences in molecular friction was accomplished in situ for nanografted patterns of different ω-functionalized *n*-alkanethiols by Joost te Riet et al., [51, 52]. Trace and retrace lateral force images were subtracted to reveal the net frictional forces to obtain quantitative frictional force measurements at the nanoscale. Images of nanografted patterns with fluorocarbon-, hydroxyl-, thiol-, amine- and acid- terminated head groups were obtained in 2-butanol under common conditions of load force and scan speed. The same cantilever was used for nanografting patterns and acquiring in situ images in liquid media. In each case, they observed that the friction of the nanografted patches was lower than that of the surrounding matrix SAM. However, nanografted patterns with functional head groups showed statistically higher friction values than nanografted patterns with methyl groups. These observations were attributed to differences in topographical roughness of the nanografted patches, the amount of disorder and defects within the patterns, as well as surface composition.

Changes in molecular-level packing, molecule chain lengths, domain boundaries, and surface chemical functionalities in nanografted SAM nanopatterns can be sensitively characterized using force modulation imaging [53]. Size-dependent changes in elasticity were detected for test platforms of nanografted SAM patterns by Price, et al., [54]. Surface patterns of octadecanethiol (ODT) of designed sizes and shapes were nanografted into *n*-alkanethiol SAMs for studies of the local mechanical properties using force modulation imaging. Certain surface features such as the edges of the domains and nanostructures or desired chemical functionalities can be selectively enhanced in the amplitude images when the driving frequency of sample modulation is tuned to the resonance frequencies of the tip-surface contact [53]. By means of tuning the driving frequency of sample modulation to certain frequencies, the resonances at the tip-surface contact are activated to sensitively reveal characteristic contrast for surface changes in molecular-level packing, molecule chain lengths, domain boundaries, and surface chemical functionalities of SAM nanopatterns. These studies demonstrate that the resonance frequency of the tip surface contact vary according to dimensions of the nanostructures. Frequency spectra of the tip surface contacts were acquired for nanografted ODT structures, from which Young's modulus was calculated using continuum mechanics models.

An approach to study metal-molecule-metal junctions based on combining approaches for nanografting and conductive probe AFM was demonstrated by Scaini, et al., [55]. Patterns of alkanethiol molecules were nanografted within a SAM of alkanethiol molecules of different chain lengths for local measurements of charge transport at the molecular level. The approach enables relative determination of the differential resistance between two molecular layers in ambient conditions; however absolute transport measurements also depend on the nature of the AFM tip-molecule contact. The tunneling decay constants of alkanethiols were measured as a function of chain lengths for octanethiol, nonanethiol and decanethiol nanopatterns relative to a matrix SAM of octadecanethiol/Au(111).

5.3 Spatially Confined Self-Assembly Mechanism of Nanografting

Both the assembly mechanism and kinetics of certain surface reactions can be sterically changed by spatial confinement with nanografting. Nanografted patterns of *n*-alkanethiols exhibit higher coverage and two-dimensional crystallinity than the matrix SAMs [56]. During the process of nanografting, thiolated molecules self-assemble within a spatially confined environment. A transient nanoscopic area of the surface is exposed by the scanning probe, which is confined by the surrounding matrix and the probe. During the nanografting process, thiol molecules present in the solution rapidly assemble onto the exposed nanometer-size area of gold substrate that is confined by the scanning tip and surrounding matrix SAM. Spatial confinement is considered to alter the pathway for the self-assembly process causing the initially adsorbed thiols to adopt a standing-up configuration directly within a nano-sized environment. The mechanism for conventional solution self-assembly occurs through a two-step process when bare gold substrates are immersed in thiol solutions, because the assembly of thiols takes place in unconstrained conditions. Initially a "lying-down" phase is spontaneously formed which subsequently transitions over time by rearrangement to a standing-up orientation [38]. In contrast, with nanografting the "lying-down" configuration is not possible because the area of the surface exposed is smaller than the molecular length, therefore the molecules assemble directly into an upright or standing orientation [57]. Self-assembly within the constrained areas proceeds with a faster reaction rate because the time lapse for a phase transition from lying-down to an upright configuration is bypassed. Thus, the kinetics of SAMs formed with nanografting occur more rapidly than during natural growth on unconstrained surfaces. The spatially confined environment was found to reduce the amount of disorder present in the resulting nanografted patterns, to produce SAMs which exhibit fewer scars or defects [51, 56].

5.3.1 Studies with Binary Mixtures of SAMs

A nanoengineering approach to regulate the lateral heterogeneity of mixed self-assembled monolayers was reported using nanografting and self-assembly chemistry [51]. Formation of segregated domains in mixed SAMs results from the interplay between reaction kinetics and thermodynamics. Considerable effort has been directed to investigate the impact of either reacting agents or surface reaction conditions such as concentration, temperature, thiol species and molar ratio of mixed components for achieving control of the resulting local domain structures. For example, kinetics-driven products for mixed SAMs with a near molecular-level mixing were favored during coadsorption of thiol mixtures at high concentration with elevated temperature [58]. Thermodynamics-driven layers of large segregated domains were observed after long immersion in dilute solutions and/or when the adsorbate chain length and termini were sufficiently different [59]. Nanografting provides additional control of the reaction mechanism for thiol self-assembly on gold, and thus affects the local domain structures that are produced from solutions of mixed SAMs.

The heterogeneity of mixed solutions of SAMs can be regulated by changing the speed of nanografting [57]. This was demonstrated both theoretically [60] and experimentally [57]. Monte Carlo simulations of nanografting were found to reproduce experimental observations concerning the variation of SAM heterogeneity with the speed of an AFM tip. Simulations by Ryu, et al. demonstrated that the faster the AFM tip displaced adsorbed molecules in a monolayer, the monolayers formed behind the tip became more heterogeneous, according to the amount of space and time available for the formation of phase-segregated domains. By varying fabrication parameters of nanografting, the lateral heterogeneity can be adjusted to produce near molecular mixing or to form segregated domains ranging from several to tens of nanometers [51].

5.4 In Situ Studies of Polymerization Reactions via Nanografting

Beyond preparing monolayer patterns of ω-functionalized *n*-alkanethiols, multilayer nanostructures can also be generated by nanografting. Depending on the concentration of thiols in the imaging media, patterns with the thickness of a bilayer were shown to form spontaneously by nanografting SAMs of certain head group chemistries [12, 61]. This is mediated by self-polymerization of molecules which have reactive groups through coupling of headgroups. Under certain conditions of high concentration, the intermolecular interactions between molecules in solution predominate, to direct the vertical self-assembly of certain α, ω-alkanedithiols to produce bilayer patterns. For SAM patterns with methyl, hydroxyl, thiol, or carboxylic acid head groups, monolayer patterns were generated when nanografting in dilute ethanol or aqueous solutions. However, as the solution concentration was increased beyond a certain threshold, nanografted patterns were formed with thicknesses corresponding to a double layer for molecules with carboxylic acid head groups or with α, ω-alkanedithiols, as reported by Kelley, et al. [12]. Nanografted patterns with methyl or hydroxyl head groups were observed to exclusively form monolayer structures for a fairly wide range of concentrations that were tested.

Designed functional groups of *n*-alkanethiols were used to attach additional organic molecules to enable site-selective surface reactions for studies of polymerization reactions at the nanoscale [62]. In the first step, nanografting was used to produce 2D nanopatterns of methyl head groups in a matrix SAM with hydroxyl head groups. The nanopatterns were then used to further construct 3D nanostructures by successive steps of an in situ reaction with organosilanes. Jun-Fu Liu et al. demonstrated transfer of 2D nanopatterns to chemically distinct 3D nanostructures with different head groups. The scheme and results for pattern transfer are shown in Fig. 5.4. A nanografted rectangular frame of octadecanethiol was inscribed within a matrix SAM of mercaptoundecanol on a gold substrate. The pattern of a frame in Fig. 5.4b measured 0.7 ± 0.2 nm taller than the matrix monolayer, in agreement with the expected theoretical dimensions. After nanografting, the AFM liquid cell was rinsed three times with decahydronaphthalene to remove any residual thiols, then a solution of octadecyltrichlorosilane ($CH_3(CH_2)_{17}SiCl_3$ or OTS) was injected into the cell for several minutes. The trichlorosilanes from the liquid media reacted with the hydroxyl terminal groups of the surrounding matrix SAM of mercaptoundecanol to form a thicker layer. However, the frame patterns did not react with OTS since the nanografted pattern with methyl head groups provided an effective resist, as shown in Fig. 5.4c. After reaction with OTS the nanografted frame is shorter than

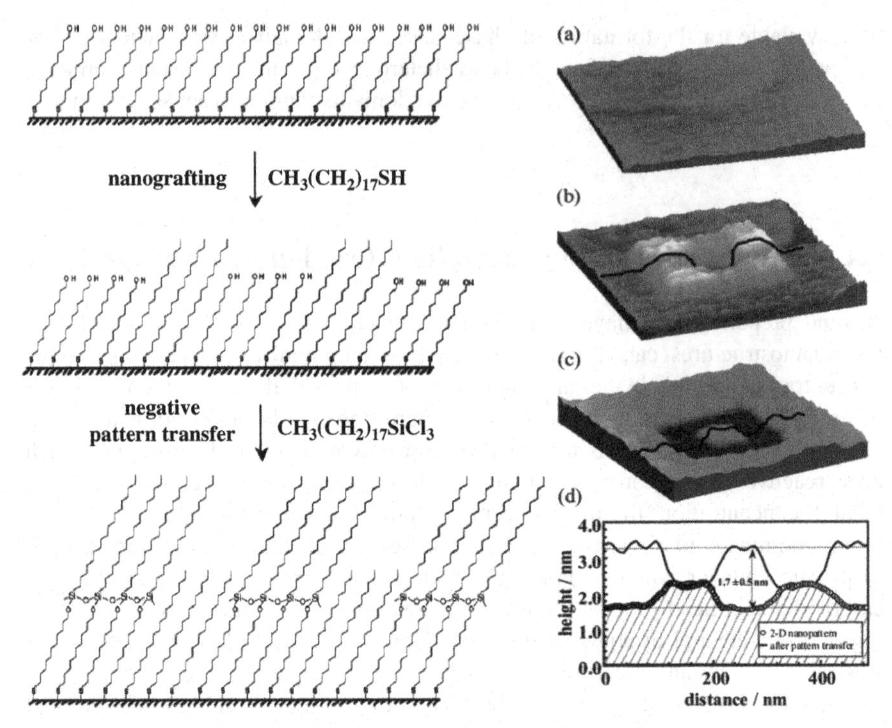

Fig. 5.4 Snapshots showing bottom-up assembly accomplished in situ with a polymerization reaction for attaching organosilanes to a hydroxyl-terminated SAM. (**a**) Initial view of a mercaptoundecanol monolayer formed on Au(111); (**b**) Nanografted frame of ODT; (**c**) Pattern is shorter than the matrix SAM after reaction with OTS; (**d**) representative cursor profile for lines in (**b**) and (**c**). (Reprinted with permission from Ref. [62]. Copyright © American Chemical Society)

the surrounding matrix film. The height changes at each step of the in situ reaction are shown with representative cursor profiles in Fig. 5.4d. The process was completed within a few minutes and the time duration for immersion in OTS was found to influence the height of siloxane structures.

Nanografting enables a critical first step for developing further protocols for designed surface reactions to construct hierarchical nanostructures with desired spacer lengths, composition and functionalities. The 2D patterns produced by nanografting provide a surface template for spatially directing the selective adsorption or binding of other molecules or nanomaterials in subsequent steps. Further examples will be presented in the next sections. The desired interfacial properties, such as lubricity, protein adhesion or resistance, and electron transfer, may be designed from the bottom-up by selection of various functional groups and designated architectures of the nanografted structures of metals, nanoparticles, protein or DNA.

5.5 Generating Patterns of Metals and Nanoparticles with Nanografting

Certain systems of metals and nanoparticles have been patterned successfully with AFM-based lithography. Nanopatterns of thiol-coated gold nanoparticles were prepared within a decanethiol SAM on Au(111) by scanning probe lithography [27]. To attach nanoparticles to gold surfaces via sulfur-gold chemisorption, surface-active gold nanoparticles were prepared with a shell of a mixed monolayer comprised of alkanethiol and alkanedithiol molecules. Local regions of a decanethiol SAM were shaved using an AFM tip under high force to expose the substrate in a solution containing nanoparticles. Unlike nanografting where surface assembly is immediate, the kinetics of larger nanomaterials such as gold nanoparticles were found to be slower and took place over longer time scales. Depending on the concentration, thiolated nanoparticles adsorbed onto the exposed areas uncovered by the AFM tip after several hours, and particles were not observed to bind to the surrounding matrix areas of the methyl-terminated decanethiol SAM. Gold nanoparticles attached to the gold substrate via sulfur-gold chemisorption. The outer shell of the nanoparticles was encapsulated with mixed thiol groups of hexanethiol and hexanedithiol molecules. Cursor measurements of the nanoparticles revealed sizes ranged from 3 to 5 nm in diameter, and patterns were formed with a single layer of nanoparticles. The slower adsorption of the nanoparticles on shaved areas of the substrate compared to nanografting of molecular patterns was attributable to differences in mobility and concentration.

5.5.1 Electroless Deposition of Metals on Nanografted SAM Patterns

Site specific reactions for electroless deposition of metals were accomplished using nanografting. Copper nanostructures formed selectively on carboxylic acid terminated SAM patterns that were nanografted within a hydroxyl-terminated resist monolayer, using electroless plating without a catalyst [26]. To accomplish in situ studies, the AFM cantilevers were coated with silane to prevent copper deposition on the probe. An example showing selective growth of copper nanostructures on nanografted patterns of 16-mercaptohexadecanoic acid (16-MHA) is displayed in Fig. 5.5. A computer script was designed to automate the nanografting process to generate patterns of different line densities within a matrix SAM of 11-mercaptoundecanol (11-MUD), which resists copper deposition. The parameters of the tip trajectory during nanografting can be used to define the thickness of copper according to the density of grafted molecules. Lower density of carboxylic acid groups resulted in differences along the gradients for deposition of copper. Changes in the surface density of 16-MHA were systematically varied by designing the probe trajectory to advance either at the edges or centers of the patterns. The difference in the molecular gradients of 16-MHA nanopatterns was evaluated by introducing

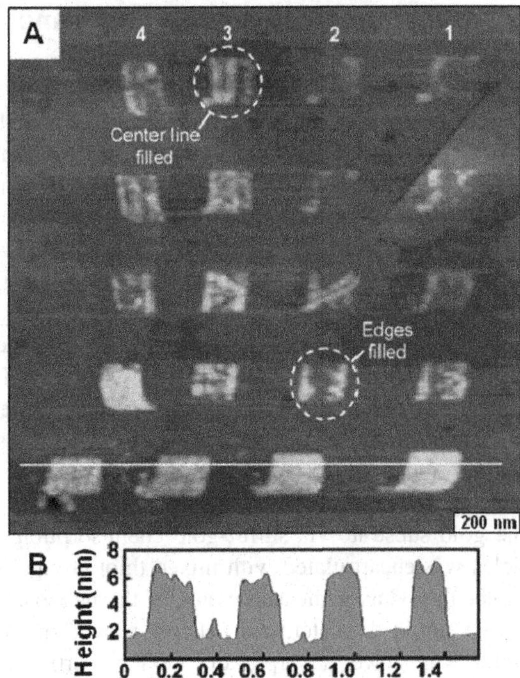

Fig. 5.5 Nanografted patterns of carboxylic acid terminated SAMs were generated with different densities for electroless deposition of copper. (**a**) View of copper nanopatterns grown on nanografted patterns written with different line densities; (**b**) cursor plot for copper structures of the bottom row. (Reprinted with permission from Ref. [26]. Copyright © American Chemical Society)

a copper solution. Metal ions (Cu^{2+}) deposited selectively in the reduced form as Cu^0 via an autocatalytic reaction on regions patterned with 16-MHA. For patterns written with lower density, less copper was observed to deposit. When the probe was traced only once (top rows) less copper deposition occurred compared to the bottom rows where the tip was swept twice along a linescan.

Systematically engineering the writing parameters for arrays of nanopatterns generated by automated nanografting offers a further useful strategy for controlling reaction conditions for bottom-up surface assembly. Essentially, the surface density of reactive moieties can be defined to further control spatial parameters of surface reactions. In addition, the writing path itself was shown to influence the initial stages of metal deposition. The general approach for patterning metals with electroless deposition could readily be extended to other metals such as platinum or nickel for construction of a range of metal structures and nanoscale metal junctions.

5.6 Nanografting with Porphyrins

An obstacle for producing patterns with nanografting has been the limitation of using thiol-based chemistries. New directions are being developed for expanding beyond preliminary model systems of chemisorbed *n*-alkanethiols on gold substrates to other chemical linkers. Porphyrins and metalloporphyrins have a macrocyclic tetrapyrrole structure, which may be functionalized with various substituents. The choice of focusing research efforts on model systems of porphyrins is highly practical, because of the associated electrical, optical and chemical properties of this functional class of molecules. More complex surface structures could be achieved with nanografting by using porphryins with thiolated substituents [35] or pyridyl functional groups [28]. Modifications of the macrocycle, peripheral groups or bound metal ions can generate a range of electrical, photoemissive or magnetic properties. The orientation of porphyrins on surfaces is determined by factors such as the nature of the peripheral substituents and their position on the macrocycle. The resulting surface structures influence the photonic and electronic properties of the systems. Also, different properties result when different metals are coordinated to the macrocycle. Porphyrin and metalloporphyrin systems are excellent materials for surface studies, due to their diverse structural motifs and associated electrical, optical and chemical properties, and thermal stability [63, 64]. The rigid planar structures and π-conjugated backbone of porphyrins convey robust electrical properties for potential molecular electronic devices.

Scanning probe studies of nanografted patterns of dipyridyl porphyrins were used to provide insight for the molecular orientation and surface assembly of porphyrins from mixed solvent media, with studies by LeJeune, et al., [65]. In situ AFM furnished local views of the assembly of porphryins with pyridyl-substituents on surfaces of Au(111). Experiments were accomplished for nanografting *n*-alkanethiols within a matrix film of 5,10-diphenyl-15,20-di-pyridin-4-yl-porphyrin (DPP) as well as for nanografting patterns of DPP within different matrix SAMs of *n*-alkanethiols. The solubility of porphyrins in ethanol, butanol or water are problematic for accomplishing in situ AFM studies, therefore a solvent mixture was used for nanografting. First the porphyrin was dissolved in a parent solution of dichloromethane, and then further diluted 100-fold in ethanol. Examples of nanografted porphyrin patterns are displayed in Fig. 5.6. Dodecanethiol (C12) was used as a matrix SAM for writing nanostructures of DPP in a solution containing 1% dichloromethane in ethanol. The overall final concentration of DPP used for nanografting was 1 micromolar.

A mosaic design of 20 oval patterns was produced by nanografting DPP within a C12 SAM, as shown in the AFM topograph of Fig. 5.6a. The patterns were produced by tracing the probe in a circular trajectory four times, so that the centers of the rings were not disturbed. The patterns were produced within 5 min using a scan speed of 0.1 μm/s. The dimensions of the oval structures of DPP measure 77 $\pm$ 3 nm from side to side, and 99 $\pm$ 6 nm from top to bottom. The dodecanethiol islands in the middle of the rings that are surrounded by a ring of DPP have an average diameter of 58 $\pm$10 nm and furnish a convenient height reference for evaluating the depth of the DPP patterns. The distance between patterns ranged between 53 and 115 nm in the vertical direction and between 44 and 200 nm horizontally. A force of 2.3 nN was applied to write patterns of porphyrins within dodecanethiol while imaging in liquid media of mixed solvents. Characteristics of the underlying Au(111) substrate such as etch pits and scar defects are apparent in the 700 $\times$ 700 nm^2 topograph, indicating that after nanografting multiple patterns the probe still maintains a sharp

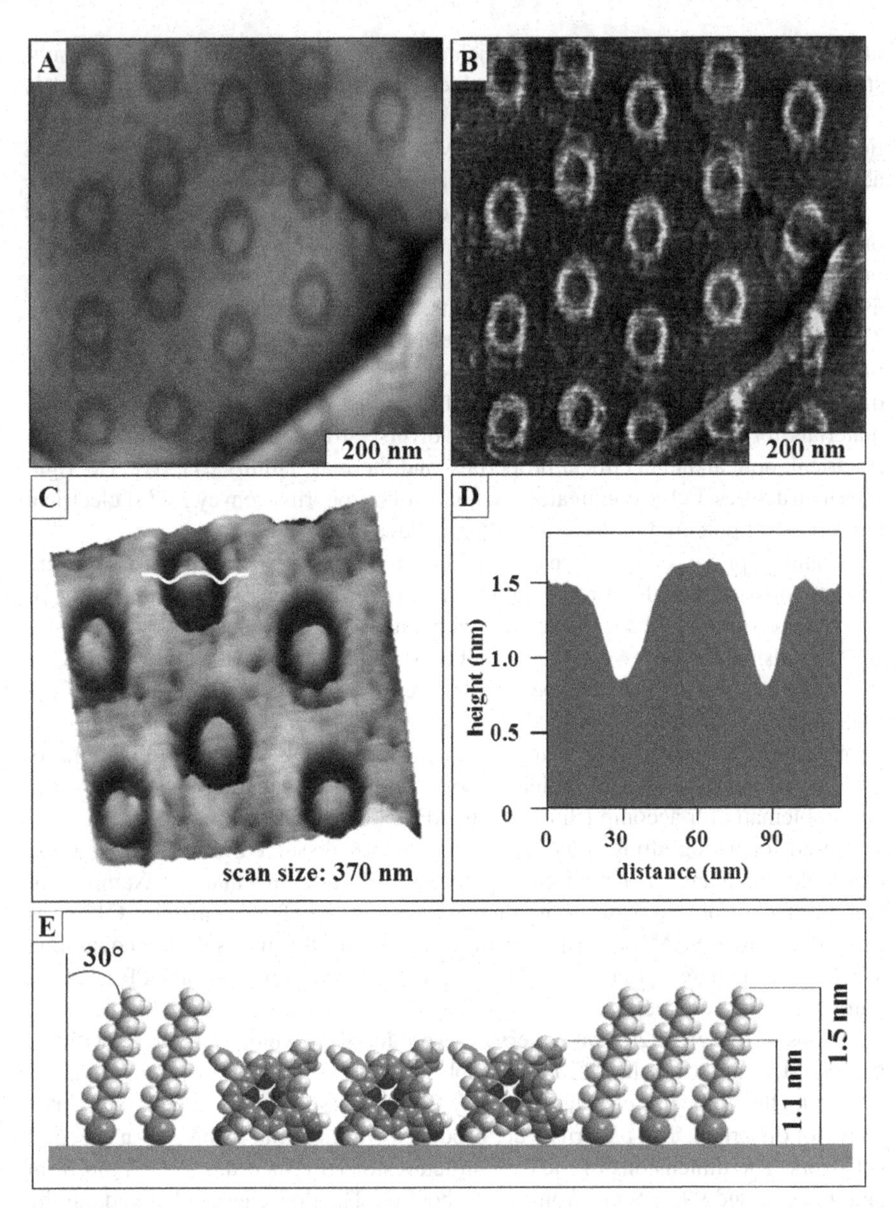

Fig. 5.6 Nanopatterns of diphenyl-dipyridyl porphyrin nanografted within dodecanethiol. (**a**) Mosaic design of 20 ring nanostructures viewed by an AFM topograph; (**b**) simultaneously acquired lateral force image; (**c**) magnified view; (**d**) cursor profile across one of the patterns traced in (**c**); (**e**) height model

geometry for accomplishing high-resolution imaging. The lateral force image (Fig. 5.6b) exhibits distinct contrast because of the different head groups of the C12 matrix and DPP nanopatterns. A zoom-in view of six ring nanopatterns is presented in Fig. 5.6c showing the fine details of the pattern shapes and height differences. The difference in height for the matrix dodecanethiol and DPP measures 0.5 ± 0.2 nm as shown by a representative line profile in Fig. 5.6d. This height difference corresponds to an upright configuration of DPP for a perpendicular orientation on Au(111) as shown by the molecular model of Fig. 5.6e.

For nanografted patterns of DPP, the heights measured from cursor profiles indicate that molecules assemble with an upright configuration with the porphyrin macrocycle oriented perpendicular to the substrate. As previously shown for nanografted molecules of *n*-alkanethiols which have a rod-like shape, planar macrocycles of DPP likewise are confined during nanografting. Constrained conditions prevent molecules of DPP from adopting a coplanar orientation on the surface to directly generate an upright configuration. The mechanical process of nanografting alters the assembly pathway providing a means to control molecular orientation of nanopatterned porphyrins on surfaces.

5.7 Nanografted Patterns of Proteins

Methods for nanoscale fabrication are becoming important for biochemical investigations, supplying tools for basic research concerning protein-protein interactions and protein function. Protein patterning is essential for the integration of biological molecules into miniature bioelectronic and sensing devices. Often, fabrication of functional nanodevices for biochemical assays requires that biomolecules be attached to surfaces with retention of structure and function. Nanoscale studies can facilitate the development of new and better approaches for immobilization and bioconjugation chemistries, which are key technologies in manufacturing surface platforms for biosensors. Nanografting provides a way to spatially control the deposition of proteins on well-defined, local areas of patterned surfaces for accomplishing in situ studies of biochemical reactions. The ability to define the chemical functionalities of nanografted patterns at nanometer length scales offers new possibilities for studies of biochemical reactions in controlled environments. Capturing AFM images in situ throughout the progressive steps of nanografting and surface patterning can disclose reaction details at a molecular level, providing direct visualization of biochemical reactions.

An overview of the different proteins that have been patterned with nanografting is summarized in Table 5.2, with spatial dimensions reaching the level of single molecule detection with protein monolayers. Spatially well-defined regions of surfaces can be nanografted with reactive or adhesive terminal groups for the attachment of biomolecules. The dimensions of many proteins are on the order of tens to hundreds of nanometers, therefore nanografting provides a way to generate patterns with appropriate sizes for defining the placement of individual proteins on surfaces. The terminal moieties of SAMs mediate the nature of protein binding,

Table 5.2 Protein studies accomplished in situ with nanografted patterns of SAMs

Biomolecule	Nanografted molecule	Pattern dimensions	Matrix SAM	Liquid media	Binding motifs	References
Antibiotin IgG	1-hexanethiol	5.2 nm × 5.2 nm	Thiolated biotin SAMs	Ethanol	Specific biotinylation	[32]
Gal	Thiolated Gal	130 nm × 110 nm	Octanethiol	Ethanol	S-Au carbohydrate ligand	[25]
GalCer	Thiolated GalCer	150 nm × 150 nm	1-decanethiol	Ethanol	S-Au carbohydrate ligand	[25]
De novo 4-helix bundle protein S-824-C	S-824-C protein	100 nm × 100 nm 200 nm × 200 nm	Octadecanethiol	Mixed aqueous buffer	S-Au single cysteine thiol	[37]
De novo maltose binding protein (MBP)	MBP	50 nm × 100 nm 100 nm × 200 nm	Undecanethiol triethylene glycol	Mixed aqueous buffer	S-Au double cysteine residues at C terminus	[36]
Lysozyme	$HS(CH_2)_2COOH$	10 nm × 150 nm 100 nm × 150 nm	Decanethiol	2-butanol	Electrostatic	[25]
Staphylococcal protein A (SpA)	Mercapto-hexadecanoic acid	100 nm × 100 nm	Octadecanethiol	Ethanol water EDC/NHS	Covalent activation chemistry	[29]
Bovine carbonic anhydrase	3-mercapto-1-propanoic acid 6-mercaptohexanol	400 nm × 400 nm	$C_{11}(EG)_6$	Water	Electrostatic	[73]
Rabbit IgG	11-mercapto-undecanoic acid	5,000 nm × 5,000 nm	Octanethiol	Buffer	Covalent	[74]

Table 5.2 (continued)

Biomolecule	Nanografted molecule	Pattern dimensions	Matrix SAM	Liquid media	Binding motifs	References
Bovine serum albumin	3-mercapto-1-propanal	200 nm × 250 nm	Hexanethiol	Buffer	Covalent	[30]
Rabbit IgG mouse anti-rabbit IgG	11-mercapto-undecanal	300 nm × 300 nm	Octadecanethiol	Buffer	Covalent	[24]
Acetylcholine esterase (AChE)	$HS(CH_2)_{11}$-$(OCH_2CH_2)_3OH$	~150 nm × 150 nm	$HS(CH_2)_{11}(OCH_2CH_2)_6O(CH_2)_{11}$-$CH(OH)CH_2OH$	Ethanol	Covalent	[87]
Insulin	$HS(CH_2)_{11}$-$(OCH_2CH_2)_3OH$	~150 nm × 150 nm	$HS(CH_2)_{11}(OCH_2CH_2)_6O(CH_2)_{11}$-$CH(OH)CH_2OH$	Ethanol	Covalent	[87]
Anti-mouse IgG	Mouse IgG	400 nm × 4,000 nm	Octadecanethiol	Ethanol	Antigen-antibody recognition	[88]
Three-helix bundle metalloproteins	C-terminal thiolated protein	NA	Octadecanethiol	Trifluoro-ethanol	S-Au	[36]
Maltose binding protein (MBP)	MBP with a double cysteine	NA	Undecanethiol triethylene glycol	Buffer	S-Au double cysteine thiol	[89]

such as through electrostatic interactions, covalent binding, molecular recognition or through specific interactions such as streptavidin-biotin recognition. The chemistry of SAM surfaces can be engineered to avoid non-specific protein adsorption for surrounding matrix monolayers, yet make specific interactions with selected proteins to be immobilized on nanografted patterns. Very few surfaces resist protein adsorption, and efforts have been directed to understand the mechanisms that contribute to protein resistance or adhesion to surfaces. Systematic studies of functionalized SAMs have been reported which evaluated the molecular characteristics that impart resistance to protein adsorption [66–71]. Depending on the protein of interest and buffer conditions, methyl-, hydroxyl- or glycol-terminated SAMs have been used effectively as matrices that resist non-specific protein adsorption.

The typical general steps of an in situ protein binding experiment with nanografting are to first graft nanopatterns of protein-adhesive *n*-alkanethiols within a resistive matrix, then rinse the liquid cell and inject a solution of proteins to bind to the SAM nanopatterns. In a final step, the activity of the immobilized proteins can be tested by introducing an antibody or protein which binds specifically to the surface-bound protein. With nanografting the same tips that are used to produce patterns are also used to characterize the morphology of nanopatterns after successive steps of protein adsorption. Unlike electron microscopy methods which require high vacuum chambers and conductive coatings for specimens, in situ AFM experiments can be accomplished under near-physiological conditions in aqueous buffered environments. With in situ nanografting, the protein patterns are not exposed to air or dried, and remain in a carefully controlled liquid environment by rinsing and exchanging solutions within the liquid cell. Sequential real time AFM images can disclose reaction details at a molecular level, revealing information about the adsorption kinetics and configurations of protein binding.

The first studies using nanografting to immobilize proteins were conducted in 1999 by Wadu-Mesthrige, et al., using protocols with either electrostatic or covalent interactions to immobilize lysozyme, rabbit immunoglobulin G (IgG) and bovine serum albumin (BSA) on SAM nanopatterns [31]. In these initial investigations, functionalized alkanethiol SAMs of carboxylic acid head groups or aldehydes were nanografted to mediate either electrostatic or covalent binding of IgG and lysozyme. Proteins were sustained on patterns despite steps of washing with buffer and surfactant solutions and were stable for at least 40 h of AFM imaging. The smallest protein feature yet produced by nanografting is a $10 \times 150\ \text{nm}^2$ line pattern containing three proteins [31].

5.7.1 Studies with Antigen-Antibody Binding Accomplished with Nanografting

The first successful AFM experiment reported that applied nanografting to study antigen-antibody binding in situ was conducted by Wadu-Mesthrige, et al., [30]. The activity of rabbit IgG immobilized covalently on an aldehyde-terminated pattern produced by nanografting was tested for reactivity toward monoclonal mouse anti-rabbit IgG. Six aldehyde-terminated nanopatterns of different sizes and arrangement

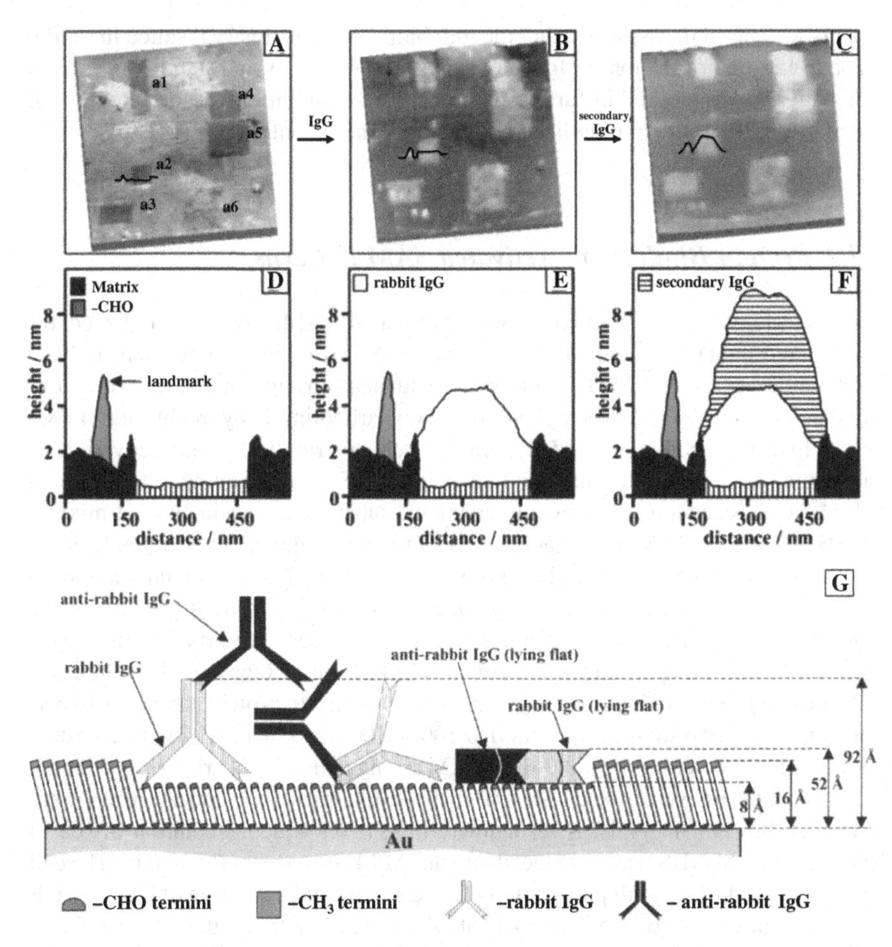

Fig. 5.7 The steps of protein binding and molecular recognition with nanografted patterns captured by AFM topographic images. (**a**) Six nanopatterns of 3-mercapto-1-propanal were written in a dodecanethiol SAM. (**b**) The image contrast changed after rabbit IgG bound covalently to the aldehyde-terminated nanopatterns. (**c**) After introducing mouse anti-rabbit IgG, the patterns display further height changes, indicating the antibody binds specifically to the protein nanopatterns. Cursor traces across pattern a2 indicate the height changes (**d**) after nanografting; (**e**) after injecting IgG; (**f**) after introducing anti-rabbit IgG. (**g**) Map for understanding the evolution of molecular height changes during the steps of this in situ experiment. (Reprinted with permission from Ref. [30])

were first grafted within a dodecanethiol SAM matrix (Fig. 5.7a). After injecting rabbit IgG and rinsing with a surfactant solution, selective adsorption of IgG was observed on all six nanopatterns (Fig. 5.7b). In the next step, mouse anti-rabbit IgG was introduced (Fig. 5.7c) revealing further increases in the heights of patterns. The changes in the height of nanopatterns before and after secondary IgG binding could be monitored in situ (cursor profiles, Fig. 5.7d–f), exhibiting thicknesses which correspond to the different surface configurations of IgG (Fig. 5.7g). Changes in pattern heights were used to assess whether the immobilization chemistry resulted in a side-on or an end-on orientation for IgG molecules. The reactivity and stability of protein nanopatterns was studied in further reports, with investigations of the retention of specific activity of the immobilized proteins for binding antibodies [24, 30].

5.7.2 Protein Binding on Activated SAM Patterns

Chemical activation of carboxylic acid terminated SAMs was accomplished for nanografted patterns of staphylococcal protein A (SpA) through covalent linkage by Ngunjiri, et al., [29]. The carboxylic acid head groups of SAMs were activated using 1-ethyl-3-(3-dimethylaminopropyl) carbodiimide hydrochloride (EDC) and N-hydroxysuccinimide (NHS) coupling chemistries [72]. The activation of carboxylic acid groups of nanografted patterns of 11-mercaptoundecanoic acid (11-MUA) was accomplished by immersing the substrate in an aqueous 1:1 mixture of NHS/EDC for 30 min to generate an activated complex with a stable reactive intermediate (N-succinimidyl ester). The resulting NHS ester interacts by a nucleophilic substitution reaction with accessible α-amine groups present on the N-termini of proteins or with ε-amines on lysine residues. The proteins bind covalently to nanografted patterns by forming a Schiff's base linkage to make complexes with the carboxylic acid groups of 11-MUA. For the in situ protein patterning experiment with SpA, 16 square nanopatterns ($100 \times 100\ nm^2$) of 11-MUA were written within a matrix octadecanethiol (ODT) SAM arranged in a 4×4 array (Fig. 5.8a–c). The nanopatterns were spaced 50 nm apart within each row, and the rows were spaced at 100 nm intervals. After nanografting, a 1:1 aqueous solution of 0.2 M EDC and 0.05 M NHS was introduced into the AFM cell to react for 30 min. The cell was then rinsed twice with phosphate-buffered saline, and a solution of 0.05 mg/mL SpA solution was introduced and incubated for 30 min. Finally, the cell was rinsed with water and ethanol to completely remove any unreacted protein. After chemical activation and protein immobilization, the same array of nanostructures was imaged in ethanol with AFM (Fig. 5.8d–f). All of the steps of nanografting, NHS/EDC activation of carboxylate groups, and protein adsorption were accomplished in situ with the same tip, and the entire experiment was completed in ~3 h. The SpA molecules were shown to bind selectively to the 11-MUA nanopatterns, forming a single layer of protein attached to nanopatterns of 11-MUA.

For in situ studies of biochemical reactions using nanografting, the most suitable immobilization chemistries for nanoscale experiments should proceed under aqueous conditions to preserve protein activity. Also, investigations should be completed using very dilute protein and reagent solutions to slow the reaction rate so that the reaction transpires over time intervals of 20–30 min. A potential technical detail is that the motion and force of the scanning tip can sweep away adsorbates or perturb the reaction environment. To address this concern, the immobilization chemistry selected for patterning must be sufficiently robust to enable continuous imaging and scanning by the tip. Imaging in liquids enables using small imaging

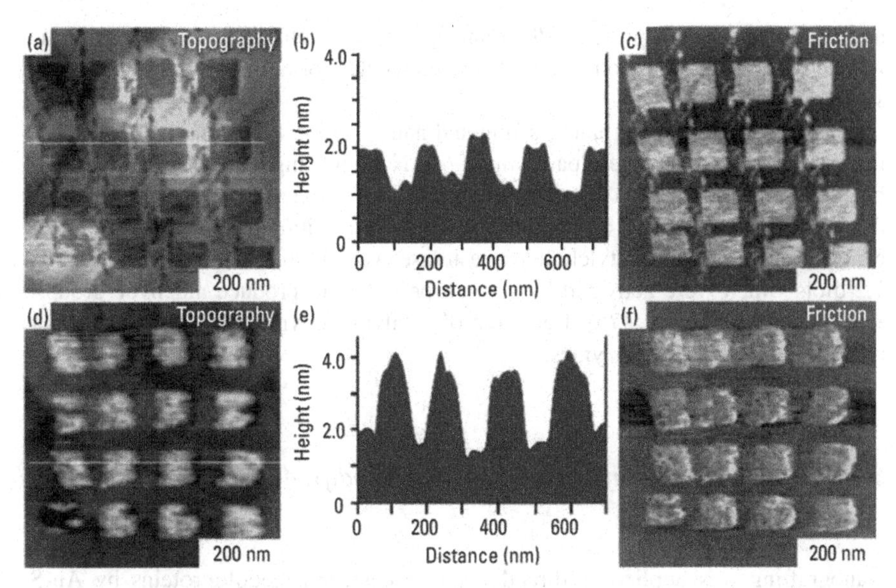

Fig. 5.8 Nanoscale protein assay of the adsorption of SpA on nanografted patterns. (**a**) An array of 11-MUA squares written in an ODT matrix SAM, (**b**) cursor plot along the white line; (**c**) corresponding lateral force image for (**a**); (**d**) same area after EDC/NHS activation and subsequent adsorption of SpA; (**e**) cursor plot along the white line in (**d**); (**f**) simultaneously acquired lateral force image for (**d**). (Reprinted with permission from Ref. [29], Copyright © American Chemical Society)

forces (0.005–0.2 nN) because the adhesive interactions between the tip and sample are minimized. An intrinsic advantage for these protocols is that small forces in the range of piconewtons to nanonewtons can be precisely controlled with AFM instruments.

5.7.3 In Situ Studies of Protein Adsorption on Nanografted Patterns

Nanografting has been applied by several investigators to write nanopatterns for studies of protein immobilization and reactivity. Zhou et al. evaluated protein adsorption at the nanoscale by comparing differently functionalized SAMs side-by-side using nanografting [3, 73]. Protein adsorption on three differently charged linkers nanografted within a hexa(ethylene glycol) terminated alkanethiol resist SAM, was monitored in situ by AFM at different pH conditions. The adsorption of proteins onto nanografted patches of 6-mercaptohexan-1-ol (MCH), *n*-(6-mercapto hexyl) pyridinium bromide (MHP), and 3-mercaptopropionic acid (MPA), was studied with lysozyme, IgG and carbonic anhydrase II. They concluded that the overall charge of protein molecules as well as the charge of local domains of the proteins plays a role in immobilization. In the same report, nanografting was applied to assemble multilayered protein G/IgG/anti-IgG nanostructures through electrostatic interactions, as an approach to orient IgG molecules for antibody-based biosensor surfaces.

Using SPL methods of nanografting and nanoshaving, Kenseth, et al. compared three approaches for protein patterning [74]. Nanografting was successfully combined with immobilization of IgG through EDC activation of 11-MUA acid and also through chemisorption of a disulfide coupling agent, dithiobis(succinimidyl undecanoate). Insulin and acetylcholinase esterase were immobilized on nanografted 1,2-diols which were activated by sodium periodate to produce aldehyde groups, reported by Jang, et al., [75]. Retention of catalytic activity was demonstrated for nanografted patterns of enzymes.

5.7.4 Direct Nanografting of Proteins Modified with Thiol Residues

Nanografting was applied to directly pattern designed metalloproteins by Au-S chemisorption by Case, et al., [36]. A 3-helix bundle protein structure with a 78 amino acid iron(II) complex was nanografted into an ethylene glycol-terminated SAM. The protein was designed to present the C-termini of three helices, terminated with D-cysteine residues for attachment to gold surfaces. The heights of nanografted patterns of this protein measured 5.3 nm, in good agreement with the dimensions predicted theoretically for the *de novo* protein to assemble in a upright orientation normal to the Au(111) substrate. A *de novo* 4-helix bundle protein was nanografted within an ODT matrix through a single cysteine thiol by Hu, et al., [37]. The protein used for these studies was engineered to have a glycine-glycine-cysteine tag at its C-terminus for attachment to the gold surface through a single cysteine thiol.

Maltose Binding Protein (MBP) was successfully patterned using nanografting by Staii, et al., [76]. The MBP protein was engineered to terminate with a double-cysteine residue for chemisorptive binding to gold surfaces. The biochemical activity of the substrate immobilized proteins was verified in situ, demonstrating that MBP function is not altered by either the immobilization process, the spatial confinement associated with the surrounding proteins, or protein-substrate interactions. The dependence of the frictional force upon the maltose concentration was used to extract the dissociation constant: $k_d = 1 \pm 0.04$ μM for this system, detecting maltose at the level of tens of attograms.

5.7.5 Reversal Nanografting

An approach for "reversal" nanografting was introduced for regulating surface heterogeneity to control protein binding [32]. As with nanografting, the reversal method also has three main steps of imaging, shaving-and-replacement, and imaging again. However, rather than directly nanografting desired termini for protein binding, the matrix SAMs are made of the binding termini, and nanografted thiols are used to isolate and separate well- defined areas of the matrix SAMs to generate ultra-small domains of protein binding sites. By controlling the shaving size and the spacing between the shaving lines, broad areas of arrays of regular nanostructures were rapidly fabricated, achieving dimensions of 5–30 nm for nanografted patterns. Reversal nanografting was demonstrated with an array of thiolated biotin nanostructures which were reacted with antibiotin IgG. Within a single experiment, reversal nanografting produced 1,089 biotin nanostructures measuring with 5.2 nm × 5.2 nm; 288 nanostructures with dimension of 12.7 nm × 12.7 nm; and 144 nanopatterns with dimensions of 10.3 nm × 31.9 nm. Thus, by changing the dimension and separation of each element of nanografted arrays the coverage and orientation of protein molecules can be regulated at the molecular level.

Although not yet practical for high throughput applications and manufacturing, combining the in situ steps of nanografting with protein immobilization enables new approaches for directly investigating changes that occur on surfaces during biochemical reactions from the bottom-up. In situ AFM investigations of protein reactions are valuable for studying antigen-antibody binding at the nanometer scale, for assessing the specificity of protein-protein binding, and for evaluating the orientation of immobilized proteins and the corresponding accessibility of ligands for binding.

5.8 Patterns of DNA Produced by Nanografting

Surface platforms of arrays of DNA patterns are used for studies with gene mapping, drug discovery, DNA sequencing and disease diagnosis. Scanning probe-based experiments offer compelling advantages and opportunities for high sensitivity, label-free detection with studies of molecular-level phenomena. Initial studies have been advanced using nanografting to prepare patterns of DNA with successive steps of enzyme digestion [33, 77], hybridization studies [78–80], as well as DNA-mediated binding of proteins [81]. A comparison of the different DNA systems and pattern dimensions produced by nanografting is provided in Table 5.3.

Individual DNA molecules can be localized within mixed patterns by diluting DNA with another alkanethiol molecule. To achieve single-molecule precision, Josephs et al., nanografted thiolated double-stranded DNA (dsDNA) with 94 base pairs from a solution containing a ~10000:1 mixture of aminoundecanethiol and dsDNA [82]. By diluting DNA molecules with another alkanethiol molecule, DNA can be positioned on a chemically well-defined, atomically flat surface and be imaged in situ. One to four dsDNA molecules were localized confined within a nanografted area to provide high precision for positioning individual DNA molecules within biochemical structures.

Table 5.3 Studies reported with nanografted patterns of DNA

System	Pattern sizes	Matrix film	Liquid media	Year	References
DNA-derivatized gold nanoparticles	100 nm × 50 μm lines	Octadecanethiol	Buffer: 1 M NaCl, 10 mM phosphate, pH 7	2001	[79]
Single stranded DNA (ssDNA) 5′-HS $(CH_2)_6$-CTAGCTCTAATCTGCTAG 5′-HS $(CH_2)_6$-AGAAGGCCTAGA	Dimensions in nm: 115 × 135; 190 × 255; 20 × 170; 15 × 150; 25 × 160	1-hexanethiol 1-decanethiol	Mixed solvent of 2-butanol/ water/ ethanol 6:1:1 (v/v/v) containing 40 μM ssDNA.	2002	[33]
Single stranded DNA 5′-HS-$(CH_2)_6(T)_{15}$ 3′-HS-$(CH_2)_6(T)_{25}$ 5′-HS-$(CH_2)_6(T)_{35}$ 5′-HS$(CH_2)_6$ ACTGCACATGGCGTG TTGCGGTGATT CGCGTTGGT	Dimensions in nm: 120 × 200; 100 × 380; 100 × 200; 250 × 250; 80 × 220; 100 × 400; 180 × 250; 40 × 250; 150 × 75	1-decanethiol	Mixed solvent of water saturated with 2-butanol and ethanol (6:1)	2005	[78]
Nanografted patterns of mercaptoethanol were used to evaluate thickness of DNA SAMs	300 × 300 nm squares of 2-mercaptoethanol	HSC_6H_{12}-5′-CCCT AACCCTAACCCTAA CCC-3′-rhodamine green 5′-GTGTTAGGT TTAGGGTTAGTG-3′	Phosphate buffered saline (pH 4.5)	2006	[50]
λ-DNA adsorbed to octadecyldimethyl-monochloro-silane (C18DMS)	100 nm × 3 μm lines of (C18DMS)	Octadecyldimethyl monochlorosilane	Nanografted patterns were incubated with λ-DNA in TE buffer (pH 7.2)	2007	[83]
Thiolated ssDNA	300 nm × 300 nm to 1 μm × 1 μm	Oligo-ethyleneglycol modified thiols	1:1 mixture of buffer and ethanol	2008	[80]

Table 5.3 (continued)

System	Pattern sizes	Matrix film	Liquid media	Year	References
ssDNA with 44 base pairs	1 μm × 1 μm	Top-oligo ethylene-glycol (EG) HS-$(CH_2)11$-$(EG)_3$-OH	Thiol-DNA containing 3:2 mixtures (v/v) of 1 M buffer and ethanol	2008	[77]
ssDNA-mediated binding of proteins thiol modified oligonucleotides	200 nm × 200 nm to 1 μm × 1 μm	Ethylene glycol- terminated alkylthiols	1:1 mixture of buffer and ethanol	2009	[81]
94 basepair thiolated double stranded DNA attached to nanografted patterns	50 nm × 50 nm	Octadecanethiol	Mixture of 11-aminoundecane thiol with DNA (10,000:1) in Tris acetate EDTA (TAE)	2010	[82]
Thiol derivatized single-stranded oligonucleotide HS-C_6H_{12}-5′-AGA TCA GTG CGT CTG TAC TAG CAC A-3′ and complementary sequence	0.5–1 μm	6-mercapto-1-hexanol	10 μM probe DNA in a 1:1 mixture (v/v) of STE-buffer and absolute ethanol	2010	[90]

5.8.1 In Situ Studies of Hybridization with Nanografted Patterns of ssDNA

Nanostructures of single stranded oligonucleotides or single stranded DNA (ssDNA) have been produced with nanografting for molecular-level studies of DNA hybridization [77–80]. Label-free hybridization of ssDNA nanostructures was accomplished for nanografted patterns of ssDNA incubated with complementary segments of designed sequences [78]. To mediate attachment to gold surfaces for nanografting, the DNA molecules were designed to contain a short thiol linker at either the $3'$ or $5'$ end. These investigations provide information about the specificity, kinetics and selectivity of surface-bound ssDNA for hybridization with complementary strands.

Label-free hybridization of nanostructures has proven to be highly selective and sensitive; as few as 50 molecules can be detected by in situ AFM studies [78]. The efficiency of the hybridization reaction at the nanometer scale depends sensitively on the packing density of DNA within the nanostructures [77, 78, 80]. The density of ssDNA molecules within nanografted patterns can be regulated by changing certain experimental parameters such as written line density and concentration. The structure of nanografted patterns and the relative surface orientation of the ssDNA molecules have been determined in situ using AFM to show that molecules of ssDNA adopt a standing upright orientation.

Nanopatterns of thiolated ssDNA were produced using nanografting by Maozi Liu, et al., [33]. Thiolated ssDNA molecules adsorb chemically onto exposed areas of gold through sulfur-gold chemisorption. The ssDNA molecules within nanopatterns adopt an upright, standing orientation on gold surfaces which were found to be accessible by enzymes. A ssDNA pattern (115×135 nm^2) of an 18-nucleotide oligomer ($5'$-HS-$(CH_2)_6$-CTAGCTCTAATCTGCTAG) was nanografted into a hexanethiol matrix, as shown in Fig. 5.9a. Nanografting and imaging of the patterns were conducted in a mixed solvent of 2-butanol/water/ethanol with a (v/v/v) ratio of 6:1:1 containing 40 μM ssDNA. The heights of the nanografted patterns were found to match well with the theoretical dimensions of an upright configuration of DNA, shown with cursor profiles. In Fig. 5.9c, a second 12-mer ssDNA ($5'$-HS-$(CH_2)_6$-AGAAGGCCTAGA) was grafted into a dodecanethiol SAM. Line patterns of ssDNA as narrow as 10 nm were produced, as shown in Fig. 5.9e. Three lines of the 12-nucleotide oligomer were nanografted within decanethiol.

Unlike natural, unconfined solution adsorption of thiolated DNA on gold surfaces, in which DNA oligomers tend to assemble with the backbone parallel to the substrate in a lying down configuration, nanografted patterns of ssDNA form a standing conformation, confined by the surrounding matrix monolayer to generate a fairly dense, close-packed structure of upright strands [33, 78]. The alkanethiol matrix SAM guides the adsorption of DNA to define the geometry and packing of grafted ssDNA molecules. Upright ssDNA molecules within the nanografted structures maintain their reactivity, as demonstrated by hybridization reactions with complementary DNA in solution. The hybridization and corresponding control experiments indicate that nanografted patterns of ssDNA exhibit high specificity and selectivity towards complementary strands.

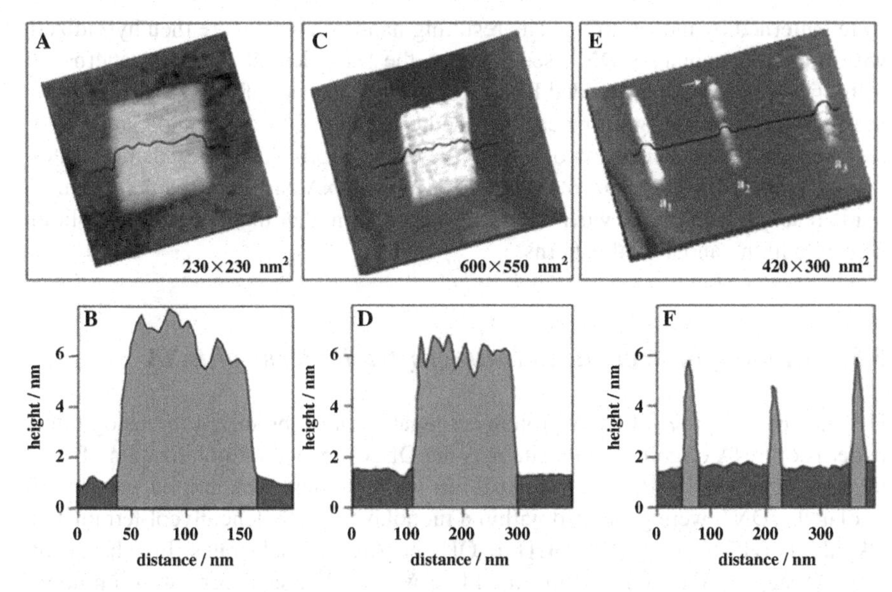

Fig. 5.9 Patterns of single-stranded DNA were nanografted into an alkanethiol SAM matrix. (**a**) Topograph of an 18-nucleotide ssDNA nanografted into a hexanethiol SAM (115×135 nm^2); (**b**) corresponding profile for the line in (**a**). (**c**) Nanografted rectangle (190×255 nm^2) of ssDNA with 12 nucleotides inscribed within a dodecanethiol matrix; (**d**) cursor profile for (**c**). (**e**) Line patterns of the ssDNA 12-mer nanografted into decanethiol; (**f**) profile for (**e**). The 18-mer and 12-mer ssDNA strands are 5′-HS-$(CH_2)_6$-CTAGCTCTAATCTGCTAG and 5′-HS-$(CH_2)_6$-AGAAGGCCTAGA, respectively. (Reproduced with permission from reference [33], Copyright © American Chemical Society)

5.8.2 Reactions with Restriction Enzymes Studied Using Nanografted Patterns of DNA

Time-dependent AFM images were acquired in situ for a nanografted pattern of the 18-nucleotide oligomer during digestion by the enzyme shown in Fig. 5.9a. The RQ1 DNase I enzyme endonucleotically degrades DNA to produce oligonucleotide fragments at the 3′ end with a hydroxyl terminal group. After nanografting steps, the ssDNA patterns were rinsed and the solvent was then replaced sequentially by ethanol, water, and finally buffer solution. Next, RQ1 DNase I was introduced and surface changes were captured in situ with high-resolution AFM images. The liquid cell experiment establishes that upright, densely-packed strands of DNA within nanografted patterns are accessible to enzyme digestion.

Studies with the cutting action of restriction enzymes were accomplished by Castronovo, et al. to better understand enzyme/DNA interactions [77]. An enzymatic reaction (*DpnII* restriction digestion) with DNA nanopatterns of variable density (surface coverage) was investigated to understand the effect of molecular crowding on the accessibility of the DNA molecules to the restriction enzyme. Single-stranded DNA molecules containing 44 base pairs (bps) with a 4 base pair recognition sequence (specific to the *DpnII* restriction enzyme) in the middle were patterned by nanografting. The resulting nanostructures were then hybridized with a complementary ssDNA sequence of the same length to yield patterns of restriction-ready double stranded DNA. The surface density of the DNA nanostructures produced by nanografting can be tuned by changing the writing parameters or by changing the concentration of the DNA when grafting. The study demonstrates that the *DpnII* restriction enzyme is sensitive to the DNA packing density; the enzymatic reaction is inhibited when the DNA density is higher than a certain threshold density within nanografted patterns.

5.8.3 Binding of Proteins to Nanografted Patterns of DNA

Hybrid nanostructures of DNA-protein conjugates can be produced for nanografted patterns of DNA oligomers with site-specific DNA-directed immobilization of proteins, as reported by Bano, et al., [81]. In the first step, nanografted patches of thiolated ssDNA were generated within a monolayer of ethylene glycol-terminated alkylthiols (HS-$(CH_2)_{11}$-$(OCH_2CH_2)_3$-OH) on Au(111) substrates. In subsequent reaction steps, proteins covalently modified with cDNA sequences were immobilized onto the $1 \times 1\ \mu m^2$ nanografted patterns. A covalent conjugate of streptavidin tethered with a DNA oligomer was found to bind to the nanografted ssDNA pattern by sequence-specific DNA hybridization. The surface was carefully rinsed with phosphate buffered saline to remove any physically adsorbed molecules and imaged with AFM between successive biochemical reaction steps. Changes in heights of the patterns enabled label-free detection of protein binding between each step of the reactions, which were likewise accomplished in multiplex experiments with control samples of streptavidin that did not have the complementary DNA tethers. The nanopatterns of DNA-protein conjugates were then used for further studies of selected protein-protein interactions with an anti-streptavidin immunoglobulin G as well as with the biomedically relevant matrix of human serum. The fabrication of nanografted arrays of multiple proteins in this study demonstrates that the interactions of biomolecular recognition mediated by DNA-protein recognition are highly specific and that bound proteins retain activity for further selective binding of proteins.

5.8.4 Using Nanografted SAM Patterns to Mediate Binding of DNA

Nanografted patterns of an aminopropyldiethoxysilane (APDES) SAM were used as sites for selective adsorption of DNA within matrices of octyldimethylmonochlorosilane (C8DMS) monolayers by Lee, et al., [83]. Line patterns of APDES that were 100 nm wide were nanografted in a C8DMS monolayer prepared on silicon dioxide substrates. After incubation in a 10 ng/μL solution of λ-DNA in buffer (pH 7.2) the heights of the nanopatterns was increased and revealed the shapes of individual DNA strands. The negatively charged DNA deposited on the positively charged amine-functionalized line patterns of aminosilanes. The negatively charged DNA molecules bound to nanografted patterns via electrostatic interactions with the positively charged amine groups of APDES, but did not bind to matrix areas terminated with methyl headgroups. These investigations provide a fundamental step toward sensitive DNA detection and construction of complex DNA architectures on surfaces.

Nanografting provides a useful protocol towards sensitive DNA detection and likely attains the most sensitive detection levels yet achievable for label-free assays. The DNA nanopatterning methodology provides a unique opportunity for engineering biostructures with nanometer precision, which benefits the advancement of technologies for DNA biosensors and biochips.

5.9 Limitations of AFM-Based Nanografting

Thus far, the capabilities for molecular manipulation by nanografting have primarily been a tool for academic research. However one may anticipate that nanografting will eventually provide commercial value for chemical or biochemical sensing or for nanotechnology. A potential disadvantage for nanografting is that over time, molecular exchange reactions take place between solution molecules and the matrix SAM for certain systems of alkanethiol matrices. Natural processes of self-exchange become an issue specifically when nanografting longer chain alkanethiols into a shorter chain matrix layer, thus it is important to use dilute ($< 0.1\ \mu M$) solutions for nanografting. Depending on the nature and age of the matrix SAM, exchange reactions can be detected within 2–4 h when molecules from solution adsorb onto defect sites and at step edges. Software addresses this problem by enabling rapid automation of the nanofabrication process. Hundreds of exquisitely regular patterns can be produced within an hour or less, leaving sufficient time to progress to further in situ steps of reactions before exchange reactions have occurred.

The serial nature of nanografting with a single probe may be a problem for applications that require higher throughput, such as at scales of millions of nanostructures. Prototype arrays of 1,024 and 55,000 AFM probes have been developed for high-throughput nanopatterning [84, 85]. At this time, nanoscale studies with AFM enable new approaches to refine and optimize parameters used to link and organize proteins and other nanomaterials on surfaces. With in situ AFM characterizations, the orientation, reactivity, and stability of molecules adsorbed on SAM nanostructures can be monitored with successive time-lapse images using liquid AFM. These investigations provide the groundwork for advancing nanotechnology toward the nanoscale and furnish molecular-level information through the visualization of surface reactions.

5.10 Future Prospectus

Nanografting provides a practical tool to precisely control the arrangement of molecules on surfaces to enable bottom-up nanofabrication of structures through successive chemical reactions. In situ AFM studies with nanografting furnish opportunities for visualization, physical measurements and precise manipulation molecules at the nanometer scale. There are multiple advantages for nanografting, particularly because experiments are accomplished using liquid media. Advantages are the ability to precisely produce nanometer-sized patterns of metals, polymers, proteins and DNA with the benefits of successively imaging and accomplishing fabrication within well-controlled environments. Because so many chemical reactions can be accomplished in solution, there are rich possibilities for studying other surface reactions, in ambient, cooled or heated conditions. The capabilities for capturing real time images throughout sequential steps of reactions offer intriguing possibilities for new studies, with directly viewing the role of temperature, reagents and solvents. Nanografting protocols provide an additional unique capability for defining spatial parameters for controlling surface coverage and confining reactions within defined boundaries. The challenge for future research directions will be to achieve greater complexity for experiments for building ever more sophisticated 3D architectures from the bottom-up.

Acknowledgements The authors received financial support from the National Science Foundation (DMR-0906873) and also from the Dreyfus Foundation for a Camille Dreyfus Teacher-Scholar award. Wilson K. Serem is an LSU doctoral candidate supported by study-leave from Masinde Muliro University, Kenya, Africa.

References

1. A. B. Chwang, E. L. Granstrom, C. D. Frisbie, Fabrication of a sexithiophene semiconducting wire: Nanoshaving with an atomic force microscope tip, *Adv. Mater.* **12**, 285–288 (2000).
2. J. Shi, J. Chen, P. S. Cremer, Sub-100 nm patterning of supported bilayers by nanoshaving lithography, *J. Am. Chem. Soc.* **130**, 2718–2719 (2008).
3. D. Zhou, A. Bruckbauer, L. Ying, C. Abell, D. Klenerman, Building three-dimensional surface biological assemblies on the nanometer scale, *Nano Lett.* **3**, 1517–1520 (2003).
4. L. G. Rosa, J. Jiang, O. V. Lima, J. Xiao, E. Utreras, P. A. Dowben, L. Tan, Selective nanoshaving of self-assembled monolayers of 2-(4-pyridylethyl)triethoxysilane, *Mater. Lett.* **63**, 961–964 (2009).
5. J. E. Headrick, M. Armstrong, J. Cratty, S. Hammond, B. A. Sheriff, C. L. Berrie, Nanoscale patterning of alkyl monolayers on silicon using the atomic force microscope, *Langmuir* **21**, 4117–4122 (2005).
6. J. C. Garno, J. D. Batteas, in *Applied scanning probe methods, vol. IV, industrial applications,* B. Bhushan, Ed. (Springer, Berlin; Heidelberg; New York, 2006).
7. Z. M. LeJeune, W. Serem, A. T. Kelley, J. N. Ngunjiri, J. C. Garno, in *Encyclopedia of nanoscience and technology, (2nd edition),* H. S. Nalwa, Ed. (American Scientific Publishers, Stevenson Ranch, CA, 2010).
8. S. Xu, G. Y. Liu, Nanometer-scale fabrication by simultaneous nanoshaving and molecular self-assembly, *Langmuir* **13**, 127–129 (1997).

9. S. Xu, S. Miller, P. E. Laibinis, G. Y. Liu, Fabrication of nanometer scale patterns within self-assembled monolayers by nanografting, *Langmuir* **15**, 7244–7251 (1999).
10. D. S. Ginger, H. Zhang, C. A. Mirkin, The evolution of dip-pen nanolithography, *Angew. Chem. Int. Ed.* **43**, 30–45 (2004).
11. K. Salaita, Y. H. Wang, C. A. Mirkin, Applications of dip-pen nanolithography, *Nat. Nanotechnol.* **2**, 145–155 (2007).
12. A. T. Kelley, J. N. Ngunjiri, W. K. Serem, J.-J. Yu, S. Lawrence, S. Crowe, J. C. Garno, Applying AFM-based nanofabrication for measuring the thickness of nanopatterns: The role of headgroups in the vertical self-assembly of ω-functionalized *n*-alkanethiols, *Langmuir* **26**, 3040–3049 (2010).
13. C. A. Hacker, J. D. Batteas, J. C. Garno, M. Marquez, C. A. Richter, L. J. Richter, R. D. vanZee, C. D. Zangmeister, Structural and chemical characterization of monofluoro-substituted oligo(phenylene-ethynylene) thiolate self-assembled monolayers on gold, *Langmuir* **20**, 6195–6205 (2004).
14. N. A. Amro, S. Xu, G.-Y. Liu, Patterning surfaces using tip-directed displacement and self-assembly, *Langmuir* **16**, 3006–3009 (2000).
15. W. T. Muller, D. L. Klein, T. Lee, J. Clarke, P. L. McEuen, P. G. Schultz, A strategy for the chemical synthesis of nanostructures, *Science* **268**, 272–273 (1995).
16. J. J. Davis, C. B. Bagshaw, K. L. Busuttil, Y. Hanyu, K. S. Coleman, Spatially controlled Suzuki and Heck catalytic molecular coupling, *J. Am. Chem. Soc.* **128**, 14135–14141 (2006).
17. J. J. Davis, K. S. Coleman, K. L. Busuttil, C. B. Bagshaw, Spatially resolved Suzuki coupling reaction initiated and controlled using a catalytic AFM probe, *J. Am. Chem. Soc.* **127**, 13082–13083 (2005).
18. H. Lee, S. A. Kim, S. J. Ahn, H. Lee, Positive and negative patterning on a palmitic acid Languir–Blodgett monolayer on Si surface using bias-dependent atomic force microscopy lithography, *Appl. Phys. Lett.* **81**, 138–140 (2002).
19. J. Gu, C. M. Yam, S. Li, C. Cai, Nanometric protein arrays on protein-resistant monolayers on silicon surfaces, *J. Am. Chem. Soc.* **126**, 8098–8099 (2004).
20. H. G. Hansma, J. Vesenka, C. Siegerist, G. Kelderman, H. Morrett, R. l. Sinsheimer, V. Elings, C. Bustamante, P. K. Hansma, Reproducible imaging and dissection of plasmid DNA under liquid with the atomic force microscope, *Science* **256**, 1180–1184 (1992).
21. A. L. Weisenhorn, P. Maivald, H. J. Butt, P. K. Hansma, Measuring adhesion, attraction, and repulsion between surfaces in liquids with an atomic-force microscope, *Phys. Rev. B* **45**, 11226–11232 (1992).
22. J. N. Ngunjiri, A. T. Kelley, Z. M. LeJeune, J.-R. Li, B. Lewandowski, W. K. Serem, S. L. Daniels, K. L. Lusker, J. C. Garno, Achieving precision and reproducibility for writing patterns of n-alkanethiol SAMs with automated nanografting, *Scanning* **30**, 123–136 (2008).
23. S. Cruchon-Dupeyrat, S. Porthun, G. Y. Liu, Nanofabrication using computer-assisted design and automated vector-scanning probe lithography, *Appl. Surf. Sci.* **175**, 636–642 (2001).
24. G. Y. Liu, N. A. Amro, Positioning protein molecules on surfaces: A nanoengineering approach to supramolecular chemistry, *Proc. Natl. Acad. Sci. USA* **99**, 5165–5170 (2002).
25. M. Liu, N. A. Amro, G.-Y. Liu, Nanografting for surface physical chemistry, *Ann. Rev. Phys. Chem.* **59**, 367–386 (2008).
26. J. C. Garno, C. D. Zangmeister, J. D. Batteas, Directed electroless growth of metal nanostructures on patterned self-assembled monolayers, *Langmuir* **23**, 7874–7879 (2007).
27. J. C. Garno, Y. Y. Yang, N. A. Amro, S. Cruchon-Dupeyrat, S. W. Chen, G. Y. Liu, Precise positioning of nanoparticles on surfaces using scanning probe lithography, *Nano Lett.* **3**, 389–395 (2003).
28. Z. M. LeJeune, M. McKenzie, E. Hao, M. G. H. Vicente, B. Chen, J. C. Garno, Surface assembly of pyridyl-substituted porphyrins on Au(111) investigated in situ using scanning probe lithography, *SPIE Proceedings* **7593**, 759311 (2010).
29. J. N. Ngunjiri, J. C. Garno, AFM-based lithography for nanoscale protein assays, *Anal. Chem.* **80**, 1361–1369 (2008).

30. K. Wadu-Mesthrige, N. A. Amro, J. C. Garno, S. Xu, G. Y. Liu, Fabrication of nanometer-sized protein patterns using atomic force microscopy and selective immobilization, *Biophys. J.* **80**, 1891–1899 (2001).
31. K. Wadu-Mesthrige, S. Xu, N. A. Amro, G. Y. Liu, Fabrication and imaging of nanometer-sized protein patterns, *Langmuir* **15**, 8580–8583 (1999).
32. Y. H. Tan, M. Liu, B. Nolting, J. G. Go, J. Gervay-Hague, G. Y. Liu, A nanoengineering approach for investigation and regulation of protein immobilization, *ACS Nano* **2**, 2374–2384 (2008).
33. M. Liu, N. A. Amro, C. S. Chow, G.-Y. Liu, Production of nanostructures of DNA on surfaces, *Nano Lett.* **2**, 863–867 (2002).
34. S. M. D. Watson, K. S. Coleman, A. K. Chakraborty, A new route to the production and nanoscale patterning of highly smooth, ultrathin zirconium films, *ACS Nano* **2**, 643–650 (2008).
35. Y.-H. Chan, A. E. Schuckman, L. M. Perez, M. Vinodu, C. M. Drain, J. D. Batteas, Synthesis and characterization of a thiol-tethered tripyridyl porphyrin on Au(111), *J. Phys. Chem. C* **112**, 6110–6118 (2008).
36. M. A. Case, G. L. McLendon, Y. Hu, T. K. Vanderlick, G. Scoles, Using nanografting to achieve directed assembly of *de novo* designed metalloproteins on gold, *Nano Lett.* **3**, 425–429 (2003).
37. J. Hu, A. Das, M. H. Hecht, G. Scoles, Nanografting de novo proteins onto gold surfaces, *Langmuir* **21**, 9103–9109 (2005).
38. S. Xu, S. J. N. Cruchon-Dupeyrat, J. C. Garno, G. Y. Liu, G. K. Jennings, T. H. Yong, P. E. Laibinis, In situ studies of thiol self-assembly on gold from solution using atomic force microscopy, *J. Chem. Phys.* **108**, 5002–5012 (1998).
39. G. E. Poirier, E. D. Pylant, The self-assembly mechanism of alkanethiols on Au(111), *Science* **272**, 1145–1148 (1996).
40. G. E. Poirier, Mechanism of formation of Au vacancy islands in alkanethiol monolayers on Au(111), *Langmuir* **13**, 2019–2026 (1997).
41. T. T. Brown, Z. M. LeJeune, K. Liu, S. Hardin, J.-R. Li, K. Rupnik, J. C. Garno, Automated scanning probe lithography with *n*-alkanethiol self-assembled monolayers on Au(111): Application for teaching undergraduate laboratories, *J. Am. Lab Automat.*, **16**, 112–125 (2011).
42. S. Xu, N. A. Amro, G. Y. Liu, Characterization of AFM tips using nanografting, *Appl. Surf. Sci.* **175**, 649–655 (2001).
43. A.-S. Duwez, Exploiting electron spectroscopies to probe the structure and organization of self-assembled monolayers: A review, *J. Electron Spectrosc. Relat. Phenom.* **134**, 97–138 (2004).
44. P. Fenter, P. Eisenberger, K. S. Liang, Chain-length dependence of the structures and phases of ch3(ch2)n-1sh self-assembled on Au(111), *Phys. Rev. Lett.* **70**, 2447–2450 (1993).
45. R. G. Nuzzo, E. M. Korenic, L. H. Dubois, Studies of the temperature-dependent phase-behavior of long-chain normal-alkyl thiol monolayers on gold, *J. Chem. Phys.* **93**, 767–773 (1990).
46. M. D. Porter, T. B. Bright, D. L. Allara, C. E. D. Chidsey, Spontaneously organized molecular assemblies. 4. Structural characterization of normal-alkyl thiol monolayers on gold by optical ellipsometry, infrared-spectroscopy, and electrochemistry, *J. Am. Chem. Soc.* **109**, 3559–3568 (1987).
47. R. G. Nuzzo, B. R. Zegarski, L. H. Dubois, Fundamental-studies of the chemisorption of organosulfur compounds on Au(111) – implications for molecular self-assembly on gold surfaces, *J. Am. Chem. Soc.* **109**, 733–740 (1987).
48. T. L. Brower, J. C. Garno, A. Ulman, G. Y. Liu, C. Yan, A. Golzhauser, M. Grunze, Self-assembled multilayers of 4,4′-dimercaptobiphenyl formed by Cu(II)-catalyzed oxidation, *Langmuir* **18**, 6207–6216 (2002).

49. M. Kadalbajoo, J.-H. Park, A. Opdahl, H. Suda, J. C. Garno, J. D. Batteas, M. J. Tarlov, P. DeShong, Synthesis and structural characterization of glucopyranosylamide films on gold, *Langmuir* **23**, 700–707 (2007).
50. D. Liu, A. Bruckbauer, C. Abell, S. Balasubramanian, D.-J. Kang, D. Klenerman, D. Zhou, A reversible pH-driven DNA nanoswitch array, *J. Am. Chem. Soc.* **128**, 2067–2071 (2006).
51. J. t. Riet, T. Smit, M. J. J. Coenen, J. W. Gerritsen, A. Cambi, J. A. A. W. Elemans, S. Speller, C. G. Figdor, AFM topography and friction studies of hydrogen-bonded bilayers of functionalized alkanethiols, *Soft Matter* **6**, 3450–3454 (2010).
52. J. T. Riet, T. Smit, J. W. Gerritsen, A. Cambi, J. Elemans, C. G. Figdor, S. Speller, Molecular friction as a tool to identify functionalized alkanethiols, *Langmuir* **26**, 6357–6366 (2010).
53. W. J. Price, P. K. Kuo, T. R. Lee, R. Colorado, Z. C. Ying, G.-Y. Liu, Probing the local structure and mechanical response of nanostructures using force modulation and nanofabrication, *Langmuir* **21**, 8422–8428 (2005).
54. W. J. Price, S. A. Leigh, S. M. Hsu, T. E. Patten, G.-Y. Liu, Measuring the size dependence of Young's modulus using force modulation atomic force microscopy, *J. Phys. Chem. A* **110**, 1382–1388 (2006).
55. D. Scaini, M. Castronovo, L. Casalis, G. Scoles, Electron transfer mediating properties of hydrocarbons as a function of chain length: A differential scanning conductive tip atomic force microscopy investigation, *ACS Nano* **2**, 507–515 (2008).
56. S. Xu, P. E. Laibinis, G. Y. Liu, Accelerating the kinetics of thiol self-assembly on gold – a spatial confinement effect, *J. Am. Chem. Soc.* **120**, 9356–9361 (1998).
57. J. J. Yu, Y. H. Tan, X. Li, P. K. Kuo, G. Y. Liu, A nanoengineering approach to regulate the lateral heterogeneity of self-assembled monolayers, *J. Am. Chem. Soc.* **128**, 11574–11581 (2006).
58. S. F. Chen, L. Y. Li, C. L. Boozer, S. Y. Jiang, Controlled chemical and structural properties of mixed self-assembled monolayers of alkanethiols on Au(111), *Langmuir* **16**, 9287–9293 (2000).
59. D. Hobara, T. Kakiuchi, Domain structure of binary self-assembled monolayers composed of 3-mercapto-1-propanol and 1-tetradecanethiol on Au(111) prepared by coadsorption *Electrochem. Commun.* **3**, 154–157 (2001).
60. S. Ryu, G. C. Schatz, Nanografting: Modeling and simulation, *J. Am. Chem. Soc.* **128**, 11563–11573 (2006).
61. J. Liang, L. G. Rosa, G. Scoles, Nanostructuring, imaging and molecular manipulation of dithiol monolayers on Au(111) surfaces by atomic force microscopy, *J. Phys. Chem C* **111**, 17275–17284 (2007).
62. J. F. Liu, S. Cruchon-Dupeyrat, J. C. Garno, J. Frommer, G. Y. Liu, Three-dimensional nanostructure construction via nanografting: Positive and negative pattern transfer, *Nano Lett.* **2**, 937–940 (2002).
63. J. M. Lim, Z. S. Yoon, J. Y. Shin, K. S. Kim, M. C. Yoon, D. Kim, The photophysical properties of expanded porphyrins: Relationships between aromaticity, molecular geometry and non-linear optical properties, *Chem. Commun.* **3**, 261–273 (2009).
64. T. Hasobe, Supramolecular nanoarchitectures for light energy conversion, *Phys. Chem. Chem. Phys.* **12**, 44–57 (2010).
65. Z. M. LeJeune, M. E. McKenzie, E. Hao, M. G. H. Vicente, B. Chen, J. C. Garno, Self-assembly of pyridyl-substituted porphyrins investigated in situ using AFM, *Abstracts Am. Chem. Soc.* **235**, 216-COLL (2008).
66. R. G. Chapman, E. Ostuni, S. Takayama, R. E. Holmlin, L. Yan, G. M. Whitesides, Surveying for surfaces that resist the adsorption of proteins, *J. Am. Chem. Soc.* **122**, 8303–8304 (2000).
67. E. Ostuni, R. G. Chapman, M. N. Liang, G. Meluleni, G. Pier, D. E. Ingber, G. M. Whitesides, Self-assembled monolayers that resist the adsorption of proteins and the adhesion of bacterial and mammalian cells, *Langmuir* **17**, 6336–6343 (2001).

68. R. E. Holmlin, X. Chen, R. G. Chapman, S. Takayama, G. M. Whitesides, Zwitterionic sams that resist nonspecific adsorption of protein from aqueous buffer, *Langmuir* **17**, 2841–2850 (2001).
69. E. Ostuni, R. G. Chapman, R. E. Holmlin, S. Takayama, G. M. Whitesides, A survey of structure-property relationships of surfaces that resist the adsorption of protein, *Langmuir* **17**, 5605–5620 (2001).
70. Y.-Y. Luk, M. Kato, M. Mrksich, Self-assembled monolayers of alkanethiolates presenting mannitol groups are inert to protein adsorption and cell attachment, *Langmuir* **16**, 9604–9608 (2000).
71. S. Herrwerth, W. Eck, S. Reinhardt, M. Grunze, Factors that determine the protein resistance of oligoether self-assembled monolayers – internal hydrophilicity, terminal hydrophilicity, and lateral packing density, *J. Am. Chem. Soc.* **125**, 9359–9366 (2003).
72. Z. Grabarek, J. Gergely, Zero-length crosslinking procedure with the use of active esters, *Anal. Biochem.* **185**, 131–135 (1990).
73. D. J. Zhou, X. Z. Wang, L. Birch, T. Rayment, C. Abell, AFM study on protein immobilization on charged surfaces at the nanoscale: Toward the fabrication of three-dimensional protein nanostructures, *Langmuir* **19**, 10557–10562 (2003).
74. J. R. Kenseth, J. A. Harnisch, V. W. Jones, M. D. Porter, Investigation of approaches for the fabrication of protein patterns by scanning probe lithography, *Langmuir* **17**, 4105–4112 (2001).
75. C.-H. Jang, B. D. Stevens, R. Phillips, M. A. Calter, W. A. Ducker, A strategy for the sequential patterning of proteins: Catalytically active multiprotein nanofabrication, *Nano Lett.* **3**, 691–694 (2003).
76. C. Staii, D. W. Wood, G. Scoles, Verification of biochemical activity for proteins nanografted on gold surfaces, *J. Am. Chem. Soc.* **130**, 640–646 (2007).
77. M. Castronovo, S. Radovic, C. Grunwald, L. Casalis, M. Morgante, G. Scoles, Control of steric hindrance on restriction enzyme reactions with surface-bound DNA nanostructures, *Nano Lett.* **8**, 4140–4145 (2008).
78. M. Liu, G.-Y. Liu, Hybridization with nanostructures of single-stranded DNA, *Langmuir* **21**, 1972–1978 (2005).
79. P. V. Schwartz, Meniscus force nanografting: Nanoscopic patterning of DNA, *Langmuir* **17**, 5971–5977 (2001).
80. E. Mirmomtaz, M. Castronovo, C. Grunwald, F. Bano, D. Scaini, A. A. Ensafi, G. Scoles, L. Casalis, Quantitative study of the effect of coverage on the hybridization efficiency of surface-bound DNA nanostructures, *Nano Lett.* **8**, 4134–4139 (2008).
81. F. Bano, L. Fruk, B. Sanavio, M. Glettenberg, L. Casalis, C. M. Niemeyer, G. Scoles, Toward multiprotein nanoarrays using nanografting and DNA directed immobilization of proteins, *Nano Lett.* **9**, 2614–2618 (2009).
82. E. A. Josephs, T. Ye, Nanoscale positioning of individual DNA molecules by an atomic force microscope, *J. Am. Chem. Soc.* **132**, 10236–10238 (2010).
83. M. V. Lee, K. A. Nelson, L. Hutchins, H. A. Becerril, S. T. Cosby, J. C. Blood, D. R. Wheeler, R. C. Davis, A. T. Woolley, J. N. Harb, M. R. Linford, Nanografting of silanes on silicon dioxide with applications to DNA localization and copper electroless deposition, *Chem. Mater.* **19**, 5052–5054 (2007).
84. M. Despont, U. Drechsler, U. Durig, W. Haberle, M. I. Lutwyche, H. E. Rothuizen, R. Stutz, R. Widmer, G. Binnig, The “Millipede” – more than one thousand tips for future AFM data storage, *IBM J. Res. Dev.* **44**, 323–340 (2000).
85. K. Salaita, Y. Wang, J. Fragala, R. A. Vega, C. Liu, C. A. Mirkin, Massively parallel dip-pen nanolithography with 55,000-pen two-dimensional arrays, *Angew. Chem. Int. Ed.* **45**, 7220–7223 (2007).
86. J.-J. Yu, J. N. Ngunjiri, A. T. Kelley, J. C. Garno, Nanografting versus solution self-assembly of α, ω-alkanedithiols on Au(111) investigated by AFM, *Langmuir* **24**, 11661–11668 (2008).

87. C. H. Jang, B. D. Stevens, R. Phillips, M. A. Calter, W. A. Ducker, A strategy for the sequential patterning of proteins: Catalytically active multiprotein nanofabrication, *Nano Lett.* **3**, 691–694 (2003).
88. N. Nuraje, I. A. Banerjee, R. I. MacCuspie, L. T. Yu, H. Matsui, Biological bottom-up assembly of antibody nanotubes on patterned antigen arrays, *J. Am. Chem. Soc.* **126**, 8088–8089 (2004).
89. C. Staii, D. W. Wood, G. Scoles, Ligand-induced structural changes in maltose binding proteins measured by atomic force microscopy, *Nano Lett.* **8**, 2503–2509 (2008).
90. I. Kopf, C. Grunwald, E. Brundermann, L. Casalis, G. Scoles, M. Havenith, Detection of hybridization on nanografted oligonucleotides using scanning near-field infrared microscopy, *J. Phys. Chem. C* **114**, 1306–1311 (2010).

Chapter 6
Nanopattern Formation Using Dip-Pen Nanolithography

Bernhard Basnar

Abstract The fabrication of bottom-up nanostructures is a crucial step for the advancement of nanotechnology. Dip-pen nanolithography has started off as a method for the transfer of small organic molecules and has matured over the years to one of the most versatile patterning techniques available in the nanoscale. Three-dimensional structures made from organic or inorganic materials on a large variety of different substrates and length scales have been fabricated. This review highlights the techniques used for the fabrication of these structures together with their practical applications. Furthermore, the physical mechanisms involved in the dip-pen process are discussed by summarizing the experimental and theoretical results obtained so far.

Keywords Dip-pen nanolithography · Atomic force microscopy · Nanochemistry · Self-assembled monolayers · Nanowires · Biosensors · Nanoparticles · Structuring · Bottom-up · Template

Abbreviations

AFM	Atomic force microscopy
CNT	Carbon nanotube
DNA	Deoxyribonucleic acid
DPN	Dip-pen nanolithography
MHA	Mercaptohexadecanoic acid
MUA	Mercaptoundecanoic acid
ODT	Octadecanethiol
SAM	Self-assembled monolayer
STM	Scanning tunneling microscope

B. Basnar (✉)
Center for Micro- and Nanostructures, Vienna University of Technology, 1040 Vienna, Austria
e-mail: basnar@tuwien.ac.at; basnar@fkeserver.fke.tuwien.ac.at

A.A. Tseng (ed.), *Tip-Based Nanofabrication*, DOI 10.1007/978-1-4419-9899-6_6,

6.1 Introduction

Sub-micron structures are of high interest in many fields of science such as electronics, photonics, catalytics, or medical applications. The interest is based on the one hand on the fact that materials of this size show novel properties due to the large surface-to-volume ratio (leading for example to higher catalytic activity) or due to quantization effects. On the other hand, fabrication of such small structures enables higher density ordering providing higher storage capacity or, for medical applications, higher number of sensors on the same area. A variety of techniques have been used towards the generation of such small structures on surfaces. These can be divided into mask- or template-based system (such as photolithography, microcontact printing, nanoimprint lithography) which provide large area modification with high throughput, beam-assisted techniques (e.g. focused ion beam and electron beam lithography) which provide the pattern flexibility for fast prototyping but posses low throughput, and self-assembly processes (e.g. aggregation, alignment, segregation or dewetting) which are also found in nature and provide potentially fast modification of large areas whilst having very limited flexibility in the size and arrangement of different structures.

The invention of the STM in 1982 [1] and the AFM in 1986 [2] gave birth to one of the most powerful surface-analytical tools available. For the first time the topography and surface related properties could be probed simultaneously with down to atomic resolution. Especially the AFM with its different measurement modes (lateral force microscopy, phase imaging, force spectroscopy, ...) [3] gave the possibility to test nanomechanical properties with high lateral resolution. Apart from the high resolution imaging of samples, the close interaction between the tip and the surface provides the foundation for surface manipulation. In its most extreme form, individual atoms can be picked up and repositioned. This was demonstrated some 20 years ago, when a STM was used for one of the most intriguing experiments in the field of nanomanipulation – the moving of individual xenon atoms on a nickel substrate to form the IBM logo [4]. However, also mechanical or electrochemical interactions between the tip of the AFM and a solid surface can cause changes to the surface in the form of scratching [5, 6] or local oxidation [7], both of which have been utilized toward the fabrication of nanostructures [8–10].

About 10 years ago, in the year 1999, a new method for the patterning of surfaces based on an atomic force microscope set-up was developed [11]. The so-called "dip-pen nanolithography" combined the precision and resolution of atomic force microscopy with the local transfer of simple organic molecules to the surface [12]. The name very well described the process of pattern generation as, similar to a quill pen, a tip was dipped into an ink solution and used for writing on a solid surface until the ink was depleted, necessitating a re-dipping into the ink-pot. For the first time, the direct transfer of a multitude of different materials in a single patterning step with complete freedom on pattern size, shape, and position became possible. Since this first demonstration, hundreds of works have been published describing routines for the fabrication of patterns made from organic, inorganic, or biological species as well as their application. Also a large number of theoretical investigations have been undertaken, providing a good understanding of the governing principles in the material transfer and pattern formation.

The present article provides an introduction to the physical principles behind DPN and depicts the different approaches available for generation of nanolithographic patterns. The sections dealing with nanostructures made from various materials give a concise overview from literature on the various structures that have been prepared with different material systems. Finally, the future direction, challenges, and possible applications for DPN are addressed.

6.2 The Physics of Dip-Pen Nanolithography

The seemingly simple task of taking a goose feather, cutting it to size, dipping it into an ink pot and writing on a piece of paper is far from trivial. Over hundreds of years methods were developed to enhance the durability of the quill feather, to obtain stable ink solutions, and to optimize the paper properties in order to obtain homogeneous ink transfer and, thus, a high quality writing method. And still today, progress is being made which can be seen in the transition from quill pen to fountain pen and ball pens and which is evidenced by the continuing stream of patent applications. Similarly, DPN is focused on the apparently easy task of transferring molecules or particles from a coated tip to a substrate of interest. Like conventional ink-pen writing, the shape of the writing tip, the writing speed, and the composition of the ink and the substrate are important. However, in addition to this, diffusional parameters such as temperature and humidity, tip wear, applied potentials and other factors can greatly influence the writing process.

The transfer of material from a tip to a substrate can be divided into several different steps. The first step in this process is the formation of a liquid bridge between the tip and the sample due to capillary condensation. This meniscus plays a crucial role as it forms the transfer path for the ink molecules. The size of this meniscus was estimated by simulations and an increase with increasing humidity or reduced distance between the tip and the sample was predicted [13–15]. Experimental confirmation for this assumption as well as data related to the evolution of the meniscus as a function of humidity and time was obtained both by environmental scanning electron microscopy [16–17] (Fig. 6.1a) as well as by STM [18].

The shape of the meniscus is also influenced by the hydrophilic/hydrophobic character of the substrate. In case of a hydrophilic surface, the small contact angle leads to a spreading of the meniscus, providing a larger volume of the meniscus which is beneficial for the material transfer. On hydrophobic surfaces, however, the contact area of the meniscus is reduced, limiting the cross-section of the water bridge and, thus, reducing the material transfer. In order to verify this process experimentally, double ink DPN experiments with MHA (hydrophilic) and ODT (hydrophobic) were performed [19]. First, a dot pattern was produced with one ink. Afterwards, a second pattern was deposited on top of the first one, using the complementary ink (Fig. 6.1b). For ODT the transfer rate was increased when performing

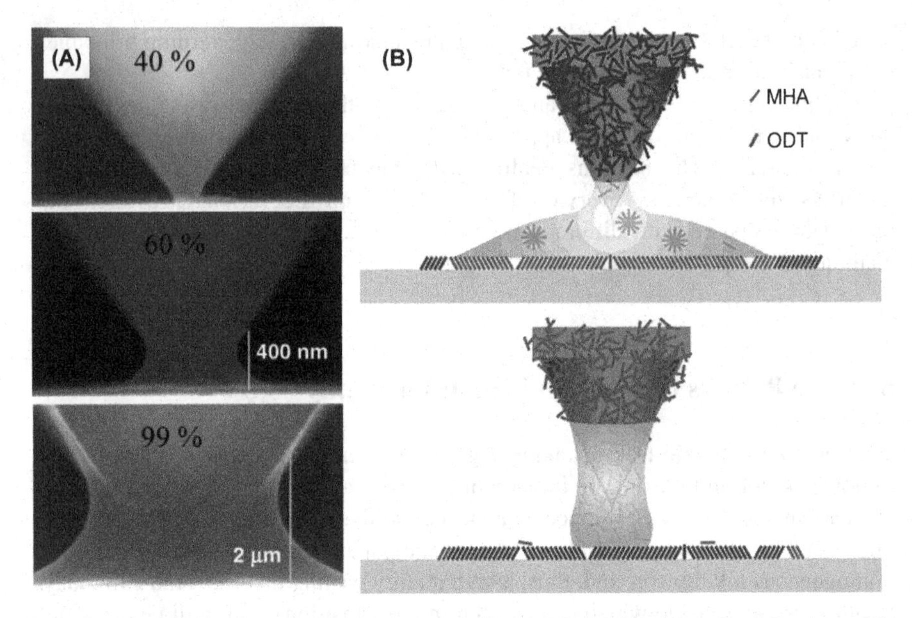

Fig. 6.1 Formation of a liquid meniscus as a function of humidity and surface chemistry: (**a**) Environmental scanning electron microscopy images of a meniscus between tip and substrate under varying humidity. (Reproduced in part with permission from [16]. Copyright 2005 American Chemical Society). (**b**) Different shape of a meniscus for hydrophilic (*top*) and hydrophobic (*bottom*) substrates, influencing the ink transfer rate. (Reproduced in part with permission from [19]. Copyright 2006 American Chemical Society)

the patterning on top of the hydrophilic MHA as compared to deposition on the bare gold substrate. For MHA the trend was the exact opposite, i.e. higher transfer rate was observed on naked gold as compared to on top of the hydrophobic ODT.

The formation of the meniscus has also pronounced consequences on the tip-sample interaction due to capillary forces. These capillary forces can cause tip abrasion [20] leading to larger radii of curvature. This, in turn changes the contact area and the size of the meniscus [13], leading to changes in the ink transport from the tip to the sample [21]. A detailed theoretical approach, based on the experimental data cited above, was developed for estimating the force [22]. Custom made modification of the radius of curvature by laser ablation allowed investigating the influence of the tip size on the deposition [23]. For tips showing radii between 40 and 500 nm a sub-proportional increase in the deposition rate was observed for increasing size of the tip, leading to larger feature sizes during the patterning process.

Whilst the meniscus provides the pathway for the ink-molecules from the tip to the surface, the material transfer itself is dependent on several experimental parameters. As was mentioned above, a higher humidity leads to a larger meniscus. This is advantageous for hydrophilic molecules (e.g. MHA), which exhibit higher transfer rates with increasing humidity [24, 25]. In the case of hydrophobic inks (e.g. ODT), this trend does not hold true anymore. Rather, the transfer rate seems to remain constant [26] or even to decrease slightly [25] with increasing humidity. This observation was attributed to the fact that the ink molecules needed to dissociate from the cantilever and dissolve in the meniscus fluid. For hydrophobic compounds, however, this transition is not favorable. In spite of the dependence of the meniscus on the humidity, even under 0% relative humidity (nitrogen-purged atmosphere) a residual water layer is present on the surface [27], which is sufficient for the formation of a meniscus, and deposition has been carried out successfully under these conditions [26, 28].

In order to elucidate the material transport further, the influence of organic vapors was investigated. Long chain alkanethiols have a substantially higher solubility in organic solvents as compared to water. Thus, if the solubility of the thiols is the rate-limiting step, then the presence of organic vapors should enhance material transport, especially if one keeps in mind that already a partial pressure of hexane of 75 torr is sufficient to form a 1 nm thick layer of hexane on gold [29]. Experiments with different organic solvents confirmed this assumption, i.e. an increase in material transport was observed for both ODT and MHA in the presence of these solvents [30]. However, no direct correlation was found between the solubility of these compounds in the various organic solvents and the actual increase in transfer rate.

A detailed investigation into the influence of ODT deposition under ethanol or water vapors provided additional insight into this phenomenon [31]. For low humidity (0%), homogeneously filled, dense lines were obtained. At 73% relative humidity, the line patterns showed dense layers only at the edge of the pattern with the middle exhibiting a low density of the molecules, suggesting that the diffusion of the ODT molecules occurs along the water/air interface of the meniscus instead of through the meniscus itself. In the presence of ethanol vapors (ethanol being a good solvent for ODT), the main part of the deposition was close to the tip apex with low density and poor resolution at the edges, indicating the formation of unordered films, especially at the edges of the pattern.

As the material transport is based on desorption, solubilization, and diffusion, a strong dependence on the temperature can be expected. Simulations investigating the deposition of ODT during short contact times were performed [32]. It was discovered, that most of the transferred material stems from the ruptured meniscus. An increase in temperature from 300 to 500 K led to an increase in the deposited amount by a factor of three. For longer contact times, the diffusion on the surface of the substrate becomes a dominating factor. In order to elucidate the diffusion behavior, experimental investigation of the material transfer of MHA in the temperature range between 10 and 30°C were undertaken. The deposition was found to be strongly influenced by the temperature, causing an increase in the diameter of dot features by up to a factor of 6 for this 20°C temperature rise [33]. Even more pronounced was the influence on the deposition of ODT, where a temperature increase of only 5°C (from 15 to 20°C) quadrupled the diameter of the patterns. Another group found both materials (MHA and ODT) to exhibit an exponential dependence of the deposition on the temperature, equivalent to an activation energy for the surface diffusion of 0.74 eV [31].

In addition to the sensitivity of the deposition on temperature and humidity, the transfer rate is still varying over the course of an experiment, even if all environmental parameters are kept constant. Reduction in the material flux will occur in case of diffusional limitations [34] or material depletion [28]. Also the surface diffusion of the patterned molecules can influence the material transfer rate [24, 35] (see further below for a more detailed discussion on the surface diffusion).

Although DPN is a method for the fabrication of nanopatterns of diverse chemical nature on a wide selection of substrates, the transfer of material from the tip to the surface is only successful if the choice of the ink fits the substrate properties. A main requirement in this respect is a similarity in the hydrophilicity/hydrophobicity of the involved materials. However, this similarity is not always given. In such cases, the use of a surfactant can be required for the effective transport of molecules to the surface. Studies on the transport of maleimide-linked biotin to mercaptosilane modified glass substrates investigated the influence of small amounts of Tween-20 (a non-ionic surfactant) in the ink solution [36]. The surfactant was found to significantly decrease the contact angle with increasing concentration, leading to a larger spreading of the meniscus and, hence, facilitating the transfer of the hydrophilic biotin derivative.

Once the ink molecule has been transferred to the surface, surface diffusion will determine the spreading and, thus, the final extension and shape of the created pattern. This diffusion can occur in two different ways, termed the "serial pushing" and the "hopping down" process (Fig. 6.2). In the serial pushing mode [37] a newly transferred molecule will displace a previously deposited molecule, pushing it outward. The displacement continues for subsequent molecules until the pattern boundary is reached. This diffusion mode is expected to happen in cases where only a small number of strong binding sites are available, as is the case for alkylsilanes on mica. In the hopping down process [38], the molecules transferred to the substrate are immediately bound strongly to the surface. New molecules thus diffuse on top of the first monolayer and "hop down" to the substrate once they reach the border of the patterned area, again becoming bound immediately. Such a system is found in the case of thiols on gold.

In many cases, a mixture of these two processes will occur, the extent of the contribution of each of these mechanisms being determined by the actual molecule-substrate binding energy and the number of binding sites. According to simulations [39, 40] low binding energies will lead to fractally shaped structures, intermediate energies will lead to patterns related to the substrate anisotropy (e.g. cross- or star-shaped patterns), whereas high binding energies lead to circular shapes. Experimental results for dodecylamine on mica [41] showed the formation of irregular shaped patterns expected for a low binding energy surface. Similar observations were obtained in the solution based deposition of octadecyltrichlorosilane on mica where this behavior was attributed to the small number of hydroxyl binding sites [42]. For the strong binding of thiols on gold, circular shaped patterns have been obtained regularly, with dual ink experiments giving proof to the hopping down mechanism [19].

In addition to the binding energy, the final shape of the patterns is also influenced by other parameters such as the "coherence length". The coherence length describes

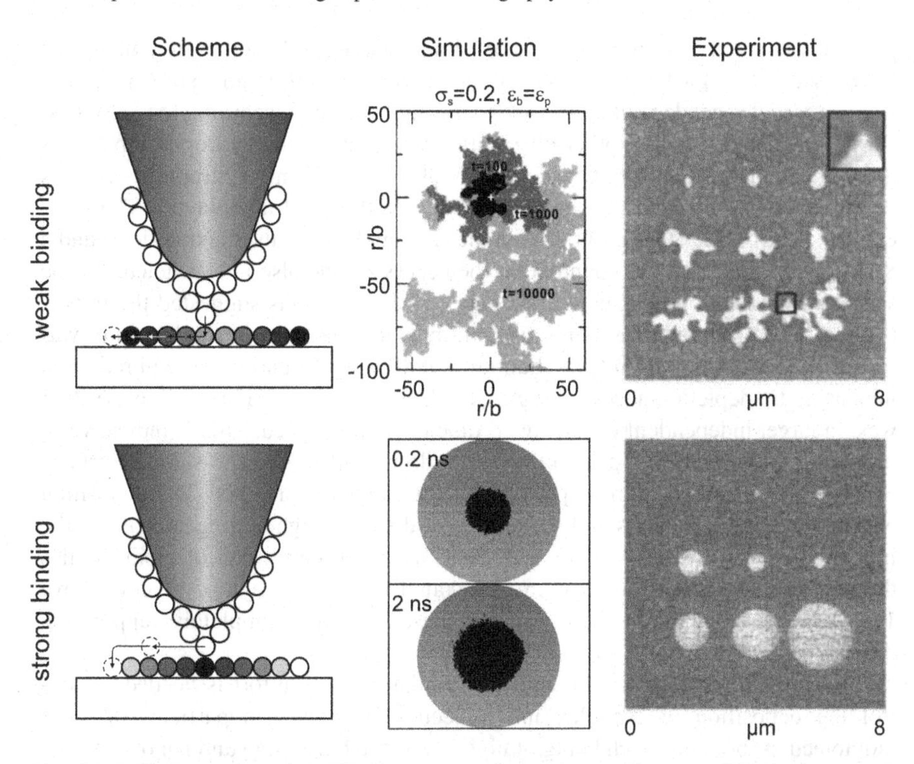

Fig. 6.2 Diffusion regimes in DPN. The *top row* depicts the pushing-type diffusion in case of weak ink-substrate binding, the bottom row the hopping down-type diffusion in case of strong ink-substrate binding. The shading of the deposited molecules depicted in the schemes relates to the time of their transfer to the surface with darker shading representing earlier transfer. Simulation for weak binding reprinted with permission from [40], Copyright 2006, American Institute of Physics; Simulation for strong binding reproduced in part with permission from [39], Copyright 2008 American Chemical Society. Experimental results reprinted with permission from [41], Copyright 2003 by the American Physical Society

the number of consecutive pushes (in the case of the serial pushing process) in the same direction induced by a single molecule transfer to the surface. This has an influence on the sensitivity of the diffusion on the substrate anisotropy [37]. Whilst for values up to 5 (i.e. on average 5 molecules will be pushed in the same direction by a single displacement event) the substrate anisotropy is negligible, already a value of 10 leads to the formation of branched structures with preferential orientation along the crystal axes of the substrate.

All these theoretical and experimental studies already provide a good understanding for the deposition mechanisms involved in DPN. Nevertheless, a full theoretical description is still missing, even in the case of such simple system like thiols on gold. Even more challenging is the investigation of other ink materials such as nanoparticles. A detailed investigation into the parameters influencing nanoparticle deposition was performed recently [43]. Gold nanoparticles with diameters ranging from 1 through 8 nm were deposited in high concentration on conventional cantilevers to allow for prolonged patterning. Different capping layers (negatively charged citrate, positively charged N,N-dimethylaminopyridine) and ink solvents (water, methanol) were used. The first finding was that for mixtures of differently sized particles, smaller particles get deposited first due to higher diffusivity according to the Stokes-Einstein equation [44]. Another observation was that for the hydrophilic particles investigated, deposition is only successful in the case of highly hydrophilic surfaces or in the case of covalent linking (e.g. through a thiolated monolayer). The presence of electrostatic attraction between the capping layer of the nanoparticles and a SAM on the substrate was found not to be necessary and also not sufficient for the successful generation of nanoparticle patterns. These results suggested the importance of the formation of a meniscus and, thus, the influence of the humidity was investigated. A minimum relative humidity of 40% was found to be required, with increasing tip depletion necessitating a further increase in humidity. The deposition was, in large, independent on the dwell time or writing speed. This is indicative of the fact that the particles, due to their size, do not diffuse on the sample surface, blocking further deposition of particles at the same sample spot. The deposition was not very homogeneous and so addition of different high boiling solvents to the ink solution was investigated (in analogy to increased deposition of biomolecules through surfactants [36]). Addition of hexanol (50%) or ethylene glycol (above 1%), however, only led to the transfer of these additives, completely suppressing the nanoparticle transfer.

This investigation shows that still substantial research effort is needed to control the deposition of complex ink molecules such as nanoparticles or large biomolecules, both of which being of high interest to biosensing and nanoelectronic applications.

6.3 Deposition Techniques

Dip-pen nanolithography is probably the most versatile of all nanopatterning techniques. This is due to the fact that it allows the formation of various nanostructures (organic, inorganic, and biological) with high resolution (a few nanometers) and high precision in potentially a single step. In order to obtain optimum results for such diverse material systems, several different methods for the generation of nanostructures have been developed. Figure 6.3 depicts the basic approaches for DPN pattern generation which are detailed in the following sections. In discussing the various approaches we will slightly expand the classical definition of DPN to all tip-based methods, where material is transferred to the surface by means of a cantilever or a nano-fountain pen.

6.3.1 Direct Transfer

The direct transfer of a chosen molecule to a specified position on the surface was the first DPN method developed. The beauty of this approach lies in its simple, one

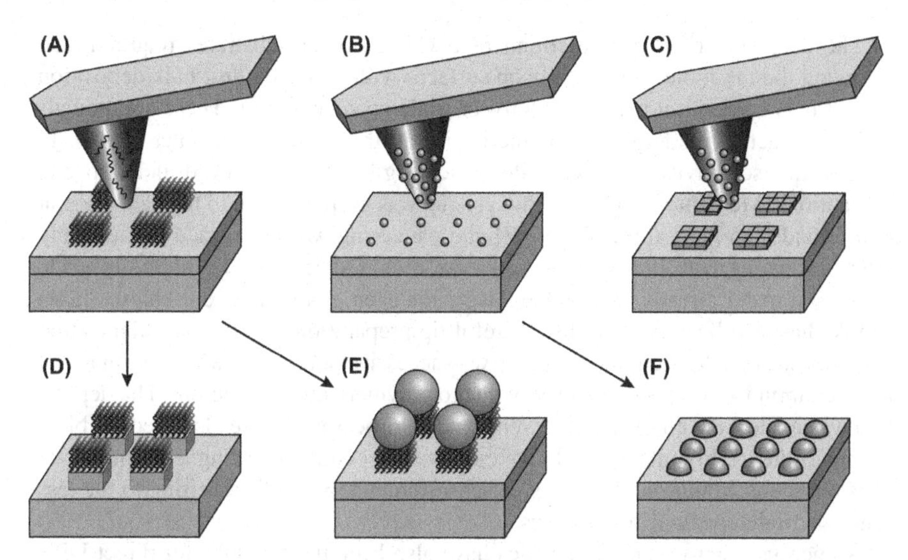

Fig. 6.3 Basic DPN techniques: The top row deals with techniques for the direct formation of nanostructures via DPN, whereas the bottom row deals with ex-situ techniques for the modification of such DPN patterns. Parts (**a**) and (**b**) show the deposition of molecular or nanoparticle patterns, respectively, by direct transfer to the surface. Part (**c**) depicts the pattern generation through an in-situ chemical reaction. Part (**d**) shows the transfer of a DPN pattern by etching. Parts (**e**) and (**f**) depict the growth of deposited nanostructures by selective binding or chemical enlargement, respectively

step lithographic patterning which avoids drawbacks like unspecific adsorption or sample contamination. As it is the most straightforward of all DPN techniques a lot of research effort has been put into developing recipes for all different kinds of material systems.

6.3.1.1 Conventional Deposition

Originally, this technique was developed using long-chain hydrocarbons with a thiol group as linker for binding to gold substrates [11]. This model system has been thoroughly investigated not only for the generation of various patterns, but also in order to understand the fundamental principles of DPN (see Section 6.2: Physics of DPN). Thiolated SAMs have several highly beneficial properties for applications in scanning probe lithography. Firstly, they have a high coefficient of diffusion from the tip to the surface, allowing for fast writing. Secondly, they form strong bonds with the underlying gold substrates which leads to minimal surface diffusion and, thus, patterns with sharp boundaries. The strong binding, together with a high degree of ordering due to hydrophobic interactions of adjacent molecules leads to dense, crystal-like patterns with well defined height. Furthermore, by varying the functional end-group of these molecules, subsequent linking of nearly any kind of material can be performed [45].

The extension of DPN patterning of SAMs to other substrates required also changing the ink molecule. For silicon surfaces with native oxides, bulk deposition of SAMs is carried out using trichloro- or trialkoxysilanes [46]. These compounds are very reactive, undergoing polymerization in the presence of water vapors. In spite of this sensitivity to water, DPN patterning has been achieved under special precautions. Mercaptopropyltrimethoxysilane was evaporated onto the cantilever at 120°C under an inert atmosphere [47] and patterning was performed immediately after tip coating with humidity being below 25%. Exceeding 30% humidity made material transfer already impossible. Using the even more reactive trichlorosilanes as inks has also been shown [48]. Careful tip preparation by previous silanization was needed and allowed patterning for two successive hours after which the material transfer abruptly stopped, supposedly due to polymerization of the ink. The deposition was made from chlorinated solvents and chlorocarbons were also used for blow drying of the freshly inked tips. In spite of the successful patterning, environmental considerations would prompt using either different solvents or reverting to the less sensitive trialkoxysilane alternatives.

Numerous other organic molecules have also been used as inks for direct DPN pattern generation. Among these are dyes [49], monomers [50], and others. As these molecules have not been that thoroughly investigated, no general parameters are available for their deposition. However, similarity in hydrophilicity or the presence of surface binding groups (e.g. electrostatic [51] or host/guest [52]) have been found to be advantageous for a fast and consistent material transfer.

Apart from simple molecules this technique of direct deposition has also been applied to the formation of nanoparticle patterns (Fig. 6.4a). A wide variety of different particles have been investigated with special emphasis on their surface modification (thiolated [53], negatively charged [54], positively charged [55], hydrophobic [56]). Particles in general do not show diffusion on the substrate surface [43], making the pattern size independent on the dwell time. However, the features can be increased in size by higher contact pressure and higher humidity. In the case of nanoparticles, the presence of strong binding groups is even more important than in the case of small organic molecules due to the higher mass and, thus, lower coefficient of diffusion.

Also biological entities can be transferred with this method. DNA was found to easily transfer to the surface. In this case, modification of one end of the DNA with thiols [57] provided good anchoring possibilities to the gold surface. More complicated molecules, such as enzymes [58], were also deposited successfully, retaining their catalytic activity which is indicative of the fact that the deposition does not cause significant changes to the secondary or tertiary structure of these proteins. Their size allows attachment through electrostatic attraction although in many cases chemical linking through histidine [59, 60], epoxy-groups [61], host-guest interaction [62], or others was preferred in order to obtain patterns which would be insensitive towards subsequent cleaning steps (e.g. rinsing).

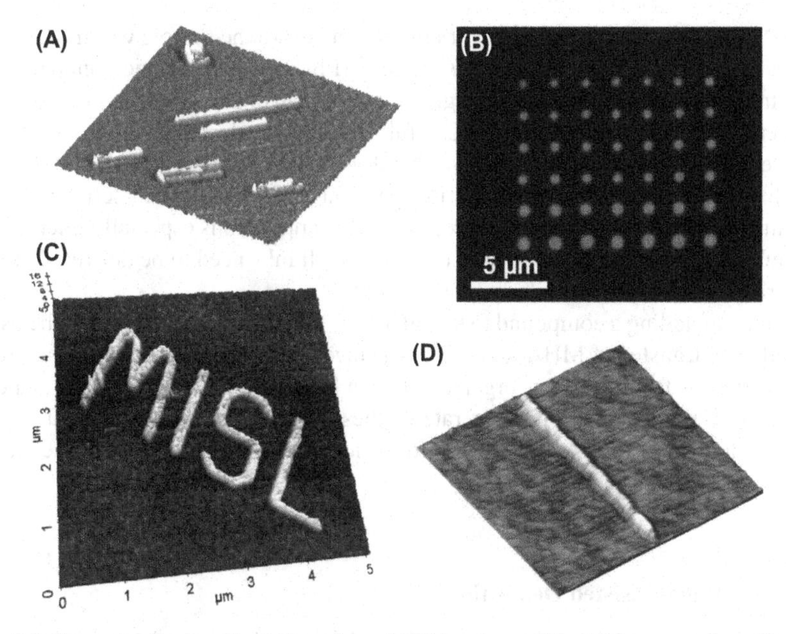

Fig. 6.4 Examples for the direct formation of DPN patterns: (**a**) AFM image of a palladium nanoparticle pattern generated by conventional deposition ([55] Reproduced by permission of the Royal Society of Chemistry). (**b**) Epifluorescence image of IgG patterns obtained by matrix-assisted deposition using agarose as carrier (Reproduced in part with permission from [63]. Copyright 2009 American Chemical Society). (**c**) AFM image of a carbon pattern generated by voltage-assisted material transfer (Reproduced in part with permission from [66]. Copyright 2009 American Chemical Society). (**d**) AFM image of an indium pattern obtained via temperature assisted deposition (Reprinted with permission from [74]. Copyright 2006, American Institute of Physics)

6.3.1.2 Additive Enhanced Deposition

The ink for DPN patterning consists, in the simplest case, only of the molecule to be patterned, either in the neat form or dissolved in a suitable solvent. However, in many cases additives are added which influence the reactivity or diffusion of the ink. In one of the first approaches, a surfactant was added to the ink solution [36]. The surfactant increased the wettability of the partially hydrophobic surface, leading to enhanced ink transfer. This addition of surfactant allows limiting the effect of the differences in hydrophilicity/hydrophobicity of ink and substrate, providing a more versatile patterning technique.

A different approach pursued was the use of a matrix for the ink molecule. By incorporating the ink molecule into the solution of a matrix material, the deposition is not dependent anymore on the diffusion rate of the ink but rather on the carrier. In this way, different ink molecules can be transferred with identical deposition rates. To this end, agarose was employed [63] which is a good matrix material for various biological compounds. The DPN patterning was performed on reactive surfaces which allowed for strong binding of the biomolecules, and the transferred proteins and DNA strands were let to react for at least 4 h before the sample was thoroughly rinsed in order to remove the agarose matrix (Fig. 6.4b). Inorganic nanoparticles were transferred using polyethylene glycol as the matrix compound [64]. Gold and magnetic nanoparticles as well as C_{60} fullerenes were deposited on a variety of surfaces and the polymer was removed by dry oxidation in oxygen plasma. It was found, that the pattern size was nearly independent on the substrate material, proving the universal nature of matrix-assisted DPN. This approach is especially interesting for multiplexed patterning of several inks, where all inks need to be delivered at the same rate, so as to yield similarly sized features.

Instead of adding a compound to the ink itself, vapors phase additives can be used as well. The transfer of MHA and ODT was investigated as a function of the presence of vapors from different organic solvents [30]. The presence of these vapors significantly increased the transfer rate of these thiols. This was attributed to the higher solubility of MHA and ODT in organic solvents as compared to the water present in the meniscus.

6.3.1.3 Voltage-Assisted Deposition

The use of voltage allows adding an additional driving force for the material transfer from tip to sample. This can be especially important for larger molecules or nanoparticles, where this approach can open up a pathway for the controlled activation and deactivation of the patterning simply by changing the applied voltage. A quite extensive set of experiments was performed investigating the deposition of organic molecules and polymers onto a variety of substrate (highly ordered pyrolithic graphite, indium tin oxide, native and passivated gold) [65]. The field-assisted deposition of fullerenes onto passivated gold allowed monolayer dots and multilayer lines to be written, with the feature size increasing with increasing bias. The patterning was successful for all the different inks under investigation. As no deposition occurred without applied potential and the influence of humidity was negligible, the authors concluded that the material transfer is not based on a meniscus (as in conventional DPN), but that it is rather akin to field ionization microscopy. This means that the ink molecules travel to the apex of the tip and get ejected due to the high field gradients. This field-assisted patterning is a promising path towards producing high resolution patterns (down to a few nanometers) by utilizing tips with a small radius of curvature.

Another interesting approach for voltage assisted deposition entailed the formation of carbon nanostructures on gold substrates [66]. To this end, the carbon was first harvested by scanning over a glassy carbon electrode. Due to the applied positive tip potential, carbon was moved from the substrate to the cantilever. Subsequent scanning of a gold surface with inverted voltage led to the redeposition of the collected carbon (Fig. 6.4c). As usual for tip-based techniques, relatively high voltages of ± 6 V were required for the material transfer. A big advantage of this technique lies in the possibility for error correction as deposited structures could be removed again from the gold surface by applying a positive tip bias.

The transfer of gold nanoparticles was also enhanced using external potentials [67]. Gold nanoparticles with a negatively charged capping layer were transferred to a silicon surface which also had been coated with a negatively charged SAM. Application of large, positive tip biases (6–12 V) led to the oxidative removal of the monolayer on the surface and formation of a field-induced oxide. Simultaneously, the larger the applied bias, the better the transfer of the nanoparticles occurred. The negatively charged layer on the surface was found to be crucial to prevent unspecific adsorption of the gold on the surface, leading to higher pattern fidelity.

As shown in the previous example, the applied voltage not only can be used for providing an additional driving force for the deposition. Rather, it can also be utilized to activate the surface by local oxidation. The approach was employed towards the deposition of trialkoxysilanes on passivated surfaces [68]. A silicon substrate was modified with octadecyltrichlorosilane to form a hydrophobic monolayer. The tip was dipped into a solution of trialkoxysilanes in toluene and scanned across the surface under an applied potential of 5–10 V. In order to facilitate the oxidation and material transfer high humidity (close to 100%) and contact forces between 1 and 10 nN were employed. Monolayer structures with 50 nm widths could be created in this way whilst scanning without applied potential showed no pattern generation.

This activation of the surface can also be utilized for the creation of biopatterns. Histidine-tagged proteins were used as an ink for deposition onto nickel substrates [69]. Histidine is known to form a complex with nickel ions, thus forming a linker between the surface and the protein. By application of relatively small potentials (between –2 and –3 V), the nickel surface was locally oxidized, providing the necessary binding sites for the histidine group. This approach was applicable for various proteins (investigated up to a size of 43 kDa) and for porphyrin, which yielded fluorescently labeled patterns.

6.3.1.4 Temperature-Assisted Deposition

As the DPN process is in most cases governed by diffusion, increasing temperatures will also increase the deposition rate [33]. Furthermore, chemical binding reactions can also be favorably influenced by higher temperature [70]. In addition to these effects, temperature can also aid the deposition of solid materials. Coating of the cantilever with solids of low melting point offers a practical way for the temperature assisted DPN fabrication of nanostructures by heating the tip. One of the biggest advantages is that by reducing the temperature, the same tip used for patterning can also be employed for characterization of these patterns without causing additional material transfer. This on/off behavior of the patterning is facilitated by the potentially low heat capacity of the cantilever, enabling fast heating and cooling.

Octadecylphosphonic acid was the first solid ink used for thermal DPN [71]. By heating the cantilever above 99°C (which is the melting point of this compound) lines and squares could be written on mica. It was found that already a temperature close to the melting point led to partial deposition. Also, the heat capacity of the cantilever led to a still noticeable deposition even after the heating current had been turned off for 2 min.

Patterns made from polymeric material can be fabricated in a similar fashion. Poly(3-dodecylthiophene), a semiconducting polymer, was deposited onto silicon by heating the cantilever up to 150°C [72]. Varying the temperature and/or the scan speed led to changes in the feature height in monolayer increments. Additionally, working close to the melting point of the polymer induced a high orientation of the polymer along the fast scan axis due to molecular combing by the tip. A diamond coated cantilever allowed fabrication of several thousand such structures without change in the tip shape and, thus, the feature size [73].

Continuous metallic lines have been produced using an indium coated cantilever [74]. Indium is very advantageous due to its good wetting properties on various surfaces and its relatively low melting point (156°C). However, high cantilever temperatures were needed to facilitate pattern formation. For a contact force of 1,000 nN temperatures of 500°C were required with lower contact forces necessitating even higher temperatures (above 1,000°C at 20 nN contact force). In spite of these technical difficulties, wires down to 100 nm width and 5 nm height with good electrical conductivity could be produced (Fig. 6.4d).

6.3.2 Patterns from In-Situ Reactions

Although, strictly speaking, also the direct material transfer described in the previous chapter often times meets the criterion of in-situ reactions through the formation of new bonds between the ink and the substrate, this chapter deals with more complex chemical reactions. The in-situ chemical conversion of materials combines the advantage of being a direct deposition technique (i.e. single step patterning without unspecific deposition) with the means to pattern a broad variety of different materials which cannot easily be patterned by direct deposition itself.

6.3.2.1 Redox Reactions

Redox reactions are the basis for galvanochemistry, which is a well known method for the fabrication of thin coatings on surfaces. The electroless or potential driven deposition of metals can be carried out with high control on the physicochemical properties of the film such as thickness or composition. The same basic chemistry can also be used in the nanoscale for DPN patterning.

One approach involves the electrochemical reduction of metal salt-based inks by the surface atoms. This has been shown for noble metals like gold on silicon [75] (Fig. 6.5a) and gold, silver, or platinum on germanium [76]. This method works without applied potentials harnessing the difference in electrochemical potential between the metal salt and the substrate material. The reaction for gold follows Eq. (1) and leads to a self-limiting growth of metallic patterns as the underlying semiconductor substrate becomes insulating due to oxide formation, blocking the transport of electrons from the substrate to the excess metal salt solution. The minimum feature size for this approach was lines of 30 nm width.

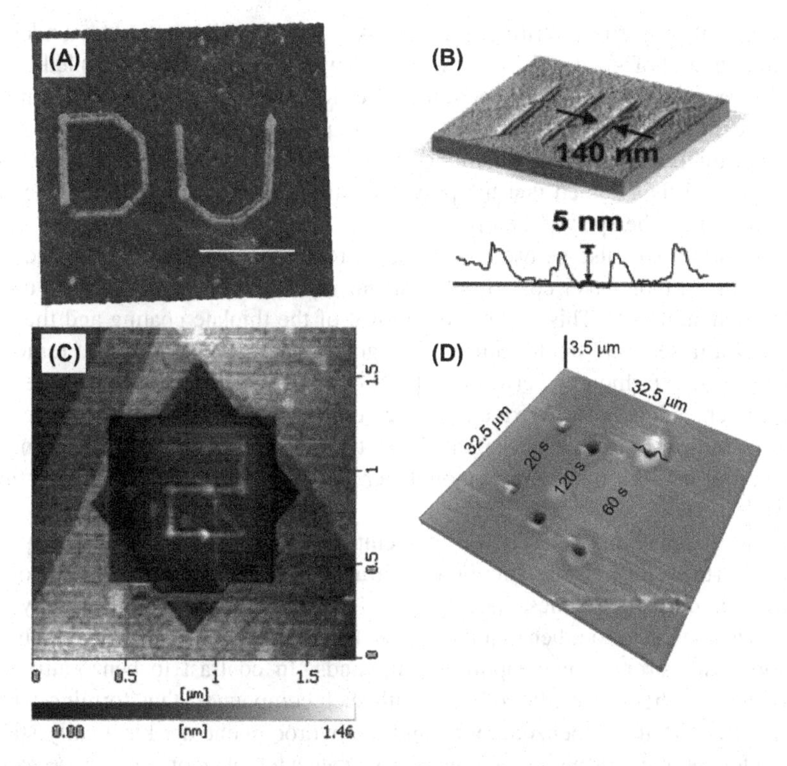

Fig. 6.5 Patterns created by in-situ reactions: (**a**) Gold pattern obtained by electroless deposition of gold salt. (Reproduced in part with permission from [75]. Copyright 2001 American Chemical Society). (**b**) Silicon oxide structures created by hydrolysis of a polymer-embedded precursor with subsequent thermal treatment. (Reproduced in part with permission from [81]. Copyright 2002 American Chemical Society). (**c**) Nanostructure obtained by etching of gold with allyl bromide. The inner pattern shape (similar to the number 2) consists of redeposited material from a previous etching and acted as etch resist during the fabrication of the etched squares. (Reproduced with permission from [83]). (**d**) Digestion of bovine serum albumin films by protease deposited with a fountain pen. (Reproduced in part with permission from [84]. Copyright 2003 American Chemical Society)

$$4HAuCl_4 + 3Si + 6H_2O \rightarrow 4Au + 3SiO_2 + 16HCl \qquad (1)$$

Metallic nanostructures can also be obtained using an applied bias. Platinum lines were deposited onto silicon by application of bias to the tip [77]. Whilst a negative bias of –10 V led to the local oxidation of the silicon substrate, a bias of +4 V yielded a platinum wire of 0.4 nm height and 30 nm width. This structure was found to be catalytically active at elevated temperatures, proving the metallic nature of the deposit.

Instead of reducing metals, potentials can also be used for the electrochemical polymerization of a monomeric ink. 3,4-Ethylenedioxythiophene is the precursor for PEDOT, a well known conducting polymer with interesting electrical and electrooptical properties. Writing with an inked tip on a silicon surface under an applied potential of –9 to –15 V (for substrates with a native oxide layer) led to the oxidative polymerization of the monomer and deposition of patterns consisting of the conducting polymer [78]. Thicker oxides required higher voltages (–130 V for a 500 nm thick oxide layer). As no significant current can be expected with such thick oxides, it is assumed that the polymerization is caused by the large electric field gradients at the apex of the tip.

Ink solutions can also be used for the local reduction of surface compounds. In an experiment, monolayer coated gold was locally oxidized by an AFM tip under applied potential [79]. This led to the removal of the thiolate coating and the formation of a raised gold oxide pattern. This gold oxide structure could be removed in an electroless fashion by performing DPN of ethanol. The ethanol acts as reducing agent, changing the gold oxide to native gold. As a consequence, the previously raised feature is converted into a small pit, the depth of which corresponding to the height of the surrounding thiol monolayer. This oxidation and reduction is fully reversible.

An ink-based oxidation scheme was employed for the patterning of hydroquinone surfaces [80]. Cerric ammonium nitrate is the ink used for the oxidation of surface-bound hydroquinones, causing a local activation of the surface. The hydroquinone is converted to a benzoquinone which can undergo Michael-type addition reactions with strongly nucleophilic compounds. In contrast to benzoquinones, hydroquinones themselves do not react with such compounds, thus forming a passivating layer. Both the benzoquinone and the hydroquinone are kinetically stable at room temperature, providing the means to create a template on the surface which can be modified at a later stage by reaction with a suitable nucleophile, such as the amine found in native proteins or amine-modified nucleotides.

6.3.2.2 Hydrolysis

The presence of water molecules in the form of a water adlayer on the surface or in the vapor phase is essential for the formation of the meniscus. In addition to this, they can also actively take part in chemical reactions by converting the ink molecules via a hydrolysis reaction.

The first work utilizing hydrolysis was developed already 3 years after the invention of DPN. Sol-gel chemistry was employed to produce patterns of aluminum oxide, tin oxide, and silicon oxide [81]. The metal chlorides were mixed with a copolymer in an ethanolic solution. The resulting sol is stable for several hours, providing sufficient time for performing DPN experiments. The chemical reaction taking place follows the general formula (Eq. 2) with M denoting the respective metal.

$$2MCl_n + nH_2O \rightarrow M_2O_n + 2nHCl \quad (2)$$

The obtained patterns consisted of the metal oxide embedded in a polymer matrix. By heating up the samples to 400°C, the polymer was decomposed, leaving behind the purely inorganic pattern (Fig. 6.5b). This was accompanied by a reduction in the pattern height by about 50% as compared to the height measured directly after the pattern deposition.

A more complex chemical reaction was utilized for the fabrication of cadmium sulfide nanostructures [82]. An ink consisting of thioacetamide and cadmium acetate was used as the patterning solution. The thioacetamide can undergo hydrolysis (Eq. 3), producing hydrogen sulfide as one of the reaction products. This compound forms a precipitate with the cadmium ions present in the ink solution (Eq. 4).

$$CH_3CSNH_2 + 2H_2O \rightarrow CH_3COONH_4 + H_2S \quad (3)$$

$$H_2S + Cd(CH_3COO)_2 \rightarrow CdS + 2CH_3COOH \quad (4)$$

The DPN process was found to be highly dependent on the experimental conditions with optimum results being achieved using negatively charged surfaces, room temperature deposition at 40% relative humidity, and slow scan speeds of 60 nm/s.

6.3.2.3 Etching

Instead of depositing ink molecules on the surface, the ink solution can also consist of an etchant. In this case, the DPN approach leads to a subtractive pattern generation similar to top-down approaches. One etch-based approach was shown by patterning gold substrates with allyl bromide [83]. This organic compound is able to dissolve gold, simultaneously forming a solid residue which gets deposited onto the cantilever. This deposit is weakly bound and can, also, be transferred to the surface in a DPN fashion, acting as etch resist. This redeposited material is chemically stable towards reaction with allyl bromide and, thus, subsequent DPN etching of the surface with allyl bromide leads to the formation of raised deposits on an etched background (Fig. 6.5c). The reactions involved in this etching process are not yet fully understood and are, thus, subject to further investigations.

The approach of subtractive patterning is also feasible for the chemical degradation of protein films using proteases. Deposition of trypsin on a protein layer consisting of bovine serum albumin was performed using fountain-pen nanolithography [84]. The deposition of small volumes of the enzyme solution led to a selective digestion of the protein film, causing the formation of dips or trenches (Fig. 6.5d). Concurrently, the incorporation of water into the protein film led to a localized swelling. Thus, the structures obtained consisted of an etched trench with raised rim, which was in good agreement with simulations. A more detailed investigation [85] allowed estimating the total turnover of the trypsin molecules (about 1,900 BSA molecules per trypsin molecule) until deactivation of the enzyme due to evaporation of the co-deposited water. As expected, deposition of larger volumes or longer dwell times led to a higher degree of hydrolysis and, thus, more pronounced etching of the protein layer due to a higher amount of enzymes transferred to the surface and longer time until evaporation of the water.

Similarly, also an oligonucleotide surface can be digested by a DNase delivered to the surface [86]. In this case, DNase I patterns were created by DPN on an oligonucleotide SAM on gold. The ink solution was phosphate buffer-based and without magnesium ions. The DNase could be visualized after deposition by AFM. Immersion of the sample into a buffer solution with magnesium ions led to the selective digestion of the oligonucleotide. High pattern fidelity was achieved and only slight widening of the patterned lines due to diffusion was observed. The magnesium ions were necessary for the functioning of the DNase, and immersion into magnesium free solutions showed no oligonucleotide degradation.

Instead of patterning oligonucleotide films, also the controlled dissection of DNA strands using enzymatic cleavage is possible. Again a buffered solution of DNase was used, this time containing magnesium [87]. Linear λ-DNA was spread out on an aminosilane treated mica surface by a molecular combing technique. Small droplets of the DNase ink were deposited along the length of the DNA. The samples were kept in an incubator at 37°C and 90% humidity for 30 min, followed by rinsing with distilled water. In 50% of all cases, the DNase spots led to the localized digestion of the DNA, breaking the long strand into smaller pieces. Control experiments again confirmed that both the DNase as well as magnesium ions were required for the reaction. This approach could prove very helpful for single molecule investigations of DNA where a selective cutting of individual domains is of great interest.

6.3.2.4 Other In-Situ Reactions

The tip-sample system can be viewed as a miniature reaction vessel. In such a reaction vessel several compounds can be present simultaneously in high concentrations (either adsorbed on the surface or present in the ink) and can be subjected to high local pressures or electric fields to induce chemical reactions (Fig. 6.6). The high concentration of the reactants as well as the pre-orientation of surface-adsorbed molecules leads to high reaction rates which are an essential requirement for DPN as the dwell-time (and thus the time for reaction) is very short. One of the first examples for an in-situ reaction was based on a local Diels Alder reaction [88], which is a 2+4 cycloaddition reaction, using a terminal alkene monolayer on the surface and a modified furan as ink. The high local concentrations together with a tip-force of 32 nN (equating to a pressure of 26 MPa) induced the cycloaddition reaction. Monomolecular lines with heights of 2.5 nm and linewidths of less than 40 nm were obtained.

A similar kind of reactions is used in the case of click chemistry based on a 1,3-dipolar cycloaddition reaction [89]. A surface was modified with an alkyne. This compound can undergo an addition reaction with a modified azide. However, DPN of such an azide showed no reaction even for dwell times of up to 10 min. By adding copper ions to the ink solution, the reaction was catalyzed by a factor of 10^7, making patterning possible even at write speeds above 1 μm/s.

The Michael addition reaction is a very versatile reaction for the formation of covalent bonds between α,β-unsaturated carbonyl compounds and thiols in solvent

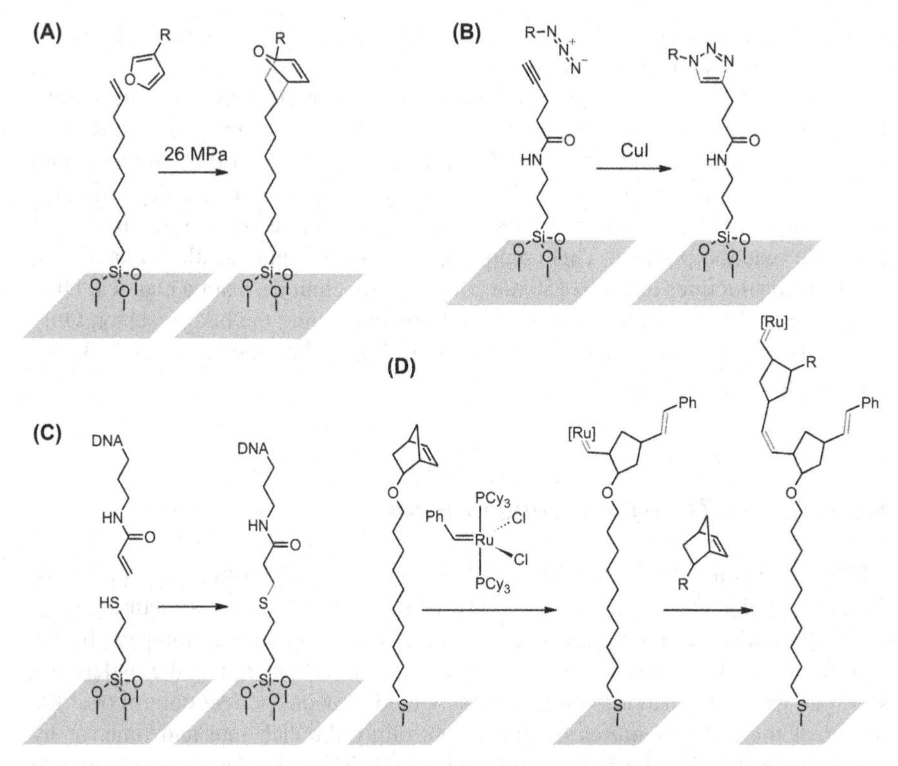

Fig. 6.6 Different DPN induced reaction schemes for the fabrication of nanostructures: (**a**) Diels-Alder reaction, (**b**) 1,3-dipolar cycloaddition, (**c**) Michael addition, and (**d**) ring-opening metathesis polymerization. The grey color marks the new bonds formed between the molecules during the reaction process

free environment [90] as is encountered during DPN. Reaction of an acrylamide-modified DNA strand with a thiol-modified silicon surface enabled the direct deposition of DNA onto a semiconducting substrate [57]. The DNA was covalently bonded to the surface and retained its ability for hybridization with a complementary DNA strand.

Also catalyzed polymerization reactions can be carried out. Ring-opening metathesis polymerization of modified norbornen molecules was investigated [91]. A substrate was modified with a monolayer of norborneyl silane and let to react with a Ruthenium based Grubbs' Catalyst. DPN of a norbornen derivative caused localized polymerization. By increasing the deposition time (i.e. the dwell time on a specific spot), the degree of polymerization also increased. Due to the fast reaction, this did not lead to a growth in the lateral direction but rather to a linear increase in the structure height over time. This increase in size was larger than was observed for surface polymerization performed in solution. As before, the reason for this observation is probably found in the high local concentration of reactants, leading to higher reaction rates. This approach allows generating, in-situ, patterns of varying heights and/or patterns of varying chemical nature simply by changing the interaction time or by using differently modified monomers.

A completely different approach towards reaction based DPN was pursued utilizing an enzyme immobilized to the cantilever [92]. Alkaline phosphatase was covalently linked to the tip of an AFM cantilever. Immersion of this cantilever into a solution containing 5-bromo-4-chloro-3-indolyl phosphate led to a dephosphorylation and, in the presence of nitro blue tetrazolium chloride, to the formation of an insoluble reaction product. This compound precipitated underneath the cantilever and formed structures down to 150 nm width. This technique is not a classical DPN technique as the tip is immersed into a solution containing the ink precursor. Only the catalytically active enzyme converts the precursor into the respective ink and transfers the ink through precipitation.

6.3.3 Pattern Transfer Through Etching

DPN, in general, creates patterns by depositing material (bottom-up approach). However, such patterns can, subsequently, be used for subtractive structuring (top-down approach) very much alike conventional photolithographic techniques. To this end, the created patterns are being used as etch resist, protecting the underlying substrate from chemical dissolution. This method provides perfect control over the height of the etched features by either controlling the etch rate and time, or by depositing material in the form of thin films of the desired thickness onto a suitable substrate before carrying out the patterning process. For metallic nanostructures created in this fashion an additional advantage is the high conductivity achievable due to the bulk nature of the thin-films.

The standard compounds used for this purpose are ODT and MHA on gold substrates. The advantage of long-chain SAMs is their high degree of order and the hydrophobic nature of their aliphatic back bone. This prevents the aqueous etch solutions to penetrate the pattern and, thus, provides good etch protection in spite of the small thickness (about 2 nm) [93] of these films. The etching of gold substrates was achieved by using a ferri/ferrocyanide solution, yielding gold patterns whose lateral size was determined by the dwell time of the ODT [94] or MHA [95] patterning. Silver patterns could be prepared with the same recipe, whilst for palladium an iron chloride-based etchant was used [96]. This latter etchant gave good results for ODT resist layers, but with MHA the pattern integrity was lost. The metal structures obtained in this way can be used as etch resist for the patterning of the underlying material. By using substrates consisting of a gold film on silicon, silicon pillars with defined lateral size (determined by the DPN patterning process) and height (determined by the etch time) were produced [94] (Fig. 6.7).

The use of ODT or MHA patterns provides a negative etch resist, i.e. the patterned area is not etched. However, also positive etch resist behavior can be achieved by carrying out an image reversal process [97]. The basis for this image reversal is found in the difference in electrochemical potential needed to desorb MHA or

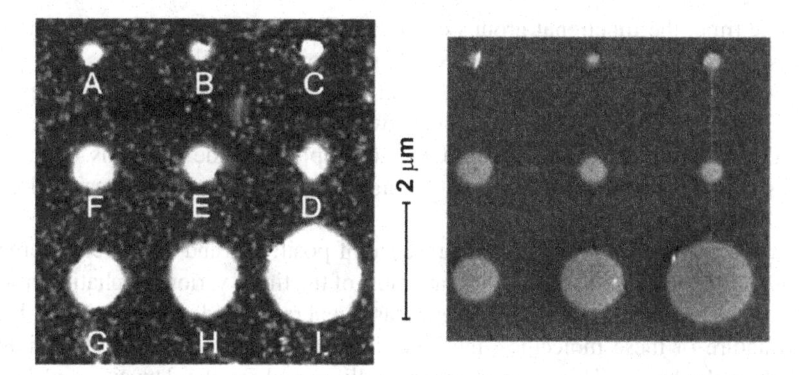

Fig. 6.7 Pattern transfer from a DPN generated structure to the substrate through etching. The *left part* shows a lateral force image of DPN generated ODT patterns on gold coated Si/SiO_2, the *right part* depicts silicon oxide pillars on a silicon substrate obtained by etching the unprotected gold with ferri/ferrocyanide and etching the unprotected silicon oxide with hydrofluoric acid. (Reproduced with permission from [94])

ODT from the metal substrate. MHA patterns are deposited by DPN on gold and the remainder of the surface is passivated by immersion into an ODT solution. The MHA can subsequently be removed by electrochemical desorption at −0.8 V without influencing the ODT molecules. Thus an inverted image is obtained where the original DPN-patterned areas are now unprotected whereas the remaining surface is coated by an ODT etch resist.

6.3.4 *Pattern Growth Through Seeding*

The formation of patterns by DPN is often times just the first step in the generation of the desired nanostructure. In this case, the pattern acts as seed or template and a subsequent modification through selective binding of nanostructured materials (biomolecules, nanoparticles, . . .) or chemical conversion (growth of nanowires, electroless deposition, . . .) can lead to new structures [98]. The advantages of this approach lie in the fact that simple molecules can be used for the templating and that three-dimensional structures are attainable which are otherwise impossible to create via DPN, especially in regards to controlling the height of the features.

6.3.4.1 Selective Binding to DPN Patterns

As direct deposition techniques are not applicable or practicable for many different material systems, the idea to deposit a linker molecule has been implemented. Such a linker allows specific binding to all different kinds of materials, from nanoparticles to organic molecules all the way to cells. One of the biggest advantages is that the linker generally is a simple organic molecule with some functional group at the end of the chain. For such molecules well established deposition routines exist. At the same time, the functional groups can be modified quite freely through surface chemistry, providing a wealth of different binding options [99].

Covalent binding was achieved in a variety of different ways. The most popular technique utilized amide bond formation between carboxylic acid or activated esters (N-hydroxysuccinimid) and amines. Other examples include reactions of nucleophiles with benzoquinone patterns [80], catalytic ring-opening metathesis [91], or Schiff base formation [100].

Electrostatic binding requires the presence of positively and negatively charged groups. The positively charged groups are most of the times various aliphatic amines whereas the negatively charged moieties are generally acid functionalities. Due to the nature of these molecules, the bond is sensitive to changes in pH or ionic strength in the buffer. This method is especially useful for the deposition of large molecules (e.g. polyelectrolytes [101], nanowires [102]) due to the large number of electrostatic bonds which provide higher stability of the linked material to the DPN template.

Even relatively weak bonds such as van-der-Waals bonds can be utilized for the linking of molecules to DPN patterns. These bonds are the method of choice in the case of hydrophobic structures without functional groups. Due to the weak nature of the bonds, again relatively large molecules are required for stable bond formation. An example for such a van-der-Waals binding is the attachment of CNTs to prepatterned MHA/ODT surfaces [103].

A different type of binding is the formation of complexes with metal ions. Nickel ions are known to form complexes with different dye molecules or proteins. Thus, the deposition of Nickel salts provides specific binding sites for such molecules [104]. Such complexes have been shown to retain their fluorescence properties (dyes) and 3D-structure (proteins) [69].

Apart from these bonding schemes, which are based on simple molecular features, biorecognition and host-guest interactions can be used as well. Prime examples for the biorecognition are the hybridization of oligonucleotide strands with complementary DNA or the antibody-antigen recognition. Molecular recognition events provide high specificity towards the target molecule [105] which is beneficial for the simultaneous modification of several different DPN patterns with a mixture of the corresponding target molecules. Such biorecognition can be applied not only for direct sensing of viruses [106] or antibodies [107], but also for the guided assembly of nanoparticles [108].

6.3.4.2 Chemical Evolution of DPN Patterns

The term "chemical evolution" is meant to comprise all techniques, where a post deposition chemical transformation or growth of the DPN pattern occurs. One direction deals with the growth of nanostructures utilizing the catalytic activity of the patterns. It is well known that metallic nanoparticles can act as catalysts for the deposition of different materials. In the easiest way, this is the simple enlargement of nanoparticles with a metal salt and a reducing agent. By choosing a sufficiently mild reduction scheme, deposition will only occur on the nanoparticle seeds without causing cross-contamination.

Potential-enhanced transfer of gold nanoparticles to a silicon substrate was used for the templating of a surface [67]. Subsequent immersion into a solution of a gold salt and a reducing agent led to the enlargement of the deposited gold nanoparticles and, thus, to the formation of gold structures on the surface. It was found that the voltage applied for the transfer of the nanoparticle seeds from the tip to the surface played a crucial role for the pattern growth during the enlargement step. A tip bias of +12 V led to homogeneous patterns whilst lower potentials led to inhomogeneous pattern formation.

Instead of externally adding a reducing agent, such a substance can also be generated in-situ. This was investigated by depositing enzymes which had been decorated with 1.4 nm-sized gold nanoparticles [58]. These enzyme-nanoparticle hybrids were patterned with DPN on silicon surfaces. For the fabrication of gold structures, the conversion of glucose by glucose oxidase was utilized, yielding hydrogen peroxide as one of the reaction products which, subsequently, acted as reducing agent for $HAuCl_4$ in the presence of the gold nanoparticles. The enlargement led to 300 nm high gold lines (Fig. 6.8a). Similarly, silver lines were produced by reduction of silver nitrate with p-aminophenol which was the reaction product from digestion of

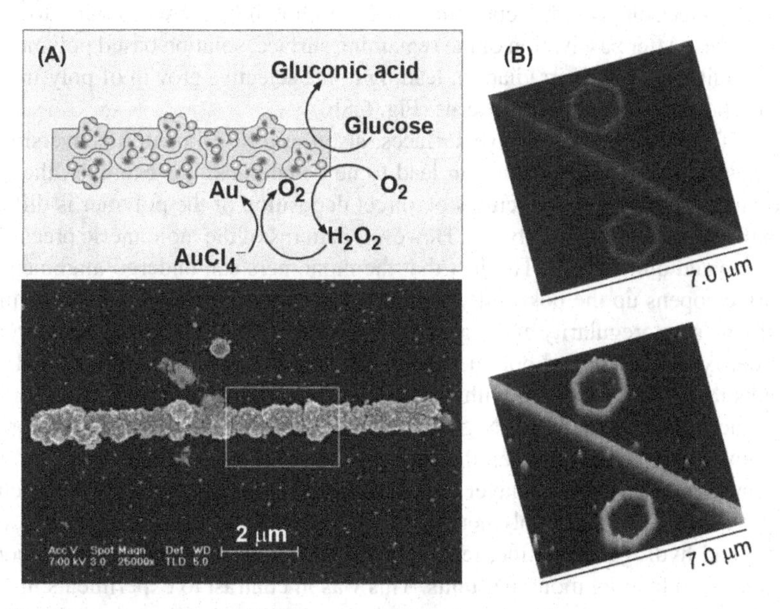

Fig. 6.8 Chemical evolution of DPN patterns: (**a**) Creation of gold nanostructures by biocatalytic enlargement of gold seeds. The *top part* shows the reaction scheme for enzyme-based reduction of the gold salt. The *bottom* image shows a SEM image of a gold nanowire created in this fashion. (Reproduced with permission from [58]). (**b**) A gold pattern created by electroless deposition of gold salt (*top*) serves as the binding site for iniferters leading to the site-selective growth of polymers on the gold pattern (*bottom*). ([111] Reproduced by permission of The Royal Society of Chemistry)

p-aminophenolphosphate by alkaline phosphatase. Here, lines of 25 nm height were obtained. The enlargement is self-limiting as the reaction stops once the enzymes are enclosed by the metal shell. This is advantageous for the controlled growth of metal structures.

Nanoparticles are not just well-known seeds for the growth of larger particles, but they are especially important in the field of nanowire growth. Controlled deposition of individual seeds in a two-dimensional array would greatly contribute to the practical application of nanowires in technological applications such as transistors or field emission arrays. In early attempts, gallium nitride nanowires were fabricated by depositing nickel nitrate on the surface of a silicon substrate [109]. After chemical vapor deposition of gallium nitride, nanowires were found to grow specifically on the patterned areas, with the number of nanowires per DPN pattern being dependent on size of the DPN pattern and, thus, on the humidity during the patterning process. Nickel chloride was used in a similar fashion to create CNTs [110]. Nanowires with 1 μm in length and 80 nm in diameter were obtained and x-ray spectroscopic characterization suggested a base-growth mechanism for these wires.

Not only metals can act as catalysts. Organic molecules can also induce reactions, as is the case for photoinitiators in polymerization reactions. Such a system was utilized for the localized growth of polymers [111]. To this end, gold patterns were created by electroless DPN deposition, and the photoinitiator was bound to these gold patterns. After passivation of the remaining surface, solution based polymerization was initiated by UV irradiation, leading to the selective growth of poly(methyl methacrylic acid) on the gold patterns (Fig. 6.8b).

Apart from catalytically active surfaces, also the post-deposition conversion of the deposited molecules or ions can lead to new materials. For example, the generation of polymeric nanostructures by direct deposition of the polymer is difficult due to the large size of the polymer. However, patterns of the monomeric precursors can be created quite easily. The fact that the monomers can undergo alignment on the surface opens up the possibility for the generation of highly oriented polymers by utilizing stereoregularity and regioselectivity. One example for such an approach is the peroxidase-catalyzed polymerization of caffeic acid [112]. Gold-coated silicon substrates were modified with a monolayer of aminothiophenol. This surface was further modified by the DPN generation of patterns of caffeic acid. The formation of hydrogen bonds between the carboxylic acid group of the caffeic acid and the amino group of the monolayer led to an alignment of the monomer molecules on the surface. Oxidative polymerization was carried out using horseradish peroxidase and hydrogen peroxide, leading to the formation of carbon-carbon bonds between the individual monomer units. This was in contrast to experiments in bulk solution, where the polymerization occurred predominantly through C-O-C bond formation.

Also magnetic structures have been obtained by converting two co-deposited salts [113]. Iron nitrate and barium carbonate were deposited onto a silicon substrate. Upon a two-step thermal annealing process, these precursors decomposed and formed a BaFe alloy with hard-magnetic properties.

6.3.5 Extensions to DPN

The art of writing has progressed from the quill pen to fountain pens and, eventually, to needle- and inkjet-printers. Similarly, also DPN has progressed yielding related techniques with novel properties. The fountain pen analogue is found in fountain pen nanolithography and still resembles a serial approach to the pattern generation. An increase in patterning speed is possible by using parallel writing techniques (analogous to the needle- and inkjet-printers) such as tip arrays or polymer pen lithography.

6.3.5.1 Fountain-Pen Nanolithography

Already with quill feathers one of the limiting drawbacks was the necessity to repeatedly immerse the feather into the ink solution. DPN suffers from the similar problem of material depletion, requiring repeated coating for large area patterning. A way to overcome this problem was the invention of the fountain-pen nanolithography [114, 115] (the DPN equivalent to regular fountain pens and ball-point pens). Fountain-pen nanolithography uses a small, micromachined capillary which is capable of transferring the ink from a small container through microfluidic channels directly to the surface. The use of such a micromachined pipette, thus, allows overcoming the limitation of ink depletion due to the large reservoirs. Additionally, the ink molecules do not need to be dissolved and do not have to pass through a humidity dependent meniscus. Rather they are constantly dissolved and pass through the aperture of the pipette, which possesses a much greater effective diameter than the meniscus. This leads to a constant stream of ink solution with constant concentration, thus avoiding the limitations due to ink diffusion. Additionally, extra driving forces in the form of membrane pumps can be used to have forced material transport [116, 117]. The fountain-pen approach is also viable to a certain degree of integration. An array of 12 pens [118] connected to two ink supplies (one ink pot for six cantilevers) was fabricated. This allowed for the simultaneous deposition of two inks in the form of dots and lines with a resolution below 100 nm.

A recent addition to fountain-pen nanolithography is the coupling of this technique with capillary electrophoresis [119]. Two electrodes are fabricated on the fountain-pen. Already the application of a small voltage of –1 V led to an increase in the concentration of the transferred bovine serum albumin by a factor of six. The electrophoretic technique should also allow for separation of different biomolecules, yielding patterns of different materials with a single ink mixture.

6.3.5.2 Tip Arrays

One of the main limitations of DPN for technological applications is the low throughput due to the serial writing capabilities. It was estimated that coating of an area of 1 cm^2 by diffusion from a single tip would take 100 years due to the speed of the material transfer (10^5 molecules/s) [31]. As the diffusion rate is a material constant (under given experimental conditions), the way to increase throughput and, thus, decrease patterning time, was the invention of parallel patterning methodologies based on tip arrays.

The use of multiple tips was spearheaded by IBM who developed a nanomechanical system ("millipede") for data storage based on arrays of individually heatable AFM cantilevers [120]. This technique allows storage densities in the Tbit/in^2 range and, due to the simultaneous readout of 4,096 tips, fast read and write times.

For DPN applications the current induced heating of the millipede approach is in many cases disadvantageous. Nevertheless, already early on multiple-pen setups were investigated [121], based on approaches developed for conventional AFM imaging and field-induced oxidation lithography [122]. An important issue in using arrays of cantilevers is the alignment of the array. An angular misalignment of a few degrees would cause some cantilevers to be suspended in the air whereas others would be pressed onto the surface at high forces. One way to reduce the influence of alignment is by receiving a feedback from each individual cantilever. However, this makes the fabrication of such arrays much more complicated and also requires significantly more hardware for data acquisition and evaluation.

For arrays consisting of conventional cantilevers, only a small pressure dependence of the DPN pattern size is given. This removes the necessity to address every single cantilever in the array individually, making the fabrication much easier. 2D-arrays based on this approach were fabricated with 55'000 cantilevers [123]. The alignment in this case was done by simply resting the whole array on the sample and then gluing it to the piezo scanner. This allowed for the generation of over 80 million dot features (with 100 nm diameter) with a single inking step.

Nevertheless, confirmation of contact for the patterning cantilevers is a desired feature to verify on-line the success of the patterning step. A method suitable for addressing individual tips in a probe array is the use of conductivity measurements [124]. Coating the probes with a metal film provides the possibility to measure the conductivity between the tip and the sample. For ODT coated tips a voltage higher than 3 V had to be applied to obtain a current signal upon contact to the substrate. The current signal provided a binary information of whether the tip was in contact or not and was not dependent on the contact force.

In all the above cases, the cantilevers were identical, except for slight variations due to the fabrication process, and were used simultaneously with the same parameters. However, arrays also lend themselves to the fabrication of cantilevers with different properties which can be addressed individually. A simple method for the individual actuation of cantilevers in an array is the use of thermal bimorph actuation [125]. Resistive heating of the cantilevers led to controlled retraction or engagement of the respective cantilever. An array of 10 such cantilevers was used to write the numbers 0–9 simultaneously with each cantilever writing a different number. Similarly, actuation can also be achieved by electrostatic attraction and repulsion between the cantilever and a counter-electrode [126]. The use of electric fields eliminates unwanted heating of the cantilevers and minimizes cross-talk between them, allowing fabrication of denser arrays.

Arrays of different cantilevers can be obtained by custom-made variation of the bluntness [127]. These are produced by creating a silicon master from a silicon-on-insulator substrate. Anisotropic etching of differently sized squares leads to inverted pyramid shapes with a flat-bottom of well defined size at the apex of the pyramid. The cantilevers are made from gold by electroplating of the master. Such an array was used to create lines of varying width of rhodamine-B by a single-pass DPN patterning. Instead of blunt tips, slight variations to this process also allow fabricating rounded tips with controlled radii of curvature [128]. Another technique yielding differently sized tip radii is the irradiation of the cantilever with short laser pulses [23]. Using 60 mJ pulses at 355 nm wavelength provided tips with radii between 40 and 500 nm.

A combination of differing tips with individual thermal actuation was also proposed [129]. A complex micromachining process yielded an array of nine scanning probe contact printing tips, five DPN tips, and three AFM tips for characterization. Each cantilever could be engaged/disengaged through resistive heating of the cantilever. This allowed simultaneous formation of DPN patterns and contact printing features with in-situ characterization without changing of the tips or cross-contamination.

Another important issue is the separate inking of cantilevers within an array. Multiple ink wells have been used both with passive tip arrays [130] as well as arrays of individually actuated tips [131] for the simultaneous inking with up to 10 different inks. A different approach utilized the precision of ink-jet printing for the controlled inking of individual cantilevers within an array [132]. Here, specialized surface modification was employed (hydrophobic back side coating and gold/MHA tip coating). This forced the ink droplet to remain in the area of the tip, giving maximum coating.

6.3.5.3 Polymer Pen Lithography

The latest addition in the realm of advanced DPN techniques is the polymer-pen lithography [133], also called molecular printing. This technique brings together the advantages of DPN and microcontact printing. Arrays of tips are made in the same fashion as in microcontact printing, i.e. a silicon master is formed by photolithography and etching. Afterwards, polydimethylsiloxane is cast onto the master, cured, and removed again from the master. In this way, 11 million pyramid-shaped tips were generated in a single array. The polydimethylsiloxane offers several important advantages to the patterning process. Firstly, this material is capable of absorbing and storing relatively large amounts of ink [134], substantially increasing the number of patterns which can be created with a single ink load. Secondly, the soft nature of the material allows controlling the feature size not just by the dwell time but also by applied force [135]. A larger force leads to a larger contact area, producing larger patterns in the same amount of time.

This large pressure dependence also requires a good leveling of the array in respect to the substrate. The first approach was based on optical alignment [133], utilizing the transparent nature of the tip material and the glass support. The tips change their reflectivity in accordance to the applied pressure. This can be used to tilt either the sample or the tip array until uniform reflectivity across the array is observed, indicating alignment. In a second approach, the applied force of the array was detected using highly sensitive scales [136]. The alignment procedure was based on the fact that the applied force is highest when the array is parallel to the surface as in this configuration the largest number of tips is actively applying pressure on the support. Thus, tilting the substrate in x- and y-direction was performed until the maximum force signal was obtained.

Finally, a suitable technique for the individual inking of the tips with different inks was developed. For this purpose, the silicon master is being used as an array of individual ink pots [137]. Ink-jet delivery of different inks into these pots was performed and the tip array was immersed into the master, coating the tips. Simultaneous patterning of three different fluorescently labeled proteins was performed with feature sizes ranging from 65 nm to over 10 μm.

These properties (large number of tips, addressable inking and high ink loading, pressure dependent pattern size, good leveling techniques) make this approach the fastest DPN technique available to date [138]. The time needed for pattern generation is close to values obtainable by soft lithography but with much higher flexibility in the pattern shape and higher resolution.

6.4 Organic Patterns

Organic molecules form the basis for most of the published data related to DPN. A small portion of these works investigated the deposition of the organic patterns themselves, whereas the majority of investigations utilized such organic patterns as templates for guided deposition, phase separation, or as etch resists, providing a simple path to more complex nanostructures.

6.4.1 Self-Assembled Monolayers

SAMs are prime candidates for the custom-made modification of surfaces in the macroscopic [139, 140] and nanoscopic realm [141]. In the case of DPN, especially thiolated monolayers on gold have been used. They are easily transferred due to their small size and good solubility, they produce highly ordered structures, and they can be fabricated with a large number of functional groups for subsequent modification [45]. Long chain thiols such as MHA and ODT have been used as the main compounds towards the investigation of the physical principles behind DPN (see Section 6.2: Physics of DPN). A special feature is the possibility to write patterns of different molecules on the same area [142]. In this way, the outline of a pattern can be written with one ink and the pattern can be filled with another ink without distorting the original pattern [121].

Extension of this technique to other substrates is important from a technological point of view. Especially semiconductor surfaces are important targets due to their applications in electronics and photonics. Both alkoxysilanes [47] and trichlorosilanes [48] have been applied for pattern generation of silicon. Their high sensitivity to humidity requires careful control of the environmental parameters and special preparatory steps like modification of the tip prior to inking.

A different kind of self-assembly is found for the biologically important lipid molecules. Due to their surfactant-like structure with a hydrophobic and a hydrophilic part they appear, in general, in a bilayer form. DPN on surfaces provides the possibility to generate multilayer structures. Phospholipids are very suitable inks for DPN as their deposition rate can be readily adjusted by the humidity and deposition in a non-covalent fashion is possible on a variety of surfaces [143]. They have been used for the fabrication of gratings based on phospholipids [144]. Biosensors based on these multilayer lipid-gratings use the diffraction change upon changes in the shape of the lipid lines based on intercalation, dewetting or spreading. In this way, the presence of streptavidin down to a concentration of 500 pM could be detected in a label-free fashion.

Although SAMs provide a practical way for direct modification of surface properties on the nanoscale, most DPN applications are related to the formation of templates from these molecules for subsequent sample modification by etching or deposition [145].

6.4.2 Other Monomeric Molecules

A close relative to SAMs, hexamethyldisilazane, was the first organic compound to be deposited directly onto semiconducting substrates [70]. This compound was a good choice due to its high affinity to the semiconductor surfaces, forming covalent bonds, and its low sensitivity to humidity.

Shortly afterwards, non-covalent attachment of dye molecules onto silicon substrates utilizing electrostatic attraction was demonstrated [49]. The use of various fluorescent dyes proved the possibility to link molecules non-covalently to a surface. In addition, optical characterization of the patterns showed that the fluorescence properties were maintained, providing confirmation that the structure of the dyes is not distorted in the deposition and binding process. Similar results were obtained for deposition on mica [146].

In many cases, though, the monomeric molecules are not directly attached to the native metal or semiconductor substrates. Rather, chemical reactions between a modified surface and the ink molecules are utilized. Some examples of this can be reviewed in Section 6.3.2.4 where various reactions from organic chemistry yield covalently bound organic nanostructures. A non-covalent technique was developed using host-guest interactions. This “molecular printboard” [147] termed technique is based on a surface modification with cyclodextrins and allowed the specific binding of various molecules to the surface.

6.4.3 Polymers

Creating patterns of functional polymers is of high interest for applications in optics, electronics, and biology, and many different techniques are being employed towards their fabrication [148]. In DPN, the biggest challenges lie in the large size and the, oftentimes, limited solubility of the polymers. For this reason, several different methods have been developed, ranging from direct deposition to on-surface synthesis.

6.4.3.1 Direct Deposition

The direct deposition of polymers is highly dependent on the mass and the number of functional groups present in the material. In order to evaluate these parameters, dendrimers of varying size and chemistry were deposited by DPN [149]. It was found that the deposition speed reduced substantially with increasing size, which also led to smaller pattern sizes. The number of functional groups on the polymer did not seem to influence the deposition significantly. The nature of the functional groups (in this study amino groups or hydroxyl groups), however, did change the deposition rate, with hydroxyl groups leading to higher deposition rates.

Covalent linking of amine-containing dendrimers was achieved through in-situ reaction with N-hydroxysuccinimide ester-modified surfaces [150]. This provides strong coupling to the surface whilst retaining free amine groups for subsequent modification of the patterns. The chemical linkage also leads to a suppression of surface diffusion, making the pattern width independent on the scan speed [151].

Polyethylene glycol is another polymer that can be patterned directly. Due to its compatibility with many different materials it has been used as an ink matrix for the assisted transfer of inorganic nanoparticles [64]. In this specific case the transfer of the polymer was not the main goal and the obtained structure underwent an oxidation step to eliminate the organic matrix. However, modified polyethylenglycol has also been used as the starting material for the fabrication of hydrogel patterns [152]. Transfer of the molecule to a surface was performed by DPN and the structure was converted to a hydrogel by photoinitiated cross-linking. Such hydrogels are of interest for biological studies, as they exhibit similar properties to cell tissue.

A way to increase deposition is to use an additional driving force in the form of electrostatic attraction. In most cases, electrically conducting polymers are doped by an ionic species, causing the polymer to exhibit a net charge. This circumstance was put to use for the electrostatic attraction of sulfonated polyaniline (negatively charged) and polypyrrole (positively charged) to a surface [51]. On positively charged substrates, obtained by modification of silicon wafers with an aminosilane, the sulfonated polyaniline could be deposited, whilst polypyrrole showed no pattern formation. For negatively charged substrates (obtained through treatment of silicon with piranha solution), the opposite behavior was observed, with polypyrrole showing good patterning properties.

In a similar fashion, non-conducting polyelectrolytes were also deposited by DPN [101]. Positively charged poly(diallyldimethyl ammonium chloride) and negatively charged poly(styrene sulfonate) were used as inks on negatively charged silicon (piranha treated) or positively charged poly(diallyldimethyl ammonium chloride) modified surfaces. The results were similar to before with opposite charges attracting each other, leading to good patterning results. An interesting extension to the direct deposition of a polymer was found in the layer-by-layer deposition of positively and negatively charged polymers for the fabrication of multilayer structures.

A completely different route for the fabrication of polymeric nanostructures was pursued by thermal DPN. Poly(3-dodecylthiophene) is a low-melting polymer which was coated onto a cantilever [72]. By resistive heating of the cantilever, the polymer could be selectively patterned on a silicon oxide surface. Controlling the temperature allowed controlling the deposition rate, effectively providing an ON/OFF switch for the deposition. Of special interest is the possibility to deposit multi-layers by multiple passes over a surface spot whilst retaining a high degree of molecular ordering in the polymeric structure (Fig. 6.9a).

6.4.3.2 Templated Deposition

Polyelectrolyte thin films can also be prepared on templated surfaces. MHA can be used for the fabrication of negatively charged patterns which can bind positively

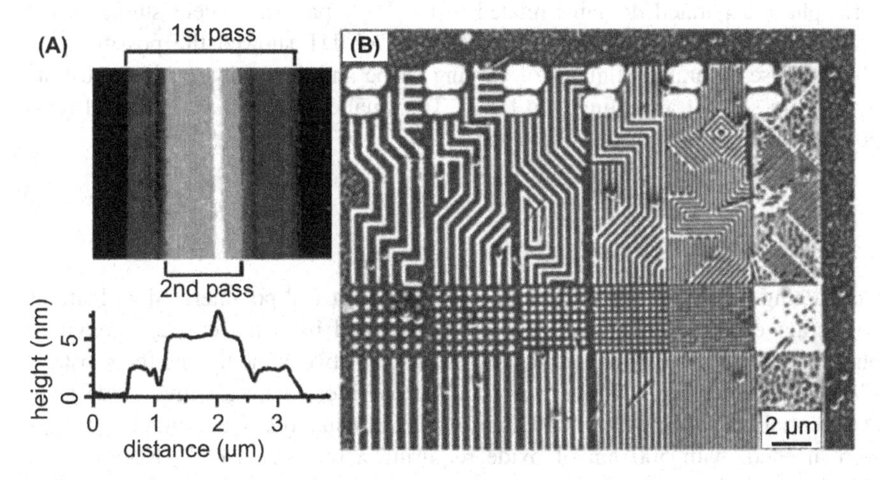

Fig. 6.9 Examples for polymeric patterns: (**a**) Pattern of poly(3-dodecylthiophene) created by temperature assisted deposition. Performing multiple passes across the same sample spot led to a linear increase in structure height, the increment being equal to the thickness of a monolayer. (Reproduced in part with permission from [72]. Copyright 2006 American Chemical Society). (**b**) Field-emission scanning electron microscope image of a phase separated poly(styrene)/poly(methyl methacrylic acid) polymer blend. The phase separation was directed by spin-coating the polymer blend onto the corresponding MHA/ODT template and performing a solvent anneal step. (Reproduced with permission from [157])

charged molecules [153]. To this end, MHA was patterned by DPN and the remaining surface was passivated. For the passivation ODT, 16-mercapto-1-hexadecanol, and a thiolated ethylene glycol were investigated with protection against unspecific absorption increasing from ODT to the thiolated ethylene glycol. Sequential deposition of poly(allyl amine) and poly(styrene sulfonate) led to a linear increase in the pattern height with the number of polyelectrolyte layers. Another polymer pattern that has been prepared consisted of poly(ethylene imine). This polymer was bound in a covalent fashion or through electrostatic binding to DPN-patterns with carboxylic acid functionality [48]. Apart from individual polymer chains, also prestructured polymers such as modified polystyrene beads have been arranged on negatively charged [154] and positively charged [155] DPN patterns utilizing electrostatic attraction.

An intriguing application for DPN is the templating of substrates towards phase separation in mixed polymers or block copolymers. Although phase separation in mixed polymers is oftentimes unwanted, controlled morphology can lead to a variety of applications. To this end patterns of MHA were drawn on gold and the remaining surface was passivated with a benzenethiol. A mixture of poly-3-hexylthiophene and polystyrene was spin-casted onto the surface [156]. It was found that on pure MHA surfaces lateral phase separation occurred, whilst on pure benzenethiol vertical phase separation was observed. The latter was found not to be stable as annealing at 140°C for 24 h also led to a laterally phase separated film. It was also found that for prepatterned surfaces a minimum feature size of the MHA patterns and a minimum content of poly-3-hexylthiophene were needed for the polymer film to form phase separated domains related to the DPN pattern. Recent studies using poly(methyl methacrylate)/polystyrene on MHA/ODT showed the possibility to obtain phase separated domains of varying shape and dimension on one substrate by using a solvent annealing step [157]. The smallest feature size obtained for a polymer domain was 50 nm (Fig. 6.9b).

6.4.3.3 In-Situ or Ex-Situ Synthesis

Conducting polymers can be created by in-situ electrical polymerization. Patterns consisting of modified polythiophene were created by using ethylenedioxythiophene as ink and application of negative voltages between the tip and the substrate [78]. The conducting polymer was obtained on silicon with a native oxide layer at potentials between –9 and –15 V. For thicker insulators, much higher voltages were needed, with 500 nm of oxide requiring a bias of –130 V. Nevertheless, electropolymerization was successful even on insulating substrates.

An ex-situ approach for the fabrication of polymeric brushes was also reported [111]. Gold nanostructures were created by in-situ reduction of gold salt on hydrogenated silicon surfaces. After oxidation of the remaining silicon surface at 60°C, a so-called iniferter (INItiator-transFER-TERminator agent) was linked to the gold structures. This compound acted as initiator for the photopolymerization of poly(methyl methacrylic acid). To this end, the modified substrate was immersed into a methyl methacrylic acid solution which was free of oxygen through bubbling with nitrogen. Irradiation with UV light caused the localized polymerization reaction on the gold nanostructures.

Using DPN for the deposition of the monomer units can also be utilized to generate novel polymeric structures. Caffeic acid was deposited onto aminothiophenol surfaces [112]. The electrostatic interaction between the carboxylic acid groups of the caffeic acid and the amino group of the surface led to a preferential orientation of the monomers. Subsequent polymerization using horseradish peroxidase and hydrogen peroxide led to the covalent linking of the monomers by direct C-C coupling of neighboring rings. This is in stark contrast to conventional polymerization where the dominating bond formation is based on a C-O-C linking via the hydroxyl groups of the caffeic acid.

6.4.3.4 Applications

Polymers can be used as sensing units. An example for a gas-sensitive system is based on doped polypyrrole which was deposited between gold electrodes on a silicon substrate with thick oxide [158]. The polymer is sensitive to the concentration of CO_2 so that exposure to this compound causes changes in the resistance of the conducting polymer. The relative resistance (R/R_0) was found to be linearly dependent on the concentration of carbon dioxide in the concentration range of 18–1,800 ppm.

A photosensitive conducting polymer can be used as photodetector. For this purpose a monomeric ink consisting of pyrrole, tetrahydrofuran, and perchloric acid was used [50]. The tetrahydrofuran acts as a modifier to reduce the speed of the acid catalyzed polymerization of pyrrole. Thus, an ink is obtained which yields good patterning results for over a week. After deposition of the ink on the surface, the polymerization continues with the acid catalyst acting as a dopant for the polymer. The polymer is light sensitive and a linear decrease of resistance with the increasing intensity squared was observed.

Of special interest is also the fabrication of stimuli-responsive structures. A polypeptide was chemically linked to patterns of MHA [159]. The polypeptide can undergo a structural change at a critical temperature, changing from hydrophilic to hydrophobic character. This transition was utilized by binding a thioredoxin fusion protein which can bind a thioredoxin-specific antibody. Above the critical temperature, this complex is stable, whereas reduction of the temperature below the critical point leads to desorption of the thioredoxin-antibody complex, restoring the native polypeptide. Such a bistable configuration is an interesting tool for biosensing, where restoring the sensing surface is an important issue.

6.5 Biological Patterns

The custom-made fabrication of highly integrated patterns made from biomolecules is an area which attracts a lot of research efforts [160, 161]. Smaller pattern sizes allow higher density sensor arrays for multiplexed detection and lower concentrations of the analyte are required. At the same time, the patterning of such structures on surfaces of biological importance (e.g. cell membranes [162]) helps bridging the gap towards in-vivo sensing schemes.

6.5.1 Proteins

The creation of protein patterns is of great importance for biochemistry, medical analytics, and biosensing. For this reason, many different techniques have been investigated for the directed deposition of proteins on various surfaces [163]. DPN patterning is an ideal candidate for this task as it allows creating such structures by direct deposition or by fabrication of binding sites. Although the large size of some protein structures makes their direct deposition sometimes quite difficult, this has nevertheless been achieved in a number of cases.

Polyethyleneglycol-modified cantilevers were coated with immunoglobuline G and patterning was achieved both by electrostatic attraction to negatively charged silicon oxide surface (through treatment with a base), as well as by covalent binding to an aldehyde-functionalized silicon oxide surface [100]. For both surfaces, the protein structure remained intact after the patterning as was evidenced by biorecognition reactions with the respective, fluorescently-labeled antibodies. Enhanced transfer was obtained by performing DPN with agarose added to the ink solution [63]. The addition of this biopolymer led to an increase in the deposition rate of up to three orders of magnitude. In order to utilize this, the proteins needed to be bound strongly to the surface in order to be able to rinse off the excess agarose, yielding pure protein nanopatterns. As before, the structure of the proteins is not degraded, maintaining their biological function and biorecognition capabilities.

Protein structures can also be prepared on nickel surfaces. Histidine possesses a high affinity towards nickel ions, providing a strong bond for surface immobilization. This selective binding was utilized for the DPN based patterning of proteins by in-situ oxidation of metallic nickel substrates [77] or by deposition on nickel oxide obtained from a pre-patterning oxidation through exposure to air [164]. Whilst in the first case the protein activity was not investigated, in the second case the proteins retained their full biofunctionality. Other surfaces amenable to direct protein deposition include native silicon oxide or tissue membranes [162]. In the latter case, the addition of surfactants to the ink or the pretreatment of the tissue membranes with surfactant proved beneficial for enhanced deposition rates.

Instead of depositing proteins on normal substrates, the positioning can also be made on selected parts of a DNA strand [165]. For this purpose, DNA is spread out on a suitable substrate. After identifying the region of interest by a survey scan with the AFM, the protein is then transferred to the chosen portion of the nucleotide strand (Fig. 6.10a). This provides the means for the selective modification or dissection of DNA strands by enzymes [87], which is of interest for DNA analysis.

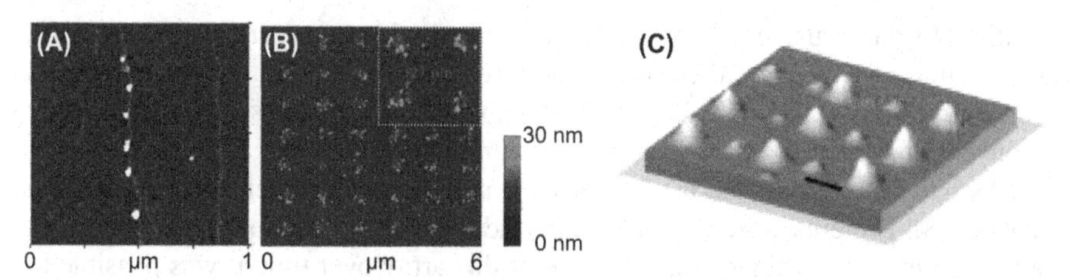

Fig. 6.10 Protein and DNA patterns obtained by DPN: (**a**) Antibody Cy3 deposited onto a DNA strand. (Reprinted from [165] with permission from Elsevier). (**b**) Detection of HIV by a multistep modification consisting of DPN of MHA, binding of anti-p24 antibody, passivation with bovine serum albumin, incubation with HIV-1, and sandwiching with anti-p24-modified nanoparticles. (Reproduced in part with permission from [106]. Copyright 2004 American Chemical Society). (**c**) DPN-patterns of two different DNA strands after hybridization with nanoparticle-modified complementary DNA. The nanoparticles had two different sizes with one size being used for DNA-1 recognition and the second size for recognition of the second DNA. (Reproduced with permission form [154])

Although direct deposition of proteins is possible and, in some cases advantageous, the more versatile technique is found in binding to a prepatterned substrate. This was originally shown for MHA patterns on gold [166] where the proteins were directly associated to the carboxylic acid functionality. Another approach was based on the deposition of nickel ions onto a nitrilotriacetic acid monolayer [104]. In this case the high affinity of histidine to nickel ions was utilized for the linking process to the surface.

An interesting application of DPN is the immobilization of proteins from a cell lysate without prior purification. The method is based on the formation of patterns made from maleimide-modified nitriloacetic acid [167]. This compound is transferred to a surface with thiol functionalization via DPN in the presence of a surfactant. The surfactant is necessary to overcome the difference in hydrophilicity between the ink molecule and the substrate. After passivation of the remaining surface by polyethylene glycol-maleimide, incubation with nickel ions was performed. This nitriloacetic acid-nickel complex is a specific binding agent for histidine-tagged proteins. Cell lysates containing a specifically expressed histidine-tagged protein with fluorescence marker (green fluorescent protein) were deposited on the sample. The tagged protein was bound to the DPN patterns and the remaining lysate constituents could be removed by washing.

Proteins are important biomarkers and, as such, interesting for biosensing applications. DPN patterns have led to interesting progress in this field due to the small pattern size and high pattern density. Leptin, a protein associated with obesity, was detected [107] based on a biotinylated maleimide which was deposited onto a mercaptopropyltrimethoxysilane monolayer. This was followed by several modification steps with the analyte, antigens, antibodies, and fluorescent labels. The leptin could be detected down to concentrations of 100 zM (i.e. 100×10^{-21} M), which is orders of magnitude better than for a conventional ELISA test. In a similar way, the detection of the HI-virus was investigated [106]. By forming, again, a sandwiched sensors of anti-p24 IgG, virus, and anti-p24 IgG decorated with nanoparticles (Fig. 6.10b), the detection of less than 50 copies of the virus RNA in plasma (corresponding to a concentration of 0.025 pg/ml) was possible, which again surpassed the sensitivity of conventional ELISA tests by orders of magnitude.

DPN patterns can also be used to investigate individual protein-protein binding events [168]. Proteins were deposited into the holes of a nanohole array. Utilizing the change in extra-ordinary transmission of this array over time it was possible to detect individual protein-protein binding events with a lateral separation of 16 μm and high temporal resolution. By using the array technique, currently over 20,000 individual, independent binding events could be monitored.

Apart from such sensing applications, specific proteins can also be used to create patterns by their enzymatic activity. Protein [84] and DNA [86] surfaces have been etched by deposition of the corresponding proteinase or DNase, respectively. For DNA, the digestion of individual strands was even observed in-situ [169], where the DNase transfer was done in intermittent contact mode whilst simultaneously measuring the sample topography. These enzymatic techniques yield a removal of material very much like a top-down approach (albeit with very high resolution and chemical specificity). Apart from etching, enzymes have also been used for the biocatalytic generation of metallic nanostructures [58]. In this case, nanoparticle-modified enzymes were transferred directly to silicon surfaces. The enzymatic conversion of a substrate yielded a product which acted as reducing agent for metal salts, leading to a localized enlargement of the nanoparticles on the enzyme. Thus, gold- and silver nanostructures could be generated.

6.5.2 DNA

DNA patterns are of high interest because of the high specificity in binding complementary strands, allowing for example to detect single base mismatches. In order to obtain such patterns different techniques for the deposition were developed. Originally, linking of modified DNA-strands to DPN patterns was used. Alkylamine-modified DNA was linked covalently to MHA patterns by formation of an amide bond [170]. Another approach used ODT or MHA as etch resist for gold [171]. After gold etching, the etch resist was removed by oxidation and thiolated DNA was linked to the gold structures. However, also the direct deposition of thiolated DNA to gold or acrylamide-modified DNA to silicon surfaces with mercaptosilane modification is possible [57], paving the way for multiplexed arrays of DNA. Matrix-assisted deposition of DNA by adding agarose to the ink solution provided enhanced transfer and less substrate sensitivity of the deposition [63].

Whilst all of the above mentioned techniques deposit DNA by DPN, DNA patterns can also be generated by localized digestion using DNase. This approach has been shown both for films of DNA [86] as well as for the selective cutting of individual DNA strands spread out on a surface [87].

DNA patterns in and of themselves are not very interesting to either science or technology. However, as soon as specific hybridization with complementary DNA comes into play, a multitude of different applications become available. DNA patterns can be used for the detection of analyte DNA in a label-free fashion directly by AFM [172] or by linking of nanoparticles for electrical characterization [173].

An area of great interest is the in-situ study of hybridization events. This requires optical solutions with sufficiently high lateral resolution and sensitivity. One way to reach this goal is the combination of DPN with surface enhanced resonance Raman scattering on nanostructured gold surfaces [174]. Limits of detection of better than 10^{-11} M were achieved whilst maintaining the full spectral information which allows distinguishing between different dye-labeled DNA strands.

A completely different application is the guided deposition of nanoparticles. Using oligonucleotide inks to pattern nanogaps between two electrodes, single particles could be selectively bound in the gap [108]. The advantage of this approach comes into play when different gaps are modified with different oligonucleotides, providing selective guiding of different DNA-capped nanoparticles. This allows performing transport measurements through single particle electrical circuits.

Another application for this guided deposition is the creation of patterns of different nanoparticles. This has been shown by the deposition of two different DNA patterns on a surface [154]. Subsequent modification with differently sized gold nanoparticles (one size being modified with one kind of complementary DNA, the other size with the second one) led to the formation of an ordered array of these particles (Fig. 6.10c).

6.5.3 Cells

Although cells are far too big to be directly deposited from a cantilever to the surface, DPN is still a highly useful technique for cell studies. An interesting application of DPN is the fabrication of suitable substrates for studies on the adhesion and proliferation of cells. Hydroquinone patterns were created on a gold-coated slide and the remaining surface was passivated with an ethyleneglycol oligomer [175]. After activation of the hydroquinone pattern by oxidation to benzoquinone, a short protein was linked to the pattern which served as anchoring point for the cell adhesion. It was found that the shape of this protein primer (linear or cyclic) significantly influenced the cell adhesion with the cyclic protein yielding significantly higher affinity. This was further elucidated investigating the immobilization on dot arrays with different pixel spacing within the array [176]. For the linear protein linker a net polarization of the cell directed towards the higher density pattern was induced, whereas no such effect was observed for the high-affinity, cyclic protein (Fig. 6.11a).

The deposition of cells on a protein layer can also be influenced by the underlying substrate [177]. Patterns of MHA, backfilled with ODT, were created and coated with zein, which is a protein obtained from corn endosperm. Subsequently attached

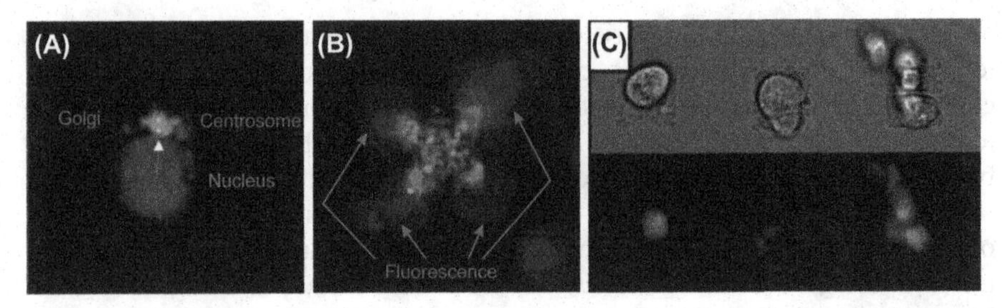

Fig. 6.11 Use of DPN patterns for single-cell investigations: (**a**) Cell polarization due to binding of the cell to an asymmetric hydroquinone pattern. (Reproduced in part with permission from [176]. Copyright 2008, American Chemical Society). (**b**) Activation of T-cells, indicated by a fluorescence signal (marked by *arrows*), through binding to modified lipid nanostructures. (Reproduced with permission from [178]). (**c**) Investigation of single cell infectivity. *Top part* shows an optical micrograph of three cells bound to virus-modified DPN patterns. The *bottom* image shows the fluorescence for the left and right cell induced by the virus infection. (Reproduced with permission from [179])

mouse fibroblasts were able to proliferate on the MHA-templated areas only. This was attributed to a different orientation of the zein molecules on MHA and ODT, respectively.

Apart from simply fixing cells to the surface, the DPN patterns have also been used to activate cells as in the example of T-cells [178]. In this case, biomimetic lipid membrane structures were prepared by DPN. Lipid monolayers were deposited onto glass slides in a non-covalent fashion. These monolayers were stable in a cell culture medium and were amenable to adsorption of cells. By chemical linking of antibodies specific for the CD3 ε chain of the T-cell receptors to these lipid membranes, an additional activation of these cells could be achieved (Fig. 6.11b).

Similarly, by creating nanopatterns of viruses, single cell infectivity can be investigated [179]. MHA was patterned on gold-coated glass and the remaining surface was passivated with a thiolated oligoethyleneglycol. Incubation with Zink ions, a polyclonal antibody, and a virus solution led to a nanoarray of virus particles which served as substrate for the binding of kidney cells. Incubation for 3 days led to an infection rate of over 90% for the immobilized cells, causing a change in the cell morphology and an increased fluorescence signal due to the virus-expressed green fluorescent protein (Fig. 6.11c).

These findings mark an important step in mimicking the immunoresponse of the human body and allow following the infection pathways of pathogens directly.

6.6 Inorganic Patterns

Inorganic structures are of high importance for electronic or optical applications. They have been investigated as storage media or for their catalytic properties. The wide variety of different material systems, from metallic, over semiconducting and magnetic, to insulating structures makes them interesting for many different research fields. DPN is a viable way for the controlled formation of such structures in the nanoscale and the different techniques have recently been reviewed in detail [180].

6.6.1 Metallic Patterns

Gold nanoparticles are the typical representative for metallic nanoparticles. Therefore, a variety of different techniques for their deposition have been developed. Hydrophobic particles on hydrophilic surfaces [56] have been patterned just as much as hydrophilic particles on hydrophobic surfaces [181]. Some recipes used covalent thiol-gold bonds [53], whilst others resorted to electrostatic attraction between acid and amine functions [54], and yet others used van-der-Waals interactions [181]. In all of these cases, the deposition was not influenced by the dwell time as the particles did not diffuse on the surface. A time-dependent patterning technique was developed by adding polyethylene glycol to the nanoparticle ink [64]. The growth of the pattern is determined by the diffusion of the polymer in which the nanoparticles are embedded. Also palladium nanoparticles have been used successfully for the generation of DPN patterns [55]. The polyvinylpyrrolidone-coated nanoparticles were solubilized and transferred to mica and copper surfaces. Whilst on mica, no lateral diffusion was observed (probably due to strong electrostatic interactions), the nanoparticles did diffuse on the copper substrate.

A different way for the deposition of nanoparticles is the fabrication of templates which are capable of forming nanoparticle conjugates. Towards this end, patterns made from SAMs with thiol- [68] or amino-functionality [182], DNA [57], or proteins [183] were deposited by DPN and the particles were associated by a subsequent immersion of the patterns into a nanoparticle solution with corresponding capping. The advantages of this approach are that only small molecules are transferred during the DPN process and that every kind of particle-surface interaction can be utilized such as covalent binding, electrostatic attraction, biorecognition, host-guest interaction, and van-der-Waals forces. Although the formation of nanoparticle patterns by a template method might not be very practical from a patterning point of view, it is a very interesting approach for enhancing the signal from a biosensing event. In this way, the sensitivity of HIV-detection [106] was enhanced, and less than 50 copies of the virus within a plasma sample could be detected.

Nanoparticle patterns are easy to fabricate. However, they generally exhibit poor conduction properties, limiting their use for nanoelectronic applications. This can be eliminated by subsequent growth of the patterns. This was utilized in the case of DNA-templated nanowires [173]. DNA patterns were generated between two electrodes and the pattern was hybridized with nanoparticles coated with the complementary DNA. The resultant structure was exposed to silver stain solution which led to the deposition of silver on the nanoparticles. Deposition of the silver could be followed by recording the resistance of the templated wire. Similarly, copper structures were obtained through electroless deposition of copper onto dendrimer-coated

Au nanoparticles which had been attached to a cyclodextrin-coated surface through host-guest interaction [184]. A self-limiting growth technique was developed by in-situ generation of the reducing agent through nanoparticle-modified enzymes [58]. Gold and silver wires were formed by the reduction of the respective salts through the enzymatic reaction product. Once the enzyme was coated with metal, the reaction stopped, controlling the size of the metallic structure.

The creation of metallic structures by deposition of nanoparticles is interesting because of the direct patterning possibility, which avoids sample contamination. Nevertheless, the first works related to metal patterns were based on etch procedures. The first structures were made from gold on silicon with ODT acting as the etch resist [94]. After the deposition of ODT, etching was performed with ferri/ferrocyanide, leading to ODT capped gold pillars. This technique was also successful using MHA as etch resist [95] and it could be extended to other metals (Ag, Pd) [96]. Silver nanostructures created in this fashion could be further transformed into gold structures by electroless deposition [185]. The deposition of the gold was accompanied by dissolution of the silver, leading to hollow gold cylinder structures from solid silver pillars (Fig. 6.12a).

In contrast to the use of an etch resist, direct etching of gold also has been described. The approach is based on the use of allyl bromide as ink which acts as an etchant for gold surfaces [83]. Selective etching was observed in the scanned areas. Simultaneously, the product of the etch process was accumulated on the tip and could be transferred to the surface in a conventional DPN fashion, yielding positive structures.

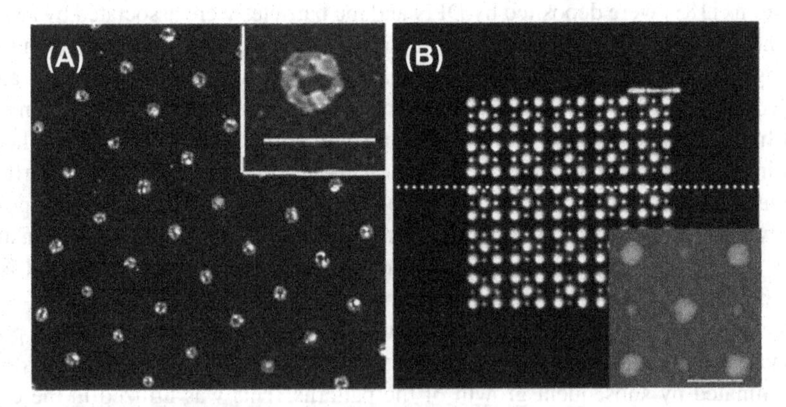

Fig. 6.12 Examples of metallic nanostructures: (**a**) Hollow gold cylinders obtained by depositing SAMs on a silver-coated silicon substrate, etching of the unprotected silver, and electroless deposition of gold with concurrent dissolution of the silver. (Reproduced in part with permission from [185]. Copyright 2004 American Chemical Society). (**b**) Silver nanopillars obtained by electrodeposition of silver on an ODT coated substrate where the initial MHA pattern was removed electrochemically. (Reproduced in part with permission from [97]. Copyright 2006 American Chemical Society)

Resist layers are not limited to etch procedures but can also be employed in cases of selective growth. MHA patterns were created and the remaining surface was passivated with ODT. Slight differences in the electrochemical potential for desorption allowed removing the MHA selectively whilst retaining the ODT passivation. Such patterns were used for etching but also for the electrochemical deposition of silver [97]. In the latter case, the height of the created silver pillars was related to the diameter of the MHA pattern, with larger patterns also yielding higher structures (Fig. 6.12b).

Electrochemistry can also be used without resist layers. Reduction of metal salt inks leads to the direct formation of metallic nanostructures. Gold, palladium, and platinum were reduced without external potentials through oxidation of the silicon [75] or germanium [75] substrates. In a different experiment, the reduction was obtained through local electrochemistry based on an external voltage that was applied between tip and substrate [77].

Conducting metal lines can also be patterned in a direct fashion. Indium-coated cantilevers were heated utilizing the low melting point of the metal [74]. Nevertheless, temperatures around 500°C and high contact forces were needed to produce metal structures on the surface. These showed metallic conductivity. Recently, a suspension of silver nanoparticles was also successfully employed towards the fabrication of metallic patterns [186]. A commercial silver nanoparticle ink with 40 nm particles was used for the preparation of nanoparticle patterns which underwent a subsequent annealing step at 150°C for 10 min. The resulting pattern showed ohmic conductance with a bulk resistivity close to that of metallic silver. A special advantage of the ink used in this case was an extremely high patterning speed (up to 1.6 mm/s) on various substrates.

6.6.2 Semiconducting Patterns

Semiconducting nanostructures are interesting due to their application in optics (waveguides, photonic crystals) and electronics. Elemental semiconductors made from group IV elements are widely available and relatively cheap. They form the basis for most of modern electronics and for passive waveguide structures due to their low price and their good processability. DPN nanostructures of such materials have first been reported by using an etch transfer step. ODT [94] or MHA [95] were deposited on a gold coated silicon surface. Subsequent etching with ferri/ferrocyanide (removal of the unprotected gold), hydrofluoric acid (etching of the silicon substrate), and aqua regia (removal of the remaining gold which served as etch mask) led to silicon structures with defined lateral and vertical dimensions. This process was extended to parallel fabrication with 55,000 cantilevers [123] and also different other etch recipes for the gold coating were investigated [187]. The use of wet chemical etching for the silicon leads to silicon structures with non-perpendicular sidewalls due to isotropic etching or preferential etching along certain crystal plains. In order to obtain perpendicular sidewalls, reactive ion etching has been successfully employed [188]. Apart from gold, also silver and palladium have been applied as etch masks for the fabrication of silicon nanostructures, with ODT and MHA again serving as the initial patterning material [96].

In contrast to the etch-approach for silicon, germanium nanostructures were created by in-situ electrochemical reduction of GeO_2 [77]. The ink, consisting of a saturated solution of GeO_2 in 0.05 M sodium hydroxide, was deposited with an applied potential of 5 V for reduction of the semiconductor, and line structures made from germanium were obtained.

In comparison to the above mentioned elemental semiconductors, compound semiconductors are advantageous in respect to the tunability of their optical or electrical properties. Their applications range from quantum dots as fluorescence markers all the way to semiconductor lasers. One example for such a compound material system is cadmium sulfide. Cadmium sulfide structures are accessible through in-situ synthesis from a chemically reactive ink solution [82] containing the precursors for cadmium and sulfide ions. Both the cadmium acetate and thioacetamide are prone to hydrolysis, providing the necessary ions for the nanoparticle formation. It was assumed that the formation already occurs on the cantilever, and that small particles, together with excess ink, are transferred to the surface. The patterning process was found to be very sensitive to humidity, scan speed, and hydrophilicity of the surface. Deposition was only observed on negatively charged surfaces at low velocities (around 100 nm/s) and medium humidity of around 50%.

Cadmium selenide nanoparticle patterns were created by using the template method with subsequent growth of the nanoparticles [48]. To this end, carboxylic acid-terminated patterns were created by DPN and the remaining surface was hydrophobically passivated. The carboxylate ion is capable of binding cadmium ions from solution through electrostatic attraction. After rinsing and drying of the substrate, exposure to hydrogen selenide led to the formation of CdSe on the carboxylate-modified surface with the carboxylic acid groups still acting as anchoring groups for the nanoparticles (Fig. 6.13a). In the same way, also PbS nanoparticle patterns were created.

Semiconducting structures are interesting also for sensing applications. The properties of tin oxide, for example, are highly dependent on the environment and have thus been harnessed for various sensing schemes. As the response time and recovery time are diffusion limited, miniaturization is a feasible way for improving the sensor performance. Native and doped tin oxide structures have been produced via a sol-gel based DPN approach [189]. Positioning of the structures between gold electrodes allowed the electrical contacting of the nanostructures in order to evaluate the gas-dependent changes in resistance (Fig. 6.13b). The response times of the gas sensors prepared by DPN were found to be about 10–20 times faster than for thin film sensors prepared from the same sol. This was attributed to the small size (both lateral and thickness) and nanoporous structure of the DPN pattern, allowing rapid diffusion of gases. By making an array of differently doped SnO_2 sensors, discrimination between different gases was possible due to the differing responses of the individual sensing spots.

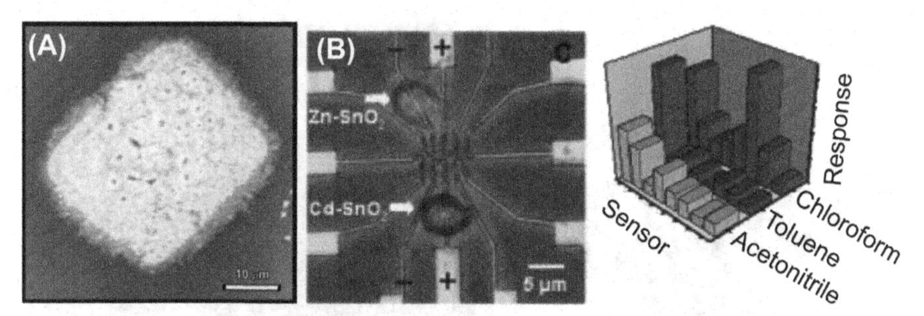

Fig. 6.13 Examples of semiconducting nanostructures generated by DPN: (**a**) CdSe pattern obtained by linking cadmium ions to a carboxylic acid-terminated DPN pattern and subsequent exposure to hydrogen selenide (Reproduced with permission from [48]). (**b**) Sensing of different gases using doped SnO2 deposits placed between electrodes by DPN. The *left part* shows an optical microscope image, the *right part* shows the response of different sensors to three different volatile organic compounds (Reproduced in part with permission from [189]. Copyright 2003 American Chemical Society)

6.6.3 Magnetic Patterns

Magnetic micro-structures have been used already for a substantial amount of time in audio- and video-tapes as well as computer hard-discs. However, the storage maximum attainable by conventional processing has nearly been reached. DPN offers the possibility to create patterns of magnetic structures with densities in the Tbit/in^2 range, which could serve as the next generation of magnetic storage devices.

Magnetic patterns have been created in a variety of different ways. Direct deposition of Fe_2O_3 nanoparticles has been performed on silicon and mica [190]. The pattern size of these particles was independent on the dwell time, indicating that the particles do not exhibit diffusion on the sample surface.

Instead of directly transferring magnetic nanoparticles to the surface, DPN can be used for the fabrication of template structures which provide binding sites for the subsequent immobilization of the particles. Iron oxide and manganese ferrite nanoparticles have been deposited on MHA modified gold via electrostatic attraction [191]. Such electrostatic binding has also been utilized for negatively charged magnetic particles on poly(allylamine) hydrochloride patterns [155]. DNA-modified particles have been linked both via electrostatic interaction with a polyelectrolyte pattern as well as through hybridization with complementary DNA molecules [192].

The use of DPN templates also provides a feasible way for the fabrication of patterns of individual nanoparticles. Carbon coated single crystal iron nanoparticles of smaller than 5 nm diameter were deposited on DPN generated MHA patterns [193]. These particles, though superparamagnetic at room temperature, exhibit a nearly 30-times higher coercivity at 10 K than do similar sized magnetite nanoparticles. Fabrication of patterns of individual particles was achieved by varying the size of the MHA templates. It was found that MHA patterns of 60–70 nm in diameter

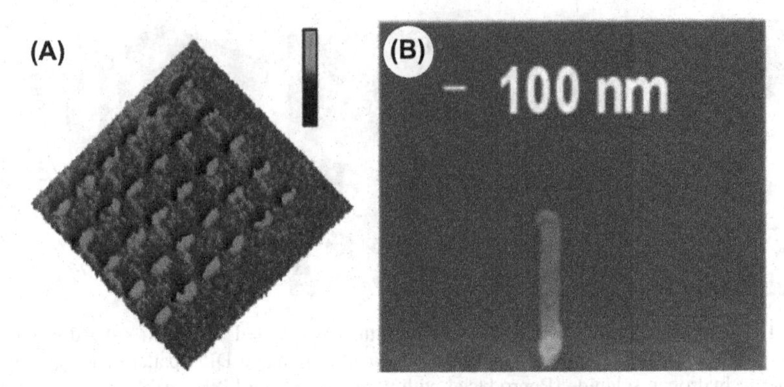

Fig. 6.14 Examples of other inorganic nanostructures: (**a**) Magnetic force microscopy image of BaFe deposits created by deposition of a mixed ink and conversion by thermal treatment (Reproduced in part with permission from [113]. Copyright 2003 American Chemical Society). (**b**) Scanning electron microscope image of a single carbon nanotube seeded by deposition of nickel chloride (Reproduced with permission from [110])

exhibited a near 100% yield of single particle immobilization, whilst smaller patterns (lower yield) and larger patterns (more than one nanoparticle per pattern) were not as favorable.

The advantage of patterns created from pre-fabricated nanoparticles is the possibility to control the size and magnetic properties before performing the patterning. However, also other approaches towards the nanopattern formation have been investigated. An interesting method was pursued in the synthesis of hard-magnetic structures by temperature induced reaction between iron nitrate and barium carbonate [113]. An ink was used containing these two salts. The co-deposition led to structures of the salts without magnetic properties. However, a two-stage thermal treatment led to the decomposition of the salts and the formation of a magnetic BaFe alloy (Fig. 6.14a).

A different approach consisted in the deposition of ferritin. Ferritin is a protein acting as iron storage container within living organism. The small spherical molecule is filled with iron oxide. Consequently, ferritin was deposited by DPN and the organic shell was removed by oxidation [194]. In this way, iron oxide nanostructures were obtained. However, the magnetic properties of these structures were not investigated.

6.6.4 Nanowires and Nanotubes

Nanowires and nanotubes are highly interesting as 1D-transistors or as sources for field-emitter devices. An important requirement for their use, however, is the fabrication of ordered arrays of these structures with high precision in both the lateral position and the dimensions. DPN has been investigated towards the fabrication of such nanowire systems. The conventional route consists of the deposition of catalytically active seeds and subsequent growth of the nanostructures on these seeds. This has been shown for GaN where nickel nitrate patterns were created with DPN and chemical vapor deposition of gallium nitride led to the growth of multiple nanowires from a single nickel nitrate pattern [109].

Apart from such semiconducting nanowires, especially CNTs have attracted a lot of interest and several studies have dealt with their DPN templated synthesis. In one approach, patterns consisting of 5 nm Cobalt particles were prepared [195]. The particles possessed an organic shell (trioctylphosphine and oleic acid) to provide sufficient mobility for the DPN process even in the dried state. Coating of the cantilever was achieved by scanning the tip across an evaporated droplet of the nanoparticle solution. The thus coated cantilever was used for the fabrication of dot patterns. It was observed that small dot sizes (sub-70 nm) only led to non-uniform, low-yield nanotube growth. For larger pattern sizes (diameters of 900 nm and above), high yield and high uniformity of the single-wall CNTs was obtained. On Si/SiO_2 substrates, the growth direction was random. Using quartz cut at a specific angle, the growth of the nanotubes could be oriented along the 100 crystal plane, leading to highly oriented CNT structures. Detailed investigations confirmed that a base-growth mechanism was responsible for the nanotube formation. Nickel ions also act as catalysts for the growth of CNTs. Nickel chloride patterns led to the formation of individual nanotubes with lengths up to 1 μm [110] (Fig. 6.14b). Also in this case, a base-growth mechanism was responsible for the nanotube formation. Compared to the Cobalt-based system, however, the yield was significantly lower.

Whilst direct growth of nanowires would be advantageous for practical applications, the above cited literature provides confirmation that control of the density and length of these wires and nanotubes is difficult to achieve. For this reason, the selective binding and alignment of harvested nanowires is an interesting alternative for several applications, such as nanotransistors. For vanadium oxide nanowires, positively charged patterns consisting of cysteamine (on gold) or aminopropyltriethoxysilane (on silicon) were created and the remaining surface was passivated with a hydrophobic monolayer [102]. Immersion into the nanowire solution led to a directed assembly of these wires on the positively charged domains. Subsequent fabrication of metal pads allowed contacting the wires for electrical characterization. The nanowires are only bound through electrostatic attraction and, thus, could be removed again by rinsing with a buffer solution.

Also CNTs can be deposited onto mixed hydrophilic/hydrophobic surfaces [103]. Dot patterns of MHA were produced and the surface was backfilled with ODT. Depositing a drop of a solution of nanotubes in dichlorobenzene onto the patterned substrate led to the accumulation of the nanotubes at the boundary of ODT and MHA. Drying of the solvent caused the nanowires to bend along this boundary, forming ring structures. This behavior was strongly dependent on the material combination used for templating, with polyethylene glycol and mercaptoundecanol as substitute for either the MHA or the ODT, respectively, showing significantly less tendency towards the nanotube ring formation. Fabrication of line-patterns led to the formation of CNT structures which exactly traced the template structure.

Although CNTs are bulky and poorly soluble, it is well known that aromatic molecules can be adsorbed on the sidewalls due to π–π stacking which allows incorporation of functional groups [196]. This has been utilized to solubilize CNTs through adsorption of alcian blue-tetrakis(methyl pyridium) chloride [197]. By doing this, an ink was obtained which allowed direct deposition of CNTs with DPN. Lines down to 110 nm in width were obtained which consisted, to the most part, of multilayers of nanotubes.

6.7 Conclusions and Outlook

Twelve years have passed since the invention of DPN, and this methodology has, indeed, proven to be one of the most versatile techniques for nanopatterning available. Hundreds of publications describe the various approaches to create patterns made of simple long-chain thiols all the way to the selective activation of individual cells for single cell studies. It has developed to be a mature technique in science with reproducible pattern generation and fast prototyping properties. Now DPN stands at the brink of transition to becoming a technologically relevant methodology. First important steps have been made by parallelization and the use of polymeric tips [138], providing the means for the simultaneous generation of millions of structures. It has, in this respect, overtaken techniques such as focused ion beam structuring, where only now multibeam approaches are being developed [198]. Because of this, the patterning times have greatly been reduced, closing in on the times needed in photolithographic or soft lithographic (microcontact printing, nanoimprint lithography) techniques (Fig. 6.15), whilst providing complete flexibility in regards to pattern shape, substrate material, and the chemical nature of the desired pattern.

Nevertheless, several challenges still need to be tackled to successfully bridge the gap from scientific tool to industrial technology. For sensor array applications, ways have to be developed to ink the thousands or millions of cantilevers each with a different ink in an addressable, error-free and cross talk-free fashion. So far, this has been demonstrated with just a handful of different inks using ink-jet printing [132], microfluidic ink wells [130], or a combination of both [137]. Also, high resolution read-out methodologies are required to make full use of the high integration capability of DPN.

Another challenge is the online monitoring of the pattern fabrication and the possibility for error correction. Online monitoring has been shown for fluorescently-labeled structures using optical microscopy [61], but no general paradigm has been developed. Simultaneously, first approaches demonstrating write/read/erase capabilities have been developed [66, 199], but again, no general approach has been found.

One of the biggest challenges still remaining for this bottom-up approach in the area of nanostructuring, however, is control of the third dimension, i.e. the height of structures. Several routes have already been investigated, including layer-by-layer growth [153], multiple-pass techniques [72], or self-limiting enlargement of

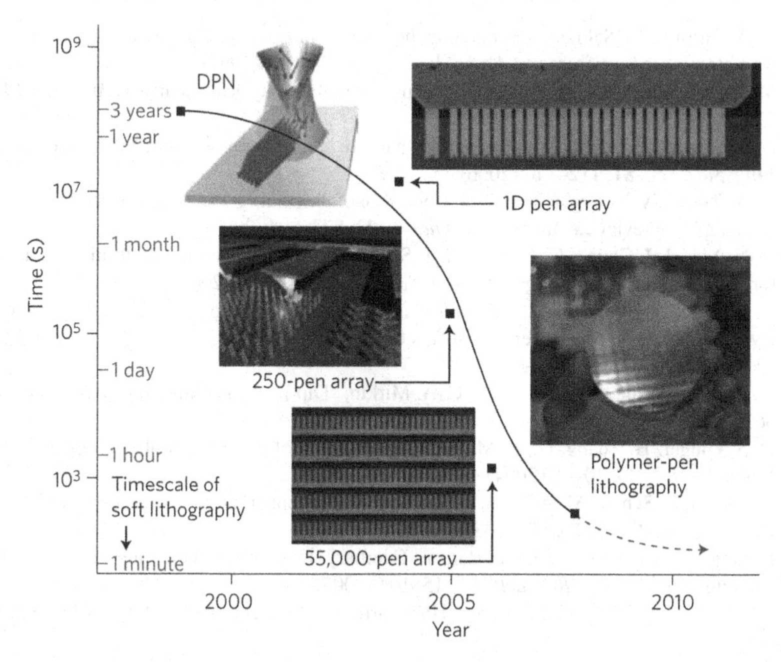

Fig. 6.15 Decrease of the patterning time as a result of advancements in DPN. (Reprinted with permission from McMillan Publishers Ltd: Nature Chem. [138], copyright 2009)

structures [58]. However, patterning time or cross-contamination both constrain the general applicability of these techniques.

Even with a number of important challenges still present, the first decade of DPN has shown how many obstacles have been overcome by novel and creative approaches. Therefore it can be expected that DPN will continue to greatly influence not only scientific research but that it will also make its way towards technological applications such as biochemical sensing or novel photonic devices.

Acknowledgments This work was financed by the Austrian Ministry of Transport, Innovation, and Technology (BMVIT) and the FFG through the Nanoinitiative project PLATON (Project No. 819654).

References

1. G. Binnig, H. Rohrer, C. Gerber, E. Weibel, Surface studies by scanning tunneling microscopy, *Phys. Rev. Lett.*, **49**, 57–61 (1982).
2. G. Binnig, C. F. Quate, C. Gerber, Atomic force microscope, *Phys. Rev. Lett.*, **56**, 930–933 (1986).
3. G. Friedbacher, H. Fuchs, Classification of scanning probe microscopies, *Pure Appl. Chem.*, **71**, 1337–1357 (1999).
4. D. M. Eigler, E. K.Schweizer, Positioning single atoms with a scanning tunnelling microscope, *Nature*, **344**, 524–526 (1990).

5. R. W. Carpick, M. Salmeron, Scratching the surface: fundamental investigations of tribology with atomic force microscopy, *Chem. Rev.*, **97**, 1163–1194 (1997).
6. M. Liu, N. A. Amro, G. Y. Liu, Nanografting for surface physical chemistry, *Ann. Rev. Phys. Chem.*, **59**, 367–386 (2008).
7. D. Stievenard, B. Legrand, Silicon surface nano-oxidation using scanning probe microscopy, *Prog. Surf. Sci.*, **81**, 112–140 (2006).
8. A. A. Tseng, A. Notargiacomo, T. P. Chen, Nanofabrication by scanning probe microscope lithography: a review, *J. Vac. Sci. Technol. B*, **23**, 877–894 (2005).
9. X. N. Xie, H. J. Chung, C. H. Sow, A. T. S. Wee, Nanoscale materials patterning by atomic force microscopy nanolithography, *Mat. Sci. Eng. R*, **54**, 1–48 (2006).
10. D. Wouters, S. Hoeppener, U. S. Schubert, Local probe oxidation of self-assembled monolayers: templates for the assembly of functional nanostructures, *Angew. Chem. Int. Ed.*, **48**, 1732–1739 (2009).
11. R. D. Piner, J. Zhu, F. Xu, S. Hong, C. A. Mirkin, "Dip-pen" nanolithography, *Science*, **283**, 661–663 (1999).
12. D. S. Ginger, H. Zhang, C. A. Mirkin, The evolution of dip-pen nanolithography, *Angew. Chem. Int. Ed.*, **43**, 30–45 (2004).
13. J. Jang, G. C. Schatz, M. A. Ratner, Liquid meniscus condensation in dip-pen nanolithography, *J. Chem. Phys.*, **116**, 3875–3886 (2002).
14. J. Jang, G. C. Schatz, M. A. Ratner, Capillary force on a nanoscale tip in dip-pen nanolithography, *Phys. Rev. Lett.*, **90**, 156104 (2003).
15. J. Jang, G. C. Schatz, M. A. Ratner, How narrow can a meniscus be? *Phys. Rev. Lett.*, **92**, 085504 (2004).
16. B. L. Weeks, M. W. Vaughn, J. J. DeYoreo, Direct imaging of meniscus formation in atomic force microscopy using environmental scanning electron microscopy, *Langmuir*, **21**, 8096–8098 (2005).
17. B. L. Weeks, J. J. DeYoreo, Dynamic meniscus growth at a scanning probe tip in contact with a gold substrate, *J. Phys. Chem. B*, **110**, 10231–10233 (2006).
18. M. G. Boyle, J. Mitra, P. Dawson, The tip-sample water bridge and light emission from scanning tunneling microscopy, *Nanotechnology*, **20**, 335202 (2009).
19. J. R. Hampton, A. A. Dameron, P. S. Weiss, Double-ink dip-pen nanolithography studies elucidate molecular transport, *J. Am. Chem. Soc.*, **128**, 1648–1653 (2006).
20. R. Agrawal, N. Moldovan, H. D. Espinosa, An energy model to predict wear in nanocrystalline diamond atomic force microscopy tips, *J. Appl. Phys.*, **106**, 064311 (2009).
21. D. M. Heo, M. Yang, H. Kim, L. C. Saha, J. Jang, Tip dependence of the self-assembly in dip-pen nanolithography, *J. Phys. Chem. C*, **113**, 13813–13818 (2009).
22. J. W. van Honschoten, N. R. Tas, M. Elwenspoek The profile of a capillary liquid bridge between solid surfaces, *Am. J. Phys.*, **78**, 277–286 (2010).
23. N. S. John, G. U. Kulkarni, Dip-pen lithography using pens of different thicknesses, *J. Nanosci. Nanotechnol.*, **7**, 977–981 (2007).
24. B. L. Weeks, A. Noy, A. E. Miller, J. J. DeYoreo, Effect of dissolution kinetics on feature size in dip-pen nanolithography, *Phys. Rev. Lett.*, **88**, 255505 (2002).
25. S. Rozhok, R. Piner, C. A. Mirkin, Dip-pen nanolithography: what controls ink transport? *J. Phys. Chem. B*, **107**, 751–757 (2003).
26. P. E. Sheehan, L. J. Whitman, Thiol diffusion and the role of humidity in "dip-pen nanolithography", *Phys. Rev. Lett.*, **88**, 156404–156107 (2002).
27. S. Rozhok, P. Sung, R. Piner, M. Lieberman, C. A. Mirkin, AFM study of water meniscus formation between an AFM tip and NaCl substrate, *J. Phys. Chem. B*, **108**, 7814–7819 (2004).
28. E. J. Peterson, B. L. Weeks, J. J. De Yoreo, P. V. Schwartz, Effect of environmental conditions on dip pen nanolithography of mercaptohexadecanoic acid, *J. Phys. Chem. B*, **108**, 15206–15210 (2004).

29. G. W. Bradberry, P. S. Vukusic, J. R. Sambles, A study of the adsorption of alkanes on a thin-metal film, *J. Chem. Phys.*, **98**, 651–654 (1993).
30. K. Salaita, A. Amarnath, T. B. Higgins, C. A. Mirkin, The effects of organic vapor on alkanethiol deposition via dip-pen nanolithography, *Scanning*, **32**, 9–14 (2010).
31. P. V. Schwartz, Molecular transport from an atomic force microscope tip: a comparative study of dip-pen nanolithography, *Langmuir*, **18**, 4041–4046 (2002).
32. C.-D. Wu, W.-H. Fang, J.-F. Lin, Formation mechanism and mechanics of dip-pen nanolithography using molecular dynamics, *Langmuir*, **26**, 3237–3241 (2010).
33. R. G. Sanedrin, N. A. Amro, J. Rendlen, M. Nelson, Temperature controlled dip-pen nanolithography, *Nanotechnology*, **21**, 115302 (2010).
34. I. Nocedal, H. Espinosa, K.-H. Kim, Ink diffusion in dip-pen nanolithography: a study in the development of nano fountain probes, *Nanoscape*, **2**, 105–111 (2005).
35. S. K. Saha, M. L. Culpepper, A surface diffusion model for dip pen nanolithography line writing, *Appl. Phys. Lett.*, **96**, 243105 (2010).
36. H. Jung, C. K. Dalal, S. Kuntz, R. Shah, C. P. Collier, Surfactant activated dip-pen nanolithography, *Nano Lett.*, **4**, 2171–2177 (2004).
37. H. Kim, J. Jang, Serial pushing model for the self-assembly in dip-pen nanolithography, *J. Phys. Chem. A*, **113**, 4313–4319 (2009).
38. J. Jang, S. Hong, G. C. Schatz, M. A. Ratner, Self-assembly of ink molecules in dip-pen nanolithography: a diffusion model, *J. Chem. Phys.*, **115**, 2721–2729 (2001).
39. D. M. Heo, M. Yang, S. Hwang, J. Jang, Molecular dynamics of monolayer deposition using a nanometer tip source, *J. Phys. Chem. C*, **112**, 8791–8796 (2008).
40. N.-K. Lee, S. Hong, Modeling collective behavior of molecules in nanoscale direct deposition processes, *J. Chem. Phys.*, **124**, 114711 (2006).
41. P. Manandhar, J. Jang, G. C. Schatz, M. A. Ratner, S. Hong, Anomalous surface diffusion in nanoscale direct deposition processes, *Phys. Rev. Lett.*, **90**, 115505 (2003).
42. H. Brunner, T. Vallant, U. Mayer, H. Hoffmann, B. Basnar, M. Vallant, G. Friedbacher, Substrate effects on the formation of alkylsiloxane monolayers, *Langmuir*, **15**, 1899–1901 (1999).
43. W. M. Wang, R. M. Stoltenberg, S. Liu, Z. Bao, Direct patterning of nanoparticles using dip-pen nanolithography, *ACS Nano*, **2**, 2135–2142 (2008).
44. A. Einstein, On the motion of small particles suspended in liquids at rest required by the molecular-kinetic theory of heat, *Ann. Phys. (Leipzig)*, **17**, 549–560 (1905).
45. X.-M. Li, J. Huskens, D. N. Reinhoudt, Reactive self-assembled monolayers on flat and nanoparticle surfaces, and their application in soft and scanning probe lithographic nanofabrication technologies, *J. Mater. Chem.*, **14**, 2954–2971 (2004).
46. S. R. Wasserman, Y.-T. Tao, G. M. Whitesides, Structure and reactivity of alkylsiloxane monolayers formed by reaction of alkyltrichlorosilane on silicon substrates, *Langmuir*, **5**, 1074–1087 (1989).
47. H. Jung, R. Kulkarni, C. P. Collier, Dip-pen nanolithography of reactive alkoxysilanes on glass, *J. Am. Chem. Soc.*, **125**, 12096–12097 (2003).
48. S. E.Kooi, L. A. Baker, P. E. Sheehan, L. J. Whitman, Dip-pen nanolithography of chemical templates on silicon oxide, *Adv. Mater.*, **16**, 1013–1016 (2004).
49. M. Su, V. P. Dravid, Colored ink dip-pen nanolithography, *Appl. Phys. Lett.*, **80**, 434–4436 (2002).
50. M. Su, M. Aslam, L. Fu, N. Wu, V. P. Dravid, Dip-pen nanopatterning of photosensitive conducting polymer using a monomer ink, *Appl. Phys. Lett.*, **84**, 4200–4202 (2004).
51. J.-H. Lim, C. A. Mirkin, Electrostatically driven dip-pen nanolithography of conducting polymers, *Adv. Mater*, **14**, 1474–1477 (2002).
52. A. Mulder, S. Onclin, M. Péter, J. P. Hoogenboom, H. Beijleveld, J. ter Maat, M. F. García-Parajó, B. J. Ravoo, J. Huskens, N. F. van Hulst, D. N. Reinhoudt, Molecular printboards on silicon oxide: lithographic patterning of cyclodextrin monolayers with multivalent, fluorescent guest molecules, *Small*, **1**, 242–253 (2005).

53. J. C. Garno, Y. Yang, N. A. Amro, S. Cruchon-Dupeyrat, S. Chen, G.-Y. Liu, Precise positioning of nanoparticles on surfaces using scanning probe lithography, *Nano Lett.*, **3**, 389–395 (2003).
54. B. Wu, A. Ho, N. Moldovan, H. D. Espinosa, Direct deposition and assembly of gold colloidal particles using a nanofountain probe, *Langmuir*, **23**, 9120–9123 (2007).
55. P. J. Thomas, G. U. Kulkarni, C. N. R. Rao, Dip-pen lithography using aqueous metal nanocrystal dispersions, *J. Mater. Chem.*, **14**, 625–628 (2004).
56. M. Ben Ali, T. Ondarçuhu, M. Brust, C. Joachim, Atomic force microscope tip nanoprinting of gold nanoclusters, *Langmuir*, **18**, 872–876 (2002).
57. L. M. Demers, D. S. Ginger, S.-J. Park, Z. Li, S.-W. Chung, C. A. Mirkin, Direct patterning of oligonucleotides on metals and insulators by dip-pen nanolithography, *Science*, **296**, 1836–1838 (2002).
58. B. Basnar, Y. Weizmann, Z. Cheglakov, I. Willner, Synthesis of nanowires using dip-pen nanolithography and biocatalytic inks, *Adv. Mater.*, **18**, 713–718 (2006).
59. G. Agarwal, L. A. Sowards, R. R. Naik, M. O. Stone, Dip-pen nanolithography in tapping mode, *J. Am. Chem. Soc.*, **125**, 580–583 (2003).
60. C.-C. Wu, H. Xu, C. Otto, D. N. Reinhoudt, R. G. H. Lammertink, J. Huskens, V. Subramaniam, A. H. Velders, Porous multilayer-coated AFM tips for dip-pen nanolithography of proteins, *J. Am. Chem. Soc.*, **131**, 7526–7527 (2009).
61. A. Noy, A. E. Miller, J. E. Klare, B. L. Weeks, B. W. Woods, J. J. DeYoreo, Fabrication of luminescent nanostructures and polymer nanowires using dip-pen nanolithography, *Nano Lett.*, **2**, 109–112 (2002).
62. M. Lee, D.-K. Kang, H.-K. Yang, K.-H. Park, S. Y. Choe, C.-S. Kang, S.-I. Chang, M. H. Han, I.-C. Kang, Protein nanoarray on Prolinker^{TM} surface constructed by atomic force microscopy dip-pen nanolithography for analysis of protein interaction, *Proteomics*, **6**, 1094–1103 (2006).
63. A. J. Senesi, D. I. Rozkiewicz, D. N. Reinhoudt, C. A. Mirkin, Agarose-assisted dip-pen nanolithography of oligonucleotides and proteins, *ACS Nano*, **3**, 2394–2402 (2009).
64. L. Huang, A. B. Braunschweig, W. Shim, L. Qin, J. K. Lim, S. J. Hurst, F. Huo, C. Xue, J.-W. Jang, C. A. Mirkin, Matrix-assisted dip-pen nanolithography and polymer pen lithography, *Small*, **6**, 1077–1081 (2010).
65. J.-F. Liu, G. P. Miller, Field-assisted nanopatterning, *J. Phys. Chem. C*, **111**, 10758–10760 (2007).
66. P. S. Spinney, S. D. Collins, R. L. Smith, Solid-phase direct write (SPDW) of carbon via scanning force microscopy, *Nano Lett.*, **7**, 1512–1515 (2007).
67. W.-K. Lee, S. Chen, A. Chilkoti, S. Zauscher, Fabrication of gold nanowires by electric-field-induced probe lithography and in situ chemical development, *Small*, **3**, 249–254.
68. Y. Cia, B. M. Ocko, Electro pen nanolithography, *J. Am. Chem. Soc.*, **127**, 16287–16291 (2005).
69. G. Agarwal, R. R. Naik, M. O. Stone, Immobilization of histidine-tagged proteins on nickel by electrochemical dip-pen nanolithography, *J. Am. Chem. Soc.*, **125**, 7408–7412 (2003).
70. A. Ivanisevic, C. A. Mirkin, "Dip-pen" nanolithography on semiconductor surfaces, *J. Am. Chem. Soc.*, **123**, 7887–7889 (2001).
71. P. E. Sheehan, L. J. Whitman, W. P. King, B. A. Nelson, Nanoscale deposition of solid inks via thermal dip pen nanolithography, *Appl. Phys. Lett.*, **85**, 1589–1591 (2004).
72. M. Yang, P. E. Sheehan, W. P. King, L. J. Whitman, Direct writing of a conducting polymer with molecular-level control of physical dimensions and orientation, *J. Am. Chem. Soc.*, **128**, 6774–6775 (2006).
73. P. C. Fletcher, J. R. Felts, Z. Dai, T. D. Jacobs, H. Zeng, W. Lee, P. E. Sheehan, J. A. Carlisle, R. W. Carpick, W. P. King, Wear-resistant diamond nanoprobe tips with integrated silicon heater for tip-based nanomanufacturing, *ACS Nano*, **4**, 3338–3334 (2010).

74. B. A. Nelson, W. P. King, A. R. Laracuente, P. E. Sheehan, L. J. Whitman, Direct deposition of continuous metal nanostructures by thermal dip-pen nanolithography, *Appl. Phys. Lett.*, **88**, 033104 (2006).
75. B. W. Maynor, Y. Li, J. Liu, Au "ink" for AFM "dip-pen" nanolithography, *Langmuir*, **17**, 2575–2578 (2001).
76. L. A. Porter Jr., H. C. Choi, J. M. Schmeltzer, A. E. Ribbe, L. C. C. Elliott, J. M. Buriak, Electroless nanoparticle film deposition compatible with photolithography, microcontact printing, and dip-pen nanolithography patterning technologies, *Nano Lett.*, **2**, 1369–1372 (2002).
77. Y. Li, B. W. Maynor, J. Liu, Electrochemical AFM "dip-pen" nanolithography, *J. Am. Chem. Soc.*, **123**, 2105–2106 (2001).
78. B. W. Maynor, S. F. Filocamo, M. W. Grinstaff, J. Liu, Direct-writing of polymer nanostructures: poly(thiophene) nanowires on semiconducting and insulating surfaces, *J. Am. Chem. Soc.*, **124**, 522–523 (2002).
79. Z. Zheng, M. Yang, B. Zhang, Reversible nanopatterning on self-assembled monolayers on gold, *J. Phys. Chem. C*, **112**, 6597–6604 (2008).
80. A. B. Braunschweig, A. J. Senesi, C. A. Mirkin, Redox-activating dip-pen nanolithography (RA-DPN), *J. Am. Chem. Soc.*, **131**, 922–923 (2009).
81. M. Su, X. Liu, S.-Y. Li, V. P. Dravid, C. A. Mirkin, Moving beyond molecules: patterning solid-state features via dip-pen nanolithography with sol-based inks, *J. Am. Chem. Soc.*, **124**, 1560–1561 (2002).
82. L. Ding, Y. Li, H. Chu, X. Li, J. Liu, Creation of cadmium sulfide nanostructures using AFM dip-pen nanolithography, *J. Phys. Chem. B*, **109**, 22337–22340 (2005).
83. Z. Zheng, M. Yang, Y. Liu, B. Zhang, Direct patterning of negative nanostructures on self-assembled monolayers of 16-mercaptohexadecanoic acid on Au(111) substrates via dip-pen nanolithography, *Nanotechnology*, **17**, 5378–5386 (2006).
84. R. E. Ionescu, R. S. Marks, L. A. Gheber, Nanolithography using protease etching of protein surfaces, *Nano Lett.*, **3**, 1639–1642 (2003).
85. R. E. Ionescu, R. S. Marks, L. A. Gheber, Manufacturing of nanochannels with controlled dimensions using protease nanolithography, *Nano Lett.*, **5**, 821–827 (2005).
86. J. Hyun, J. Kim, S. L. Craig, A. Chilkoti, Enzymatic nanolithography of a self-assembled oligonucleotide monolayer on gold, *J. Am. Chem. Soc.*, **126**, 4770–4771 (2004).
87. B. Li, Y. Zhang, S.-H. Yan, J.-H. Lü, M. Ye, M.-Q. Li, J. Hu, Positioning scission of single DNA molecules with nonspecific endonuclease based on nanomanipulation, *J. Am. Chem. Soc.*, **129**, 6668–6669 (2007).
88. S. Matsubara, H. Yamamoto, K. Oshima, E. Mouri, H. Matsuoka, Fabrication of nanostructure by Diels-Alder reaction, *Chem. Lett.*, **31**, 886–887 (2002).
89. D. A. Long, K. Unal, R. C. Pratt, M. Malkoch, J. Frommer, Localized "click" chemistry through dip-pen nanolithography, *Adv. Mater.*, **19**, 4471–4473 (2007).
90. B. Movassagh, P. Shaygan, Michael addition of thiols to α,β-unsaturated carbonyl compounds under solvent-free conditions, *ARKIVOC*, 130–137 (XII/2006).
91. X. G. Liu, S. W. Guo, C. A. Mirkin, Surface and site-specific ring-opening metathesis polymerization initiated by dip-pen nanolithography, *Angew. Chem. Int. Ed.*, **42**, 4785–4789 (2003).
92. L. Riemenschneider, S. Blank, M. Radmacher, Enzyme-assisted nanolithography, *Nano Lett.*, **5**, 1643–1646 (2005).
93. J. Yang, J. Han, K. Isaacson, D. Y. Kwok, Effects of surface defects, polycrystallinity, and nanostructure of self-assembled monolayers for octadecanethiol adsorbed onto Au on wetting and its surface energetic interpretation, *Langmuir*, **19**, 9231–9238 (2003).
94. D. A. Weinberger, S. Hong, C. A. Mirkin, B. W. Wessels, T. B. Higgins, Combinatorial generation and analysis of nanometer- and micrometer-scale silicon features via dip-pen" nanolithograpy and wet chemical etching, *Adv. Mater.*, **12**, 1600–1603 (2000).

95. H. Zhang, S. W. Chung, C. A. Mirkin, Fabrication of sub-50-nm solid-state nanostructures on the basis of dip-pen nanolithography, *Nano Lett.*, **3**, 43–45 (2003).
96. H. Zhang, C. A. Mirkin, DPN-generated nanostructures made of gold, silver, and palladium, *Chem. Mater.*, **16**, 1480–1484 (2004).
97. K. S. Salaita, S. W. Lee, D. S. Ginger, C. A. Mirkin, DPN-generated nanostructures as positive resists for preparing lithographic masters or hole arrays, *Nano Lett.*, **6**, 2493–2498 (2006).
98. M. Woodson, J. Liu, Functional nanostructures from surface chemistry patterning, *Phys. Chem. Chem. Phys.*, **9**, 207–225 (2007).
99. X. M. Li, J. Huskens, D.N. Reinhoudt, Reactive self-assembled monolayers on flat and nanoparticle surfaces, and their application in soft and scanning probe lithographic nanofabrication technologies, *J. Mater. Chem.*, **14**, 2954–2971 (2004).
100. J.-H. Lim, D. S. Ginger, K.-B. Lee, J. Heo, J.-M. Nam, C. A. Mirkin, Direct-write dip-pen nanolithography of proteins on modified silicon oxide surfaces, *Angew. Chem. Int. Ed.*, **42**, 2309–2312 (2003).
101. M. Yu, D. Nyamjav, A. Ivanisevic, Fabrication of positively and negatively charged polyelectrolyte structures by dip-pen nanolithography, *J. Mater. Chem.*, **15**, 649–652 (2005).
102. S. Myung, M. Lee, G. T. Kim, J. S. Ha, S. Hong, Large-scale "surface-programmed assembly" of pristine vanadium oxide nanowire-based devices, *Adv. Mater.*, **17**, 2361–2364 (2005).
103. Y. Wang, D. Maspoch, S. Zou, G. C. Schatz, R. E. Smalley, C. A. Mirkin, Controlling the shape, orientation, and linkage of carbon nanotubes with nano affinity templates, *Proc. Nat. Acad. Sci. USA*, **103**, 2026–2031 (2006).
104. C.-C. Wu, D. N. Reinhoudt, C. Otto, A. H. Velders, V. Subramaniam, Protein immobilization on Ni(II) ion patterns prepared by microcontact printing and dip-pen nanolithography, *ACS Nano*, **4**, 1083–1091 (2010).
105. D. C. Kim, D. J. Kang, Molecular recognition and specific interactions for biosensing applications, *Sensors*, **8**, 6605–6641 (2008).
106. K. B. Lee, E. Y. Kim, C. A. Mirkin, S. M. Wolinsky, The use of nanoarrays for highly sensitive and selective detection of human immunodeficiency virus type 1 in plasma, *Nano Lett.*, **4**, 1869–1872 (2004).
107. S. Lee, S. Lee, Y.-H. Ko, H. Jung, J. D. Kim, J. M. Song, J. Choo, S. K. Eo, S. H. Kang, Quantitative analysis of human serum leptin using a nanoarray protein chip based on single-molecule sandwich immunoassay, *Talanta*, **78**, 608–612 (2009).
108. S. W. Chung, D. S. Ginger, M. W. Morales, Z. Zhang, V. Chandrasekhar, M. A. Ratner, C. A. Mirkin, Top-down meets bottom-up: dip-pen nanolithography and DNA-directed assembly of nanoscale electrical circuits, *Small*, **1**, 64–69 (2005).
109. J. Li, C. Lu, B. Maynor, S. Huang, J. Liu, Controlled growth of long GaN nanowires from catalyst patterns fabricated by "dip-pen" nanolithographic techniques, *Chem. Mater.*, **16**, 1633–1636 (2004).
110. S.-W. Kang, D. Banerjee, A. B. Kaul, K. G. Megerian, Nanopatterning of catalyst by dip pen nanolithography (DPN) for synthesis of carbon nanotubes (CNT), *Scanning*, **32**, 42–48 (2010).
111. S. Zapotoczny, E. M. Benetti, G. J. Vancso, Preparation and characterization of macromolecular "hedge" brushes grafted from Au nanowires, *J. Mater. Chem.*, **17**, 3293–3296 (2007).
112. P. Xu, H. Uyama, J. E. Whitten, S. Kobayashi, D. L. Kaplan, Peroxidase-catalyzed in situ polymerization of surface orientated caffeic acid, *J. Am. Chem. Soc.*, **127**, 11745–11753 (2005).
113. L. Fu, X. Liu, Y. Zhang, V. P. Dravid, C. A. Mirkin, Nanopatterning of "hard" magnetic nanostructures via dip-pen nanolithography and a sol-based ink, *Nano Lett.*, **3**, 757–760 (2003).

114. A. Lewis; Y. Kheifetz, E. Shambrodt, A. Radko, E. Khatchatryan, C. Sukenik, Fountain pen nanochemistry: atomic force control of chrome etching, *Appl. Phys. Lett.*, **75**, 2689–2691 (1999).
115. M.-H. Hong, K. H. Kim, J. Bae, W. Jhe, Scanning nanolithography using a material-filled nanopipette, *Appl. Phys. Lett.*, **77**, 2604–2606 (2000).
116. Y.-K. Lee, S.-H. Lee, Y.-J. Kim, H. Kim, A novel passive membrane pumping nano fountain-pen, *Proc. 2nd IEEE Conf. Nano/Micro Engineered and Molecular Systems*, Jan 18–21, 2006, Zhuhai, China, pp. 1012–1017
117. K. Hwang, V.-D. Dinh, S.-H. Lee, Y.-J. Kim, H.-M. Kim, Analysis of line width with nano fountain pen using active membrane pumping, *Proc. 2nd IEEE Conf. Nano/Micro Engineered and Molecular Systems*, Jan 16–19, 2007, Bangkok, Thailand, pp. 759–763
118. N. Moldovan, K.-H. Kim, H. D. Espinosa, A multi-ink linear array of nanofountain probes, *J. Micromech. Microeng.*, **16**, 1935–1942 (2006).
119. Y. Lovski, A. Lewis, C. Sukenik, E. Grushka, Atomic-force-controlled capillary electrophoretic nanoprinting of proteins, *Anal. Bioanal. Chem.*, **396**, 133–138 (2010).
120. P. Vettiger, G. Cross, M. Despont, U. Drechsler, U. Dürig, B. Gotsmann, W. Häberle, M. A. Lantz, H. E. Rothuizen, R. Stutz, G. K. Binnig, The “millipede – nanotechnology entering data storage, *IEEE Trans. Nanotechnol.*, **1**, 39–55 (2002).
121. S. Hong, C. A. Mirkin, A nanoplotter with both parallel and serial writing capabilities, *Science*, **288**, 1808–1811 (2000).
122. S. C. Minne, J. D. Adams, G. Yaralioglu, S.R. Manalis, A. Atalar, C. F. Quate, Centimeter scale atomic force microscope imaging and lithography, *Appl. Phys. Lett.*, **73**, 1742–1744 (1998).
123. K. Salaita, Y. Wang, J. Fragala, R. A. Vega, C. Liu, C. A. Mirkin, Massively parallel dip-pen nanolithography with 55 000-pen two-dimensional arrays, *Angew. Chem. Int. Ed.*, **45**, 7220–7223 (2006).
124. J. Zou, D. Bullen, X. Wang, C. Liu, C. A. Mirkin, Conductivity-based contact sensing for probe arrays in dip-pen nanolithography, *Appl. Phys. Lett.*, **83**, 581–583 (2003).
125. X. Wang, D. A. Bullen, J. Zou, C. Liu, C. A. Mirkin, Thermally actuated probe array for parallel dip-pen nanolithography, *J. Vac. Sci. Technol. B*, **22**, 2563–2567 (2004).
126. D. Bullen, C. Liu, Electrostatically actuated dip pen nanolithography probe arrays, *Sens. Actuators A*, **125**, 504–511 (2006).
127. M. K. Yapici, J. Zou, A novel micromachining technique for the batch fabrication of scanning probe arrays with precisely defined tip contact areas, *J. Micromech. Microeng.*, **18**, 085015 (2008).
128. M. K. Yapici, J. Zou, Microfabrication of colloidal scanning probes with controllable tip radii of curvature, *J. Micromech. Microeng.*, **19**, 105021 (2009).
129. X. Wang, C. Liu, Multifunctional probe array for nano patterning and imaging, *Nano Lett.*, **5**, 1867–1872 (2005).
130. D. Banerjee, N. A. Amro, S. Disawal, J. Fragala, Optimizing microfluidic ink delivery for dip pen nanolithography, *J. Microlith., Microfab., Microsyst.*, **4**, 023014 (2005).
131. B. Rosner, T. Duenas, D. Banerjee, R. While, N. Amro, J. Rendlen, Functional extensions of dip pen nanolithography^{TM}: active probes and microfluidic ink delivery, *Smart Mater. Struct.*, **15**, S124–S130 (2006).
132. Y. Wang, L. R. Giam. M. Park, S. Lenhert, H. Fuchs, C. A. Mirkin, A self-correcting inking strategy for cantilever arrays addressed by an inkjet printer and used for dip-pen nanolithography, *Small*, **4**, 16666–1670 (2008).
133. F. Huo, Z. Zheng, G. Zheng, L. R. Giam, H. Zhang, C. A. Mirkin, Polymer pen lithography, *Science*, **321**, 1658–1660 (2008).
134. H. Zhang, R. Elghanian, N. A. Amro, S. Disawal, R. Eby, Dip pen nanolithography stamp tip, *Nano Lett.*, **4**, 1649–1655 (2004).

135. X. Liao, A. B. Braunschweig, Z. Zheng, C. A. Mirkin, Force- and time-dependent feature size and shape control in molecular printing via polymer-pen lithography, *Small*, **6**, 1082–1086 (2010).
136. X. Liao, A. B. Braunschweig, C. A. Mirkin, "Force-feedback" leveling of massively parallel arrays in polymer pen lithography, *Nano Lett.*, **10**, 1335–1340 (2010).
137. Z. Zheng, W. L. Daniel, L. R. Giam, F. Huo, A. J. Senesi, G. Zheng, C. A. Mirkin, Multiplexed protein arrays enabled by polymer pen lithography: addressing the inking challenge, *Angew. Chem. Int. Ed.*, **48**, 7626–7629 (2009).
138. A. B. Braunschweig, F. Huo, C. A. Mirkin, Molecular printing, *Nat. Chem.*, **1**, 353–358 (2009).
139. A. Ulman, *An introduction to ultrathin organic films*, Academic Press, San Diego, CA, 1991.
140. S. Onclin, B. J. Ravoo, D. N. Reinhoudt, Engineering silicon oxide surfaces using self-assembled monolayers, *Angew. Chem. Int. Ed.*, **44**, 6282–6304 (2005).
141. S. Krämer, R. R. Fuierer, C. B. Gorman, Scanning probe lithography using self-assembled monolayers, *Chem. Rev.*, **103**, 4367–4418 (2003).
142. S. Hong, J. Zhu, C. A. Mirkin, Multiple ink nanolithography: toward a multiple-pen nano-plotter, *Science*, **286**, 523–525 (1999).
143. S. Lenhert, P. Sun, Y. Wang, H. Fuchs, C. A. Mirkin, Massively parallel dip-pen nanolithography of heterogeneous supported phospholipids multilayer patterns, *Small*, **3**, 71–75 (2007).
144. S. Lenhert, F. Brinkmann, T. Lauer, W. Walheim, C. Vannahme, S. Klinkhammer, M. Xu, S. Sekula, T. Mappes, T. Schimmel, H. Fuchs, Lipid multilayer gratings, *Nat. Nanotechnol.*, **5**, 275–279 (2010).
145. H. Zhang, K.-B. Lee, Z. Li, C. A. Mirkin, Biofunctionalized nanoarrays of inorganic structures prepared by dip-pen nanolithography, *Nanotechnology*, **14**, 1113–1117 (2003).
146. H. Zhou, Z. Li, A. Wu, G. Wei, Z. Liu, Direct patterning of rhodamine 6G molecules on mica by dip-pen nanolithography, *Appl. Surf. Sci.*, **236**, 18–24 (2004).
147. J. Huskens, M. A. Deij, D. N. Reinhoudt, Attachment of molecules at a molecular printboard by multiple host-guest interactions, *Angew. Chem. Int. Ed.*, **41**, 4467–4471 (2002).
148. Z. Nie, E. Kumacheva, Patterning surfaces with functional polymers, *Nat. Mater.*, **7**, 277–290 (2008).
149. R. McKendry, W. T. S. Huck, B. Weeks, M. Fiorini, C. Abell, T. Rayment, Creating nanoscale patterns of dendrimers on silicon surfaces with dip-pen nanolithography, *Nano Lett.*, **2**, 713–716 (2002).
150. G. H. Degenhart, B. Dordi, H. Schönherr, G. J. Vancso, Micro- and nanofabrication of robust reactive arrays based on the covalent coupling of dendrimers to activated monolayers, *Langmuir*, **20**, 6216–6224 (2004).
151. R. B. Salazar, A. Shovsky, H. Schönherr, G. J. Vancso, Dip-pen nanolithography on (bio)reactive monolayer and block-copolymer platforms: deposition of lines of single macromolecules, *Small*, **2**, 1274–1282 (2006).
152. P. L. Stiles, Direct deposition of micro- and nanoscale hydrogels using dip pen nanolithography (DPN), *Nat. Methods*, **7**, (2010), doi:10.1038/nmeth.f.309.
153. S. W. Lee, R. G. Sanedrin, B.-K. Oh, C. A. Mirkin, Nanostructured polyelectrolyte multilayer thin films generated via parallel dip-pen nanolithography, *Adv. Mater.*, **17**, 2749–2753 (2005).
154. L. M. Demers, C. A. Mirkin, Combinatorial templates generated by dip-pen nanolithography for the formation of two-dimensional particle arrays, *Angew. Chem. Int. Ed.*, **40**, 3069–3071 (2001).
155. D. Nyamjav, A. Ivanisevic, Properties of polyelectrolyte templates generated by dip-pen nanolithography and microcontact printing, *Chem. Mater.*, **16**, 5216–5219 (2004).
156. D. C. Coffey, D. S. Ginger, Patterning phase separation in polymer films with dip-pen nanolithography, *J. Am. Chem. Soc.*, **127**, 4564–4565 (2005).

157. J. Chiota, J. Shearer, M. Wei, C. Barry, J. Mead, Multiscale directed assembly of polymer blends using chemically functionalized nanoscale-patterned templates, *Small*, **5**, 2788–2791 (2009).
158. Q. Tang, S.-Q. Shi, Preparation of gas sensors via dip-pen nanolithography, *Sens. Actuators B*, **131**, 379–383 (2008).
159. J. Hyun, W.-K. Lee, N. Nath, A. Chilkoti, S. Zauscher, Capture and release of proteins on the nanoscale by stimuli-responsive elastin-like polypeptide "switches", *J. Am. Chem. Soc.*, **126**, 7330–7335 (2004).
160. K. Ariga, T. Nakanishi, T. Michinobu, Immobilization of biomaterials to nano-assembled films (self-assembled monolayers, Langmuir-Blodgett films, and layer-by-layer assemblies) and their related functions, *J. Nanosci. Nanotechnol.*, **6**, 2278–2301 (2006).
161. W. Senaratne, L. Andruzzi, C. K. Ober, Self-assembled monolayers and polymer brushes in biotechnology: current applications and future perspectives, *Biomacromol.*, **6**, 2427–2448 (2005).
162. M. A. Kramer, H. C. Park, A. Ivanisevic, Dip-pen nanolithography on SiOx and tissue-derived substrates: comparison with multiple biological inks, *Scanning*, **32**, 330–34 (2010).
163. K. L. Christman, V. D. Enriquez-Rios, H. D. Maynard, Nanopatterning proteins and peptides, *Soft Matter*, **2**, 928–939 (2006).
164. J.-M. Nam, S. W. Han, K.-B. Lee, X. Liu, M. A. Ratner, C. A. Mirkin, Bioactive protein nanoarrays on nickel oxide surfaces formed by dip-pen nanolithography, *Angew. Chem. Int. Ed.*, **43**, 1246–1249 (2004).
165. B. Li, Y. Zhang, J. Hu, M. Li, Fabricating protein nanopatterns on a single DNA molecule with dip-pen nanolithography, *Ultramicroscopy*, **105**, 312–315 (2005).
166. K.-B. Lee, S.-J. Park, C. A. Mirkin, J. C. Smith, M. Mrksich, Protein nanoarrays generated by dip-pen nanolithography, *Science*, **295**, 1702–1705 (2002).
167. K. H. Kim, J. D. Kim, Y. J. Kim, S. H. Kang, S. Y. Jung, H. Jung, Protein immobilization without purification via dip-pen nanolithography, *Small*, **8**, 1089–1094 (2008).
168. J. Ji, J.-C. Yang, D. N. Larson, Nanohole arrays of mixed designs and microwriting for simultaneous and multiple protein binding studies, *Biosens. Bioelectron.*, **24**, 2847–2852 (2009).
169. J. Lü, M. Ze, N. Duan, B. Li, Enzymatic digestion of single DNA molecules anchored on nanogold-modified surfaces, *Nanoscale Res. Lett.*, **4**, 1029–1034 (2009).
170. L. M. Demers, S.-J. Park, T. A. Taton, Z. Li, C. A. Mirkin, Orthogonal assembly of nanoparticle building blocks on dip-pen nanolithographically generated templates of DNA, *Angew. Chem. Int. Ed.*, **40**, 3071–3073 (2001).
171. H. Zhang, Z. Li, C. A. Mirkin, Dip-pen nanolithography-based methodology for preparing arrays of nanostructures functionalized with oligonucleotides, *Adv. Mater.*, **14**, 1472–1474 (2002).
172. A. Baserga, M. Vigano, C. S. Casari, S. Turri, A. L. Bassi, M. Levi, C. E. Bottani, Au-Ag template stripped pattern for scanning probe investigations of DNA arrays produced by dip pen nanolithography, *Langmuir*, **24**, 13212–13217 (2008).
173. S. Li, S. Szegedi, E. Goluch, C. Liu, Dip pen nanolithography functionalized electrical gaps for multiplexed DNA detection, *Anal. Chem.*, **80**, 5899–5904 (2008).
174. R. J. Stokes, J. A. Dougan, D. Graham, Dip-pen nanolithography and SERRS as synergetic techniques, *Chem. Commun.*, 5734–5736 (2008).
175. D. K. Hoover, E.-J. Lee, E. W. L. Chan, M. N. Yousaf, Electroactive nanoarrays for biospecific ligand mediated studies on cell adhesion, *Chem. Bio. Chem.*, **8**, 1920–1923 (2007).
176. D. K. Hoover, E. W. L. Chan, M. N. Yousaf, Asymmetric peptide nanoarray surfaces for studies of single-cell polarization, *J. Am. Chem. Soc.*, **130**, 3280–3281 (2008)
177. Q. Wang, W. Xian, S. Li, C. Liu, G. W. Padua, Topography and biocompatibility of patterned hydrophobic/hydrophilic zein layers, *Acta Biomater.*, **4**, 844–851 (2008).

178. S. Sekula, J. Fuchs, S. Weg-Remers, P. Nagel, S. Schuppler, J. Fragala, N. Theilacker, M. Franzreb, C. Wingren, P. Ellmark, C. A. K. Borrebaeck, C. A. Mirkin, H. Fuchs, S. Lenhert, Multiplexed lipid dip-pen nanolithography on subcellular scales for the templating of functional proteins and cell culture, *Small*, **4**, 1785–1793 (2008).
179. R. A. Vega, C. K.-F. Shen, D. Maspoch, J. G. Robach, R. A. Lamb, C. A. Mirkin, Monitoring single-cell infectivity from virus particle nanoarrays fabricated by parallel dip-pen nanolithography, *Small*, **3**, 1482–1485 (2007).
180. B. Basnar, I. Willner, Dip-pen-nanolithographic patterning of metallic, semiconductor, and metal oxide nanostructures on surfaces, *Small*, **5**, 28–44 (2009).
181. D. Prime, S. Paul, C. Pearson, M. Green, M. C. Petty, Nanoscale patterning of gold nanoparticles using an atomic force microscope, *Mat. Sci. Eng. C*, **25**, 33–38 (2005).
182. R. J. Barsotti Jr., F. Stellacci, Chemically directed assembly of monolayer protected gold nanoparticles on lithographically generated patterns, *J. Mater. Chem.*, **16**, 962–965 (2006).
183. K. B. Lee, E. Y. Kim, C. A. Mirkin, Protein nanostructures formed via direct-write dip-pen nanolithography, *J. Am. Chem. Soc.*, **125**, 5588–5589 (2003).
184. C. M. Bruinink, C. A. Nijhuis, M. Péter, B. Dordi, O. Crespo-Biel, T. Auletta, A. Mulder, H. Schönherr, G. J. Vancso, J. Huskens, D. N. Reinhoudt, Supramolecular microcontact printing and dip-pen nanolithography on molecular printboards, *Chem. Eur. J.*, **11**, 3988–3996 (2005).
185. H. Zhang, R. Jin, C. A. Mirkin, Synthesis of open-ended, cylindrical Au-Ag alloy nanostructures on a Si/SiO_x surface, *Nano Lett.*, **4**, 1493–1495 (2003).
186. S.-C. Hung, O. A. Nafday, J. R. Haaheim, F. Ren, G. C. Chi, S. J. Pearton, Dip pen nanolithography of conductive silver traces, *J. Phys. Chem. C*, **114**, 9672–9677 (2010).
187. K. Salaita, S. W. Lee, X. Wang, L, Huang, T.M. Dellinger, C. Liu, C. A. Mirkin, Sub-100 nm, centimeter-scale, parallel dip-pen nanolithography, *Small*, **1**, 940–945 (2005).
188. H. Zhang, N. A. Amro, S. Disawal, R. Elghanian, R. Shile, J. Fragala, High-throughput dip-pen-nanolithography-based fabrication of Si nanostructures, *Small*, **3**, 81–85 (2007).
189. M. Su, S. Li, V. P. Dravid, Miniaturized chemical multiplexed sensor array, *J. Am. Chem. Soc.*, **125**, 9930–9931 (2003).
190. G. Gundiah, N. S. John, P. J. Thomas, G. U. Kulkarni, C. N. R. Rao, S. Heun, Dip-pen nanolithography with magnetic Fe_2O_3 nanocrystals, *Appl. Phys. Lett.*, **84**, 5341–5343 (2004).
191. X. Liu, L. Fu, S. Hong, V. P. Dravid, C. A. Mirkin, Arrays of magnetic nanoparticles patterned via "dip-pen" nanolithography, *Adv. Mater.*, **14**, 231–234 (2002).
192. D. Nyamjav, A. Ivanisevic, Templates for DNA templated Fe_3O_4 nanoparticles, *Biomaterials*, **26**, 2749–2757 (2005).
193. Y. Wang, W. Wei, D. Maspoch, J. Wu, V. P. Dravid, C. A. Mirkin, Superparamagnetic sub-5 nm Fe@C nanoparticles: isolation, structure, magnetic properties, and directed assembly, *Nano Lett.*, **8**, 3761–3765 (2008).
194. E. Bellido, R. De Miguel, J. Sese, D. Ruiz-Molina, A. Lostao, D. Maspoch, Nanoscale positioning of inorganic nanoparticles using biological ferritin arrays fabricated by dip-pen nanolithography, *Scanning*, **32**, 35–41 (2010).
195. B. Li, C. F. Goh, X. Zhou, G. Lu, H. Tantang, Y. Chang, C. Xue, F. Y. C. Boey, H. Zhang, Patterning colloidal metal nanoparticles for controlled growth of carbon nanotubes, *Adv. Mater.*, **20**, 4873–4878 (2008).
196. E. Granot, B. Basnar, Z. Cheglakov, E. Katz, I. Willner, Enhanced bioelectrocatalysis using single-walled carbon nanotubes (SWCNTs) polyaniline hybrid systems in thin-films and microrod structures associated with electrodes, *Electroanalysis*, **18**, 26–34 (2006).
197. A. Baba, F. Sato, N. Fukuda, H. Ushijima, K. Yase, Micro/nanopatterning of single-walled carbon nanotube-organic semiconductor composites, *Nanotechnology*, **20**, 085301 (2009).

198. S. Zaitsev, A. Svintsov, C. Ebm, S. Eder-Kapl, H. Loeschner, E. Platzgummer, G. Lalev, S. Dimov, V. Velkova, B. Basnar, 3D ion multi-beam processing with CHARPAN PMLP tool and with a single ion beam FIB tool, optimized with the 'IonRevSim' software, *in Alternative Lithographic Technologies*, F. M. Schelenberg and B. B. La Fontaine (Ed), Proceedings of SPIE Vol. 7271, The International Society for Optical Eng., 2009
199. J.-W. Jang, D. Maspoch, T. Fujigaya, C. A. Mirkin, A "molecular eraser" for dip-pen nanolithography, *Small*, **3**, 600–605 (2007).

Chapter 7
Nanofabrication of Functional Nanostructures by Thermochemical Nanolithography

Debin Wang, Vamsi K. Kodali, Jennifer E. Curtis, and Elisa Riedo

Abstract Nanofabrication is the process of building functional structures with nanoscale dimensions, which can be used as components, devices, or systems with high density, in large quantities, and at low cost. Since the invention of scanning tunneling microscopy (STM) and atomic force microscopy (AFM) in 1980s, the application of scanning probe based lithography (SPL) techniques for modification of substrates and creation of functional nanoscale structures and nanostructured materials has been widespread, resulting in the emergence of a large variety of methodologies. In this chapter, we review the recent development of a thermal probe based nanofabrication technique called thermochemical nanolithography (TCNL). We start with a brief review of the evolution of the thermal AFM probes integrated with resistive heaters. We then provide an overview of some established nanofabrication techniques in which thermal probes are used, namely thermomechanical nanolithography, the Millipede project, and thermal dip-pen nanolithography. We discuss the heat transfer mechanisms of the thermal probes in the thermal writing process of TCNL. The remainder of the chapter focuses on the use of TCNL on a variety of systems and thermochemical reactions. TCNL has been successfully used for fabrication of functional nanostructures that are appealing for various applications in nanofluidics, nanoelectronics, nanophotonics, and biosensing devices. Finally, we close this chapter by discussing some future research directions where the capabilities and robustness of TCNL can be further extended.

Keywords Atomic force microscopy (AFM) · Scanning probe microscopy (SPM) · Nanofabrication · Nanomanufacturing · Nanolithography · Nanopatterning · Thermochemical nanolithography (TCNL) · Thermomechanical nanolithography · Thermal dip-pen nanolithography (tDPN) · Millipede · Wettability · Conjugated polymer · Graphene · Graphene oxide

Abbreviations

AFM	Atomic force microscopy
CD	Cluster of differentiation
CF	Covalent functionalization

E. Riedo (✉)
School of Physics, Georgia Institute of Technology, Atlanta, GA 30332-0430, USA
e-mail: elisa.riedo@physics.gatech.edu

A.A. Tseng (ed.), *Tip-Based Nanofabrication*, DOI 10.1007/978-1-4419-9899-6_7,

DPN	Dip-pen nanolithography
GO	Graphene oxide
ICAM	Inter-cellular adhesion molecule
IgG	Immunoglobulin G
MEMS	Microelectromechanical system
MR	Molecular recognition
PKC-θ	Protein kinase C-θ
PMC-MA	Poly(3-(4-[(E)-3-methoxy-3-oxoprop-1-enyl]phenoxy)propyl 2-methacrylate)
PMMA	Poly(methyl methacrylate)
PPV	Poly(*p*-phenylene vinylene)
SAM	Self-assembled monolayers
SOI	Silicon-on-insulator
SPL	Scanning probe lithography
SPM	Scanning probe microscopy
STM	Scanning tunneling microscopy
TCNL	Thermochemical nanolithography
tDPN	Thermal dip-pen nanolithography
THP-MA	Poly(tetrahydro-2H-pyran-2-ylmeth-acrylate)
T_g	Temperature of glass transition in organic polymer materials

7.1 Introduction

Nanofabrication is the process of making functional structures with arbitrary topographic and/or chemical patterns having nanoscale dimensions [1–4]. Nanofabrication has been widely implemented for improving microelectronic devices and information technology, for example, to increase the density of components, to lower their cost, and to increase their performance per device and per integrated circuit [5]. Other areas of application beyond information processing and data storage include photonics, sensor technologies, and novel materials [6–7].

The methods used to produce nanoscale structures and nanostructured materials are commonly categorized as the "top-down" or the "bottom-up" approach [3]. The "bottom-up" approach uses interactions between molecules or colloidal particles to self-assemble discrete nanoscale structures in two- and three-dimensions. The "top-down" approach uses various lithography methods to pattern materials with nanometer resolution. This approach includes serial and parallel techniques for patterning features. The dominant "top-down" techniques are photolithography [8] and particle beam lithography [9]. The limitations of these conventional approaches, such as high capital and operational costs, restricted planar-only fabrication, and incompatibility with biological materials, have motivated the development of unconventional fabrication techniques: soft lithography [10], self-assembly [11], edge lithography [12], and scanning probe lithography (SPL) [4, 13].

SPL is a set of lithographic methods that utilizes scanning probes, in which a nanoscopic tip is attached to a microscopic cantilever, to create a variety of functional nanostructures and nanomaterials. SPL stands out amongst other nanofabrication methods due to its revolutionary capability to achieve controllable nanomanufacturing. SPL is able to manufacture nanostructures, namely nanowires, nanotubes, and quantum dots, with nanoscopic control over the shape, size, and position of each individual nanostructure. The development of SPL began soon after the inception of scanning probe microscopy (SPM), namely scanning tunneling microscopy (STM) [14] in 1982 and atomic force microscopy (AFM) [15] in 1986. SPM is capable of providing functionalized cantilevers and tips to manipulate environments at sub-micrometer scale on the substrate surface, generating optical fields, high temperature fields, and high electric, and/or magnetic fields. Therefore, a variety of SPL techniques for controlled nanomanufacturing have evolved, ranging from the subtle movement of atoms using STM [16], the formation of local deformations in soft substrates using high-contact force AFM [17], to the local application of inks and the local oxidation of suitable substrates [18–19]. With the application of SPL techniques, control over position and direction in the range of 5–50 nm is evident. Moreover, due to the very recently developed multi-probe systems [20] and automated scanning probe equipment [21], the patterning of large areas has become accessible. Not only can physical patterns on substrates be created, but also nanoscale chemical reactions on substrates have been demonstrated, which allow the combination of techniques from both the top-down and the bottom-up approaches. With the high cost of conventional photolithography and particle beam lithography techniques, SPL is a particularly useful alternative for low volume manufacturing and prototyping in conventional processes.

Among all the existing SPL techniques, dip-pen nanolithography (DPN) stands out as a versatile technique that offers high resolution and registration with direct-write patterning capabilities [22]. DPN functions by facilitating the direct transport of molecules to surfaces, much like the transfer of ink from a macroscopic dip-pen to paper. By depositing several different kinds of molecules on the same substrate, DPN can pattern a range of desired chemistries with spatial control without exposing the substrate to harsh solvents, chemical etching, and/or extreme electrical field gradients. DPN is compatible with a variety of inks, including organic molecules [23], organic [24] and biological [25] polymers, colloidal particles [26], and metal ions [27]. DPN can be used to pattern substrates ranging from metals to insulators [28]. The intrinsic linear writing speed of DPN depends on molecular transport between the probe tip and the surface, and is thus limited by mass diffusion. Like all SPL techniques, DPN is inherently a serial lithography process. In order to increase the throughput, development of microfluidic arrays of addressable ink wells is now underway [29]. However, automation of ink delivery remains the ultimate challenge of Microelectromechanical System (MEMS) integration of DPN technology.

In this book chapter, we review the recent results on a thermal AFM probe based nanofabrication technique called thermochemical nanolithography (TCNL). We begin with a brief review of the development of the thermal probes integrated with resistive heaters, followed by an overview of some established nanofabrication techniques involving thermal probes, including Thermomechanical Nanolithography, the Millipede project, and thermal Dip-Pen nanolithography. The remainder focuses on the development of TCNL. We discuss the heat transfer mechanisms of the thermal probes in the writing process. We review the use of TCNL on a variety of systems and thermochemical reactions. We show that TCNL can be used for the fabrication of functional nanostructures that are appealing for various applications in nanofluidics, nanoelectronics, nanophotonics, and biosensing devices. Finally, we close this chapter by discussing some future research directions where the capabilities and robustness of TCNL can be further extended.

7.2 Nanofabrication with Thermal AFM Probes

7.2.1 Thermal Probes

Thermal AFM probes were originally developed for scanning thermal microscopy with the aim of providing surface topography as well as local heat-related information, including temperature and thermal conductivity [30–32]. The first kind of these probes, designed for simultaneous topographical and thermal imaging, were made from a Wollaston wire consisting of a Ag sheath and a Pt core [33]. A second type of thermal cantilever was developed with a thin film Ni/W resistor sandwiched between layers of polyimide [34]. The polyimide cantilever material provided low thermal conductivity for thermal isolation from the ambient medium, low stiffness for imaging soft biological samples, and low electrical conductivity for electrical isolation of the heater from the ambient medium. Resistive heaters have eventually been incorporated into probe cantilevers with lateral resolution equivalent to state-of-the-art silicon probes [35]. When current flows through the probe legs, resistive heating near the probe tip can raise the tip temperature to over 1,000°C.

The principle of thermal sensing is based on the fact that the thermal impedance between the resistive heater and the polymer substrate changes as a function of distance between them [36–37]. The air between the resistive heater and the sample substrate transport heat from the cantilever to the substrate. The heat transport through the air varies according to the change of the distance between the cantilever and the substrate, resulting in a change of the cantilever temperature. Since the electrical resistance of the probe depends on its temperature, the change of the electrical resistance ($\Delta R/R$) can be used as a readback sensing signal to track the topography of the substrate surface. The speed of this thermal sensing process is limited by the thermal time constant of the heater that is of the order of a few microseconds. King et al. modeled and predicted that the thermal impedance sensing can provide more than one order of magnitude improved performance in sensitivity and resolution over piezoresistive-strain sensing due to the stronger thermal effects in semiconductors [31, 38–39]. Kim et al. [40] and Park et al. [41] recently demonstrated the topographical imaging by the thermal probes could be operated in AFM contact mode as well as in AFM tapping mode, respectively.

7.2.2 Thermomechanical Nanolithography

The data density of magnetic disks will likely plateau in the range of 100–200 Gb·in^{-2} due to the superparamagnetic effect [42]. Because of its ability to form and detect nanometer-scale structures, thermomechanical based SPL has been considered a promising candidate technology for advanced data storage development. Mamin et al. from IBM Research pioneered the development of thermomechanical nanolithography [43]. In this pilot work, they used a pulsed infrared laser as the heat source to heat the AFM tip to create arrays of 150 nm wide pits on a poly(methyl methacrylate) (PMMA) substrate. However, the necessity of an external laser required difficult optical alignment and inhibited further development of data storage applications. Four years later, they proposed the use of a piezoresistive cantilever in which electrical power provided an alternative heat source [44]. With this modified design, they were able to produce an array of sub-100 nm wide pits with 150 nm spacing on a polycarbonate substrate, corresponding to a data density of roughly 30 Gb·in^{-2}.

During a standard thermomechanical nanolithography process, writing is performed by a thermal AFM probe that comprises a resistive heater for heating the tip as well as a capacitive platform for applying loading force through electrostatic interaction. In a stand-by position, the tip sits a few hundred nanometers above the substrate while heated by a voltage applied across the cantilever legs. For the writing, an electrostatic force pulse is provided by applying a voltage between the substrate and the cantilever for a duration of a few microseconds. The pulse brings the hot tip into contact with the polymer surface and softens the surrounding materials, resulting in a thermomechanical indentation. The presence or absence of the indentations corresponds to logical 1s or 0s, respectively.

In addition to its writing capability, the same thermal cantilevers can be used for imaging and reading because a second heater is placed beside the tip. When the second heater operates at a relatively low temperature and the distance between heater and sample is reduced as the tip moves into a bit indentation, the temperature and electrical resistance of the heater will decrease because the heat transport through air will be more efficient. Thus, changes in electrical resistance of the continuously heated probe are monitored while the probe is scanned over data bits, providing a means of detecting the bits.

The latest thermomechanical nanolithography work by Gotsmann et al. has shown the capability of fabricating three-dimensional nanostructures by using self-amplified depolymerization polymers [45–46]. For such polymers, the breaking of a single bond induces the spontaneous depolymerization of the entire polymer chain. The decomposition reaction is fast in comparison with the mechanical motion of the tip. Therefore, nanostructures can be written less than a couple of microseconds and the writing capability can be extended to the third dimension in one patterning step. They have demonstrated arbitrary three-dimensional nanostructures with ~40 nm lateral and ~1 nm vertical resolution.

7.2.3 Thermal Dip-Pen Nanolithography

Sheehan et al. initiated the research on chemical patterning with thermal AFM probes by developing a technique known as thermal DPN (tDPN) [47]. The rise of the temperature of the tip causes solid ink to melt and wet the tip. The advantage of this approach is twofold. It can turn on and off the ink flow at will, whereas prior DPN techniques apply ink to the surface as long as the tip stays in contact with it. In addition, the rate of the ink diffusion is tunable by controlling the tip temperature. In their pilot work, Sheehan et al. has demonstrated that octadecylphosphonic acid, a kind of solid ink immobile at room temperature, can self-assemble on mica after heated to 100°C melting temperature by the tip [47]. Single lines were written less than 100 nm wide. Their recent work has demonstrated that the nanostructures of poly(*N*-isopropylacrylamide), a stimulus-responsive polymer whose surface wettability can be modified upon heating or cooling, can be directly and reproducibly written from melting on an epoxysilane SAM-functionalized silicon oxide substrate [48]. These nanostructures reversibly bind and release proteins when actuated through a hydrophilic-hydrophobic phase transition. In a separate paper, they have shown that tDPN uses a thermal AFM probe as a "nano-soldering iron" and can deposit nanoscale electrical metal connections using an analogy to conventional soldering iron (Fig. 7.1) [49]. Indium metal nanowires less than 80 nm wide have been demonstrated. The capability of direct writing of continuous electrical nanowires can be used for fabrication of nanocircuits, developing nanoelectronic prototypes, and even in situ inspection and repair of nanocircuits.

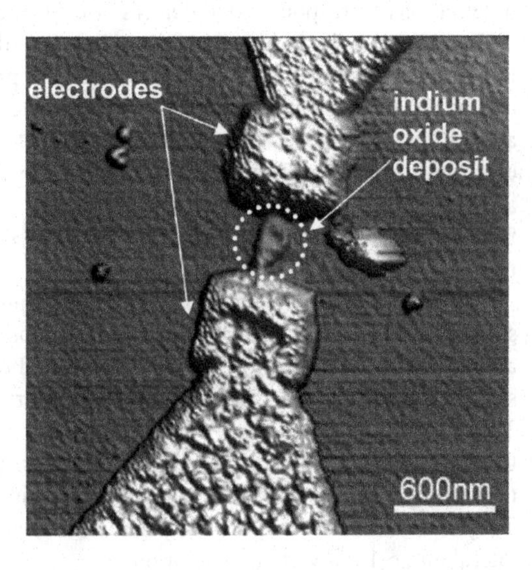

Fig. 7.1 Topographical AFM image of a continuous structure deposited with in across a 500 nm wide gap between pre-fabricated gold electrodes [49]. Copyright 2006, American Institute of Physics

7.2.4 Parallel Patterning with Probe Arrays

The throughput of SPL technologies is limited by the intrinsic serial writing fashion. A practical approach to SPL for large-volume, parallel production may emerge by simultaneously writing patterns with multiple probes. In fact, various designs of AFM probe arrays have been developed for applications in nanofabrication, as well as in parallel imaging, force spectroscopy, and bio/chemical sensing. Aeschimann et al. designed a 4×4 array of piezoresistive cantilever probes that allows for imaging cells in their native conditions and performing force spectroscopy measurements on them [50]. Arrays of probes with selective coating can used as sensing devices because absorption of bio/chemical molecules can be transduced into nanomechanical motion of the cantilevers. This mechanism has been used for investigation of DNA hybridization and receptor-ligand binding by Fritz et al. [51], the interaction of DNA-binding proteins by Huber et al. [52], and the detection of prostate-specific antigens over a range of concentrations by Wu et al. [53].

The thermal AFM probes were the first type of scanning probes that have been fabricated in two-dimensional arrays [20]. Their unique read/write dual functionality makes them particularly desirable for data storage application. The use of thermomechanical sensing mechanism circumvents the complexities of other tip-height feedback control systems, such as optical beam reflection, optical interferometry, or the electronics for piezoresistive sensing. The first 2D probe array was made possible by Lutwyche et al. [54]. They showed that a 5 × 5 probe array with successful application to parallel imaging of a test sample consisting of 200 nm pitches etched into silicon. A couple of years later, Vettiger et al. presented the first operational "Millipede" prototype design. A large 32 × 32 probe array effectively functioned in a parallel fashion. Write/read storage operation in a thin PMMA medium was demonstrated at data densities from 100 to 400 $\text{Gb}\cdot\text{in}^{-2}$ [20]. In a following paper, Pantazi et al. presented a small scale Millipede prototype system of a storage device. Experimental results of multiple sectors, recorded with a probe array of up to 64 × 64 free-standing cantilevers demonstrated the functionality of the prototype system [55]. The operation of the array of 4,096 cantilever achieved ultra-high data storage density of above 840 $\text{Gb}\cdot\text{in}^{-2}$ and a raw bit error rate of merely 10^{-4} [37].

Several parallel-probe strategies have been developed for large-scale DPN nanofabrication. Zhang et al. designed a linear array of 32 passive (non-actuated) probes that can be used to write SAM features with 60 nm resolution [56]. Prototypes of active parallel-probe arrays based on thermoelectric actuation by Wang et al. [57] or electrical conduction by Zou et al. [58] have been demonstrated to control each individual probe. Salaita et al. reported a 2D array of 55,000 probes fabricated to achieve extremely large area replication with DPN. Wang et al. recently demonstrated a multifunctional probe array for patterning and imaging [57]. The probe array comprises of three kinds of probes that are devoted to perform DPN, scanning probe contact printing, and generic AFM microscopy, respectively. With this design, functional DPN nanostructures can be generated with high throughput and can be imaged without cross contamination and changing probes.

For all the probe array operations, cantilever deflection sensing mechanism is always a major concern. Optical sensing mechanism using diode lasers can only be used for arrays of a small number of probes [57]. Arrays of a large number of probes involve integration of different deflection sensing methods, such as piezoelectric sensing, capacitive sensing, piezoresistive sensing, and thermal impedance sensing [59]. Piezoresistive sensing has been widely used because of its high sensitivity and ease of fabrication and implementation.

7.3 Heat Transfer Mechanisms in the Thermal Probes

The thermal AFM probes used for TCNL nanofabrication (Fig. 7.2) are made using a standard silicon-on-insulator (SOI) process following a documented fabrication process developed by King et al. [60]. The process starts with an SOI wafer of orientation <100>, n-type doping at $2 \times 10^{14}/\text{cm}^3$ with a resistivity of approximately 4 Ω·cm. A cantilever tip is formed using an oxidation sharpening process and typically has a radius of curvature of 20 nm and a height of 1.5 μm. The probes were made electrically active by selectively doping different parts of the cantilever through a two-step process. First, a low-dosage blanket ion implantation was performed on the entire cantilever and furnace-annealed in order to establish an essentially uniform background doping level ($10^{17}/\text{cm}^3$, phosphorous n-type). The cantilever was then subjected to a heavy implantation step during which a region around the tip (width 8 μm) is masked off ($10^{20}/\text{cm}^3$, phosphorous n type). The masked region became a relatively lightly doped region at the free end of the cantilever, *i.e.*, the resistive heater. The cantilever was electrically connected to the base via highly conducting legs (110 μm long and 15 μm wide). With the cantilever dimensions and temperature-dependent resistivity, the heater accounts for more than 90% of the electrical resistance of entire cantilever.

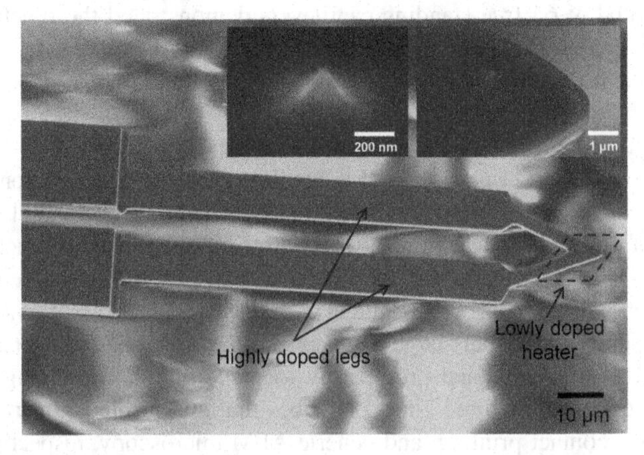

Fig. 7.2 SEM micrograph of a thermal AFM probe. The tip sits above the integrated resistive heater (shown as inset). Adapted from [66]

7.3.1 Heat Generation in the Resistive Heater

The feasibility of sub-20 nm local chemical reaction achieved by TCNL is supported by previous research on local temperature gradients near the contact between a nanoscale AFM tip and a substrate surface [61–62]. The temperature at the tip-substrate contact is significantly higher than the temperature elsewhere in the substrate. In the region close to the location of the contact, the temperature drops sharply within a few nanometers range, in both lateral and vertical directions.

Figure 7.3 depicts a schematic of the temperature control system used for a thermal AFM probe in a TCNL process. A power supply is used to provide electrical power to the thermal probe with a heating voltage (V_O). The electric circuit has a sense resistor (R_S) connected to the cantilever in series to protect the probe by limiting the current at high power as well as to sense the current (I_S). A multimeter is used to measure the voltage drop across the sense resistor (V_S). By recording V_S and V_O, dissipation power P_H can be obtained as:

$$P_H = (V_0 - V_S)\frac{V_S}{R_S}. \tag{1}$$

Since the electrical resistivity of the doped silicon is a strong function of temperature, the cantilever resistance (R_H) varies non-linearly with temperature (Fig. 7.4). However, it can be experimentally obtained using the following relationship:

$$R_H = R_S\frac{V_0 - V_S}{V_S}. \tag{2}$$

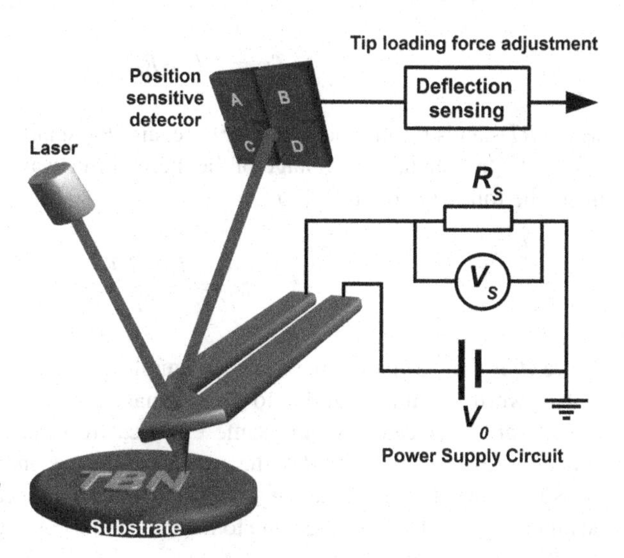

Fig. 7.3 Schematic of TCNL temperature control and deflection sensing systems

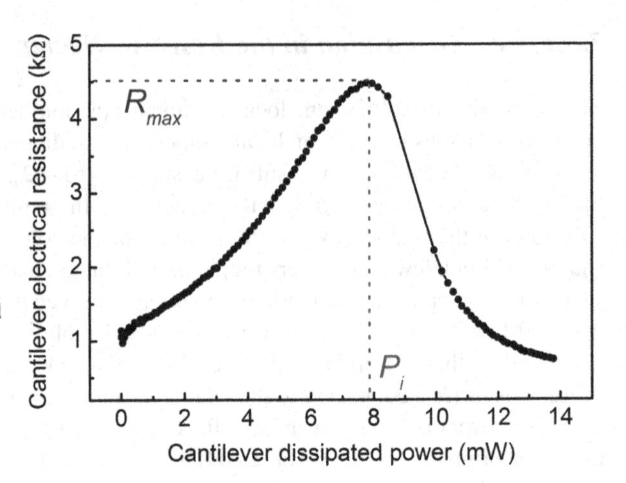

Fig. 7.4 Variation of the electrical resistance of a TCNL integrated heater as a function of dissipated power. Competition between increased electron scattering and intrinsic carrier generation at elevated temperatures results in a peak resistance of 4.5 kΩ at 550°C where $P_i = 7.9$ mW. Adapted from [66]

The cantilever resistance increases with temperature near room temperature, because the carrier mobility in the doped silicon cantilever decreases with temperature. However, the intrinsic carriers in the silicon increase with increasing temperature. At the intrinsic temperature (T_i) of approximately 550°C, the thermally generated intrinsic carriers become the dominant parameter affecting cantilever resistance, thus the electrical resistance reaches a maximum (R_{max}) and begins to drop quickly with increasing temperature. This thermal runaway effect has been well studied for cantilever heating at steady state [63–64].

Given that a linear relationship can be found between heater temperature (T_H) and dissipated power of the thermal probe (P_H) [65–66], T_H can be estimated using the following equation:

$$T_H = RT + R_{th} \cdot P_H, \tag{3}$$

where RT stands for the room temperature, and R_{th} stands for the thermal resistance constant. One can take advantage of the thermal runaway effect to obtain the R_{th} by using the following relationship:

$$R_{th} = \frac{T_i - RT}{P_i}, \tag{4}$$

where P_i is the dissipation power of the cantilever at the intrinsic temperature.

It is worthy to note that due to the thermal runaway at higher power levels, the heated cantilever can have the same electrical resistance at two different heater temperatures, which precludes the use of resistance monitoring for temperature measurements. Therefore, using cantilever power as a measure of temperature has advantages over the resistance due to its one-to-one correspondence [64].

7.3.2 Heat Transfer at the Tip-Sample Interface

The temperature rise at the tip-sample interface (T_{int}) is of greater interest than the heater temperature (T_H), because T_{int} is generally much lower than T_H. The difference between them depends on the thermal resistances of a number of components within the tip-sample thermal circuit, as shown in Fig. 7.5. The tip-sample thermal circuit includes thermal resistances of the tip (R_{tip}), tip-surface interface (R_{int}), and sample spreading resistance (R_{spread}). Heat also flows directly from the cantilever heater to the substrate through the ambient air, with thermal resistance of R_{gap}. The dominant mode of heat transfer is through the ambient air, as R_{gap} is approximately an order of magnitude smaller than the total thermal resistance through the tip. However, the temperature rise on the sample surface due to the heat transfer through air ($T_{surface}$) is much less significant than the tip-sample interface temperature (T_{int}) due to the thermal conductivities of ambient air and the sample material. Such attribute of the thermal AFM probes make them particularly useful for performing highly localized thermal processing of materials.

The thermal resistance of the tip (R_{tip}) originates from the conductance of the phonons within the silicon tip and from the layer of native oxide covering it. The thermal conductivity is given by [65]:

$$k = \frac{1}{3} C \cdot v \cdot \left(\Lambda_0^{-1} + d^{-1}\right)^{-1}, \quad (5)$$

where k is the thermal conductivity, C is the volumetric heat capacity, v is the average phonon speed, Λ_0 is the temperature-dependent phonon mean free path in the bulk material, and d is the diameter of the structure. It is obvious that the resistivity

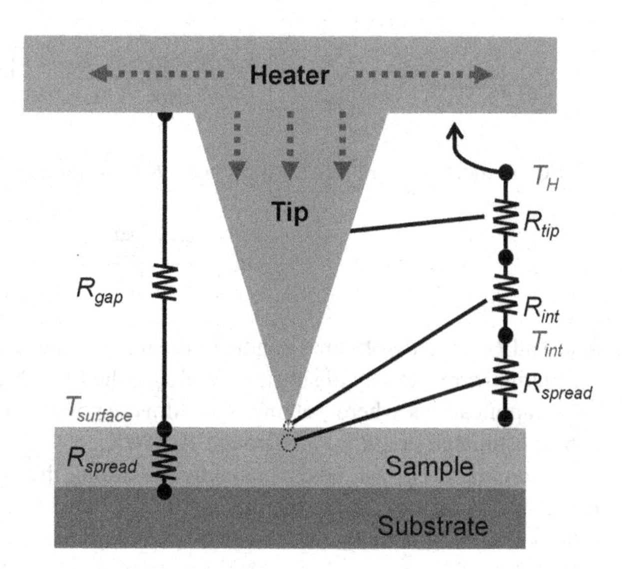

Fig. 7.5 Thermal circuit for heat flow of a heated TCNL thermal cantilever, the tip, and the contact with sample. Adapted from [66]

inside the tip increases with respect to bulk silicon because of enhanced phonon scattering with boundary surfaces, and reduction of the cross section area towards the tip apex (d changes from 1 μm at tip base to 10 nm at the tip-surface contact). The integral expression of R_{tip} for varying d is given by [67]:

$$R_{tip} = \int_{2rL/D}^{L} \frac{dz}{k(z)A(z)} = \frac{3}{2Cva}\left[1 - \left(\frac{2a}{D}\right)^2\right] + \frac{4}{2k_{tip}a}\left(1 - \frac{2a}{D}\right), \tag{6}$$

where L is the tip length, z is the height from the tip apex to tip base, D is the tip base diameter, a is the contact radius, k_{tip} is the bulk thermal conductivity of the tip, k_z is the height-dependent thermal conductivity as shown in Eq. (5). The above integral yields a thermal resistance of the order of 10^6 K/W due to phonon scattering. It is important to note that over 90% of R_{tip} occurs at the first 10% of the tip length. The end of the tip thus governs heat transport through the entire structure.

At the tip-sample interface, thermal resistance occurs due to phonon scattering at the interface. The interfacial resistance can be estimated as [68]:

$$R_{\text{int}} = \frac{r_{\text{int}}}{\pi a^2},$$

where r_{int} is the thermal boundary resistance. Experimental measurements of r_{int} for solid-solid contacts near room temperature typically give a range of 10^{-9} m^2 K/W, which is smaller than typical bulk thermal contact resistances. It is expected that R_{int} is in the range of 10^7 K/W for contact radius a of ~5 nm.

The contact of the probe tip with a flat sample surface can be approximated as a circular heat source in contact with a flat, homogeneous semi-infinite surface that has a spreading resistance given by [69]:

$$R_{spread} = \frac{1}{\pi k_{sam} a} \int_{0}^{\infty} \left[\frac{1 + K\exp(-2\zeta t_{sam}/a)}{1 - K\exp(-2\zeta t_{sam}/a)}\right] J_1(\zeta)\sin(\zeta)\frac{d\zeta}{\zeta^2}, \tag{7}$$

where the thermal conductivity parameter K is defined as

$$K = \frac{1 - \frac{k_{sub}}{k_{sam}}}{1 + \frac{k_{sub}}{k_{sam}}}, \tag{8}$$

k_{sam} and t_{sam} are the thermal conductivity and thickness of the sample film, k_{sub} is the thermal conductivity of substrate, and J_1 is the Bessel function. For most thermal probe applications where polymer thin films deposited on glass slides are used, it can be estimated that R_{spread} is around 10^8 W/K.

In summary, T_{int} can be estimated by evaluating the proportion of each thermal resistance component in the thermal circuit. We define heating efficiency (c) as the ratio between T_{int} and T_H, which can be expressed as:

$$c = \frac{T_{\text{int}}}{T_H} = \frac{R_{spread}}{R_{tip} + R_{\text{int}} + R_{spread}}. \tag{9}$$

In the cases where polymer films deposited glass substrates are used, c can be estimated in the range of 60–70%.

7.3.3 Reaction Kinetics in the Nanofabrication Processes

Reaction kinetics plays an important role in choosing a working heater temperature during TCNL operations. Even though T_{int} is as much as 60–70% of T_H in the presence of the phonon scattering in the tip apex and the tip-surface interface, experimental results indicate an even smaller value of heating efficiency. For instance, the thermogravimetric analysis (TGA) of a carbamate copolymer shows that the thermal deprotection of amine groups occurs at 160°C. But the deprotection during TCNL operations starts when T_H rises to around 410°C, which corresponds to a heating efficiency of merely 35% [70]. This is due to the fact that the thermal reaction involved in a TCNL writing process is much faster than that occurred in TGA measurements. For a moderate TCNL writing speed of 10 μm/s and a normal loading force of 100 nN, we estimate that the local heating time is 1.7 ms, which corresponds to a local temperature ramping rate of 10^7°C/min. This is six orders of magnitude faster than a typical TGA temperature ramping rate (~10°C/min). Therefore, it requires a significantly high temperature (around 350°C) to initiate the amine deprotection. Decreasing the writing speed (down to a few nm·sec^{-1}) is a judicious choice to increase the heating efficiency and reduce the heater temperature. Increasing normal loading force can also increase the heating efficiency because it can increase the tip-surface contact area and increase the dwell time of the thermal contact. However, a large loading force can induce undesirable physical damage to the sample surface.

7.4 Thermochemical Nanolithography of Functional Nanostructures

TCNL employs the thermal probes to induce well-defined chemical reactions in order to change the surface functionality of a material. A wealth of thermally activated chemistries can feasibly be employed to change the subsequent reactivity, surface energy, solubility, conductivity, *etc.*, of the material. TCNL can be used for fabrication of functional nanostructures that are appealing for various applications in nanofluidics, nanoelectronics, nanophotonics, and biosensing devices. This technique offers advantages over the aforementioned nanofabrication techniques in terms of the combination of high speed, high resolution, material flexibility, potential towards parallelization, and the versatility of working under ambient conditions.

The distance of the tip from the surface and the temperature of the tip can be modulated independently, and the tip does not have to indent the surface as in thermomechanical nanolithography. In addition, chemical changes can be written very quickly through rapid scanning of the substrate or the tip, as no mass is transferred from the tip to the surface as in DPN (writing speed is limited only by the heat transfer rate) [22]. Judicious choice of the physical properties of a material (e.g., glass transition temperature, T_g, for polymer materials) may afford a system wherein chemical changes can be performed either separately from, or accompanied by, topographical modification. Furthermore, the use of a material that can undergo multiple chemical reactions at significantly different temperatures renders the possibility of a multi-state system, wherein different functionalities can be addressed at different temperatures.

7.4.1 High-Speed, Sub-15 nm Feature Size Nanolithography

TCNL allows for simultaneous direct control of the local chemistry and topography of thin polymer films. Pioneering work demonstrating that the thermal probe can write sub-15 nm hydrophilic features on a hydrophobic polymer surface at the rate of 1.4 mm/s was conducted by Szoszkiewicz et al. [71]. The polymer used in this study was poly(tetrahydro-2H-pyran-2-ylmeth-acrylate)$_{80}$poly(3-(4-[(E)-3-methoxy-3-oxoprop-1-enyl]phenoxy)propyl 2-methacrylate)$_{20}$ (or p(THP-MA)$_{80}$p(PMC-MA)$_{20}$ for short). Upon heating at 160°C ± 30°C, the tetrahydropyran (THP) protection group was released and carboxylic acid was formed, which rendered the pristine hydrophobic surface to a hydrophilic surface.

Szoszkiewicz et al. have shown a high-resolution patterning of controlled chemical patterns on a surface. As shown in Fig. 7.6, panels A and A' shows topography and friction images of a series of hydrophilic lines with high density of 2×10^7 lines/m (corresponding to 260 Gb·in^{-2}). Panels B, B', and B" show topography and friction images of chemically written "GIT". The cross-section of letter "G" in the friction image demonstrates the highest spatial resolution of the chemical pattern is as small as 12 nm.

The very small feature size achievable is attributed to the large temperature gradients in the polymer in the vicinity of the tip. The fundamental limit to writing speed in TCNL is the thermal diffusivity of the substrate material as opposed to mass diffusivity, which limits deposition-based approaches. For example, the mass diffusivity of small molecules typically used in DPN is $\sim 10^{-10}$ m^2/s [72], while the thermal diffusivity of the organic substrates used in the present work is $\sim 10^{-7}$ m^2/s [73]. Thus, the speed of the TCNL technique is currently limited by accessible AFM scanning speeds. Calculation and modeling suggest that the maximum patterning speed can be estimated as about 30 mm/s and is, therefore, much faster than any comparable chemical nanopatterning technology.

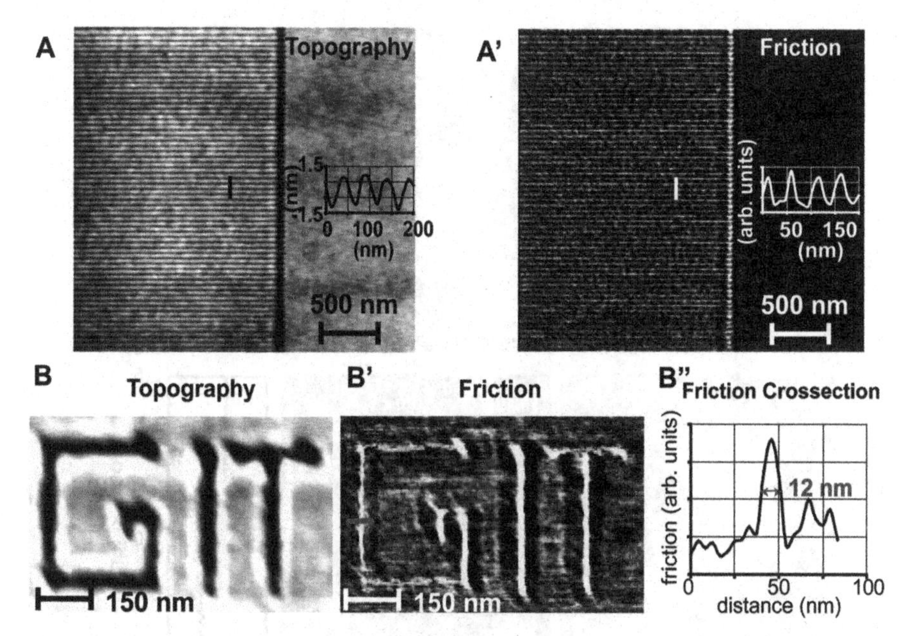

Fig. 7.6 High-resolution patterning of nanostructures by TCNL. (**a**) 3 μm × 3 μm AFM topography image and (**a'**) corresponding friction image of a cross-linked p(THP-MA)80 p(PMC-MA)20 film showing a high-density line pattern written chemically on the left side. (**b**) AFM topography and corresponding friction image (**b'**) of a modified copolymer film, with the indentation depth kept within 3 nm. The resulting friction cross-section (**b"**) shows about 12 nm half-width within the modified zone (in the letter G); topographical changes are minimal. Reprinted in part with permission from [71]. Copyright 2007 American Chemical Society

7.4.2 Local Wettability Modification of Polymer Surfaces

Combined write-read-erase-rewrite or write-read-overwrite capabilities are not only important for data storage applications, they also give the flexibility required by complex multiple-step manufacturing processes. Previous attempts to develop these rewriting/overwriting capabilities relied on reversible light-induced chemical reactions [74], oxidation/reduction reactions [75], and electrochemical deposition/removal of metallic particles [76]. Each of these methods has its disadvantages, such as the low resolution, the slow writing speed, the need for a conductive substrate and the lack of control over the water meniscus in electrochemical processes.

Wang et al. recently have demonstrated the ability to in situ write-read-overwrite chemical patterns on the surface of a p(THP-MA)$_{80}$p(PMC-MA)$_{20}$ copolymer film with no need of a probe change [77]. As shown in Fig. 7.7, the copolymer surface was first heated to 70 ± 20°C, below the THP deprotection temperature, by means of the thermal probe. No change in topography or friction was detected on the heated area. After heating a 1.5 μm × 1.5 μm square to 110 ± 20°C a corresponding pattern in the friction image is observed. The topography shows no depletion inside

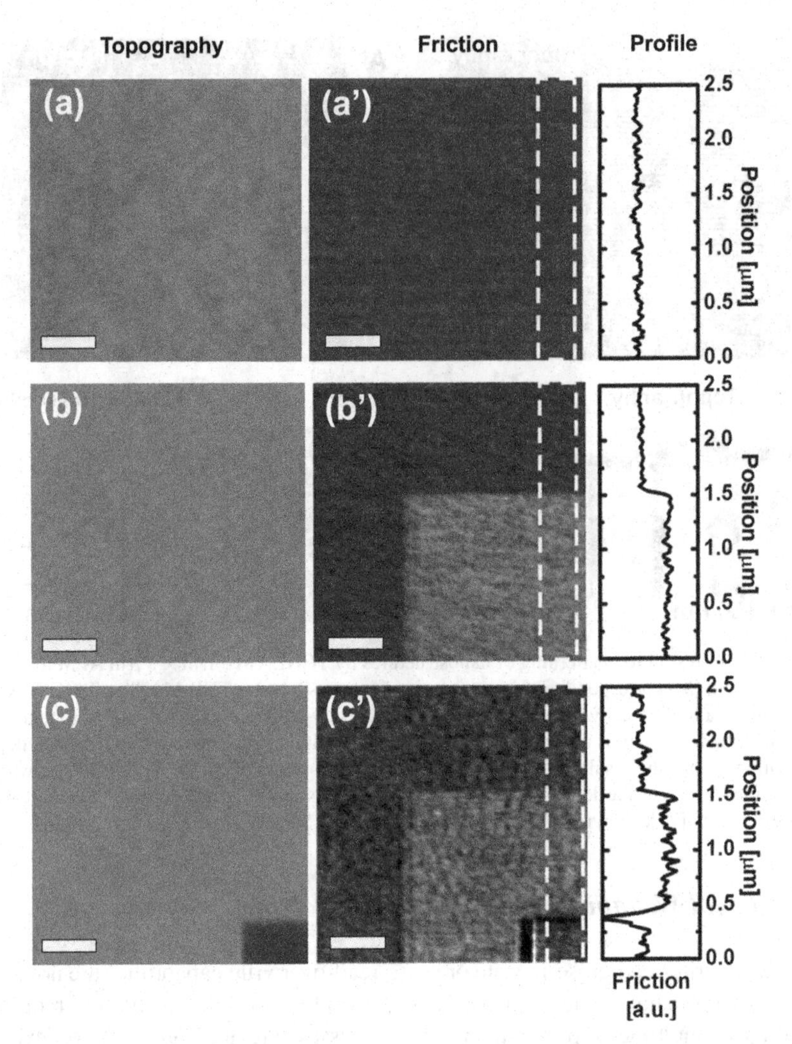

Fig. 7.7 Topography and friction images of a p(THP-MA)80 p(PMC-MA)20 copolymer surface before and after TCNL two-step modification of surface wettability. (**a**) and (**b**): topography and friction images of a pristine polymer surface. (**c**) and (**d**): topography and friction images of the surface after a first-step modification by TCNL at 110 ± 20°C over an area of 1.5 μm × 1.5 μm. (**e**) and (**f**): topography and friction images of the surface after a second-step modification was made on top of the first-step modification region by TCNL at 190 ± 20°C over an area of 0.65 μm × 0.4 μm. The friction profiles are averages over the areas delimited by the *dashed lines*. The scale bars are 500 nm long

the square. The friction increase in the written pattern suggests that the THP groups were deprotected, leaving the area covered with hydrophilic carboxylic acid groups. A second smaller 0.65 μm × 0.4 μm square pattern was overwritten inside the hydrophilic pattern by further heating to 190 ± 20°C. The friction image shows that the surface becomes again hydrophobic (lower friction). The corresponding topography depletion is 6 nm deep. This change is consistent with the formation of anhydrides [78].

7.4.3 Multifunctional Nano-Templates for Assembling Nano-Objects

Recently, Wang et al. demonstrated that nanoscale patterning of different chemical species in independent nanopatterns can be achieved by the iterative application of TCNL to inscribe amine patterns followed by their chemical conversion to other functional groups [70]. Due to the unique chemical stability of the patterns, the resultant substrates can be used for covalent and molecular-recognition based attachment of nano-objects using chemical protocols. In particular, it allows for the attachment of proteins and DNA to the chemical nanopatterns and for the creation of co-patterns of multiple distinctively bioactive proteins.

Upon heating between 150 and 220°C, a methacrylate copolymer containing tetrahydropyran carbamate groups can be thermally deprotected to unmask primary amines. The amine groups can then be further converted to aldehydes, thiols, biotins, and maleimides, and can be used, in a second stage, to attach different classes of nano-objects, such as proteins, nucleic acids, and potentially many others, by standard functionalization methods (Fig. 7.8). This new TCNL/covalent functionalization (CF)/molecular recognition (MR) approach is conceptually straightforward and patterns can be written at high resolution.

An example of nanoarrays of fibronectin proteins made by TCNL/CF/MR method is shown in Fig. 7.9. Three such features are shown as topographical and phase images taken after fibronectin or GA staining. The topographical data revealed that the TCNL holes are filled with proteins. The phase images are also consistent with the deposition of proteins in the holes. Fibronectin phase features as small as 40 nm have been measured which correspond to roughly one or two fibronectin molecules exposed at the surface.

In complex molecular systems, proteins work cooperatively to initiate biological events, for example in the adhesion plaques formed during cell adhesion or the patterning of signaling and adhesion proteins in the formation of the T-cell immunological synapse [79–81]. Wang et al. have shown that patterned proteins remain bioactive and can initiate cell activity. Figure 7.10a demonstrates the bioactivity of the biotin-bound anti-cluster-of-differentiation-3 (anti-CD3). The ability of anti-CD3 is verified by binding to a secondary antibody, anti-Immunoglobulin G (anti-IgG). Anti-CD3 is known to stimulate specific cell signaling pathways when interacting with Jurkat cells, an immortalized line of T lymphocyte cells (T-cells) that are used to investigate T-cell signaling and immune synapse formation. In a cellular assay to demonstrate the bioactivity of bound anti-CD3 (Fig. 7.10b), immunofluorescence cell staining of protein kinase C-θ (PKC-θ) in a cell interacting with a triangle shaped anti-CD3 micropattern shows the halo of the PKC-θ echoing

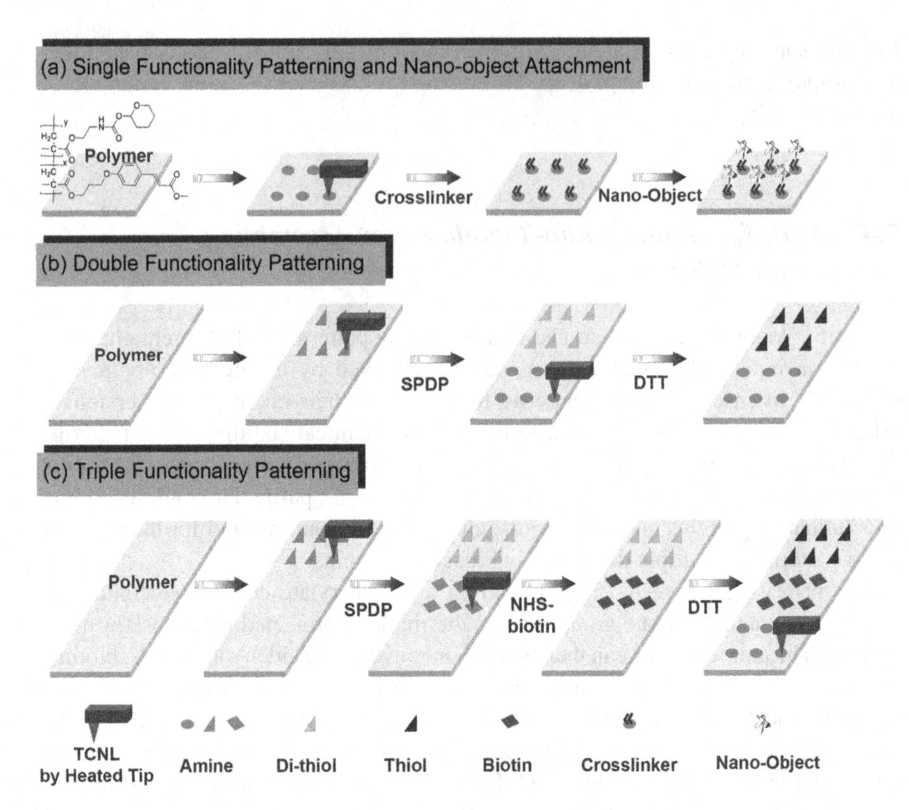

Fig. 7.8 Schematic of patterning multifunctional nanotemplates by TCNL followed by covalent functionalization. (**a**) A single nano-object pattern is created through three steps: TCNL, crosslinker incubation, and nano-object incubation. (**b**) A double functionality pattern of thiols (*triangles*) and amines (*circles*) is created through two rounds of TCNL and incubation processes. (**c**) A triple functionality pattern of thiols (*triangles*), biotins (*diamonds*), and amines (*circles*) is created through three rounds of TCNL and incubation processes [70]. Copyright Wiley-VCH Verlag GmbH & Co. KGaA. Reproduced with permission

the triangular shape of the anti-CD3 micropattern. Figure 7.10c shows a concentric set of independently immobilized proteins (inner square = anti-CD3, outer square = Inter-Cellular Adhesion Molecule (ICAM-1)) and a separate surface patterned with a two by two array comprised of two anti-CD3 triangles and of two ICAM-1 diamonds. Minimal crosstalk and non-specific binding to the untreated polymer surface is present.

Since the polymer can planarize a substrate, the technique could be applicable to glass substrates and any oxide film to which the polymer can be crosslinked. Furthermore, it is important to note that the surfaces can be pre-patterned and stored for later bio/nano functionalization (at least weeks later). Thus, the multi-protein/nano-object patterning can take place under native conditions in a second laboratory without the TCNL equipment or expertise in nanolithography. These

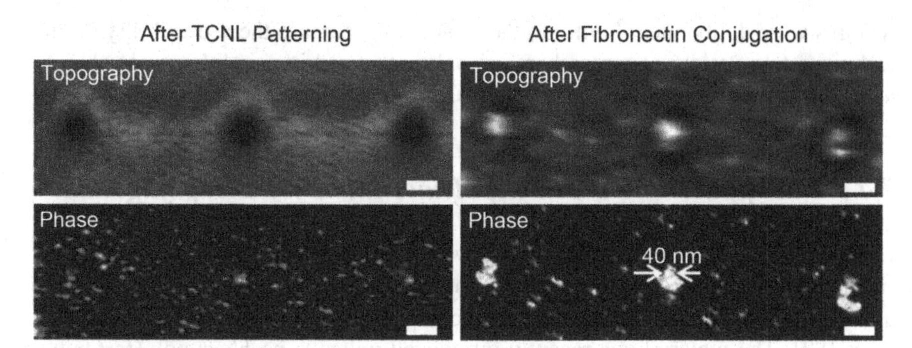

Fig. 7.9 Fibronectin nanoarrays. AFM topography and phase images of a TCNL nanoarray before and after fibronectin attachment. The topography z-range in (a) is 20 nm Scale bars: 100 nm [70]. Copyright Wiley-VCH Verlag GmbH & Co. KGaA. Reproduced with permission

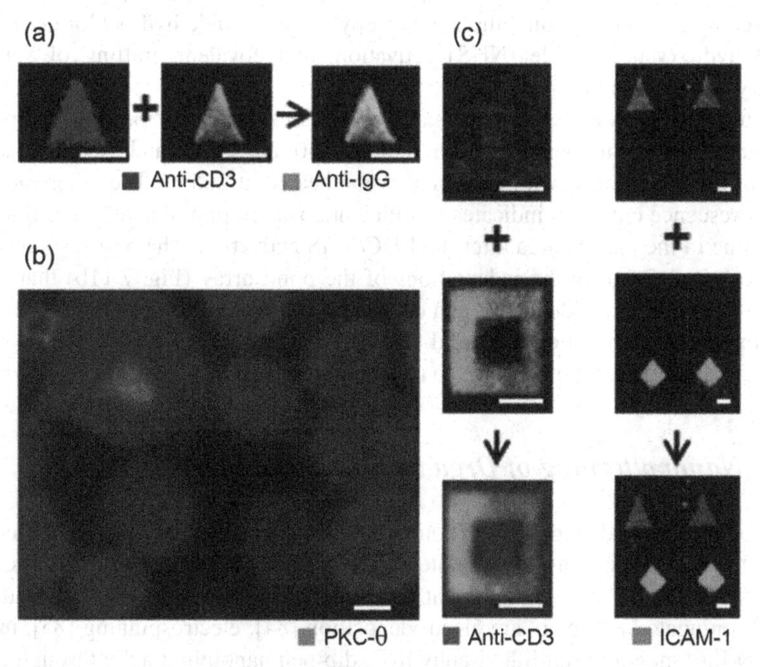

Fig. 7.10 Epi-fluorescence of anti-CD3 bioactivity and two-protein co-patterning. (**a**) Alexa350 labeled biotinylated anti-human CD3 bound to the TCNL amine pattern by means of NHS-biotin and streptavidin shows bioactive molecular recognition of FITC-labeled IgG. (**b**) Jurkat cells, immunostained for PKC-θ, lying on a triangular anti-CD3 pattern. The PKC-θ accumulation above the cell–pattern contact site can be observed. (**c**) Anti-CD3 (inner squares on the *left panel* and triangles on the *right panel*) and ICAM-1 (outer hallow squares on the *left panel* and the diamonds on the *right panel*) are closely co-patterned on a single surface [70]. Scale bars: 5 μm. Copyright Wiley-VCH Verlag GmbH & Co. KGaA. Reproduced with permission

features were deliberately built into the protocol to increase the accessibility of the technique to a variety of researchers not only interested in nanolithography, but in areas of biochemistry, nanoscience, and nanobiotechnology more broadly.

7.4.4 Nanopatterning Block Copolymer Films for Bioconjugation

Duvigneau et al. reported on an alternative thermochemical nanopatterning material that can obtain defined formation of pending carboxylic acid groups that allow for subsequent site-specific bioconjugation on polymer thin film surfaces [82]. The material is a recently introduced polystyrene-*block*-poly(*tert*-butyl acrylate) (PS-*b*-P*t*BA) block copolymer [83], whose *tert*-butyl ester moieties can be activated in the presence of acid or heat, resulting in the formation of deprotected carboxylic acid groups on surfaces. The bioconjugation can be achieved by 1-ethyl-3-(3-dimethylaminopropyl)carbodiimide hydrochloride (EDC) and *N*-hydroxysuccinimide (NHS) activation, and covalent grafting of various primary amine containing molecules.

Figure 7.11 shows fluorescence images of squares on PS-*b*-P*t*BA films prepared by raster scanning in contact mode a 30 μm $\times$ 30 μm square and by indenting an array of 25 $\times$ 25 points with a thermal probe heated at 265°C. The emergence of the fluorescence emission indicates an efficient covalent immobilization of fluoresceinamine in the pattern area after the EDC/NHS activation. The best resolution in this work is defined by the indentations of the point array (Fig. 7.11b) that were $\sim$370 nm $\times$ 580 nm wide and 30 nm deep. With optimized control of temperature, tip-sample contact duration time, and feedback setting, the resolution limits on this block copolymer film are expected to reach attainable minimum feature size.

7.4.5 Nanopatterning of Organic Semiconductors

Nano-patterning and nano-fabrication of conjugated polymers on various length scales have attracted considerable interest for nanoelectronics, nanophotonics, and biosensing. Among the methods that have been reported to date for the patterning of conjugated polymers are electrodeposition [84], electrospinning [85], inkjet printing [86], nanoimprint lithography [87], dip-pen nanolithography through electrostatic interaction or electrochemical reaction [88–89], scanning near-field optical lithography [90–91], thermochemical nanopatterning [92], and edge lithography [93]. More recently, Wang et al. reported the direct writing of nanostructures of poly(*p*-phenylene vinylene) (PPV), a widely studied electroluminescent conjugated polymer, by nanoscale heating a sulfonium salt precursor, poly(p-xylene tetrahydrothiophenium chloride) is realized by locally heating in ambient conditions with a resistively heated AFM cantilever at 240°C [94]. The precursor thin film is obtained by drop-casting precursor solution on a glass or silicon (111) substrate.

Fluorescence and AFM topography images of the PPV nanowires are shown in Fig. 7.12. These nanostructures were made at a writing speed of 20 μm/s, with a

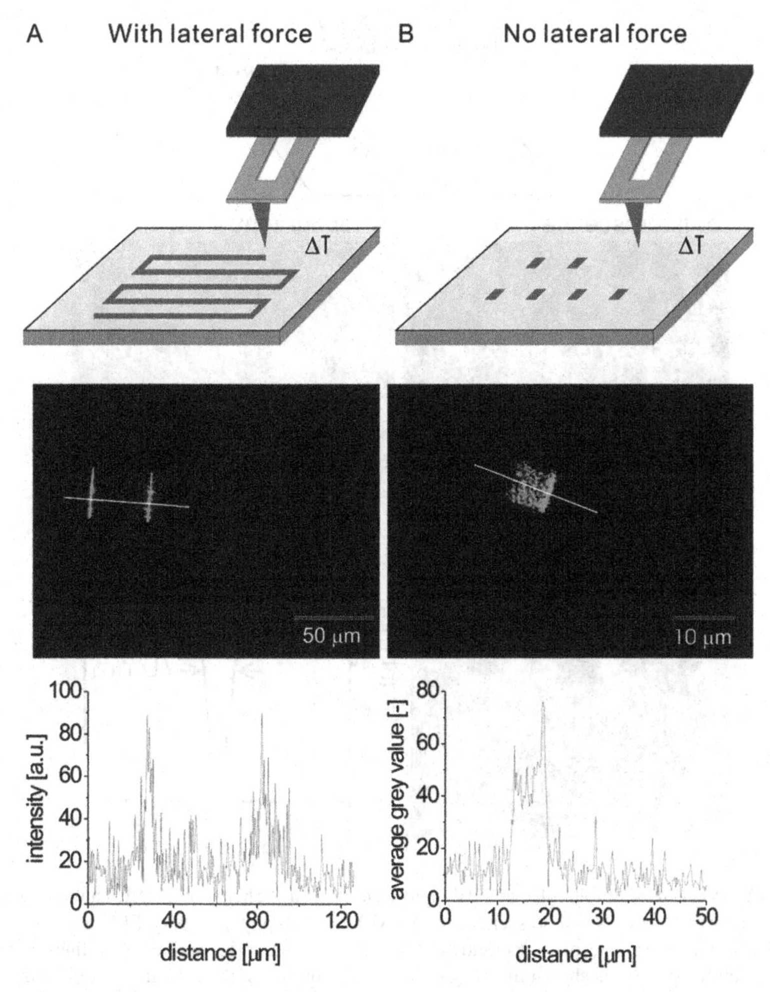

Fig. 7.11 EDC/NHS activation of thermochemical nanopatterns on PS-*b*-P*t*BA films. Schematic of scanning thermal microscopy approaches (*top*), fluorescence microscopy images (*middle*), and cross sections (*bottom*) of squares on PS-*b*-P*t*BA films prepared (**a**) by raster scanning in contact mode a 30 μm × 30 μm square and (**b**) by indenting an array of 25 points × 25 points (x-y separation 250 nm) with a thermal probe heated at 265°C. Reprinted in part with permission from [82]. Copyright 2008 American Chemical Society

normal load of 30 nN, and cantilever temperature ranging between 240 and 360°C. The nanostructures started to show a visible fluorescent contrast at 240°C. The contrast became clearer as the heating temperature was raised to 360°C. The corresponding AFM topography image reveals TCNL capability of fabricating PPV nanostructures with a high spatial resolution of 70 nm.

To study the quality of conjugated polymer nanostructures made by TCNL, Wang et al. studied a series of micropatterns (20 μm × 20 μm in size) made at 240°C,

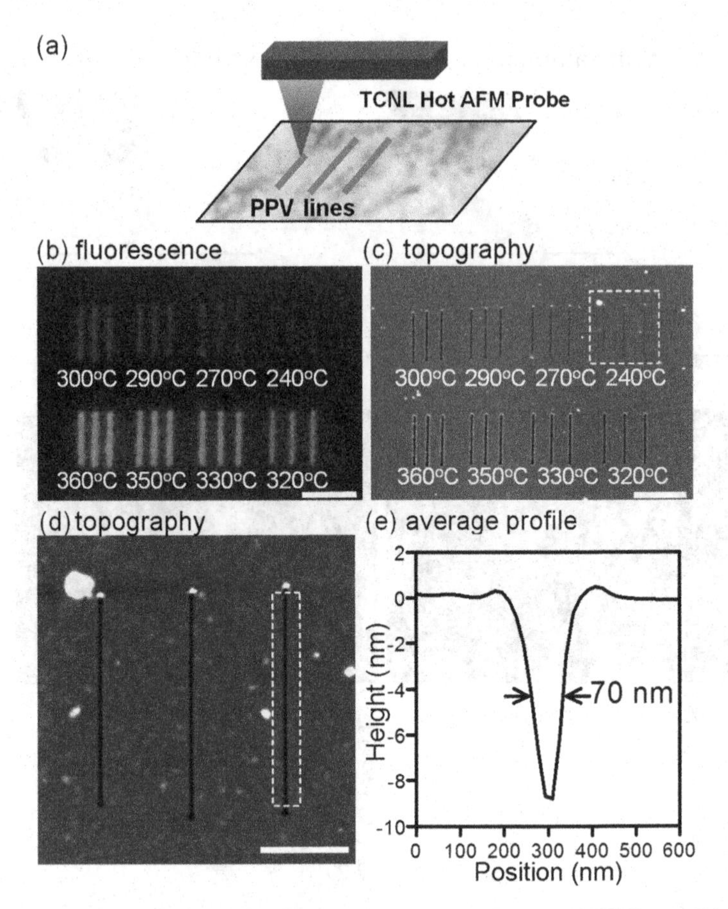

Fig. 7.12 TCNL nanolithography of PPV nanostructures. (**a**) Scheme of TCNL nanolithography of PPV nanostructures. (**b**) Fluorescence and (**c**) AFM topography images of PPV nanostructures made by TCNL at a range of temperatures, 240–360°C. A zoom-in view of PPV lines made at 240°C as outlined in (**c**) is shown in (**d**). (**e**) The average profile of the PPV trench outlined in (**d**) shows that the width (FWHM) of the line is as narrow as 70 nm. The thickness of this precursor film is 100 nm [94]. Scale bars: (**b**) and (**c**): 5 μm, (**d**): 2 μm. Copyright 2009, American Institute of Physics

280°C, and 320°C, respectively, with Raman spectroscopy. Data revealed that the quality of these PPV patterns in ambient conditions is comparable to that of a PPV sample prepared by a standard thermal annealing of a precursor polymer in vacuum conditions (here referred to as $PPV_{reference}$) as shown in Fig. 7.13. The most distinctive characteristics of the Raman spectra after the complete conversion of the precursor film into PPV is the large intensity enhancement and the shift in frequency of the Raman peaks associated with the C-C vibrations to lower frequencies. The Raman peak positions of the TCNL patterns written at three different tip temperatures are in between those of the precursor polymer and those of the $PPV_{reference}$

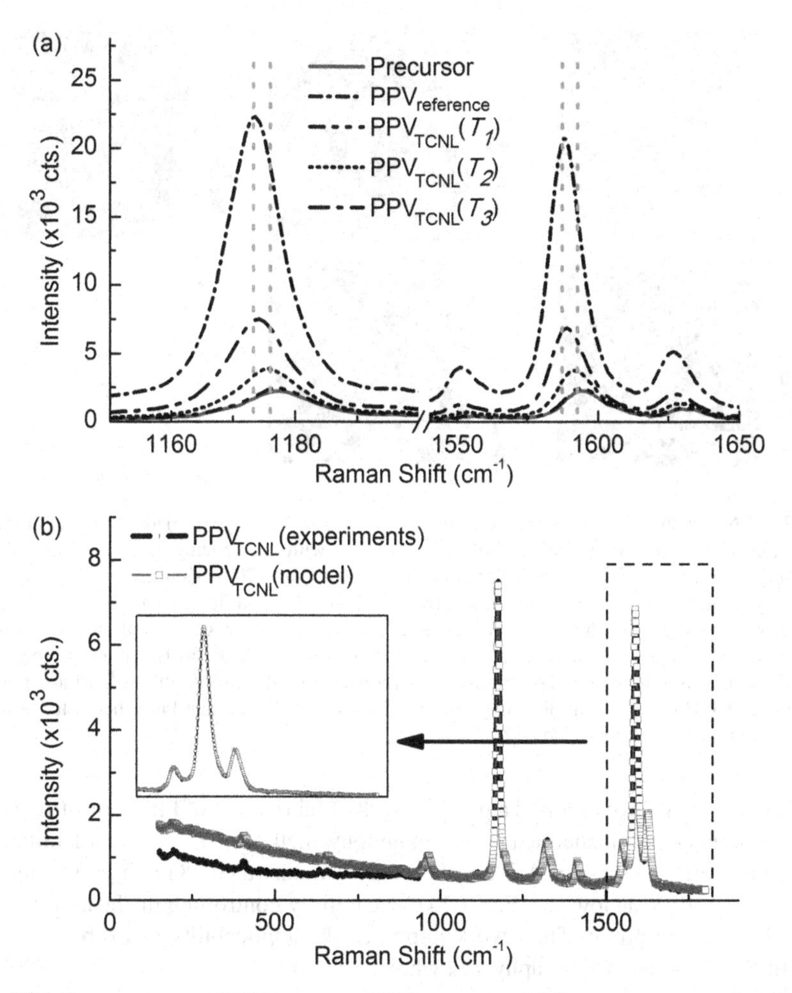

Fig. 7.13 Raman spectra of TCNL converted PPV structures. (**a**) Raman spectra as a function of the temperature used during TCNL, $T_1 = 240°C$, $T_2 = 280°C$, $T_3 = 320°C$, respectively. (**b**) Comparison between the experimental Raman spectra obtained from the PPV_{TCNL} pattern and the modeled Raman spectra [94]. Copyright 2009, American Institute of Physics

polymer. As the tip temperature used to perform TCNL increases, the Raman intensity of the written patterns increases and the peak positions shift to those of the $PPV_{reference}$ film. In addition, they modeled these nanostructures as a blend of precursor and converted PPV. The fitting yielded a blending ratio of 73% precursor and 27% PPV for the TCNL PPV nanostructures produced at 300°C on a precursor film with 1.4 μm thickness. This indicated that a precursor thickness of around 320 nm has been converted to the reference grade of PPV in a single application of TCNL at 300°C. The conversion ratio can be controlled by varying the probe temperature and linear writing speed, which is an advantage of the TCNL process.

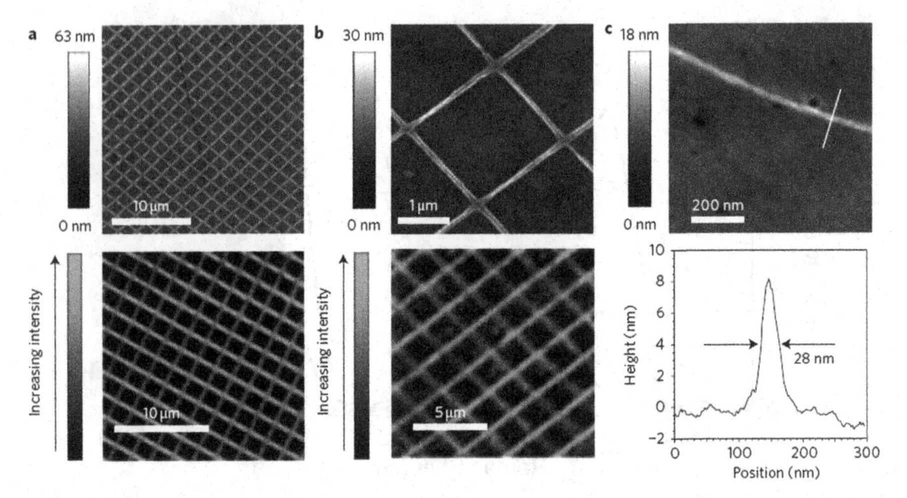

Fig. 7.14 Nanopatterning of organic semiconductors. (**a**), AFM image (*top*) and confocal fluorescence image (*bottom*) of a square grid of PPV structures with line spacing of 2 μm produced by scanning the probe in two perpendicular directions at 230°C and 10 μm/s. Each line represents a double scan (trace and retrace) of the probe. (**b**), AFM and confocal fluorescence images taken of a similar set of structures drawn at 5 μm/s and 250°C. (**c**), AFM image (*top*) of an isolated line drawn in PPV (single scan). The lines were drawn at a temperature of 250°C and a scan speed of 5 μm/s on a 15-nm-thick film. The cross-section (*bottom*) reveals a line width (full-width at half-maximum, FWHM) of 28 nm. Reprinted by permission from Macmillan Publishers Ltd: Nature Nanotechnology [92], copyright (2009)

Fenwick et al. have reported a parallel work of fabrication of PPV nanostructures via a modified thermochemical nanolithography method [92]. In lieu of utilizing the aforementioned thermal AFM probes, they demonstrated 28 nm high-resolution patterning of PPV nanowires (Fig. 7.14) by carefully controlling the heat profile of a Wollaston wire probe. Their work improved the applicability and robustness of thermochemical nanolithography as a whole.

7.4.6 Nanoscale Tunable Reduction of Graphene Oxide

Reduction of graphene oxide (GO), an insulating material with a transport gap larger than 0.5–0.7 eV at room temperature, has been identified as a promising route for translating the interesting fundamental properties of graphene into technologically viable devices [95–97], such as transparent electrodes [97], chemical sensors [98], and MEMS resonators [99]. In particular, transport measurements have shown that GO undergoes an insulator-semiconductor-semimetal transition as it is reduced back to graphene [95, 100]. Wei et al. has recently reported that graphene oxide, an oxidized form of carbon material graphene, can be controllably reduced by TCNL [101]. It allows for local tuning of the electrical properties of the material with nanoscopic resolution. This method provides a solution to pattern insulating material with conducting nanostructures, which is a major progress that may speed up its implementation in nanometer scale circuitry. The locally reduced graphene oxide (rGO) regions by thermal probes are up to four orders of magnitude more conductive than untreated GO. Conductive nanoribbons with resolution down to 12 nm could be produced in oxidized epitaxial graphene films. This procedure may enable conducting graphene circuits to be directly written on insulating graphene oxide sheets.

A zigzag rGO nanowire written with a single line scan at $T_H \sim 1{,}060°C$ on GO is shown in Fig. 7.15. Panel A is an image of the electrical current measured between a conductive platinum AFM tip and each point of the surface, showing no current on the GO surface and a current enhancement of about 100 pA in the rGO nanoribbons. These current values are consistent with the presence of 12 nm wide and several nanometers thick rGO nanoribbons presenting a vanishingly small Schottky barrier; and a resistive SiC substrate (resistivity of about 10^5 Ω·cm). The topographical image (Panel B and black graph in Panel C) indicates that the reduction produces a shallow indentation of 1 nm. This can be a result of loss of oxygen-rich functional groups and flattening of the material caused by the conversion of sp^3 carbon bonds into sp^2 carbon bonds.

Wei et al. showed that variable reduction of GO could be achieved by controlling the temperature of the AFM tip. Graphene has a low friction coefficient [102]

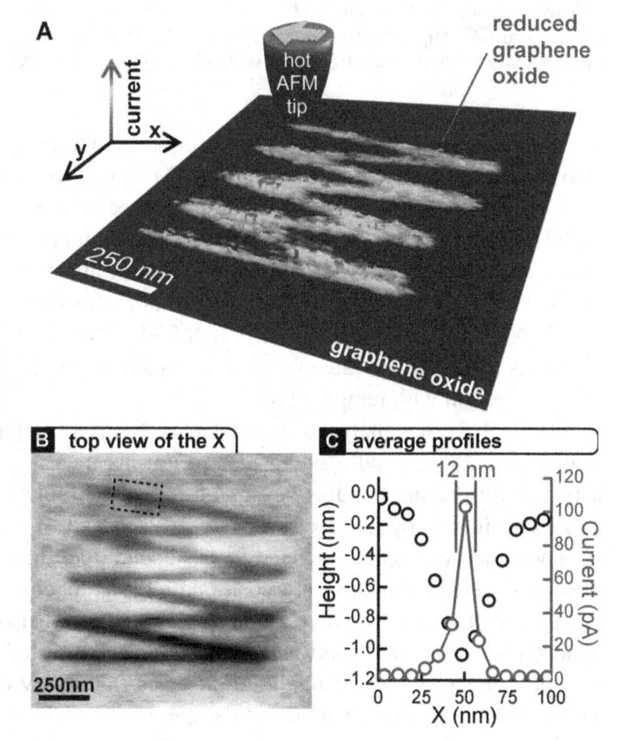

Fig. 7.15 Local thermal reduction of a GO film: current and topographical images. (**a**) 3D CAFM current image (taken with a bias voltage of 2.5 V between tip and substrate) of a zigzag shaped nanowire formed after TCNL was performed on GO at $T_H \sim 1{,}060°C$ with a linear speed of 0.2 μm/s and a load of 120 nN. (**b**) Corresponding topography image taken simultaneously with (**a**). (**c**) Averaged profiles of current and height of the cross sections that are indicated as *dashed lines* in (**a**) and (**b**) [101]. Reprinted with permission from AAAS

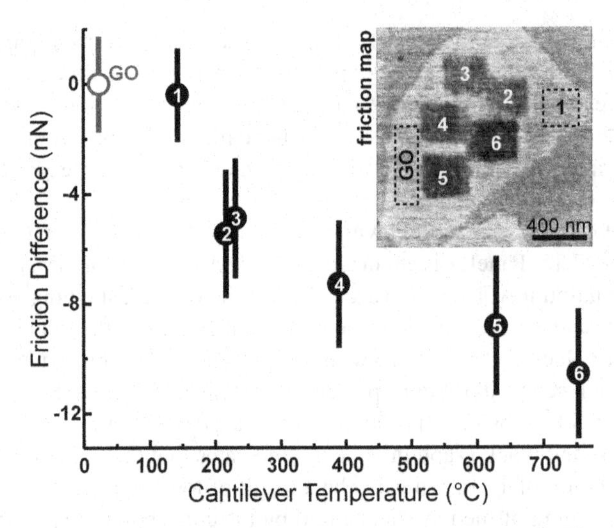

Fig. 7.16 The rate of thermal reduction depends on the tip temperature. The plot shows the decrease in lateral force on an AFM tip at room temperature as it scans over several squares previously reduced by TCNL at different temperatures. The inset is a room temperature friction image of the GO sheet on which a heated tip was previously rastered twice over six square areas, at a speed of 4 μm/s. In square 1, the tip was heated during TCNL to $T_H \sim 100°C$ yielding no apparent reduction while at temperatures $T_H > 150°C$ the rastered areas (2–6) were thermally reduced. Reduced GO, which like bulk graphite behaves as a lubricant, shows lower friction than the original GO. Higher temperatures accelerate the thermal reduction of GO and thereby more rapidly lower friction [101]. Reprinted with permission from AAAS

while oxides typically have higher friction coefficients. Thermal reduction should also reduce friction as the high friction GO is replaced with lower friction graphene. Figure 7.16 shows the strong correlation between the cantilever temperature during TCNL processing and the lateral force on a room temperature tip scanned over previously reduced squares. Reduction begins at or above 130°C which is comparable to the results of Wu et al. and Mattevi et al., who showed that reduction starts at 100°C presumably after the desorption of adventitious water [3, 103]. Higher temperatures increased the rate of reduction, as shown by the roughly linear decrease in relative friction with temperature.

Wei et al. further analyzed an isolated TCNL-rGO_{epi} nanoribbon (Fig. 7.17) with a length of 25 μm and a width of 100 nm, as measured by AFM. I-V data was acquired by placing conductive tips on top of two micron-size squares of rGO_{epi} fabricated in situ by an electron beam at each end of the nanoribbon. Two point transport measurements indicated a resistance larger than 2 gigaohm when the tips were positioned on an arbitrary position on the graphene surface (very large barrier at the contact) and a drop in resistance from 120 MΩ (between the 2 squares with no nanoribbon) to 20 MΩ (between the 2 squares connected by the nanoribbon). The transport changed from insulating to metallic (linear I-V curves) in presence of the TCNL-rGO nanoribbon between the squares.

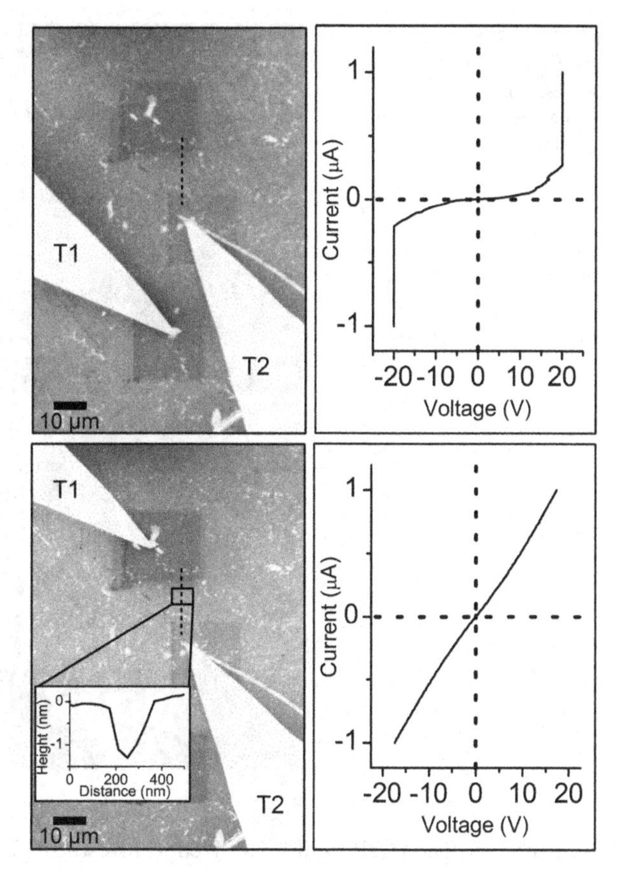

Fig. 7.17 Micro 2-point electrical transport measurement. (*Left*) SEM images of the configuration used for 2-point transport measurements when the tips are positioned between 2 rGO_{epi} squares without (*top*), and with (*bottom*) TCNL-rGO_{epi} nanoribbon in between. The AFM cross section of the nanoribbon is shown as inset of the bottom SEM image. (*Right*) *I*/*V* curves obtained measuring current between 2 rGO_{epi} squares with no nanoribbon in between (*top curve*), and between 2 rGO_{epi} squares with a nanoribbon in between (*bottom curve*) [101]. Reprinted with permission from AAAS

7.5 Conclusions

In this review chapter, we have presented the development of a thermal AFM probe based nanofabrication technique – TCNL. Although we reviewed a limited number of applications of TCNL on creating functional nanostructures for nanofluidic, nanobiosensing, nanophotonic, and nanoelectronic devices, the capabilities developed for this subset of nanostructures should be extendable to numerous other nanostructures and nanomaterials. It is important to note that the thermal probes can be used for in situ chemical characterization of nanostructures in friction force microscopy mode when the probe is not heated. One can detect and characterize the nanostructures made by TCNL without changing the probes, which would otherwise involve tremendous effort of probe alignment and delicate design of visual marks. This ability to perform in situ detection is a strong motivation to consider TCNL as a valuable tool for tip-based nanofabrication.

Efforts are underway to expand the applicability and improving the robustness of TCNL technology. Given that the thermal probes can be operated in water [104], we

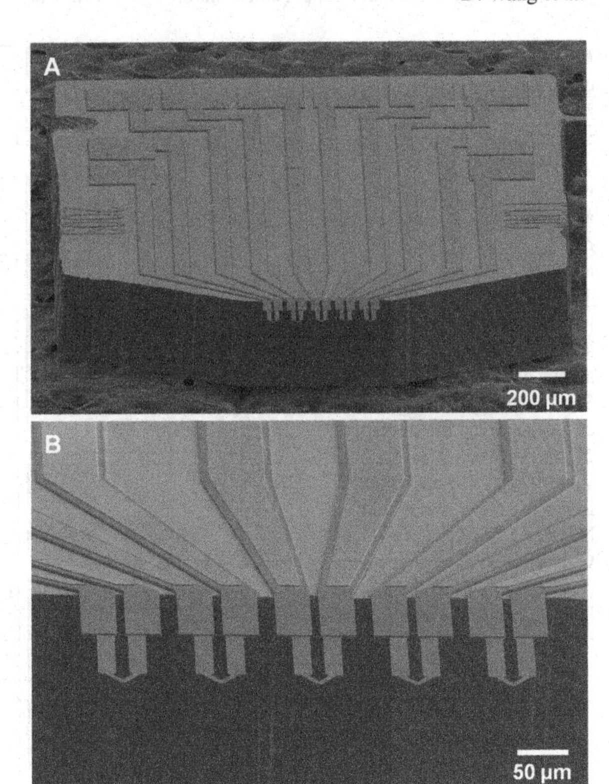

Fig. 7.18 SEM images of a 1×5 array of TCNL thermal probes. Courtesy of Dr. William P. King of University of Illinois at Urbana-Champaign

are confident that TCNL will be utilized to investigate *in vitro* chemical reactions in liquid environments. The thermal probes have shown exceptional resistance to wear, deformation, and fouling with the integration of ultrananocrystalline diamond tips with the thermal cantilevers [105]. In addition, the scalability and throughput of the TCNL technology can be significantly improved by the use of arrays of thermal probes (Fig. 7.18). The writing and reading of each probe in the array can be independently addressed. This probe-array design will lead to fabrication and integration of large 1D or 2D arrays of multifunctional microcantilevers suitable for parallel TCNL operations. Furthermore, the throughput and addressable area can also be significantly improved by application of the supramolecular nanostamping technique to replicate large areas of nanoscopic patterns [106].

Acknowledgements The authors would like to thank Dr. Seth R. Marder and his research group for the fruitful discussions and the preparation of TCNL polymer samples. We would also like to acknowledge Dr. William P. King and his research group for their continuing support on thermal AFM probes. This work was supported by National Science Foundation (CMDITR program DMR 0120967, MRSEC program DMR 0820382, and DMR-0706031), Department of Energy (DE-FG02-06ER46293 and PECASE), and Georgia Institute of Technology (Georgia Tech Research Foundation, COE Cutting Edge Research Award, and COPE graduate fellowship).

References

1. H. M. Saavedra, T. J. Mullen, P. P. Zhang, D. C. Dewey, S. A. Claridge, P. S. Weiss, Hybrid strategies in nanolithography, *Rep. Prog. Phys.* **73**, 036501 (2010).
2. Y. N. Xia, J. A. Rogers, K. E. Paul, G. M. Whitesides, Unconventional methods for fabricating and patterning nanostructures, *Chem. Rev.* **99**, 1823 (1999).
3. B. D. Gates, Q. B. Xu, M. Stewart, D. Ryan, C. G. Willson, G. M. Whitesides, New approaches to nanofabrication: Molding, printing, and other techniques, *Chem. Rev.* **105**, 1171 (2005).
4. A. A. Tseng, A. Notargiacomo, T. P. Chen, Nanofabrication by scanning probe microscope lithography: A review, *J. Vac. Sci. & Technol. B* **23**, 877 (2005).
5. R. F. Service, Can chip devices keep shrinking? *Science* **274**, 1834 (1996).
6. G. M. Whitesides, The 'right' size in nanobiotechnology, *Nat. Biotechnol.* **21**, 1161 (2003).
7. S. A. Maier, M. L. Brongersma, P. G. Kik, S. Meltzer, A. A. G. Requicha, H. A. Atwater, Plasmonics – a route to nanoscale optical devices, *Adv. Mater.* **13**, 1501 (2001).
8. M. D. Levenson, N. S. Viswanathan, R. A. Simpson, Improving resolution in photolithography with a phase-shifting mask, *IEEE Trans. Electron Dev.* **29**, 1828 (1982).
9. D. R. Medeiros, A. Aviram, C. R. Guarnieri, W. S. Huang, R. Kwong, C. K. Magg, A. P. Mahorowala, W. M. Moreau, K. E. Petrillo, M. Angelopoulos, Recent progress in electron-beam a resists for advanced mask-making, *IBM J. Res. Dev.* **45**, 639 (2001).
10. Y. N. Xia, E. Kim, X. M. Zhao, J. A. Rogers, M. Prentiss, G. M. Whitesides, Complex optical surfaces formed by replica molding against elastomeric masters, *Science* **273**, 347 (1996).
11. G. Krausch, R. Magerle, Nanostructured thin films via self-assembly of block copolymers, *Adv. Mater.* **14**, 1579 (2002).
12. M. P. Zach, K. H. Ng, R. M. Penner, Molybdenum nanowires by electrodeposition, *Science* **290**, 2120 (2000).
13. D. Wouters, U. S. Schubert, Nanolithography and nanochemistry: Probe-related patterning techniques and chemical modification for nanometer-sized devices, *Angew. Chem. Int. Ed.* **43**, 2480 (2004).
14. G. Binnig, H. Rohrer, C. Gerber, E. Weibel, Tunneling through a controllable vacuum gap, *Appl. Phys. Lett.* **40**, 178 (1982).
15. G. Binnig, C. F. Quate, C. Gerber, Atomic force microscope, *Phys. Rev. Lett.* **56**, 930 (1986).
16. D. M. Eigler, E. K. Schweizer, Positioning single atoms with a scanning tunneling microscope, *Nature* **344**, 524 (1990).
17. R. M. Nyffenegger, R. M. Penner, Nanometer-scale surface modification using the scanning probe microscope: Progress since 1991, *Chem. Rev.* **97**, 1195 (1997).
18. R. D. Piner, J. Zhu, F. Xu, S. H. Hong, C. A. Mirkin, Dip-pen nanolithography, *Science* **283**, 661 (1999).
19. J. A. Dagata, J. Schneir, H. H. Harary, C. J. Evans, M. T. Postek, J. Bennett, Modification of hydrogen-passivated silicon by a scanning tunneling microscope operating in air, *Appl. Phys. Lett.* **56**, 2001 (1990).
20. P. Vettiger, G. Cross, M. Despont, U. Drechsler, U. Durig, B. Gotsmann, W. Haberle, M. A. Lantz, H. E. Rothuizen, R. Stutz, G. K. Binnig, The "Millipede" – nanotechnology entering data storage, *IEEE Trans. Nanotechnol.* **1**, 39 (2002).
21. R. Neffati, A. Alexeev, S. Saunin, J. C. M. Brokken-Zijp, D. Wouters, S. Schmatloch, U. S. Schubert, J. Loos, Automated scanning probe microscopy as a new tool for combinatorial polymer research: Conductive carbon black/poly(dimethylsiloxane) composites, *Macromol. Rapid Commun.* **24**, 113 (2003).
22. D. S. Ginger, H. Zhang, C. A. Mirkin, The evolution of dip-pen nanolithography, *Angew. Chem. Int. Ed.* **43**, 30 (2004).
23. S. H. Hong, J. Zhu, C. A. Mirkin, Multiple ink nanolithography: Toward a multiple-pen nano-plotter, *Science* **286**, 523 (1999).

24. A. Noy, A. E. Miller, J. E. Klare, B. L. Weeks, B. W. Woods, J. J. DeYoreo, Fabrication of luminescent nanostructures and polymer nanowires using dip-pen nanolithography, *Nano Lett.* **2**, 109 (2002).
25. L. M. Demers, D. S. Ginger, S. J. Park, Z. Li, S. W. Chung, C. A. Mirkin, Direct patterning of modified oligonucleotides on metals and insulators by dip-pen nanolithography, *Science* **296**, 1836 (2002).
26. J. C. Garno, Y. Y. Yang, N. A. Amro, S. Cruchon-Dupeyrat, S. W. Chen, G. Y. Liu, Precise positioning of nanoparticles on surfaces using scanning probe lithography, *Nano Lett.* **3**, 389 (2003).
27. L. A. Porter, H. C. Choi, J. M. Schmeltzer, A. E. Ribbe, L. C. C. Elliott, J. M. Buriak, Electroless nanoparticle film deposition compatible with photolithography, microcontact printing, and dip-pen nanolithography patterning technologies, *Nano Lett.* **2**, 1369 (2002).
28. D. A. Weinberger, S. G. Hong, C. A. Mirkin, B. W. Wessels, T. B. Higgins, Combinatorial generation and analysis of nanometer- and micrometer-scale silicon features via "dip-pen" nanolithography and wet chemical etching, *Adv. Mater.* **12**, 1600 (2000).
29. T. Thorsen, S. J. Maerkl, S. R. Quake, Microfluidic large-scale integration, *Science* **298**, 580 (2002).
30. A. Majumdar, Scanning thermal microscopy, *Annu. Rev. Mater. Res.* **29**, 505 (1999).
31. W. P. King, T. W. Kenny, K. E. Goodson, Comparison of thermal and piezoresistive sensing approaches for atomic force microscopy topography measurements, *Appl. Phys. Lett.* **85**, 2086 (2004).
32. W. P. King, Design analysis of heated atomic force microscope cantilevers for nanotopography measurements, *J. Micromech. Microeng.* **15**, 2441 (2005).
33. A. Hammiche, H. M. Pollock, M. Song, D. J. Hourston, Sub-surface imaging by scanning thermal microscopy, *Meas. Sci. Technol.* **7**, 142 (1996).
34. M. H. Li, Y. B. Gianchandani, Microcalorimetry applications of a surface micromachined bolometer-type thermal probe, *J. Vac. Sci. Technol. B* **18**, 3600 (2000).
35. H. J. Mamin, B. D. Terris, L. S. Fan, S. Hoen, R. C. Barrett, D. Rugar, High-density data storage using proximal probe techniques, *IBM J. Res. Dev.* **39**, 681 (1995).
36. M. A. Lantz, G. K. Binnig, M. Despont, U. Drechsler, A micromechanical thermal displacement sensor with nanometre resolution, *Nanotechnology* **16**, 1089 (2005).
37. A. Pantazi, A. Sebastian, T. A. Antonakopoulos, P. Baechtold, A. R. Bonaccio, J. Bonan, G. Cherubini, M. Despont, R. A. DiPietro, U. Drechsler, U. Duerig, B. Gotsmann, W. Haeberle, C. Hagleitner, J. L. Hedrick, D. Jubin, A. Knoll, M. A. Lantz, J. Pentaralkis, H. Pozidis, R. C. Pratt, H. Rothuizen, R. Stutz, M. Varsamou, D. Wiesmann, E. Eleftheriou, Probe-based ultrahigh-density storage technology, *IBM J. Res. Dev.* **52**, 493 (2008).
38. W. P. King, T. W. Kenny, K. E. Goodson, G. Cross, M. Despont, U. Durig, H. Rothuizen, G. K. Binnig, P. Vettiger, Atomic force microscope cantilevers for combined thermomechanical data writing and reading, *Appl. Phys. Lett.* **78**, 1300 (2001).
39. J. Lee, W. P. King, Improved all-silicon microcantilever heaters with integrated piezoresistive sensing, *J. Microelectromech. Syst.* **17**, 432 (2008).
40. K. J. Kim, K. Park, J. Lee, Z. M. Zhang, W. P. King, Nanotopographical imaging using a heated atomic force microscope cantilever probe, *Sensor. Actuat. A-Phys.* **136**, 95 (2007).
41. K. Park, J. Lee, Z. M. Zhang, W. P. King, Topography imaging with a heated atomic force microscope cantilever in tapping mode, *Rev. Sci. Instrum.* **78**, (2007).
42. J. J. M. Ruigrok, R. Coehoorn, S. R. Cumpson, H. W. Kesteren, Disk recording beyond 100 Gb/in^2: Hybrid recording? *J. Appl. Phys.* **87**, 5398 (2000).
43. H. J. Mamin, D. Rugar, Thermomechanical writing with an atomic force microscope tip, *Appl. Phys. Lett.* **61**, 1003 (1992).
44. H. J. Mamin, Thermal writing using a heated atomic force microscope tip, *Appl. Phys. Lett.* **69**, 433 (1996).
45. A. W. Knoll, D. Pires, O. Coulembier, P. Dubois, J. L. Hedrick, J. Frommer, U. Duerig, Probe-based 3-D nanolithography using self-amplified depolymerization polymers, *Adv. Mater.* **22**, 3361–3365 (2010).

46. D. Pires, J. L. Hedrick, A. De Silva, J. Frommer, B. Gotsmann, H. Wolf, M. Despont, U. Duerig, A. W. Knoll, Nanoscale three-dimensional patterning of molecular resists by scanning probes, *Science* **328**, 732 (2010).
47. P. E. Sheehan, L. J. Whitman, W. P. King, B. A. Nelson, Nanoscale deposition of solid inks via thermal dip pen nanolithography, *Appl. Phys. Lett.* **85**, 1589 (2004).
48. W. K. Lee, L. J. Whitman, J. Lee, W. P. King, P. E. Sheehan, The nanopatterning of a stimulus-responsive polymer by thermal dip-pen nanolithography, *Soft Matter* **4**, 1844 (2008).
49. B. A. Nelson, W. P. King, A. R. Laracuente, P. E. Sheehan, L. J. Whitman, Direct deposition of continuous metal nanostructures by thermal dip-pen nanolithography, *Appl. Phys. Lett.* **88**, 033104 (2006).
50. L. Aeschimann, A. Meister, T. Akiyama, B. W. Chui, P. Niedermann, H. Heinzelmann, N. F. De Rooij, U. Staufer, P. Vettiger, Scanning probe arrays for life sciences and nanobiology applications, *Microelectron. Eng.* **83**, 1698 (2006).
51. J. Fritz, M. K. Baller, H. P. Lang, H. Rothuizen, P. Vettiger, E. Meyer, H. J. Guntherodt, C. Gerber, J. K. Gimzewski, Translating biomolecular recognition into nanomechanics, *Science* **288**, 316 (2000).
52. F. Huber, M. Hegner, C. Gerber, H. J. Guntherodt, H. P. Lang, Label free analysis of transcription factors using microcantilever arrays, *Biosens. Bioelectron.* **21**, 1599 (2006).
53. G. H. Wu, R. H. Datar, K. M. Hansen, T. Thundat, R. J. Cote, A. Majumdar, Bioassay of prostate-specific antigen (PSA) using microcantilevers, *Nat. Biotechnol.* **19**, 856 (2001).
54. M. Lutwyche, C. Andreoli, G. Binnig, J. Brugger, U. Drechsler, W. Haberle, H. Rohrer, H. Rothuizen, P. Vettiger, G. Yaralioglu, C. Quate, 5X5 2D AFM cantilever arrays a first step towards a Terabit storage device, *Sensor. Actuat. A-Phys.* **73**, 89 (1999).
55. M. Despont, U. Drechsler, R. Yu, H. B. Pogge, P. Vettiger, Wafer-scale microdevice transfer/interconnect: Its application in an AFM-based data-storage system, *J. Microelectromech. Syst.* **13**, 895 (2004).
56. M. Zhang, D. Bullen, S. W. Chung, S. Hong, K. S. Ryu, Z. F. Fan, C. A. Mirkin, C. Liu, A MEMS nanoplotter with high-density parallel dip-pen manolithography probe arrays, *Nanotechnology* **13**, 212 (2002).
57. X. F. Wang, C. Liu, Multifunctional probe array for nano patterning and imaging, *Nano Lett.* **5**, 1867 (2005).
58. J. Zou, D. Bullen, X. F. Wang, C. Liu, C. A. Mirkin, Conductivity-based contact sensing for probe arrays in dip-pen nanolithography, *Appl. Phys. Lett.* **83**, 581 (2003).
59. J. Lee, W. P. King, Array of Microcantilever Heaters with Integrated Piezoresistors, in *Proceedings of the 7th International IEEE Conference on Nanotechnology*, Hong Kong, 135–140, (2007).
60. J. Lee, T. Beechem, T. L. Wright, B. A. Nelson, S. Graham, W. P. King, Electrical, thermal, and mechanical characterization of silicon microcantilever heaters, *J. Microelectromech. Syst.* **15**, 1644 (2006).
61. L. Shi, A. Majumdar, Thermal transport mechanisms at nanoscale point contacts, *J. Heat Trans-T ASME* **124**, 329 (2002).
62. S. K. Saha, L. Shi, Molecular dynamics simulation of thermal transport at a nanometer scale constriction in silicon, *J. Appl. Phys.* **101** (2007).
63. B. W. Chui, M. Asheghi, Y. S. Ju, K. E. Goodson, T. W. Kenny, H. J. Mamin, Intrinsic-carrier thermal runaway in silicon microcantilevers, *Nanosc. Microsc. Therm.* **3**, 217 (1999).
64. B. A. Nelson, W. P. King, Modeling and simulation of the interface temperature between a heated silicon tip and a substrate, *Nanosc. Microsc. Therm.* **12**, 98 (2008).
65. B. Bhushan, H. Fuchs, *Applied Scanning Probe Methods IV: Industrial Applications* (Springer, Berlin, 2006).
66. D. Wang, Thermochemical Nanolithography Fabrication and Atomic Force Microscopy Characterization of Functional Nanostructures, Georgia Institute of Technology (2010).
67. G. Chen, Phonon heat conduction in nanostructures, *Inter. J. Therm. Sci.* **39**, 471 (2000).

68. C. Hu, M. Kiene, P. S. Ho, Thermal conductivity and interfacial thermal resistance of polymeric low k films, *Appl. Phys. Lett.* **79**, 4121 (2001).
69. M. M. Yovanovich, J. R. Culham, P. Teertstra, Analytical modeling of spreading resistance in flux tubes, half spaces, and compound disks, *IEEE T. Compon. Pack. T* **21**, 168 (1998).
70. D. Wang, V. K. Kodali, W. D. Underwood, J. E. Jarvholm, T. Okada, S. C. Jones, M. Rumi, Z. Dai, W. P. King, S. R. Marder, J. E. Curtis, E. Riedo, Thermochemical nanolithography of multifunctional nanotemplates for assembling nano-objects, *Adv. Funct. Mater.* **19**, 1 (2009).
71. R. Szoszkiewicz, T. Okada, S. C. Jones, T. D. Li, W. P. King, S. R. Marder, E. Riedo, High-speed, sub-15 nm feature size thermochemical nanolithography, *Nano Lett.* **7**, 1064 (2007).
72. S. Xu, G. Y. Liu, Nanometer-scale fabrication by simultaneous nanoshaving and molecular self-assembly, *Langmuir* **13**, 127 (1997).
73. B. Klehn, U. Kunze, Nanolithography with an atomic force microscope by means of vector-scan controlled dynamic plowing, *J. Appl. Phys.* **85**, 3897 (1999).
74. W. Fudickar, A. Fery, T. Linker, Reversible light and air-driven lithography by singlet oxygen, *J. Am. Chem. Soc.* **127**, 9386 (2005).
75. I. Turyan, U. O. Krasovec, B. Orel, T. Saraidorov, R. Reisfeld, D. Mandler, Writing-reading-erasing on tungsten oxide films using the scanning electrochemical microscope, *Adv. Mater.* **12**, 330 (2000).
76. K. Seo, E. Borguet, Nanolithographic write, read, and erase via reversible nanotemplated nanostructure electrodeposition on alkanethiol-modified Au(111) in an aqueous solution, *Langmuir* **22**, 1388 (2006).
77. D. Wang, R. Szoszkiewicz, M. Lucas, E. Riedo, T. Okada, S. C. Jones, S. R. Marder, J. Lee, W. P. King, Local wettability modification by thermochemical nanolithography with write-read-overwrite capability, *Appl. Phys. Lett.* **91**, 243104 (2007).
78. M. A. Lantz, B. Gotsmann, U. T. Durig, P. Vettiger, Y. Nakayama, T. Shimizu, H. Tokumoto, Carbon nanotube tips for thermomechanical data storage, *Appl. Phys. Lett.* **83**, 1266 (2003).
79. M. Arnold, E. A. Cavalcanti-Adam, R. Glass, J. Blummel, W. Eck, M. Kantlehner, H. Kessler, J. P. Spatz, Activation of integrin function by nanopatterned adhesive interfaces, *Chemphyschem* **5**, 383 (2004).
80. E. Zamir, B. Geiger, Molecular complexity and dynamics of cell-matrix adhesions, *J. Cell Sci.* **114**, 3583 (2001).
81. A. Grakoui, S. K. Bromley, C. Sumen, M. M. Davis, A. S. Shaw, P. M. Allen, M. L. Dustin, The immunological synapse: A molecular machine controlling T cell activation, *Science* **285**, 221 (1999).
82. J. Duvigneau, H. Schonherr, G. J. Vancso, Atomic force microscopy based thermal lithography of poly(tert-butyl acrylate) block copolymer films for bioconjugation, *Langmuir* **24**, 10825 (2008).
83. C. L. Feng, G. J. Vancso, H. Schonherr, Reactive μCP on ultrathin block copolymer films: Investigation of the μCP mechanism and application to sub-μm (Bio)molecular patterning, *Langmuir* **23**, 1131 (2007).
84. M. H. Yun, N. V. Myung, R. P. Vasquez, C. S. Lee, E. Menke, R. M. Penner, Electrochemically grown wires for individually addressable sensor arrays, *Nano Lett.* **4**, 419 (2004).
85. J. Kameoka, D. Czaplewski, H. Q. Liu, H. G. Craighead, Polymeric nanowire architecture, *J. Mater. Chem.* **14**, 1503 (2004).
86. E. Tekin, E. Holder, D. Kozodaev, U. S. Schubert, Controlled pattern formation of poly [2-methoxy-5(2 ′-ethylhexyloxyl)-1,4-phenylenevinylene] (MEH-PPV) by ink-jet printing, *Adv. Funct. Mater.* **17**, 277 (2007).
87. B. Dong, N. Lu, M. Zelsmann, N. Kehagias, H. Fuchs, C. M. S. Torres, L. F. Chi, Fabrication of high-density large-area conducting-polymer nanostructures, *Adv. Funct. Mater.* **16**, 1937 (2006).
88. J. H. Lim, C. A. Mirkin, Electrostatically driven dip-pen nanolithography of conducting polymers, *Adv. Mater.* **14**, 1474 (2002).

89. B. W. Maynor, S. F. Filocamo, M. W. Grinstaff, J. Liu, Direct-writing of polymer nanostructures: Poly(thiophene) nanowires on semiconducting and insulating surfaces, *J. Am. Chem. Soc.* **124**, 522 (2002).
90. E. Betzig, J. K. Trautman, Near-field optics- microscopy, spectroscopy, and surface modification beyond the diffraction limit, *Science* **257**, 189 (1992).
91. R. Riehn, A. Charas, J. Morgado, F. Cacialli, Near-field optical lithography of a conjugated polymer, *Appl. Phys. Lett.* **82**, 526 (2003).
92. O. Fenwick, L. Bozec, D. Credgington, A. Hammiche, G. M. Lazzerini, Y. R. Silberberg, F. Cacialli, Thermochemical nanopatterning of organic semiconductors, *Nat. Nanotechnol.* **4**, 664 (2009).
93. D. J. Lipomi, R. C. Chiechi, M. D. Dickey, G. M. Whitesides, Fabrication of conjugated polymer nanowires by edge lithography, *Nano Lett.* **8**, 2100 (2008).
94. D. Wang, S. Kim, W. D. Underwood, A. J. Giordano, C. L. Henderson, Z. T. Dai, W. P. King, S. R. Marder, E. Riedo, Direct writing and characterization of poly(p-phenylene vinylene) nanostructures, *Appl. Phys. Lett.* **95**, 233108 (2009).
95. X. S. Wu, M. Sprinkle, X. B. Li, F. Ming, C. Berger, W. A. de Heer, Epitaxial-graphene/graphene-oxide junction: An essential step towards epitaxial graphene electronics, *Phys. Rev. Lett.* **101**, 026801 (2008).
96. D. A. Dikin, S. Stankovich, E. J. Zimney, R. D. Piner, G. H. B. Dommett, G. Evmenenko, S. T. Nguyen, R. S. Ruoff, Preparation and characterization of graphene oxide paper, *Nature* **448**, 457 (2007).
97. G. Eda, G. Fanchini, M. Chhowalla, Large-area ultrathin films of reduced graphene oxide as a transparent and flexible electronic material, *Nat. Nanotechnol.* **3**, 270 (2008).
98. J. T. Robinson, F. K. Perkins, E. S. Snow, Z. Q. Wei, P. E. Sheehan, Reduced graphene oxide molecular sensors, *Nano Lett.* **8**, 3137 (2008).
99. J. T. Robinson, M. Zalalutdinov, J. W. Baldwin, E. S. Snow, Z. Q. Wei, P. Sheehan, B. H. Houston, Wafer-scale reduced graphene oxide films for nanomechanical devices, *Nano Lett.* **8**, 3441 (2008).
100. G. Eda, C. Mattevi, H. Yamaguchi, H. Kim, M. Chhowalla, Insulator to semimetal transition in graphene oxide, *J. Phys. Chem. C* **113**, 15768 (2009).
101. Z. Wei, D. Wang, S. Kim, S. Y. Kim, Y. Hu, M. K. Yakes, A. R. Laracuente, Z. T. Dai, S. R. Marder, C. Berger, W. P. King, W. A. deHeer, P. E. Sheehan, E. Riedo, Nanoscale tunable reduction of graphene oxide for graphene electronics, *Science* **328**, 1373 (2010).
102. J. H. Burroughes, D. D. C. Bradley, A. R. Brown, R. N. Marks, K. Mackay, R. H. Friend, P. L. Burns, A. B. Holmes, Light-emitting-diodes based on conjugated polymers, *Nature* **347**, 539 (1990).
103. E. Kim, Y. N. Xia, G. M. Whitesides, Polymer microstructures formed by molding in capillaries, *Nature* **376**, 581 (1995).
104. J. Lee, W. P. King, Liquid operation of silicon microcantilever heaters, *IEEE Sens. J.* **8**, 1805 (2008).
105. P. C. Fletcher, J. R. Felts, Z. T. Dai, T. D. Jacobs, H. J. Zeng, W. Lee, P. E. Sheehan, J. A. Carlisle, R. W. Carpick, W. P. King, Wear-resistant diamond nanoprobe tips with integrated silicon heater for Tip-based nanomanufacturing, *ACS Nano* **4**, 3338 (2010).
106. O. Akbulut, J. M. Jung, R. D. Bennett, Y. Hu, H. T. Jung, R. E. Cohen, A. M. Mayes, F. Stellacci, Application of supramolecular nanostarnping to the replication of DNA nanoarrays, *Nano Lett.* **7**, 3493 (2007).

Chapter 8
Proton-fountain Electric-field-assisted Nanolithography (PEN)

Andres La Rosa and Mingdi Yan

Abstract This chapter describes the implementation of Proton-fountain Electric-field-assisted Nanolithography (PEN) as a potential tool for fabricating nanostructures by exploiting the properties of stimuli-responsive materials. The merits of PEN are demonstrated using poly(4-vinylpyridine) (P4VP) films, whose structural (swelling) response is triggered by the delivery of protons from an acidic fountain tip into the polymer substrate. Despite the probably many intervening factors affecting the fabrication process, PEN underscores the improved reliability in the pattern formation when using an external electric field (with voltage values of up to 5 V applied between the probe and the sample) as well as when controlling the environmental humidity conditions. PEN thus expands the applications of P4VP as a stimuli-responsive material into the nanoscale domain, which could have technological impact on the fabrication of memory and sensing devices as well as in the fabrication of nanostructures that closely mimic natural bio-environments. The reproducibility and reversible character of the PEN fabrication process offers opportunities to also use these films as test bed for studying fundamental (thermodynamic and kinetic) physical properties of responsive materials at the nanoscale level.

Keywords Responsive materials · P4VP · Nanolithography · Swelling · Polymer film · pH responsive · Erasable patterns · PEN · Biomimetic materials · DPN · Hydrogels · Osmotic pressure · Entropy of mixing · Protonation

8.1 Introduction

8.1.1 PEN as a Method for Creating Erasable Nanostructures

The applications of tip-based nanolithography techniques that create patterns by anchoring molecules *onto* a surface of proper chemical affinity – as is the case in dip-pen nanolithography (DPN) [1, 2] – can be expanded by, alternatively, triggering the formation of nanostructures out of stimuli-responsive material

A. La Rosa (✉)
Department of Physics, Portland State University, Portland, OR 97201, USA
e-mail: andres@pdx.edu

A.A. Tseng (ed.), *Tip-Based Nanofabrication*, DOI 10.1007/978-1-4419-9899-6_8,

substrates. Since stimuli-responsive properties may have a reversible character, the alternative nanofabrication approach could have concomitant implications in bio-technology (for creating switching gates that allow manipulating the transport, separation, and detection of bio-molecules, or for fabricating soft-material nanostructures that closely mimic natural bio-environments) as well as in emerging nano-electronics technologies (for fabricating low-cost and low-voltage operation integrated logic circuits in flexible substrates). The potential technological implications that can be brought by harnessing the fabrication of nanostructures out of stimuli responsive materials underlines the interest for developing Proton-fountain Electric-field-assisted Nanolithography (PEN). In PEN the formation of nanostructures is triggered by the localized injection of protons into the substrate, with the charge-transfer from a sharp tip into the substrate being better controlled by the application of an external electric field. The development of PEN is thus conceived within the context of emerging developments in materials science and molecular engineering [3] that pursue the design of devices that rely on the transduction of environmental signals.

8.1.2 PEN in the Context of Emerging Biomimetic Engineering

Inspired by the multi-functional inner working properties of living cell membranes [4], including the surprising sensitivity of their dynamic response to the mechanical properties of surrounding material [5], a current focus in biomimetic materials constitutes the development of versatile synthetic thin films that can selectively respond to a variety of signal interactions (mechanical, chemical, optical, changes in environmental conditions, etc) [6]. In one approach, the complex synthetic hierarchy needed to eventually mimic *nature* is conceived as a combination of functional-domains separated by stimuli-responsive polymer thin films regulating the interactions between the domain compartments [7]. In another approach, the cell is conceived not just as a chemical but also as a mechanical device [8], for it is found that the cell membrane is very sensitive to the mechanical properties of its surrounding matrix (affecting their growth, differentiation, migration, and, eventually, apoptosis) [9, 10], which has triggered an interest in the development of, for example, synthetic polymer scaffold for regenerative medicine [11, 12]. Both approaches, mentioned above, emphasize the need for harnessing the fabrication of synthetic thin film responsive materials.

The different approaches to biomimetic materials have resulted in the design of a variety of responsive building blocks (gels [13], brushes [14], hybrid systems with inorganic particles [15]) that respond selectively to different (pH [16], temperature [17, 18], optical [15, 19], and magnetic [20]) external stimuli. Following the "bottom-up" route, functional materials have been prepared based on self-assembly of polymeric supramolecules [21]. Progress following the alternative "top-down" approach includes the fabrication of stimulus responsive polymer brushes [22], growth of polymers from previously DPN-patterned templates [23], and chain polymerization of monomolecular layer by local stimulation using a STM tip [24, 25] (followed up by investigation of their working principle [26]). PEN falls in the top-down category approach. In the next section, we concentrate our description on hydrogels [27], since the latter describes closer the experimental results obtained in current applications of PEN.

8.2 Underlying Working Mechanisms of Swelling in Hydrogels

This section provides a succinct summary of the main theoretical results underlying the working mechanisms involved in the swelling of hydrogels, where the concept of entropy plays a key role. In particular, it is worth to highlight the peculiar theoretical framework brilliantly introduced, time ago, by Paul J. Flory [28, 29] for analyzing polymer solutions; although his models have been refined, the essence of his clever approach is still used. Given the expected complexity of these polymer structures, it results interesting to realize the conceptual similarities between (a) the analysis of a much simpler liquid-vapor system in equilibrium [30, 31], and (b) the analysis of the more complex (hydrogel) polymer solution [32] (see Fig. 8.1 below). For comparison and illustrative purposes both analyses will be presented here.

A hydrogel refers to a flexible (typically) hydrophilic cross-linked polymer network and a fluid filling the interstitial spaces of the network. The entire network holds the liquid in place thus giving the system a solid aspect. But contrary to other solid materials, these wet and soft systems are capable of undergoing very large

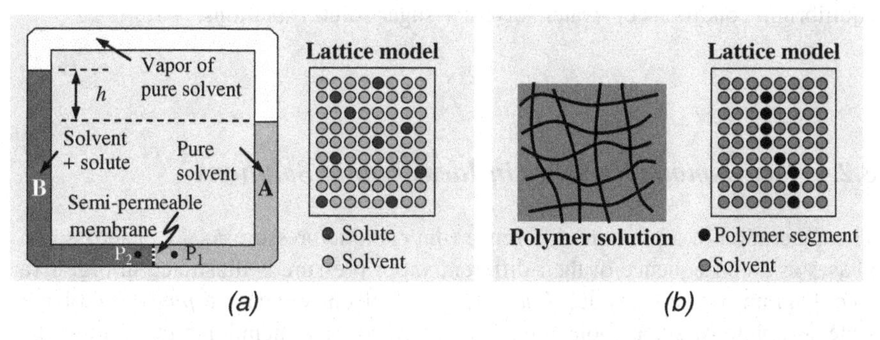

Fig. 8.1 Two different thermodynamic systems studied using similar lattice model analysis. (**a**) *Left:* Because of their different vapor pressures, a solution (water solvent + sugar solute molecules) and a pure water solvent (separated by a membrane permeable only to solvent molecules) generate an osmotic pressure ρgh (ρ is the density of the solution). *Right:* Solvent and solute molecules considered to reside in a hypothetical lattice, for entropy calculation purposes. The diagram on the *left* has been reproduced from Huang [30, p. 47] and reprinted with permission of John Wiley & Sons, Inc. (**b**) *Left:* Polymer gel system. *Right:* Polymer chain considered as solute immersed in solvent, where all molecules are considered to reside in a hypothetical liquid lattice. Figures adapted from Flory [32], Copyright @ 1953 Cornell University and Copyright @ 1981 Paul J. Flory; used by permission of the publisher, Cornell University Press

deformation (greater than 100%). Understanding the dynamic behavior of hydrogels is worthwhile to pursue due to their widespread implications. In particular, the role of gels in living organism can not be exaggerated. As it is well put by Osada and Gong [33], living organisms are largely made of gels (mammalian tissues are aqueous materials largely composed of protein and polysaccharide networks), which enables them to transport ions and molecules very effectively while keeping its solidity.

What drives the swelling in a hydrogel? One of the potential mechanisms can be described in terms of the osmotic pressure (an entropic driven phenomenon), which help us understand how the additional entropy of mixing (afforded by an increase in the system's volume due to the absorption of water by the polymer network) is counteracted by a restoring force (also of entropic origin, since a lager dimension afford less polymer configurations) from the network itself. This osmotic pressure refers to the same type of phenomenon underlying the lower vapor-pressure displayed by an ideal solution (solvent + non-volatile solute) when compared to the vapor-pressure of a pure solvent. Since the latter constitutes a much simpler and familiar process, we conveniently include its (brief) description in the next paragraph. More specifically, we address the dynamics involved when a solution (water solvent + sugar solute) and pure water solvent are separated by a semi-permeable membrane, which can help us gain knowledge about the osmotic pressure concept. The entropy of mixing involved in this phenomenon is described in the framework of a lattice model (one in which solute and solvent molecules are considered to reside on the sites of a hypothetical lattice, the latter used as a resource that facilitates the calculation of the solution's entropy). This approach will allow us to get familiar with lattice models, which are frequently used to describe polymer solutions (hydrogels). In short, we try to view the dynamics of hydrogels through the same prism used to view the equilibrium conditions of (water solvent + sugar solute) solutions.

8.2.1 The Osmotic Pressure in Ideal Liquid Solutions

The generation of a pressure difference (the osmotic pressure π_{osmotic}) across two phases as a consequence of their different vapor pressure is illustrated in Fig. 8.1a (see diagram on the left side). *Phase-A* (pure solvent water) and *phase-B* (solvent water + solute of sugar molecules) are separated by a membrane that allows the passage of water molecules but not the larger solute molecules [30]. It is an experimental fact that a dynamic equilibrium (i.e. equal rate, in both directions, of water molecules passing across the membrane) is reached when a hydrostatic pressure difference (the osmotic pressure π_{osmotic}) is established between the two phases. (Conceptually, the underlying mechanism at play here is that a water molecule in *phase-B* contributes greater to the total entropy than when inside the pure water *phase-A*; hence a net flow towards the former increases the total entropy.) The question to address is how to quantify this pressure difference.

8.2.1.1 Chemical Potentials of an Ideal Solution and a Pure Solvent

A formal description of the osmotic pressure [30, 31] takes into account the fact that under initial conditions of equal pressure P, the chemical potential $\mu^{A}_{\text{water},P}$ of water in *phase-A* (pure solvent water) is greater than the chemical potential $\mu^{B}_{\text{water},P}$ of water in *phase-B* (as it will be justified in the next paragraph and the next section below). Hence, when the phases are separated by a membrane permeable only to water, this constituent will not be in equilibrium and a net passage of water molecules from *phase-A* to *phase-B* is expected. As more water passes to *phase-B* the pressure increases and so does the chemical potential. Equilibrium with respect to the water constituent will then be established when $\mu^{B}_{\text{water},P_2}$ (the chemical potential of water in *phase-B*, at the increasing pressure P_2) becomes equal to $\mu^{A}_{\text{water},P_1}$ (the chemical potential of water in *phase-A*, at pressure P_1) [31]. The thus developed pressure difference $(P_2 - P_1)$ is referred to as the osmotic pressure, π_{osmotic}.

To find a relationship between the change in chemical potential and the osmotic pressure, let's resort first to the extensive properties of the thermodynamic potentials [34] (namely, when the amount of matter is changed by a given factor, they change by the same factor). A particular important relationship is obtained when this property is applied to the Gibbs free energy $G = G\,(T, P, N)$. In effect, being the temperature T and pressure P intensive quantities, G has to have the form

$$G = Nf(T, P),$$

where N is the number of particles of the analyzed system.

Since $\mathrm{d}G = -S\,\mathrm{d}T + V\,\mathrm{d}P + \mu\,\mathrm{d}N$ and $\mu = (dG/dN)_{T,P}$, the extensive property $G = Nf(T, P)$ implies that μ is only a function of T and P; that is,

$$\mu = G/N = f(T, P).$$

Accordingly, μ is the Gibbs free energy per molecule, and it is a quantity independent of N.

Thus,

$$d(G/N) = d\mu = -(S/N)dT + (V/N)dP,$$

which implies,

$$\frac{d\mu}{dP} = V/N.$$

This expression is pertinent to the quantification of the osmotic pressure. In effect, it reflects the change in chemical potential due to an increase in pressure, $\Delta\mu = (V/N)\,\Delta P$ (where it has been assumed that the volume does not change with pressure). Using $v \equiv (V/N)$, one obtains $\mu^{B}_{\text{water},P_2} - \mu^{A}_{\text{water},P_1} = v(P_2 - P_1)$, or,

$$\mu^{A}_{\text{water},P_1} - \mu^{B}_{\text{water},P_2} = v\pi_{osmotic} \tag{1}$$

8.2.1.2 Lattice Model for Calculating the Entropy of Mixing in Ideal Liquid Solutions

An explicit calculation of the chemical potential difference, in terms of the number of constitutive molecules, can be derived by starting (at the most fundamental level) from a relatively simple combinatorial analysis of dissimilar solvent (water) and solute (sugar) molecules allowed to reside on the sites of a hypothetical lattice (see the diagram at the right in Fig. 8.1a) [32], the latter introduced basically to facilitate the calculation the system's entropy of mixing ΔS_{mix}. In this lattice framework, the increase in the system entropy resulting from mixing N solvent and n solute molecules is given by $\Delta S_{\mathrm{mix}}(N,n) = k \ln[(N+n)!/N!\, n!]$. For large values of N and n one can use the well known Stirling's approximation that gives $\ln n! = n \ln n - n \approx n \ln n$, or $n! \approx n^n$. Using this approximation, ΔS_{M} adopts the form $k \ln[(N+n)^{(N+n)}/N^N n^n]$, or

$$\Delta S_{\mathrm{mix}}(N,n) = -k\,(N \ln f_N + n \ln f_n) \tag{2}$$

where $f_N \equiv N/(N+n)$ and $f_n \equiv n/(N+n)$ are the mole fractions of solvent (water) and solute (sugar) in the solution, respectively; k is the Boltzmann constant.

The Gibbs free energy has the general form $G = E + PV - TS = \mu N$. When applied to the system in *phase-B* (of N solvent and n solute molecules) at pressure P_2 and temperature T, one obtains,

$$G^{phase-B}(T,P_2,N,n) = G^{\mathrm{pure}}_{\mathrm{water}}(T,P_2,N) + G^{\mathrm{pure}}_{\mathrm{water}}(T,P_2,n) - T\Delta S_{\mathrm{mix}}(N,n) \tag{3}$$

where the first two terms on the right hand side are the Gibbs energy of the corresponding components in their pure state (we are assuming an ideal solution, so the components do not interact).

The chemical potential of the water component in *phase-B* will then be given by,

$$\begin{aligned}
\mu^{phase-B}_{\mathrm{water}}(T,P_2,N,n) &= \frac{dG^{phase-B}}{dN}(T,P_2,N,n) \\
&= \frac{dG^{\mathrm{pure}}_{\mathrm{water}}}{dN}(T,P_2,N) - T\frac{d}{dN}\Delta S_{\mathrm{mix}} \\
&= \frac{dG^{\mathrm{pure}}_{\mathrm{water}}}{dN}(T,P_2,N) + kT\frac{d}{dN}(N \ln f_N + n \ln f_n) \\
\mu^{phase-B}_{\mathrm{water}}(T,P_2,N,n) &= \mu^{\mathrm{pure}}_{\mathrm{water}}(T,P_2,N) + kT \ln f_N
\end{aligned} \tag{4}$$

For *phase-A*, which is constituted by pure water at pressure P_1,

$$\mu^{phase-A}_{\mathrm{water}}(T,P_1,N,n=0) = \mu^{\mathrm{pure}}_{\mathrm{water}}(T,\, P_1,\, N) \tag{5}$$

At equilibrium, $\mu_{\text{water}}^{phase-B}(T,P_2,N,n) = \mu_{\text{water}}^{phase-A}(T,P_1,N,n = 0)$. Therefore, expressions (4) and (5) lead to,

$$\mu_{\text{water}}^{\text{pure}}(T,P_2,N) - \mu_{\text{water}}^{\text{pure}}(T,P_1,N) = -kT\ln f_N \tag{6}$$

Replacing (6) in (1) gives,

$$\pi_{\text{osmotic}} = -\frac{kT}{\nu}\ln f_N = -\frac{kT}{V/N}\ln f_N \tag{7}$$

Incidentally, since we are working with solution where $f_n << 1$ and thus $\ln f_N = \ln(1 - f_n) = -f_n$, useful equivalent formulas can be obtained,

$$\pi_{\text{osmotic}} = \frac{kT}{V/N} f_n \text{ (for a dilute solution)} \tag{8}$$

or, by expanding further f_n, one gets $\pi_{\text{osmotic}} = \frac{kT}{V/N}\frac{n}{n+N} \approx kT\frac{n}{V}$.

$$\pi_{\text{osmotic}} \approx kT\frac{n}{V} = RT\frac{moles\ of\ solute}{V} = \frac{RT}{MW}\frac{mass\ of\ solute}{V} \tag{9}$$

where n is the number of solute molecules in a volume V, and MW is the solute's molecular weight.

8.2.2 Lattice Model for Describing Ideal Polymer Solutions

The basic results displayed by expressions (2) and (7) constitute a proper starting step for describing more complicated (and apparently unrelated) cases like, for example, a cross-linked polymer network interacting with a pool of water, i.e. a hydrogel. The elegant twist in the coming description lies in considering the polymer network as the solute[1] [32, 35]. That is, the trend of analysis using a hypothetical lattice conveniently remains the same; hence the observed similarity between the lattice displayed in Fig. 8.1a (used to analyze a liquid solution system) and the lattice in Fig. 8.1b (used to analyze a polymer solution) [36, 37]. In the latter, one polymer molecule is considered to be a chain of x segments, each segment (arbitrarily) considered equal in size to a solvent molecule. The objective becomes calculating the entropy of the polymer solution resulting from the different configurations that can be arranged with n_1 molecules of solvent and n_2 polymer molecules in a lattice containing $(n_1 + x\,n_2)$ cells.

[1] Lattice models of polymer solutions are widely used for their simplicity and computational convenience. Their use for predicting solution properties of polymers solutions dates back to the 1940s.

8.2.2.1 Lattice Model for Calculating the Entropy of Mixing

For the case of a polymer solution the calculation of the entropy can be conceived by counting first the different permutations associated with a given configuration of the polymer molecules (assuming no solvent molecules were present), and then adding the configurations resulting from their mixing with the solvent molecules (polymer segments and solvent molecules replacing one another in the liquid lattice). While the former is expected to contribute more effectively in a process of polymer fusion, here the interest focuses mainly on the entropy of mixing. The latter takes, quite surprisingly again, a very simple form [29, 32],

$$\Delta S_{\text{mixing}} = -k(n_1 \ln v_1 + n_2 \ln v_2) \tag{10}$$

where v_1 and v_2 are the volume fractions of solvent and solute respectively,

$$v_1 = n_1/(n_1 + xn_2)$$

$$v_2 = xn_2/(n_1 + xn_2)$$

Notice the similarity between expressions (2) and (10), except that volume fractions appear in the latter formula (mixing of molecules of different size) instead of mole fraction in the former (mixing of molecules of the same size).

8.2.2.2 The Heat Energy of Mixing ΔE_M, the Excluded Volume Effect, and the Helmholtz Free Energy ΔF_M

Given the fact that the dynamics of a polymer solution depends not only on the entropy but also on the energy of the system (the Helmholtz free energy $F = E - TS$ has to be minimum) this latter aspect is addressed in this section. In fact, the interactions between the water and the polymer molecules in a hydrogel make the polymer network a highly non-ideal thermodynamic system, where the cross-linked network structure plays an important role in determining the equilibrium aspects of the gel. In principle, any realistic model then has to take into account the intermolecular interactions due to the close proximity of the molecules, although some approximation can be applied depending on the temperature range being considered. On one hand, at relatively low temperatures a net attractive interaction between the monomers prevails, resulting in a net negative energy of the polymer system. On the other hand, at higher temperatures the repulsive interaction between the monomers when they are at very short distance (implicitly reflecting the fact that a monomer can not supplant the space already occupied by another monomer, a phenomenon better known in the jargon of polymer science as "excluded volume effect" or "excluded volume interaction") [38] will lead to a net positive energy of the polymer system. Below we provide some expressions that quantify this energy contribution.

In the lattice model, where each cell is able to accommodate either a solvent molecule or a polymer segment, the heat of mixing results from the replacement of some of the contacts between like-species (1-1 or 2-2) with unlike-constituents (1-2). If Δw_{12} represents the change in energy when one of these replacements occurs, then the heat associated with the formation of a particular configuration having p_{12} unlike-neighbors will be equal to $\Delta E_M = p_{12}\, \Delta w_{12}$.

For the calculation of p_{12}, it is plausible to assume that the probability a particular site adjacent to a polymer segment is occupied by a solvent molecule to be proportional to the volume fraction v_1 of the solvent. On the other hand, if the number of cells which are first neighbor to a given cell is z (here z is expected to be on the order of 6–12), then zx is the number of contacts per polymer molecule, and zxn_2 would be proportional to the total number of contacts. p_{12} turns out to be then proportional to $(zxn_2)v_1$. On the other hand, using the definition $v_2 = xn_2/(n_1 + xn_2)$, one obtains $n_1 v_2 = xn_2\, n_1/(n_1 + xn_2) = xn_2\, v_1$. Hence, $\Delta E_M = p_{12}\, \Delta w_{12} \sim (zxn_2)v_1 \Delta w_{12} = (z\, v_2\, n_1)\Delta w_{12} = (z\, \Delta w_{12})\, n_1\, v_2$. This result is typically expressed as,

$$\Delta E_M = kT\ \chi_1\ n_1\ v_2 \tag{11}$$

where the quantity $kT\chi_1 \equiv z\ \Delta w_{12}$ characterizes (for a given solute) the interaction energy per solvent molecule.

As mentioned above, at relatively low temperatures the net contribution from this term has a negative value. However, at relatively high temperatures, it was pointed out early on by Flory that the interpretation of χ_1 should be expanded as to include also other potential interactions between neighboring components that could have concomitant (positive value) contribution to the Helmholtz free energy. The new interpretation relates χ_1 to the number of pair-molecules collisions [38], which should be proportional also to the number of pair contacts developed in the solution, just as for the heat exchange.

Using (10) and (11), the Helmholtz free energy of mixing is given by [32],

$$\begin{aligned} \Delta F_{\text{mix}} &= \Delta E_{\text{mix}} - \text{T}\, \Delta S_{\text{mix}} \\ &= kT(\chi_1\, n_1\, v_2 + n_1 \ln v_1 + n_2 \ln v_2) \end{aligned} \tag{12}$$

8.2.3 Swelling of Neutral Polymer Networks

Notice that the above formulation basically describes the mixing of two liquids; the view of a polymer network held up by its cross-links has not appeared yet. The calculation given in (12), however, is a good initial step towards calculating the structural entropy of network formation. The latter was originally done in a very clever way by Paul Flory and John Rehner in their seminal papers [28, 29] (latter improved by Flory) [36], who modeled the cross-linked network as a system composed of v polymer chains of the same contour length (a chain counted as a polymer thread between two cross-link intersections), with the location of the cross-link points positions defined, on average, on the vertex of a regular tetrahedron [28]. Any external distortion of the network (swelling, for example, due to the mixing of the polymers with a solvent) would be monitored by the distortion of this average tetrahedron cell.

In the Flory's description, the first step consists of calculating the different configuration that results from the dilution of v (short) polymer chains (prior to the cross-linking) with n solvent molecules. This is given by expression (10), with v playing the role of n_2,

$$\Delta S_{\mathrm{D}} = -k\left(n_1 \ln \frac{n_1}{n_1 + xv} + v \ln \frac{v_1}{n_1 + xv}\right) \tag{13}$$

This is followed by a more elaborated calculation of the additional entropy corresponding to the different configurations that lead to the formation of a network of tetrahedron cells in a sea of solvent molecules.

$$\Delta S_{\mathrm{v'}} = \Delta S_{\mathrm{D}} + \Delta S_{\text{network of tetrahedron cells}} \tag{14}$$

Here the sub index V' stands for the final total volume of the swollen polymer network due to the mixing of the polymer with the solvent molecules. The objective here, however, is to calculate the net change in entropy $\Delta S_{\mathrm{swelling}}$ due just to the swelling; that is,

$$\Delta S_{\mathrm{swelling}} = \Delta S_{v'} - \Delta S_{v} \tag{15}$$

where ΔS_V stands for the entropy of the network when no solvent molecules are present. The result, in terms of the volume fraction $v_2 = x\,v/(n_1 + xv)$, is given by [36],

$$\Delta S_{\mathrm{swelling}} = \underbrace{-kn_1 \ln(1-v_2)}_{\text{Entropy of mixing polymers and solvent}} - \underbrace{\frac{3}{2} kv[(1/v_2)^{2/3} - 1] - \frac{1}{2} kv \ln v_2}_{\text{elastic entropy arising from the deformation of the network}} \tag{16}$$

This expression reflects the contribution to the entropy from two different sources. Before dilution with the solvent molecules, the configuration of the network corresponds to one of maximum entropy, hence any deformation (due to swelling) would lead to a configuration of comparatively lower entropy. The latter then competes against the tendency for an entropy increase caused by the addition of solvent (and volume) that favors the creation of new configurations. Expression (15) embodies then the physical mechanism underlying the swelling process in an uncharged hydrogel. Using (11) and (15), the change in free energy is given by, $\Delta F = \Delta E_{\mathrm{M}} - T\,\Delta S_{\mathrm{swelling}}$. Equilibrium is governed by the condition $\Delta F = 0$.

8.2.4 Swelling of Ionic Polymer Networks

Another potential channel for causing a polymer network to swell is the existence of charge centers, or molecular groups, which can strongly interact among themselves and with other ions contained in the solvent. That is the case in a poly(4-vinylpyridine) (P4VP), whose pyridyl groups can react with hydronium ions H_3O^+ thus forming positively charged nitrogen centers (N^+). The situation is depicted in Fig. 8.2. If the fixed pyridinium cations were the only ions present there would be an exceedingly large electrostatic repulsion, but such interaction is partially screened by the presence of counterions resulting from the dissociation of the phosphate salts in water,

$$\begin{aligned}
&NaH_2PO_4 \rightarrow Na^+ + H_2PO_4^- \\
&H_2PO_4^- + H_2O \rightleftharpoons H_3O^+ + HPO_4^{2-} \quad K = 6.31 \times 10^{-8} \\
&HPO_4^{2-} + H_2O \rightleftharpoons H_3O^+ + PO_4^{3-} \quad K = 3.98 \times 10^{-13}
\end{aligned} \tag{17}$$

where K stands for the corresponding dissociation constants.

Notice in Fig. 8.2 that the equilibrium between the swollen ionic polymer network and its surroundings resembles the situation depicted in Fig. 8.1a where a membrane prevents the solute sugar molecules from entering the pure solvent region. In this case, the polymer acts as a membrane, preventing the charged ions from freely diffusing into the outer solution, establishing a higher concentration of mobile ions inside the network than in the outside (mainly because of the attraction of the fixed ions). There thus exists an associated osmotic pressure arising from the difference in mobile ion concentration. Consequently, in addition to the swelling caused by the entropic mixing of polymer and water solvent, the fixed charges and counterions produce an additional driving force for the network to swell.

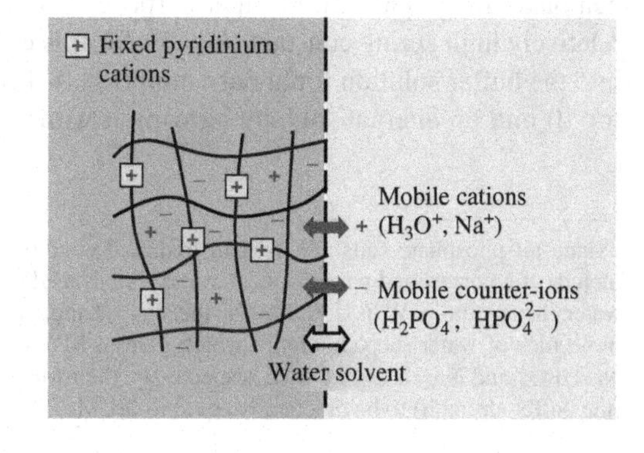

Fig. 8.2 Schematic description of an exchange of ions and solvent between a P4VP polymer network and its surrounding electrolyte. Adapted from Flory [32], Copyright @ 1953 Cornell University and Copyright @ 1981 Paul J. Flory; used by permission of the publisher, Cornell University Press

8.3 Fabrication Procedure

8.3.1 Preparation of the P4VP Responsive Material

In a typical procedure, a solution of P4VP (molecular weight ca. 160,000) in n-butanol (10 mg/ml) is prepared with the reagents used as received. Silicon wafers with a native oxide layer are cut into square pieces ~ 1 cm $\times$ 1 cm, and subsequently cleaned either by sonication in isopropyl alcohol for 15 min, or, alternatively, by immersion into piranha solution for 60 min at 80°C followed by thorough cleaning in hot water. (Caution: the piranha solution reacts violently with many organic solvents.) Subsequently, the P4VP solution is spin-coated onto the wafers at 2,000 rpm for 60 s. For crosslinking purposes, the sample is irradiated with a 450-W medium-pressure Mercury lamp (measured intensity of 5 mW/cm^2) for about 5 min. The irradiated films are then soaked in n-butanol for 24 h to remove the unbound polymer. One way to estimate the thickness of the resulting film is to use an ellipsometer. In that case, a value of 1.54 for the refractive index of the P4VP is used in the calculation [39]. This overall procedure gives film thickness in the 60–100 nm range.

8.3.2 Preparation of the Acidic Fountain Tip

A source of hydronium ions H_3O^+ (or, if desired, hydroxide ions OH^- as well) is prepared out of phosphate buffered solutions, which have the remarkable property that can be diluted and still keep the same concentration of H_3O^+. Different pH values can be obtained by dissolving corresponding quantity ratios of sodium dihydrogen phosphate (NaH_2PO_4) and sodium hydrogen phosphate (Na_2HPO_4) in distilled water. For example, mixing 13.8 g/l and 0.036 g/l of the two salts, respectively, gives 0.1 M buffer solution of pH equal to 4.0[2]. This acidic solution serves as the source of hydronium ions which, upon penetrating a P4VP film, protonate the P4VP's pyridyl groups, as suggested in Fig. 8.3 [40]. To achieve the protonation in very localized and targeted regions, however, PEN currently uses a sharp atomic force microscope (AFM) tip as an ion delivery vehicle in a similar fashion to dip-pen nanolithography [2]. For that purpose, as outlined in Fig. 8.4 , an AFM tip of relatively high spring constant ($k = 40$ N/m) is coated by simply soaking the probe into the buffer solution for about 1 min (Fig. 8.4a) and then allowing it to dry in air for 10 min or, alternatively, by blowing it with nitrogen (Fig. 8.4b). Such fountain

[2]Since all phosphate salts are used in hydrated condition, the molecular weight (MW) should include the corresponding portion of water. For NaH_2PO_4 we should include one molecules of water, hence the MW is 137.99. On the other hand, for the Na_2HPO_4 we should consider 7 molecules of water (heptahydrate), which gives a MW of 268.07. Hence, if 13.8 and 0.036 g of NaH_2PO_4 and Na_2HPO_4 are used respectively, then we can quote the concentration of NaH_2PO_4 (the buffer strength) to be practically equal to 0.1 M.

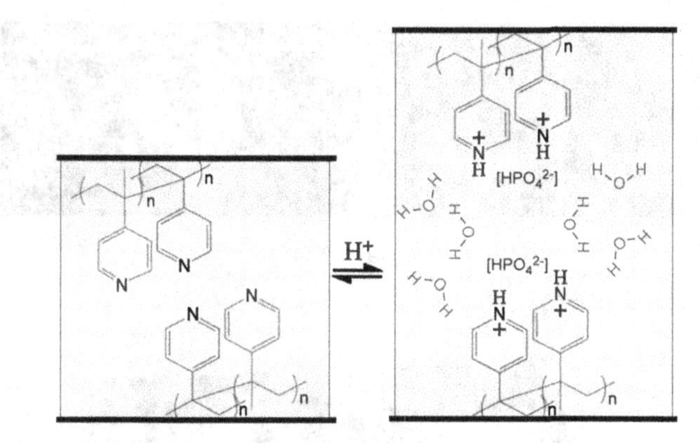

Fig. 8.3 Schematic illustration of the reversible swelling mechanism in P4VP, suggested here to be a consequence of the corresponding electrostatic interaction upon the protonation of the pyridyl groups. Illustration adapted from Maedler et al. [40] and reproduced with permission of the *SPIE*

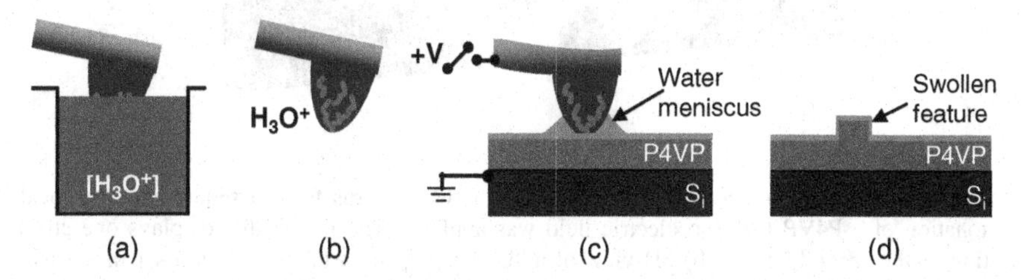

Fig. 8.4 Schematic procedure for attaining spatially-localized protonation of a P4VP film resting on a silicon substrate. After dipping an AFM tip into an acidic buffer solution, the probe is pressed against a P4VP film. The transfer of *hydronium* ions from the tip to the film is aided by the presence of a naturally-formed water meniscus (experiment performed at relative humilities above 40%). The protonation of the P4VP's pyridyl groups triggers the swelling response of the film. The bias voltage helps the process to be more reliable and allows controlling the height of the features

tip constitutes the probe for delivering hydronium ions very locally over targeted sites on a responsive material substrate.

8.3.3 Procedure for the Local Protonation

The fountain-tip is mounted on the head-stage of an AFM system,[3] whose electronic station comprises lithography software for controlling the lateral scanning of the tip along pre-determined paths, and the capability to apply a bias voltage to the probe. The accumulated experimental results suggest that, upon bringing the probe into contact with the polymer film (contact force of $\sim$1 μN), both (i) the surrounding water meniscus that naturally forms between the tip and polymer film,

[3]AFM XE-120 from Park Systems Inc.

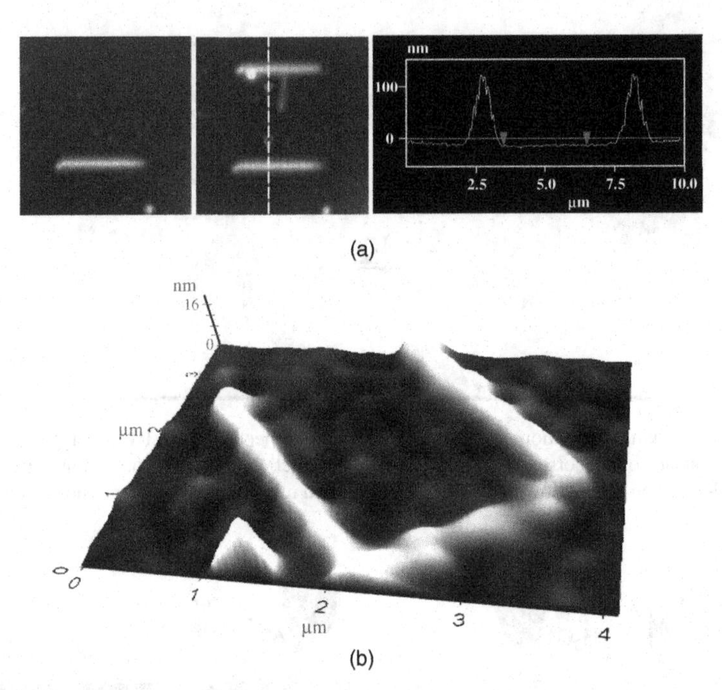

Fig. 8.5 (**a**) *Top-view images* (10 μm × 10 μm) of two patterns formed sequentially by local protonation of a P4VP film (no electric field was applied). The line profile displays one cross section of the *second image*. (**b**) 3D-view of a "U" shape pattern fabricated on a 4 μm × 4 μm region of another P4VP film (contact force = 1 μN and 5 V tip-sample bias voltage)

and (ii) the buffer concentration gradient, facilitate the transport of hydroium ions from the tip into the polymer (Fig. 8.4c). The subsequent protonation of the P4VP's pyridyl groups (as suggested in Fig. 8.3) gives rise to a net electrostatic repulsion that causes the corresponding polymer region to swell (Fig. 8.4d). However initially successful (see Fig. 8.5a), this simple contact method unfortunately did not guarantee the pattern formation reproducibly. Subsequently we discovered that pattern formation occurred more consistently when applying an electric field between the probe and the substrate.

8.3.4 Pattern Formation

First, with the "ink" loaded tip, an atomic force microscope (AFM) is set in "tapping imaging-mode," where the probe cantilever undergoes oscillations perpendicular to the sample's surface. To obtain an initial knowledge of the "blank paper" (a UV cross-linked P4VP polymer film), an image is taken at a relatively fast lateral scanning rate (5 μm/s). The use of high rates prevents transferring the buffer molecules into the substrate. Subsequently, the microscope is switched to "contact imaging-mode" for pattern formation under physical parameters controlled by the operator, namely contact forces of the order of 1 μN, "writing" speeds up to 400 nm/s, and fixed bias voltages up to 5 V. Features of different planar morphologies can be generated with pre-programmed software designs, which guide the voltage-controlled XY lateral scanning of the tip while an electronic feedback-control keeps the probe-sample contact force constant. In our case, the sample rests on a XY piezo scanner stage that is equipped with strain-gauge sensors for overcoming piezoelectric hysteresis via another internal feedback control.[4] The tip is held by an independent piezoelectric z-stage, thus conveniently decoupling the sample's horizontal XY scanning motion from the probe's vertical z-displacements. Finally, the microscope is switched back to the tapping mode for topography imaging in order to verify whether the patterns have been formed.

Figure 8.5a shows two sequentially acquired images of the first polymer structures created in our labs by exploiting the responsive characteristics of P4VP. The cross-linked P4VP film swelled only in the areas where the phosphate "ink" was delivered, forming two narrow-line terraces. These two images demonstrate the ability for sequentially creating an initial pattern, then imaging the resulting sample topography with the same ink-loaded tip, and subsequently creating additional patterns. The line profile in Fig. 8.5a reveals that both features are approximately 100 nm in height and 500 nm in width. The relatively large dimensions of these features (compared to the finer patterns we have recently created in our laboratories) may be attributed to the relatively large amount of ink initially attached to the tip (prior to the writing process). That is, after dipping the tip into the buffer, no nitrogen was blown out in front of the "pen" to let it dry; the latter became afterwards a standard practice in our laboratory as a way to evaluate the reproducibility of the fabrication method under similar conditions as possible. The line profile also reveals the hydrogel characteristics of the patterns since the 100 nm constitutes a substantial swelling compared to the initial 80 nm thickness of the polymer film. Unfortunately, the reproducibility of these initial experiments was very poor from day to day, or month to month. At the time when the structures in Fig. 8.5a were created, no electric field was used in the experimental setting.

The reproducibility of the fabrication process greatly improves when an electric field is applied between the silicon tip and the silicon substrate (see setting in Fig. 8.4c). Typical bias voltages are up to 5 V, which when applied through a 50 nm film sets a strong electric field $\sim 10^6$ V/cm (still leaving the film apparently undamaged). Provided that the humidity is above 40%, the patterns are consistently fabricated with this PEN method. Figure 8.5b shows an 8 nm height "*U*" shape structure fabricated under the new procedure, by applying a 5 V bias voltage and under 1 μN contact force. The smaller height of the structures (compared to the ones in Fig. 8.5a) could be attributed to the minimal coating ink on the probe (nitrogen gas is blown on the tip right after dipping it into the buffer solution) thus a lower number of hydronium ions diffused through the polymer network. The lateral dimension of the line-features in both cases presented in Fig. 8.5a, b is limited by

[4] ibid.

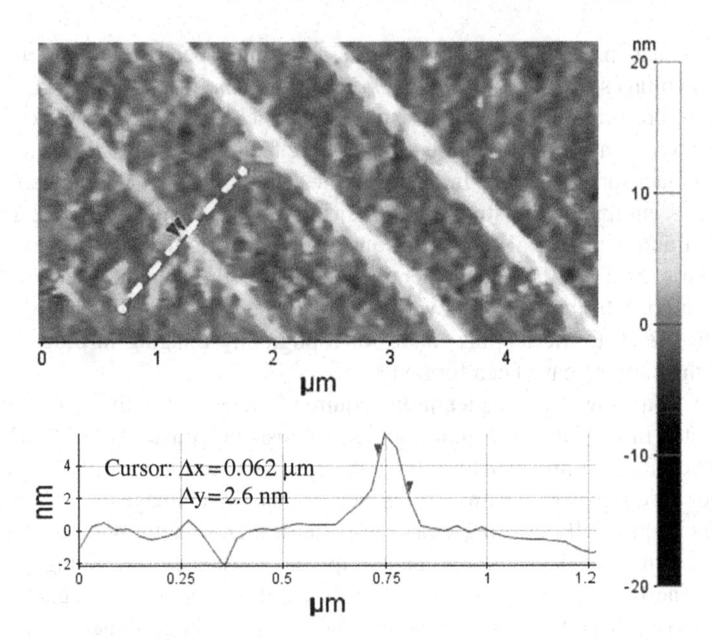

Fig. 8.6 Three line features fabricated at 45% humidity using (form *left* to *right*) contact forces of 0.6, 0.9, 1.0 μN and bias voltages of 5 V, 5 V, and 4 V respectively. The line profile across the line structure at the left displays a line feature ~6 nm tall and ~60 nm wide [41]

the diffusion of hydronium ions into the polymer network. Still, under smaller contact forces and control of the dwelling time, line features as thin as 60 nm can be created, as shown in Fig. 8.6 [41, 42].

The effect of applying an electric field appears straightforward in PEN (Fig. 8.7), helping drive the positive hydronium ions into the polymer network. To provide some context, here we mention other more sophisticated situations where the use of an electric field turns out to be also valuable. For example, an electric field is used to trigger specific conformational transitions switching between straight (hydrophobic) and bent (hydrophobic) states of low-density self-assembled monolayer (SAM) of 16-mercaptohexadecanoic acid on gold, which changes the wettability of the surface while maintaining unaltered the system's environment [43]. Also, nanometer-scale hydrogels, produced by surface grafting of a polymerizable monomer onto Au, undergo reproducible changes in thickness when a potential is applied across the film [44]. More strikingly, an infinitesimal change in electric potential across a polyelectrolyte gel produces volume collapse in polyelectrolyte gels [45]. A broader account on the effects of electric fields on gels has been given by Osada and Gong [33].

The ability of PEN to create patterns having a variety of vertical dimensions in the nanometer range can be capitalized to study molecular interactions in these hydrogel systems with very much detail, including, for example, phase transitions. Polymer gels are known to exist in two phases (swollen and collapsed) where the

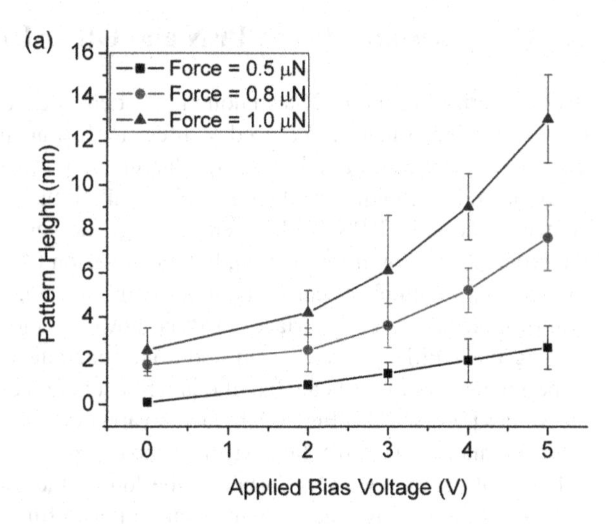

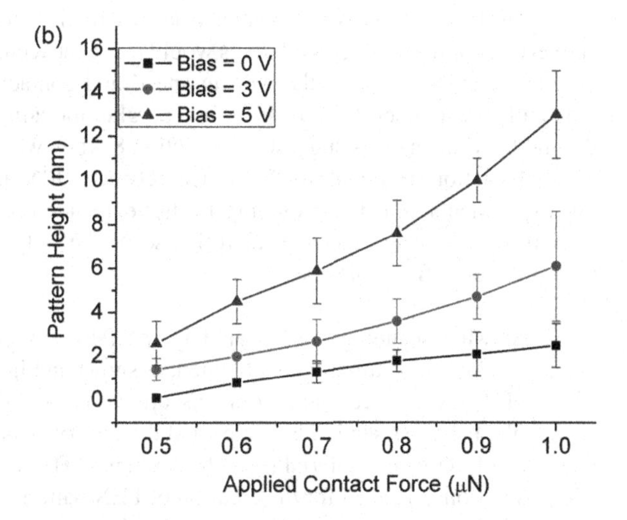

Fig. 8.7 (**a**) Pattern height (nm) vs. applied bias voltage (V) at various fixed contact forces. (**b**) The pattern height (nm) vs. applied contact forces (μN) under different constant bias voltages. Reprinted, with permission, from Wang et al. [42], Copyright @ 2010, American Chemical Society

volume transitions between the phases may occur either continuously or discontinuously [46, 47]. These transitions have been studied by monitoring the swelling as a function of different parameters (solvent quality, temperature [48], pH [49], visible light radiation) [50, 51], which occur as a result of a competition between intermolecular forces (repulsive forces are usually electrostatic whereas attraction is mediated by hydrogen bonding [52], van der Waals interactions, or radiation pressure [53]) that act to expand or shrink the polymer. However, these investigations do not disclose the microscopic view of the structure of gels. While neutron scattering has been used to elucidate these interactions at mesoscopic scales [54], here we have an opportunity to complement these studies with nanoscale-sized individual gels fabricated via PEN.

8.4 Comparison Between PEN and Other DPN Techniques

A characteristic of the PEN technique is the dependence of the pattern dimensions on the positive contact force used to press the tip against the polymer film. The greater the applied external force, the larger the pattern size. This differs greatly from the DPN technique in which the line-width of the pattern is independent of the contact force [55]. DPN works even under gentle negative external forces (where the probe pulls away from the film, but the adhesion forces keep the probe attached to the sample) simply because its underlying mechanism implies just the deposition of molecules on the surface. In PEN, however, the buffer molecules (initially coating the AFM-tip) have to, in addition, penetrate into the polymer substrate. Experimental results indicate that the latter occurs less efficiently with weaker contact forces [56]; see also Fig. 8.7. In fact, no apparent features are formed for contact forces smaller than 0.5 μN (at a writing speed of 80 nm/s), even when applying bias voltages of up to 5 V [42]. Certainly, the longer the dwelling time per pixel (i.e. slower writing speeds), the more hydronium ions diffuse into the polymer thus giving rise to larger patterns [56] which would also allow reducing the contact force. However, extremely long dwell times would make the technique less attractive, thus, a trade off exists between the writing speed and contact force for applying PEN efficiently. (Implementation of PEN in a parallel modality, in order to increase the throughput efficiency, is suggested in Section 8.5 below.)

Confirmation of a contrast difference between DPN and PEN is revealed from test experiments aimed at providing further evidence that the pattern formation in PEN results from the mechanical response of the polymer film and not just the deposition of buffer molecules:

First, when scanning a non-coated tip on a P4VP film under 3 μN contact force, no elevated features were obtained, as shown in Fig. 8.8a [56]. Only scratches of 2 nm deep resulted from this operation, as evidenced by the line profile; the cross section indicates that the observed small protuberances are the result removed material from the scratches. This result suggests that a buffer solution is needed for the creation of PEN patterns.

Second, an argument could be made about whether or not, when using a coated tip, the patterns could result from the physical deposition of buffer molecules. To refute this argument, this time the test experiment was carried out using a buffer-coated tip scanned at low speed over a non-responsive polymer, polystyrene. The result presented in Fig. 8.8b shows no pattern formation, thus providing favorable evidence that features created were not just the deposition of buffer molecules on a polymer surface, but the swelling response of the substrate itself when using responsive materials.

Third, the results given above also favor the hypothesis of protonation. But would the hydronium ions be the only ones diffusing into the polymer network? To test this hypothesis C. Maedler et al. investigated the swollen structures [57]. Armed with Kelvin Probe Force Microscopy as a tool [58], and following up studies of deposition of charges in silicon [57], they

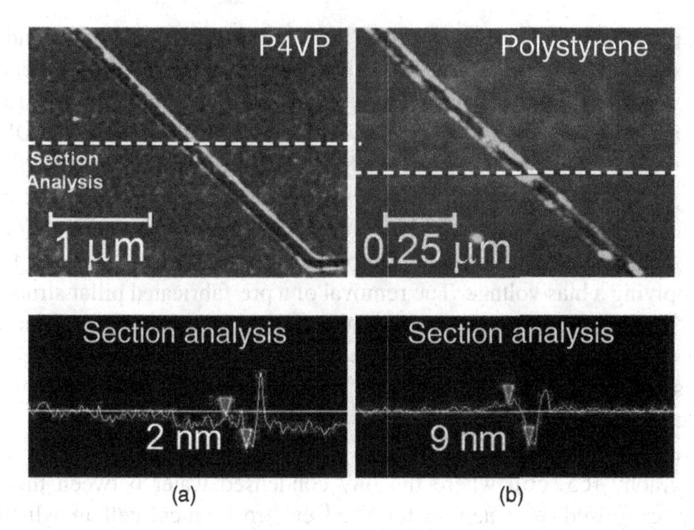

Fig. 8.8 (**a**) Scanning along a diagonal path with a contact force of 3 μN on a P4VP film, using an uncoated tip, produces only scratches, as revealed by the cross section line profile. (**b**) Image of the resulting topography produced by a line scan on a polystyrene film (a non responsive material) using a contact force of 2 μN and a coated tip. A cross section line profile reveals only a scratch indent in the substrate. Reprinted, with permission, from Maedler et al. [56], copyright 2008, American Institute of Physics

corroborated to have the sensitivity for monitoring the presence of a net charge in the polymer structures, if any. They found none. This implies that not only the hydronium ions penetrate the polymer, but they bring with them their corresponding counterions (see expression (17) above), thus keeping the charge neutrality of the sample.

DPN and PEN agree on the role played by the humidity. PEN pattern formation in P4VP films under different humidity levels has been well documented by C. Maedler [40, 56] and X. Wang [42]. No features formation occurs below 40% of relative humidity, suggesting that a water meniscus [59] naturally created between the probe and the polymer film plays a key role. In this regard, there is a coincidence of arguments with the DPN community that supports the hypothesis that a water meniscus is a fundamental component for the technique to work [60]. There is evidence that a liquid bridge exists even at zero humidity conditions [61]. (However, some authors have concluded that a meniscus is not involved in DPN experiments with certain molecules like 1-octadecanethiol [62]).

Finally, the reversibility of PEN contrasts with the, in general, non-reversible character of DPN. An exception to this notion is observed, however, in one DPN application where a voltage biased tip is used to create positive and negative charged nanopatterns on 1-hexadecanethiol (HDT) self-assembled monolayers (SAMs) on Au. The positive nanopatterns are gold oxide, which can be reduced back to gold by ethanol ink via DPN [63]. This development followed the "molecular eraser" [64] where a tip with a negative bias voltage (relative to the electrically grounded substrate that contain a monolayer of MHA 16-mercaptohexadecanoic acid) causes the molecules to desorb. Thus, regions of the alkanethiol can be selectively removed, which constitutes the pattern. The recessed areas can be refilled via DPN thus erasing the pattern. On the other hand, PEN instead is an intrinsically reversible process, using a base ink to erase the pillars previously formed by the acidic ink. The reversibility of the process was elegantly implemented by X. Wang et al. [42] using the same tip but inked with a basic phosphate buffer solution of pH 8.3, this time without applying a bias voltage. The removal of a pre-fabricated pillar structure was performed by keeping the tip in contact and stationary on top of the pillar. Results showed the progressive attenuation of the height of the structure. The basic buffer neutralizes the pyridinium converting it to the neutral pyridyl group and causing the film to "de-swell".

On the other hand, PEN differs from Electrochemical AFM Dip-Pen Nanolithography [65, 66], where the tiny condensed water between the tip and the substrate is used as a nanometer-sized electrochemical cell in which metal salts (coating the probe) are dissolved, reduced into metals electrochemically, and deposited onto the surface. This method was used first to deposit Pt on a silicon substrate. The process involves coating an AFM cantilever tip with H_2PtCl_6 and applying a positive voltage to the tip relative to the electrically grounded substrate (the latter constituted by silicon with its native oxide). Despite the oxide, the conductivity is sufficient for the reduction (electrons gain) of the precursor ions $PtCl_6{}^{2-}$. During the process, H_2PtCl_6 dissolved in the water meniscus is electrochemically reduced from Pt(IV) to Pt(0) metal at the (cathode) silicon surface and deposits as Pt according to the reduction reaction $PtCl_6{}^{2-} + 4e \rightarrow Pt + 6Cl^-$. (The DC voltages are kept in the 1–4 V; higher voltages would tend to oxidize the silicon substrate and form SiO_2 instead of Pt nanostructures.) Electrochemical AFM Dip-Pen Nanolithography is therefore a technique were the deposited materials constitute the nanostructure, similar to DPN but different than PEN.

PEN is also different from Constructive Nanolithography (CNL) [67, 68] and its similar Electro Pen Nanolithography (EPN) [69]. Contrasting Electrochemical AFM Dip-Pen Nanolithography (described above), whose working principle is based on *reducing* (gaining of electron) a precursor ion, CNL and EPN exploit an *oxidation* (loss of electron) process instead. In one application of CNL, a silicon substrate is first coated with a 2.5 nm thick monolayer of octadecyltrichlorosilane (OTS, an organic molecule with a methyl-terminated,18-carbon alkyl chain), which serves as the initial blanket surface to be oxidized on specific regions by the apex of a conductive AFM tip. The oxidation process is implemented by applying a positive bias voltage to the substrate relative to the tip (the latter kept at electrical ground). As a result, the oxidized areas (initially hydrophobic, neutral, and inert) become hydrophilic, negative-charged, and chemically active, forming the multi-functional templates for post, or in-situ, surface generation of organic (insulator), metal, or semiconductor nanocomponent features, including the implementation of a wetting-driven patterning [70, 71]. The post pattern generation process of CNL is elegantly alleviated in EPN where the electrically biased conductive AFM tip is also coated with ink (trialkoxysilane and quaternary ammonium salts), which are in situ transferred during the scanning process as the local oxidation occurs. The virtue of CNL and EPN is that the transfer of molecules occurs only on the regions that have been oxidized, which can be as small as 25 nm, thus bypassing the diffusion effect that limit the lateral resolution of DPN. CNL and EPN therefore have characteristics similar to DPN, hence different than PEN, for their patterns are constituted by anchoring molecules on the substrate, although a chemically modified substrate.

PEN is more alike to Chemical Lithography (ChemLith) [72]. This technique exploits the fact that photoresist materials change their solubility upon an acid-catalyzed chemical reaction. As an alternative to the diffraction limited photolithography method (where the photoacid generator is mixed in the resist formula and the acid is generated by photon-initiated reactions), ChemLith delivers the catalyzing acid proton source to the desired position on a negative resist film via either a nanoimprinting method or by using an sharp stylus. In a post bake step, the resist molecules are catalyzed to cross-link with each other, and thus become insoluble in the final development step. Thus ChemLith nanopatterns result from a local protonation process (similar to PEN) and a post baking step (the latter renders the process irreversible, unlike PEN).

As mentioned at the beginning of this chapter, PEN joins other efforts for developing responsive materials (including polymer nano-composites [73], soft hydrogels [3, 7] and chemically functionalized metal nanoparticles [15, 74], which may also be driven by proton sources as well as the use of bias voltages), but putting emphasis into the nanoscale size regime. This PEN development occurs in parallel to other electrochemistry SPM-based efforts for fabricating energy storage devices [75], and, more generally, for attaining reliable surface modification using SPM [76, 77].

8.5 Perspectives

We have introduced Proton-fountain Electric-field-assisted Nanolithography (PEN) as a technique for fabricating erasable nanostructures that closely mimic natural bioenvironments. Its distinct feature, contrasting other DPN-based techniques, resides in the fact that the pillar patterns are made of the substrate material itself (a polymer film). The pattern formation results from the delivery of hydronium ions through an acidic-fountain tip in contact with the substrate, causing the polymer film to swell at targeted locations. In addition, the patterns can be erased with a basic-ink loaded tip. Such a reversible character of the fabrication process could be relevant to a wide range of potential applications, ranging from microelectronics (the swollen and no-swollen states could represent the ones and zeros in a memory device) to biotechnologies (variable size gates that open and close compartment in micro- and nano-fluidic devices used for separation of molecules or chemical reagents). Further, its capability for controlling the dimension of the patterns with nanometer precision, via an electric field, offers an opportunity to use these films as testbed for studying fundamental (thermodynamic and kinetic) physical properties of responsive materials at the nanoscale level. This is concomitant to the far reaching goal towards developing thin film-based responsive materials that can selectively respond to a variety of external stimuli (mechanical, chemical, optical, and changes in environmental conditions).

Despite its current progress, PEN would benefit from further development in order to place the technique in more solid grounds. For example, it would be convenient to study with more detail the effects of humidity (at different levels) in conjunction with the application of a range of bias voltages; until now their effects have been reported only separately. The objective would be to understand, optimize, and to know better, the dynamics involved in the transportation of hydronium-ions from the tip to inside the polymer network. This study would also help to elucidate why relatively strong forces ($\sim\mu$N) are needed for a rapid pattern formation to occur in PEN. Apparently, the first monolayers of water [78, 79], found naturally adsorbed on the polymer film, is the first barrier that the ions from the tip face before rapidly diffusing into the polymer film. Functionalizing the responsive polymer film with, in turn, hydrophobic or hydrophilic layers would help to contrast this hypothesis. In passing, the latter proposed experiment could also lead to the formation of the thinnest line feature under DPN, for by writing on an hydrophobic-functionalized responsive-material film, only the region in contact with the tip would create a richer water content bridge for the hydronium ions to penetrate into the polymer network. Certainly the thinnest line width would be determined by the diffusion of the ions while inside the polymer; still, the degree of cross-link of the starting material polymer film could be used to influence this diffusion, which offers another variable to control the ultimate resolution in PEN.

To gain versatility on the PEN implementation, and with a perspective on application for building memory devices (the reversibility of the PEN allowing the implementation of memory level changes), it would be worth to explore the use of solid state sources of protons [72, 80, 81]. Such an application could benefit from current efforts for developing solid acid materials as electrolytes in fuel cells [82, 83]. In that direction of technological applications, it would be worth to also explore the implementation of PEN in a parallel format; that is, to develop capability for fabricating many patterns at once. For this application, PEN can capitalize on the fact that relatively strong forces are needed to fabricate the patterns, which provides some leverage to implement a stamping type fabrication modality. For example, metallic (master) features can be fabricated on a flat (glass) substrate, all of them interconnected as to be able to apply a bias voltage. After spin-coating a layer of buffer solution on the metallic master features, the resulting wet stamp would be pressed against a P4VP film. By applying a bias voltage, the protonation would be more effective on the regions defined by the metallic features, hence the patterns from the master stamp will be replicated onto the polymer substrate by its swelling reaction. Further, this methodology could become very versatile, since different metallic regions on the master stamp could be electrically addressed at will; thus a given mask would be used for fabricating different patterns according to a programmable voltage pattern.

Finally, it is worth to point out that the intrinsic hydrogel nature of the P4VP patterns fabricated with PEN constitutes an advantage in applications that involve the replication of structures that closely mimic natural bio-environments. This capability is particularly relevant to a recent trend in research that conceives the cell not only as a chemical factory but also as a sensitive mechanical device. Incidentally, it has recently been reported that stem cells do not regenerate efficiently in vitro environment unless the surrounding medium is made out of flexible gels [84]. Hence, further developments on PEN should benefit the implementation of these attractive bio-engineering applications.

References

1. K. Salaita, Y. Wang, and C. A. Mirkin, "Applications of dip-pen nanolithography," *Nat. Nanotechnol.* **2**, 145 (2007).
2. R. D. Piner, J. Zhu, F. Xu, S. Hong, and C. A. Mirkin, "Dip-Pen nanolithography," *Science* **283**, 661 (1999).
3. I. Tokarev and S. Minko, "Stimuli-responsive hydrogel thin films," *Soft Matter* **5**, 511 (2009).
4. L. Anson, "Membrane protein biophysics," *Nature* **459**, 343 (2009).
5. C. Ainsworth, "Stretching the imagination," *Nature* **456**, 696 (2008).
6. B. Bhushan, "Biomimetics: lessons from nature-an overview," *Phil. Trans. R. Soc. A* **367**, 1445 (2009).
7. I. Tokarev, M. Motornov, and S. Minko;, "Molecular-engineered stimuli-responsive thin polymer film: a platform for the development of integrated multifunctional intelligent materials," *J. Mater. Chem.* **19**, 6932 (2009).
8. C. Cofield, "Cell is mechanical device," *Am. Phys. Soc., APS News*, Series II, **19**, 4 (June 2010).
9. C. Wu, Y. Li, J. H. Haga, R. Kaunas, J. Chiu, F. Su, S. Usami, and S. Chien, "Directional shear flow and Rho activation prevent the endothelial cell apoptosis induced by micro patterned anisotropic geometry," *PNAS* **104,** 1254 (2007).
10. C. S. Chen, M. Mrksich, S. Huang, G. M. Whitesides, and D. E. Ingber, "Geometric control of cell life and death," *Science* **276**, 1425 (1997).
11. M. A. Greenfield, J. R. Hoffman, M. Olvera de la Cruz, and S. I. Stupp, "Tunable mechanics of peptide nanofiber gels," *Langmuir* **26**, 3641 (2010).
12. M. M. Stevens and J. H. George, "Exploring and engineering the cell surface interface," Science **310**, 1135 (2005).
13. S. Maeda, Y. Hara, T. Sakai, R. Yoshida, and S. Hashimoto, "Self-walking gel," *Adv. Mater.* **19**, 3480 (2007).
14. T. K. Tam, M. Ornatska, M. Pita, S. Minko, and E. Katz, "Polymer brush-modified electrode with switchable and tunable redox activity for bioelectronic applications," *J. Phys. Chem. C* **112**, 8438 (2008).
15. D. Wang, I. Lagzi, P. J. Wesson, and B. A. Grzybowski, "Rewritable and pH-sensitive micropatterns based on nanoparticle 'Inks'," *Small* **6**, 2114 (2010).
16. J. Ruhe, M. Ballauff, M. Biesalski, P. Dziezok, F. Grohn, D. Johannsmann, N. Houbenov, N. Hugenberg, R. Konradi, S. Minko, M. Motornov, R. R. Netz, M. Schmidt, C. Seidel, M. Stamm, T. Stephan, D. Usov, and H. Zhang, "Polyelectrolyte brushes" in: "Polyelectrolytes with Defined Molecular Architecture," M. Schmidt Ed. *Adv. Polym. Sci.* **165**, 79 (2004).
17. C. Liu, H. Qin, and P. T. Mather, "Review of progress in shape-memory polymers," *J. Mater. Chem.* **17**, 1543 (2007).
18. R. Yoshida, K. Uchida, Y. Kaneko, K. Sakai, A. Kikuchi, Y. Sakurai, and T. Okano, "Comb-type grafted hydrogels with rapid deswelling response to temperature changes," *Nature* **374**, 240 (1995).

19. T. Suzuki, S. Shinkai, and K. Sada, "Supramolecular crosslinked linear poly(trimethylene iminium trifluorosulfonimide) polymer gels sensitive to light and thermal stimuli," *Adv. Mater.* **18**, 1043 (2006).
20. B. A. Evans, A. R. Shields, R. Lloyd Carroll, S. Washburn, M. R. Falvo, and R. Superfine, "Magnetically actuated nanorod arrays as biomimetic cilia," *Nano Lett.* **7**, 1428 (2007).
21. O. Ikkala and G. Brinke, "Functional materials based on self assembly of polymeric supramolecules," *Science* **295**, 2407 (2002).
22. M. Kaholek, W. K. Lee, B. LaMattina, K. C. Caster, and S. Zauscher, "Fabrication of stimulus-responsive nanopatterned polymer brushes by scanning-probe lithography," *Nano Lett.* **4**, 373 (2004).
23. X. Liu, S. Guo, and C. A. Mirkin, "Surface and site-specific ring-opening metathesis polymerization initiated by dip-pen nanolithography," *Angew. Chem. Int. Ed.* **42**, 4785 (2003).
24. Y. Okawa and M. Aono, "Nanoscale control of chain polymerization," *Nature* **409**, 683 (2001).
25. S. P. Sullivan, A. Schnieders, S. K. Mbugua, and T. P. Beebe Jr., "Controlled polymerization of substituted diacetylene self-organized monolayers confined in molecule corrals," *Langmuir* **21**, 1322 (2005).
26. Y. Okawa, D. Takajo, S. Tsukamoto, T. Hasegawa, and M. Aono, "Atomic force microscopy and theoretical investigation of the lifted-up conformation of polydiacetylene on a graphite substrate," *Soft Matter* **4**, 1041 (2008).
27. P. Calvert, "Hydrogels for soft machines," *Adv. Mater.* **21**, 743 (2009).
28. P. J. Flory and J. Rehner, "Statistical mechanics of cross-linked polymer networks I. Rubberlike elasticity," *J. Chem. Phys.* **11**, 512 (1943).
29. P. J. Flory and J. Rehner, "Statistical mechanics of cross-linked polymer networks II. Swelling," *J. Chem. Phys.* **11**, 521 (1943).
30. K. Huang, "Statistical Mechanics," Wiley, New York, 2nd Ed. (1987).
31. F. P. Chinard and T. Enns, "Osmotic pressure," *Science* **124**, 472 (1956).
32. P. J. Flory, "Principles of Polymer Chemistry," Cornell University Press, Oxford (1969).
33. Y. Osada and J. Gong, "Soft and wet materials: Polymer gels," *Adv. Mater.* **10**, 827 (1998).
34. L. D. Landau and E. M. Lifshitz, "Statistical Physics", Elsevier, New York, 3rd Ed., Part 1, pp. 72, 267 (2006).
35. P. J. Flory, "Thermodynamics of high polymer solutions," *J. Chem. Phys.* **10**, 51 (1942).
36. P. J. Flory, "Statistical mechanics of swelling of network structures," *J. Chem. Phys.* **18**, 108 (1950).
37. W. Hu and D. Frenkel, "Lattice-model study of the thermodynamic interplay of polymer crystallization and liquid–liquid demixing," *J. Chem. Phys.* **118**, 10343 (2003).
38. A. Yu. Grosberg and A. R. Khokhlov, "Giant Molecules. Here, There and Everywhere," Academic Press, New York (1997).
39. D. Woo, "Spectroscopic Ellipsometry Studies of Polymers on Silicon Wafer," Thesis for the Master degree in Physics, Portland State University, Portland, OR (2009).
40. C. Maedler, H. Graaf, S. Chada, M. Yan, and A. La Rosa, "Nano-structure formation driven by local protonation of polymer thin films", *Proc. SPIE*, **7364**, 736409-1 (2009).
41. X. Wang, "Characterization of Mesoscopic Fluid-Like Films with the Novel Shear-Force/Acoustic Microscopy," Thesis for the Master degree in Physics, Portland State University, Portland, OR (2010).
42. X. Wang, X. Wang, R. Fernandez, L. Ocola, M. Yan, and A. La Rosa, "Electric field-assisted dip-pen nanolithography on poly(4-vinyl pyridine) films," *ACS Appl. Mater. Interface* **2**, 2904–2909 (2010).
43. J. Lahann, S. Mitragotri, T. Tran, H. Kaido, J. Sundaram, I. S. Choi, S. Hoffer, G. A. Somorjai, and R. Langer, "A reversibly switching surface," *Science* **299**, 371 (2003).
44. I. S. Lokuge and P. W. Bohn, "Voltage-tunable volume transitions in nanoscale films of poly(hydroxyethyl methacrylate) surfaces grafted onto gold," *Langmuir* **21**, 1979 (2005).

45. T. Tanaka, I. Nishio, S. Sun, and S. Ueno-Nishio, "Collapse of gels in an electric field," *Science*, **218**, 467 (1982).
46. M. Annakaand and T. Tanaka, "Multiple phases of polymer gels," *Nature* **355**, 430 (1992).
47. A. Matsuyama, "Volume phase transitions of smectic gels," *Phys. Rev. E* **79**, 051704 (2009).
48. W. Xue, and I. W. Hamley, "Thermoreversible swelling behaviour of hydrogels based on *N*-isopropylacrylamide with a hydrophobic comonomer," *Polymer* **43**, 3069 (2002).
49. A. Richter, G. Paschew, S. Klatt, J. Lienig, K. Arndt, and H. P. Adler, "Review on hydrogel-based pH sensors and microsensors," *Sensors* **8**, 561 (2008).
50. T. Tanaka, D. Fillmore, S.-T. Sun, I. Nishio, G. Swislow, and A. Shah, "Phase transitions in ionic gels," *Phys. Rev. Lett.* **45**, 1636 (1980).
51. A. Suzuki and T. Tanaka, "Phase transition in polymer gels induced by visible light," *Nature* **346**, 345 (1990).
52. F. Ilmain, T. Tanaka, and E. Kokufuta, "Volume transition in a gel driven by hydrogen bonding," *Nature* **349**, 400 (1991).
53. S. Juodkazis, N. Mukai, R. Wakaki, A. Yamaguchi, S. Matsuo, and H. Misawa, "Reversible phase transitions in polymer gels induced by radiation forces," *Nature* **408**, 178 (2000).
54. M. Shibayama and T. Tanaka, "Small-angle neutron scattering study on weakly charged temperature sensitive polymer gels," *J. Chem. Phys.* **97**, 6842 (1992).
55. S. Hong and C. A. Mirkin, "A nanoplotter with both parallel and serial writing capabilities," *Science* **288**, 1808 (2000).
56. C. Maedler, S. Chada, X. Cui, M. Taylor, M. Yan, and A. La Rosa, "Creation of nanopatterns by local protonation of P4VP via dip pen nanolithography." *J. Appl. Phys.* **104**, 014311 (2008).
57. C. Maedler, "Applying Different Modes of Atomic Force Microscopy for the Manipulation and Characterization of Spatially Localized Structures and Charges," Diploma Thesis for the academic degree of Diplom Physiker; Faculty of Natural Sciences Institute of Physics, Chemnitz University of Technology (2009).
58. M. Nonnenmacher, M. P. O'Boyle, and H. K. Wickramasinghe, "Kelvin probe force microscopy," *Appl. Phys. Lett.* **58**, 2921 (1991).
59. M. Schenk, M. Futing, and R. Reichelt, "Direct visualization of the dynamic behavior of water meniscus by scanning electron microscopy," *J. Appl. Phys.* **84**, 4880 (1998).
60. L. M. Demers, D. S. Ginger, Z. Li, S.-J. Park, S.-W. Chung, and C. A. Mirkin, "Direct patterning of modified oligonucleotides on metals and insulators by dip-pen nanolithography," *Science* **296**,1836 (2002).
61. S. Rozhok, P. Sun, R. Piner, M. Lieberman, and C. A. Mirkin, "AFM study of water meniscus formation between an aFM tip and NaCl substrate," *J. Phys. Chem. B* **108**, 7814 (2004).
62. P. E. Sheehan and L. J. Whitman, "Thiol diffusion and the role of humidity in 'dip pen nanolithography'," *Phys. Rev. Lett.* **88**, 156104 (2002).
63. Z. Zheng, M. Yang, and B. Zhang, "Reversible nanopatterning on self-assembled monolayers on gold," *J. Phys. Chem. C* **112**, 6597 (2008).
64. J.-W. Jang, D. Maspoch, T. Fujigaya, and C. A. Mirkin, "A 'molecular eraser' for dip-pen nanolithography," *Small* **3**, 600 (2007).
65. Y. Li, B. W. Maynor, and J. Liu, "Electrochemical AFM 'Dip-pen' nanolithography," *J. Am. Chem. Soc.* **123**, 2105 (2001).
66. F. C. Simeone, C. Albonetti, and M. Cavallini, "Progress in micro- and nanopatterning via electrochemical lithography," *J. Phys. Chem. C* **113**, 18987 (2009).
67. R. Maoz, E. Frydman, S. R. Cohen, and J. Sagiv, "'Constructive nanolithography': inert monolayers as patternable templates for in-situ nanofabrication of metal-semiconductor-organic surface structures: a generic approach," *Adv. Mater.* **12**, 725 (2000).
68. Z. Zheng, M. Yang, and B. Zhang, "Constructive nanolithography by chemically modified tips: nanoelectrochemical patterning on SAMs/Au," *J. Phys. Chem. C* **114**, 19220 (2010).
69. Y. Cai and B. M. Ocko, "Electro pen nanolithography," *J. Am. Chem. Soc.* **127**, 16287 (2005).
70. D. Chowdhury, R. Maoz, and J. Sagiv, "Wetting driven self-assembly as a new approach to template-guided fabrication of metal nanopatterns," *Nano Lett.* **7**, 1770 (2007).

71. A. Zeira, D. Chowdhury, S. Hoeppener, S. T. Liu, J. Berson, S. R. Cohen, R. Maoz, and J. Sagiv, "Patterned organosilane monolayers as lyophobic–lyophilic guiding templates in surface self-assembly: monolayer self-assembly versus wetting-driven self-assembly," *J. Langmuir* **25**, 13984 (2009).
72. P. Yao, G. J. Schneider, J. Murakowski, and D. W. Prather, "Chemical lithography," *J. Vac. Sci. Technol. B* **24**, 2553 (2006).
73. K. Shanmuganathan, J. R. Capadona, S. J. Rowan, and C. Weder, "Stimuli-responsive mechanically adaptive polymer nanocomposites," *ACS Appl. Mater. Interface* **2**, 165 (2010).
74. R. Klajn, P. J. Wesson, K. J. M. Bishop, and B. A. Grzybowski, "Writing self-erasing images using metastable nanoparticle 'Inks'," *Angew. Chem. Int. Ed.* **48**, 7035 (2009).
75. S. V. Kalinin and N. Balk, "Local Electrochemical functionality in energy storage materials and devices by scanning probe microscopies: status and perspectives," *Adv. Mater.* **22**, E193–E209 (2010).
76. A. A. Tseng, A. Notargiacomo, and T. P. Chen, "Nanofabrication by scanning probe microscope lithography: a review," *J. Vac. Sci. Technol. B* **23**, 877 (2005).
77. A. A. Tseng, A. Notargiacomo, T. P. Chen, and Y. Liu, "Profile uniformity of overlapped oxide dots induced by atomic force microscopy," *J. Nanosci. Nanotechnol.* **10**, 4390 (2010).
78. K. B. Jinesh and J. W. M. Frenken, "Capillary condensation in atomic scale friction: how water acts like a glue," *PRL* **96**, 166103 (2006).
79. P. J. Feibelman, "The first wetting layer on a solid," *Physics Today* **63**, 34 (2010).
80. E. Kapetanakis, A. M. Douvas, D. Velessiotis, E. Makarona, P. Argitis, N. Glezos, and P. Normand, "Hybrid organic–inorganic materials for molecular proton memory devices," *Org. Electron.* **10**, 711 (2009).
81. E. Kapetanakis, A. M. Douvas, D. Velessiotis, E. Makarona, P. Argitis, N. Glezos, and P. Normand, "Molecular storage elements for proton memory devices," *Adv. Mater.* **20**, 4568 (2008).
82. S. M. Haile, D. A. Boysen, C. R. Chisholm, and R. B. Merle, "Solid acids as fuel cell electrolytes," *Nature* **410**, 910 (2001).
83. T. Norby, "The promise of protonics," *Nature* **410**, 877 (2001).
84. P. M. Gilbert, K. L. Havenstrite, K. E. G. Magnusson, A. Sacco, N. A. Leonardi, P. Kraft, N. K. Nguyen, S. Thrun, M. P. Lutolf, and H. M. Blau, "Substrate elasticity regulates skeletal muscle stem cell self-renewal in culture," *Science* **329**, 1078 (2010).

Chapter 9
Development of High-Throughput Control Techniques for Tip-Based Nanofabrication

Haiming Wang and Qingze Zou

Abstract In this chapter, we will discuss recent developments of advanced control techniques for nanoscale precision motion in general and probe-based nanofabrication (PBN) in specific. First, from the control perspective viewpoint, the advantages and challenges in parallel-probe based approach will be discussed to clarify the needs of high-speed PBN, particularly for areas such as nanoscale rapid prototyping and self-assembly based nanomanufacturing using chemical evaporation deposition (CVD). Then secondly, control challenges encountered in high-speed PBN will be discussed to introduce three main approaches to address these challenges: the robust-control based approach, the system-inversion based approach, and the iterative control approach. The basic idea and the main results obtained in these three approaches will be comparatively discussed. We finish our discussion with a few remarks.

Keywords High-speed probe-based nanofabrication · Piezoelectric actuator nanopositioning control · Stable-inversion · Preview-based control · Iterative learning control · Robust control · Feedforward-feedback control · Cross-axis dynamics coupling · Mechanical-scratching

Abbreviations

CCF-ILC	Current-cycle-feedback ILC
CVD	Chemical evaporation deposition
DOF	Degree-of-freedom
HOPG	Highly oriented pyrolytic graphite
IIC	Inversion-based Iterative Control
ILC	Iterative learning control
LQR	Linear quadratic optimization
MAIIC	Multi-axis Inversion-based Iterative Control
MIIC	Model-less Inversion-based Iterative Control
MIMO	Multi-input-multi-output
PBN	Probe-based nanofabrication

Q. Zou (✉)
Department of Mechanical and Aerospace Engineering, Rutgers, The State University of New Jersey, Piscataway, NJ 08854, USA
e-mail: qzzou@rci.rutgers.edu

A.A. Tseng (ed.), *Tip-Based Nanofabrication*, DOI 10.1007/978-1-4419-9899-6_9,

PID	Proportional-Integral-Derivative
SISO	Single-input-single-output
SPM	Scanning probe microscope
STM	Scanning tunneling microscope

9.1 Introduction: High-Throughput PBN – A Control Perspective

Compared to other nanofabrication techniques, PBN has its own unique advantages. Unlike conventional micro-/nano-patterning techniques based on photolithography that are expensive, restricted to lateral patterning and semiconductor materials only, and not compatible with fragile materials including organic materials (and biological materials in specific) [1, 2], PBN provides a relatively low cost, inherently simple and reliable approach to fabricate nanoscale patterns and structures with a wide variety of materials (e.g., [1–5]). Numerous PBN methods have been developed [1, 4, 5] that utilize various mechanisms to fabricate/modify surface patterns and structures, physically and/or chemically. Particularly, PBN is unique in fabricating functional structures at local sites with unprecedented spatial resolution. Moreover, PBN on platforms like scanning probe microscope (SPM) also combines fabrication and characterization function together to allow convenient in situ and online pattern/structure fabrication and characterization [6]. These advantages combined demonstrate that PBN has great potentials as a unique nanofabrication tool.

9.1.1 The Need for Nanopositioning in Probe-Based Nanofabrication

Precision positioning is central to probe-based nanofabrication, particularly to migrate this technology from the early proof-of-concept stage to the product stage in industrial applications. As almost all existing PBN processes are based on x-y-z 3-degree-of-freedom (3DOF) motions, the precision positioning is needed in both the lateral x-y axes and the vertical z-axis motions.

9.1.1.1 Importance of Lateral x-y Axes Nanopositioning in PBN

One of the most important objectives in PBN processes is to drive, with submicrometer to nanometer accuracy, the probe to follow a given desired path on lateral plan. Fail to track the desired path, thereby, directly results in distortions in the nanopattern fabricated. As an illustrative example, Fig. 9.1 shows a star pattern fabricated through mechanical scratch on a gold-coated silicon sample using an AFM probe. Clearly, due to the lateral positioning errors, the fabricated star pattern is largely distorted from the desired shape. As a result, these positioning-error-caused pattern distortions will affect the performance and function, and even lead to malfunction and failure of the final nano-devices (if large).

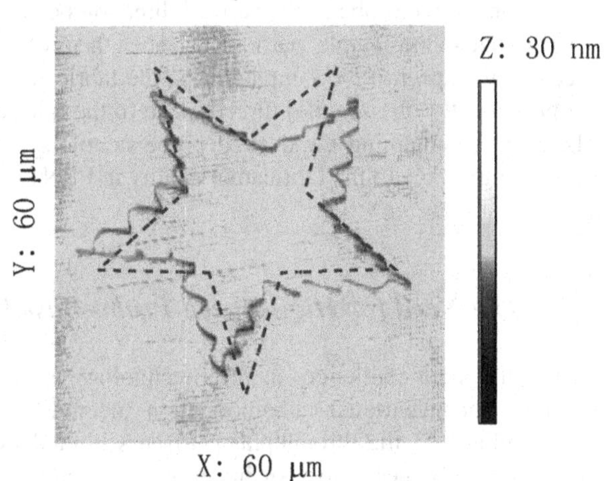

Fig. 9.1 A star pattern fabricated through mechanical scratch on a gold coated silicon sample using an AFM probe at high-speed (average lateral speed: 4.5 mm/s.) when the adverse effects on high-speed PBN process were not compensated for, where the desired star pattern is denoted by *dashed-line* [7]

We also note, compared to other non-probe-based nanofabrication techniques, one of the main advantages of PBN technique is its ability to fabricate nanoscale patterns of almost any arbitrary shape at relatively low cost. The low-cost is particularly attractive when only a small amount of copies are needed (for example, for test and evaluation during the design stage). On the contrary, the cost of other nonprobe-based nanofabrication processes directly depends on the complexity of the pattern/devices to be fabricated. For example, in the lithography technique, the cost of mask dramatically increases with the complexity of the mask itself. Thus, maintaining the lateral *x*-*y* axes precision positioning of the probe with respect to the sample is critical in PBN processes.

9.1.1.2 Importance of Vertical *z*-Axis Nanopositioning in PBN

The importance of vertical *z*-axis precision positioning to PBN processes lies in the fact that the use of probe opens the door to numerous physical and chemical effects for local surface modification (e.g., [1, 4]), almost all of these effects largely depend on the probe-sample distance (spacing). For example, when using the laser-assisted probe-based nanofabrication (e.g., [8, 9]), both the thermal expansion and the radiation effects of the probe involved are very sensitive to the probe-sample distance. When using tunneling-current-induced local oxidization of materials (such as silicon) for nanofabrications (e.g., [10, 11]), the tunneling current induced is exponentially dependent on the probe-sample spacing. Therefore, the variation of the tip-sample spacing directly results in the irregularity of the width and the depth in the pattern fabricated. As these nanopatterns are either used as parts for building nano-devices or as masks for following fabrications, these undesired pattern variations directly affect the intended functions and performance. Moreover, precision positioning in the vertical *z*-axis is also needed to prevent the modification or damage of the probe, the sample, or both during the PBN processes. When the oscillations of the probe in the vertical direction become large, the probe can crash onto or poke in the sample, particularly when the probe is moving at relatively high-speed with respect to the sample during the fabrication process. Such damages (of the probe and/or the sample) directly lead to the failure of the fabrication process. Therefore, maintaining the desired probe-sample spacing (i.e., the desired z-axis positioning) is key to the fabrication quality in PBN processes.

9.1.2 The Need for High-Speed Probe-Based Nanofabrication

One of the main challenges in PBN technology is its low throughput. Compared to other unconventional nanofabrication methods such as nanoimprinting and nanomolding [2], the throughput of PBN is limited by its serial nature of operation – the probe needs to visit in serial each sample location where the required local modification is needed. High throughput, with no loss of fabrication quality and reliability, is the key to transforming PBN from technique stunts in individual laboratory to efficient tools in industries [1, 2].

High-speed operation is needed in PBN technique for practical applications. The efforts to increase the throughput of PBN can be categorized into two groups – the parallel-probe approach and the increase of operation speed [2, 6, 12]. Central to the parallel-probe approach is the use of an array of probes (one or two dimensions) to simultaneously fabricate numerous copies of one pattern or structure. Such a parallel-probe concept has been realized in exemplary work such as the millipede system of IBM [13, 14] and the parallel-probe "dip-pen" system [15]. In the millipede project, it has been demonstrated that millions of "dots" can be reliably written or read simultaneously [13, 14]. Massive fabrication of more complicated patterns/structures using millipede, however, has yet been demonstrated. Although parallel fabrication of more complicated patterns has been achieved with the "dip-pen" technique [15], the complexity of the entire system has to be substantially increased – to realize the parallel-tip mechanism, particularly when each tip needs to be individually addressed/manipulated. The increase of system complexity inevitably results in cost increase and reliability and robustness issues. Moreover, despite the great efforts that have been made, maintaining fabrication quality and uniformity among all the patterns/structures fabricated by the parallel probes still remains as a daunting challenge. The challenge arises from the need to maintain the uniformity among all the probes, including the uniformity of not only the physical characteristics, but also the actuation and sensing of all the probes. Moreover, challenge to maintain the required uniformity among the fabricated patterns also comes from the lack of highly implementable control techniques that can efficiently compensate for the probe variations. Additionally, the operation of the parallel "dip-pen" system is still slow. Therefore, limits exist in parallel probe approach to achieve the desired high-throughput fabrication in PBN.

High-speed operation is not only complement to parallel-probe approach for high-throughput PBN, but also particularly desirable in applications where parallel-probe approach is not suitable. Note that techniques to achieve high-speed operation in single-probe PBNs are applicable to parallel-probe PBNs, as challenges to achieve high-speed operations in single-probe PBNs also exist in parallel-probe ones. Moreover, the power of parallel-probe fades in areas such as nanoscale rapid prototyping [16, 17] as only a small number of copies are needed for characterization and evaluation [16]. Thus, high-speed PBNs, particularly high-speed single-probe PBNs, are essential to accelerate the entire synthesis and design process. Another area where single-probe rather than parallel-probe PBN is needed is the pattern guided self-assembly nanofabrication [2, 4, 18, 19], where PBN technique is utilized to plant the "paths" or "seeds" on the substrate a priori to guide the growth of nano-patterns/structures. As currently such a planting process is slow and becomes the bottleneck of the entire assembly process, high-speed PBNs (not parallel-probe) are crucial to the efficiency of the entire fabrication process. Therefore, high-speed operation is indispensable to arriving at high-throughput PBN.

9.1.3 Challenges to High-Speed PBN

The fabrication speed of PBN is limited by the oscillation of the probe relative to the sample surface that can be excited at high-speed. Challenges, thereby, lie in the need to avoid such probe oscillations with no loss of fabrication range and/or quality (i.e., positioning precision).

9.1.3.1 Challenges in Nanopositioning Control for PBN

Currently, piezoelectronic actuators are widely used for nanopositioning control in areas including PBN. Therefore, nanopositioning in high-speed PBN not only shares similar challenges in controlling piezoeletrical actuators for precision positioning as in other applications, but also possesses issues especially for the probe-based operation itself.

Nanopositioning Challenges in General

Piezoeletric actuators allow precision motion to be achieved at nano- to atomic-level resolution. The precision motion, however, is limited by the adverse effects of piezo-actuators when the motion is at high-speed and over a relatively large operation range. These adverse effects include creep [20], vibration dynamics [21, 22], and hysteresis behavior [22, 23] of piezo-actuators. The creep effect occurs at relatively low speed operation (substantially lower than the bandwidth of the piezo-actuator), whereas when the operation is at high-speed, vibration dynamics of piezo-actuators can be easily excited and result in large probe-to-sample oscillations. The vibration dynamics of piezoactuators tends to be characterized by a low gain margin, as manifested by the sudden phase-drop accompanied with the sharp resonant peak(s) in the magnitude part in the Bode plot (see one example shown in Fig. 9.2). Probe-to-sample positioning errors can also be generated by the nonlinear hysteresis behavior of piezo-actuators, particularly when the operation range approaches to the full displacement size of the piezo-actuator, as demonstrated in Fig. 9.3. Additionally, the vibration and the hysteresis effects are augmented together and result in even larger positioning errors when the operation is at both high-speed and large range [20, 22, 24–26]. Readers are referred to [27] for a recent review on nanopositioning control issues.

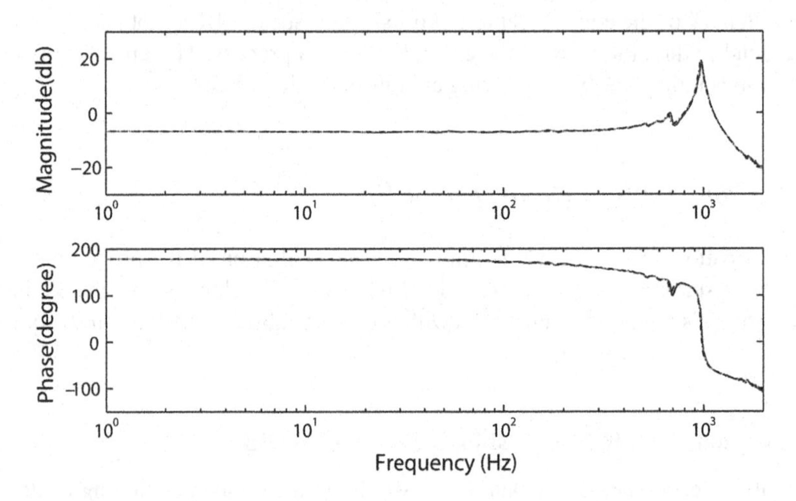

Fig. 9.2 The bode plot (frequency response) of a piezotube scanner measured in experiments

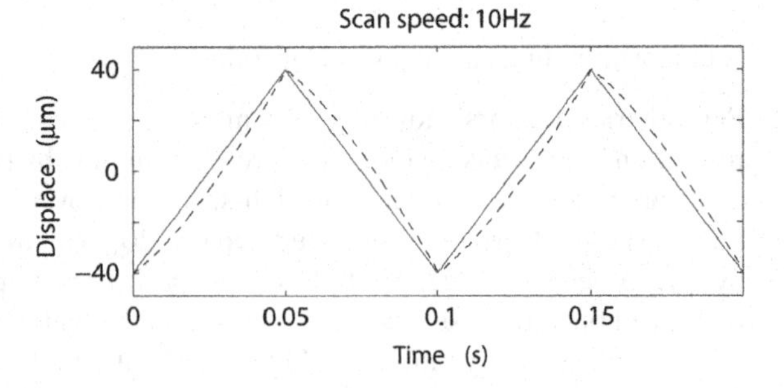

Fig. 9.3 The output tracking of a piezotube scanner with no hysteresis compensation (*dashed-line*), where the solid line denotes the desired triangle trajectory

PBN-Related Nanopositioning Control Issues

Key to achieving high-speed PBN is to account for the system dynamics changes and uncertainties. Although usually care has been taken to ensure that the PBN system is operated in a well-maintained environment (e.g., acoustic and external vibration are very small and can be ignored), the dynamics of PBN system can be altered from day-to-day operations due to actions as simple as replacing the probe (or changing the operation condition or sample material). The variation of PBN system dynamics also arises from factors such as the evolution of the materials (chemically or physically) during the fabrication process and the variations of the environment and the operation condition (e.g., thermal drift) [28]. Such operation-related dynamics changes tend to become more significant in the high frequency range, as illustrated by Fig. 9.4 where the Bode plots of a piezoelectric scanner under different conditions are shown. As a result, the positioning precisions during high-speed operations are limited. Another key issue is to account for the cross-axis coupling effects [29–31] such as the vertical motion of the probe caused by the lateral scanning. Such coupling effects are caused by the inevitable imperfectness of piezo-actuators [29], small misalignment of the sensor [7], and/or inherent coupling between motions in different axes (e.g., the "bowing" effect when using piezotube scanners to generate lateral scanning [32]). Moreover, the cross-axis coupling effect is dynamic dependent and becomes more serious during high-speed operations. This can be clearly seen from Fig. 9.5, where the x-to-z coupling dynamics becomes much larger around the resonant peak. Such large coupling dynamics can also result in extraneous oscillations of the probe relative to the sample.

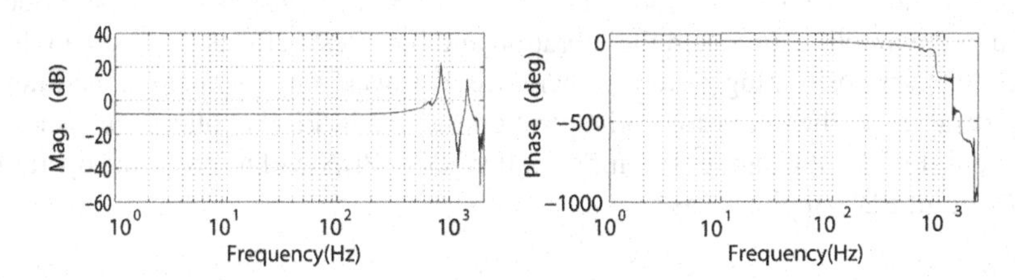

Fig. 9.4 The frequency responses of a piezotube scanner that were measured under different conditions (different input amplitude and offsets) [22]

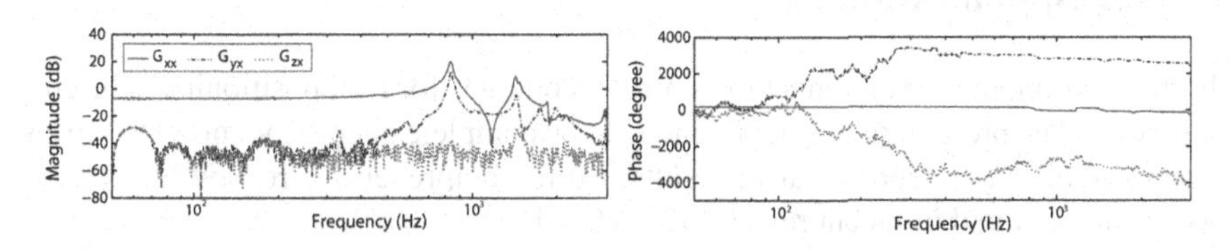

Fig. 9.5 The experimental measured frequency axis of a piezotube scanner for lateral x-axis positioning, along with the frequency responses of the coupling dynamics from the x-axis input to the y-axis and the z-axis output, respectively [87]

9.1.3.2 High-Bandwidth Approach to High-Speed Nanopositioning and Its Limits

The operation speed of PBN processes can be improved by substantially increasing the bandwidth of the nanopositioning system. The basic idea is that by using a piezo-actuator with much higher bandwidth (e.g., over two orders higher), i.e., the first resonant frequency of the piezo actuator becomes much higher, the PBN system can be moved at much higher speed without exciting the dynamics of the system. Such an approach has been explored for nanopositioning needs, particularly for high-speed SPM imaging [33–39]. As demonstrated in these works, the lateral scanning speed can be increased 10–100 folds when the bandwidth of the piezo-actuator was increased by similar amount (see a recent review article [40] for a relation between the scan rate and the first resonant frequency, based on an numerical fitting of recent published results).

The fundamental limit of the above approach to speed improvement via bandwidth-increase is that the displacement (operation) range is dramatically reduced. Such a limit is because in general, the physical size of the piezo-actuator has to be reduced to increase its bandwidth. For example, the first resonant frequency of a piezotube scanner is inversely proportional to the square of the length of the piezotube scanner. The reduction of piezo-actuator size inevitably results in the reduction of the displacement range. For piezotube scanners, the lateral scan range is inversely proportional to the first resonant frequency of the scanner (e.g., readers are referred to [40] for a detailed discussion). Currently the lateral scan range of the ultrahigh-speed SPMs above [33–38] is below 10 microns – such a small displacement range is not competent for most PBN operation needs. Although the displacement range can be increased (i.e., amplified) through the design of flexural-stage for the piezo-actuator [38, 41–43], positioning precision might be reduced through the amplification, and the vibration and hysteresis effects still exist in these extended nanopositioning systems. Therefore, hardware improvement alone cannot, in general, meet the needs for high-speed operations in PBN applications, whereas compensation for the nanopositioning challenges in PBN will improve the operation speed of all PBN applications.

9.2 Current Development of Control Techniques for Nanopositioning

In this section, the development of control techniques for nanopositioning is briefly reviewed. The presentation is not intended to be complete but instead, provide a brief overview of major representative results. More comprehensive review of relevant work can be found in recent reviews [24–26, 44].

9.2.1 Proportional-Integral-Derivative Control

9.2.1.1 PID Control for Nanopositioning

Proportional-Integral-Derivative (PID) control is probably still the most widely used control tool in industry, including the commercial nanopositioning systems. The PID feedback controller $G_c(s)$ can be generally written as (see Fig. 9.6)

$$G_c(s) = K_p \left(1 + \frac{s}{T_s} + T_d s\right) \quad (1)$$

where K_p, T_s, and T_d are the proportional, integral, and derivative gains, respectively. To account for other adverse effects such as noise and saturation problem, the above standard PID controller can be extended to include low-pass filters, as discussed in many undergraduate control textbooks (e.g., [45, 46]). The PID controller can be designed and implemented without modeling the system dynamics, while identification of system dynamics requires in-depth knowledge in dynamics and control from the user, and is still far from being automatic. Instead, protocols based on practical experiences have been developed to guide the tuning of the PID parameters (K_p, T_s, and T_d) through experiments (e.g., see textbook [45, 46]). Moreover, PID control is relatively robust against system dynamics uncertainty and adverse effects of noise and disturbances, and rich experience has been accumulated in the implementation of PID controllers. Therefore, PID control are still widely used in commercial nanopositioning systems for various submicro- to nano-sciences and engineering needs, including the PBNs (e.g., [47, 48]).

9.2.1.2 Limits of PID Control for High-Speed PBN

It has been well recognized [49–52] that PID control approach is limited by its inability to adequately compensate for the dynamics effects of piezoactuators during high-speed nanopositioning. Such a limit exists as the bandwidth of the PID nanopositioning control system is dictated by the small gain margin of piezo-actuators [49–52] and thereby becomes small, i.e., the allowable feedback gain – before the control system becomes unstable – is small. The gain margin of piezo-actuators is, in turn, limited by the rapid phase drop occurring around the resonant peaks of the piezo-actuator dynamics that are related to the lightly structure damping. Although the gain margin of piezo-actuators might be increased by using a notch-filter as a pre-filter [21, 23], the performance of PID control for high-speed tracking

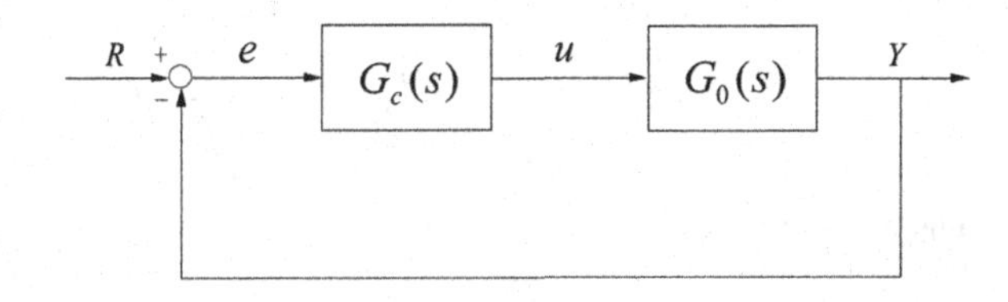

Fig. 9.6 The block diagram of feedback control

is still limited. Moreover, as a feedback control technique, PID approach is inherently causal, whereas even though the entire desired trajectory (specified by the pattern/device to be fabricated) usually is known a priori, such a knowledge of the desired trajectory cannot be fully exploited by a causal controller (i.e., a causal controller cannot utilize the future information of the desired trajectory and/or the tracking to adjust the control input at current time instant). These inefficacies of PID control for PBNs motivate the development of more advanced control tools, as reviewed below.

9.2.2 Robust-Control-Based Approach

Feedback controller design methodologies based on robust-control theory have been developed to address the limits of PID control for precision positioning in nanofabrication. H_∞-based robust control theory [53–55] provides a systematic frame work to incorporate the control performance, stability, and robustness against dynamics uncertainty and disturbances in the controller design. Particularly, the obtained controller has guaranteed robustness as the optimal feedback controller is sought to minimize the tracking error for given bound of system uncertainty (thereby "robust"). Next we briefly describe the design of robust feedback controller in the context of nanopositoning – to convey the main idea.

The H_∞-based controller can be designed by using the *mixed sensitivity* method (see textbook [53]). As schematically represented in Fig. 9.7, the feedback controller $G_c(s)$ is designed by imposing the performance, stability, and robustness requirements through the following norm criteria:

$$\left\| \begin{matrix} W_P(j\omega)S(j\omega) \\ W_T(j\omega)T(j\omega) \\ W_u(j\omega)G_c(j\omega)S(j\omega) \end{matrix} \right\|_\infty \leq 1 \tag{2}$$

where "$\|G(j\omega)\|_\infty$" denotes the H_∞ norm of the transfer function $G(j\omega)$ [53], (e.g., for single input single output systems, the H_∞ norm is the supermum of the transfer function gain over all frequencies), $S(j\omega)$ and $T(j\omega)$ are the sensitivity and complementary sensitivity of the closed-loop system, respectively,

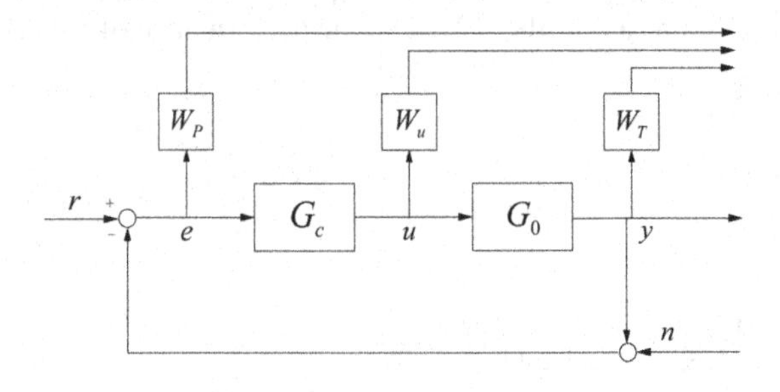

Fig. 9.7 The block diagram of robust control

$$S(j\omega) = \frac{1}{1 + G_c(j\omega)G_0(j\omega)} \qquad (3)$$
$$T(j\omega) = 1 - S(j\omega)$$

and $W_P(j\omega)$, $W_T(j\omega)$, and $W_u(j\omega)$ are frequency-dependent weighting functions, respectively. Specifically, the weighting function on the sensitivity, $W_P(j\omega)$, is designed to impose tracking performance requirement on the bandwidth, the weighting on the complementary sensitivity, $W_T(j\omega)$, is designed to impose robustness requirement against dynamics uncertainty and disturbances on the roll off of the loop transfer function at high frequency, and the weighting on the controller out, $W_u(j\omega)$, is designed to impose input amplitude requirement on saturation limit. The H_∞-based feedback controller can be designed by using commercially available tools such as Matlab (the robust control and μ-synthesis toolboxes). The design of these weighting functions demands the knowledge of the system dynamics and operation condition. Particularly, the nonlinear hysteresis effect of piezoactuators on nanopositioning control is treated in the H_∞-control as model uncertainty to the linear model of the system, $G_0(j\omega)$, thereby can be accounted for through the design of weighting functions. References [50, 51] present the design of H_∞-based feedback controller for single-axis nanopositioning, which has been extended to multi-axis nanpositioning recently in [56], and further, to the integration with feedforward control in the 2DOF-control design in [57]. Readers are also referred to Ref. [58] for a comparison study of H_∞-based robust control with other robustification controller design, and Ref. [44] for a recent review of this line of work.

9.2.2.1 Advantages of Robust Control for Nanopositioning

Recent works [50, 51, 58, 57] have demonstrated that the H_∞-based feedback control can significantly improve the control performance in nanopositioning applications over the PID control, particularly during high-speed motions. The H_∞-based robust control technique has been utilized to improve the SPM imaging speed in both the lateral scanning speed [50, 51, 58], and the tracking of sample topography in the vertical z-axis [57, 59, 60]. For PBN applications, similar approach has also been applied to the well-known IBM millipede systems [13, 14, 61]. In Refs. [62, 63], the robust control technique has been applied to the control of the thermal-sensing-actuation based 2D cantilever array for probe-based data storage applications. It is demonstrated that the H_∞-based robust control approach can effectively account for the sensor noise effect on the lateral x-y nanopositoning, the determining factor to the positioning accuracy of the system.

9.2.2.2 Limits of Robust-Control for High-Speed PBN

The performance of H_∞-based feedback control is governed by the so called *Bode's integral* [53, 64], which implies that a reduction of the feedback sensitivity in one frequency range must be paid off by an increase of the sensitivity in another range, regardless the type of feedback controllers. Thus, small sensitivity (i.e., good robustness) and large bandwidth (i.e., good tracking performance) cannot be achieved simultaneously, i.e., in feedback control the tracking precision must be traded-off with the robustness against dynamics uncertainty. Such a precision-robustness trade-off of feedback control can be alleviated by using feedforward control approach or augmenting an feedforward controller to the feedback loop in the so called two-degree-of-freedom (2DOF) control framework (see Section 9.2.5 later). However, the H_∞-based feedforward control [57] cannot fully exploit the a priori knowledge of the desired trajectory available in PBN applications – as the H_∞-based control technique is also inherently causal. Moreover, the tracking precision might be improved by re-designing the feedback controller whenever the dynamics of the PBN system changes significantly (i.e., the feedback controller does not need to be designed to have good robustness), however, such a re-design of feedback controller is not practical in PBN applications, as the H_∞ robust controller design requires in-depth knowledge of the user in robust control theory, and is still far from being automatic. On the contrary, the iterative learning control approach reviewed later (see Section 9.2.4) provides a much easier and more convenient method to account for the dynamics changes that utilizes the nature of repetitive operations involved in many PBN processes.

9.2.3 Inversion-Based Feedforward Control Approach

In the past decade, the inversion-based feedforward control techniques have been developed for nanopositioning applications (see recent review paper [40] and the references therein). The basic idea of the inversion-based techniques is rather straightforward: For a given desired trajectory to track (e.g., given by the desired geometry path in PBN), $y_d(\cdot)$, the needed control input (for tracking $y_d(\cdot)$ exactly) can be obtained by (see Fig. 9.8)

$$u_{ff}(\cdot) = G^{-1}[y_d(\cdot)], \tag{4}$$

where "$G^{-1}[\cdot]$" denotes the inverse of the input-to-output mapping $G[\cdot]: \; u(\cdot) \rightarrow y(\cdot)$. As the desired trajectory $y_d(\cdot)$ tends to be known a priori in PBN, and the mapping $G[\cdot]$ to model the system dynamics can be modeled experimentally, the inversion based feedforward approach is ideal to achieve high-speed PBN, provided that (1) the mapping $G[\cdot]$ is invertible, i.e., the inverse feedforward input $u_{ff}(\cdot)$ remains bounded; and (2) the mapping $G[\cdot]$ can be accurately modeled. These two constraints, as related to the nonminimum-phase characteristics of the system

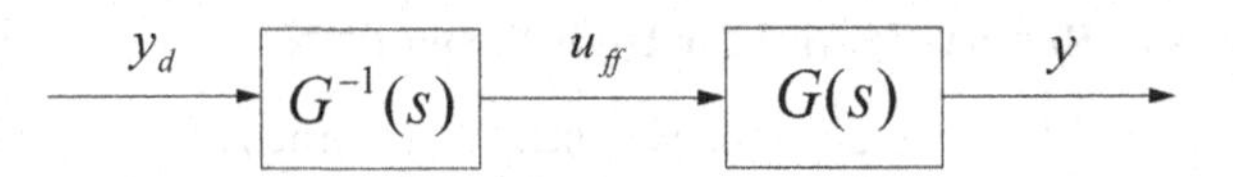

Fig. 9.8 The block diagram of inversion-based feedforward control

dynamics and the uncertainty effect on the system inverse, respectively (Note that the piezoactuator-based positioning system in PBN tends to have nonminimum-phase zeros [21, 51]), are addressed in the development of the stable-inversion theory.

9.2.3.1 The Stable-Inversion for Nonminimum-Phase Systems

Consider that the input-to-output mapping $G[\cdot]$ that models the PBN dynamics in state-space form as

$$\begin{aligned} \dot{x}(t) &= Ax(t) + Bu(t) \\ y(t) &= Cx(t), \end{aligned} \tag{5}$$

then the inverse input $u_{ff}(\cdot)$ (4) is obtained in two steps: (1) Apply the following state transformation $T: \Re^n \to \Re^n$ to (5),

$$\begin{bmatrix} \xi(t) \\ \eta_s(t) \\ \eta_u(t) \end{bmatrix} = Tx(t), \tag{6}$$

such that part of the system state in the transformed coordinate, $\xi(t)$, are the output and its derivatives

$$\xi(t) = \left[y_1, \dot{y}_1, \ldots, \frac{d^{r_1-1}y_1}{dt^{r_1-1}}, y_2, \dot{y}_2, \ldots, \frac{d^{r_2-1}y_2}{dt^{r_2-1}}, \ldots, y_p, \dot{y}_p, \ldots, \frac{d^{r_p-1}y_p}{dt^{r_p-1}} \right]^T, \tag{7}$$

where y_is are in $y = [y_1, y_2, \ldots, y_p]$, and r_is in $r = [r_1, r_2, \ldots, r_p]$ is the vector relative degree of system (5) (e.g., [65]); And (2) differentiate the output in (5) until the output appears and substitute the state transformation (6) into the system dynamics. With these two steps, the inverse feedforward input $u_{ff}(\cdot)$ can be obtained as

$$u_{ff}(t) = \mathrm{M}_Y Y_d(t) + \mathrm{M}_s \eta_s(t) + \mathrm{M}_u \eta_u(t), \tag{8}$$

where

$$Y_d(t) = \begin{bmatrix} \xi_d(t) \\ y_d^{(r)}(t) \end{bmatrix} \tag{9}$$

is the desired output and its derivatives with $\xi_d(t)$ as similarly defined in (7), and $\left[\eta_s(t) \; \eta_u(t) \right]^T$ are the stable and the unstable internal state, respectively. The dynamics of the internal state (called the internal dynamics) is obtained by representing system (5) in the transformed coordinate (6) and substituting the system state (8). Then the inverse input $u_{ff}(t)$ is obtained by solving the internal dynamics as follows,

$$\eta_s(t) = \int_{-\infty}^{t} e^{A_s(t-\tau)} B_s y_d(\tau) d\tau \tag{10}$$

$$\eta_u(t) = -\int_{t}^{\infty} e^{A_u(\tau-t)} B_u y_d(\tau) d\tau \tag{11}$$

Note that in the above (11), the bounded solution to the unstable internal dynamics is noncausal! Thus, for nonminimum-phase systems, the implementation of the stable-inversion technique in applications including high-speed PBN requires the entire trajectory to be completely specified – the scenario in majority PBN applications. For PBN applications where online pattern/structure modification is needed, a preview-based approach [66–68] has been developed that utilizes a previewed finite future desired trajectory to obtain the inverse input $u_{ff}(t)$. The basic concept is to truncate the noncausal integral in (11) to some finite preview time T_p. With such a truncation, the output will track the online-generated desired trajectory (specified by the online-generated desired pattern/structure) with a delay equaling to the preview time T_p. The amount of preview time T_p that guarantees the required tracking precision can be quantified [66, 67], and furthermore, can be minimized by using the previewed desired output within the preview-time window to optimally estimate the unknown unstable internal state at the future boundary time instant [68], $\eta_u(t+T_p)$. The readers are referred to [66–68] for details, and [69–71] for the extension to nonlinear systems.

9.2.3.2 Compensation for System Uncertainty Effect

The optimal-inversion technique accounts for the system dynamics uncertainties/variations by providing a frequency-wise trade-off between the tracking precision and the input energy in a linear quadratic optimization (LQR) framework. Specifically, the optimal inverse input $u_{opt}(\cdot)$ that minimizes the following cost function in frequency domain,

$$J(u(\cdot)) = \int_{-\infty}^{\infty} \{u^*(j\omega)R(j\omega)u(j\omega) + [y_d(j\omega) - y(j\omega)]^* Q(j\omega)[y_d(j\omega) - y(j\omega)]\} \tag{12}$$

is obtained as

$$u_{opt}(j\omega) = G^*(j\omega)Q(j\omega)[R(j\omega) + G^*(j\omega)Q(j\omega)G(j\omega)]^{-1} y_d(j\omega) \tag{13}$$

Where "*" denotes the complex conjugate transpose. It can be shown that the H_∞ robust feedforward controller is a special case of the optimal inverse controller restricted to the causal case (Readers are referred to [40] for details). Alternatively, the dynamics uncertainty effect on the inverse input can also be accounted for by seeking the inverse input that minimizes the uncertainty effect for given bound of dynamics uncertainty. This idea has been explored in the recently-developed robust inversion approach. The readers are referred to [70] for details. More generally, the dynamics uncertainty effect can be accounted for by combining the inversion-based feedforward control with a feedback control loop [21, 72]. Different configurations of the inversion-based feedforward-feedback control have been comparatively discussed in a recent review (see [40] and the references therein).

9.2.3.3 Implementation to Nanopositioning Applications

The inversion-based feedforward control approach has been successfully implemented by Prof. Devasia and his students in nanopositioning applications (see [40] and the references therein). Particularly, the optimal-inversion technique has been implemented to achieve high-speed lateral scanning in SPM-based imaging [20, 67, 73], the preview-based inversion approach to achieve precision online tracking has been demonstrated in [67] for nanopositioning. Figure 9.9 compares the scanning tunneling microscope (STM) image of a standard highly oriented pyrolytic graphite (HOPG) sample by using the preview-based optimal-inversion technique and that with no dynamics compensation. The distortion of the HOPG image in Fig. 9.9a visually demonstrated the vibration effect on the nanopositioning of the probe relative to the sample during high-speed scanning, whereas evidently the probe-vibration-caused distortion was substantially reduced with the application of the optimal-inversion technique. The integration of the inversion-based feedforward with feedback control method has also been illustrated in [21, 72]. As shown in Fig. 9.10, high-speed tracking of a periodic triangle trajectory (relative to the bandwidth of the piezo-actuation system) can be achieved by using the inversion-based feedforward-feedback control. Note that precision tracking of such a periodic trajectory is needed for lateral scanning in SPM imaging, as well as in many PBN applications. Finally, the inversion-based feedforward approach to compensate for the hysteresis effect has been presented in [20, 23].

9.2.4 Iterative Learning Control

In PBN applications, the desired trajectory – specified by the pattern/structure to be fabricated – tends to be known a priori, and furthermore, needs to be tracked

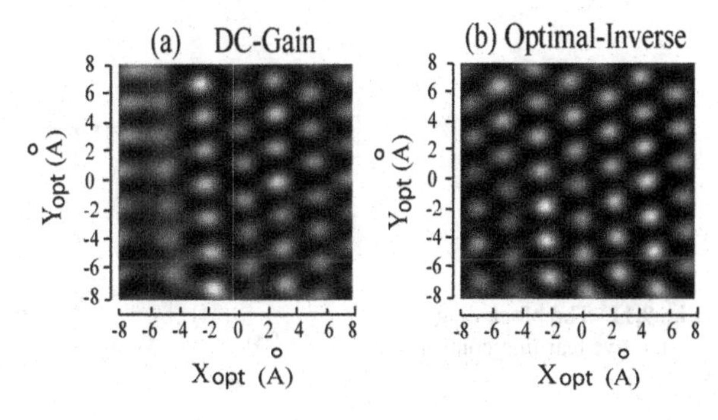

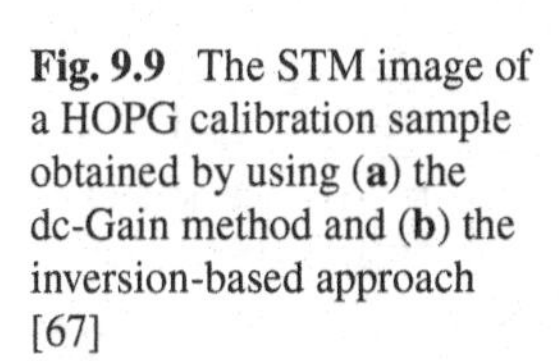
Fig. 9.9 The STM image of a HOPG calibration sample obtained by using (**a**) the dc-Gain method and (**b**) the inversion-based approach [67]

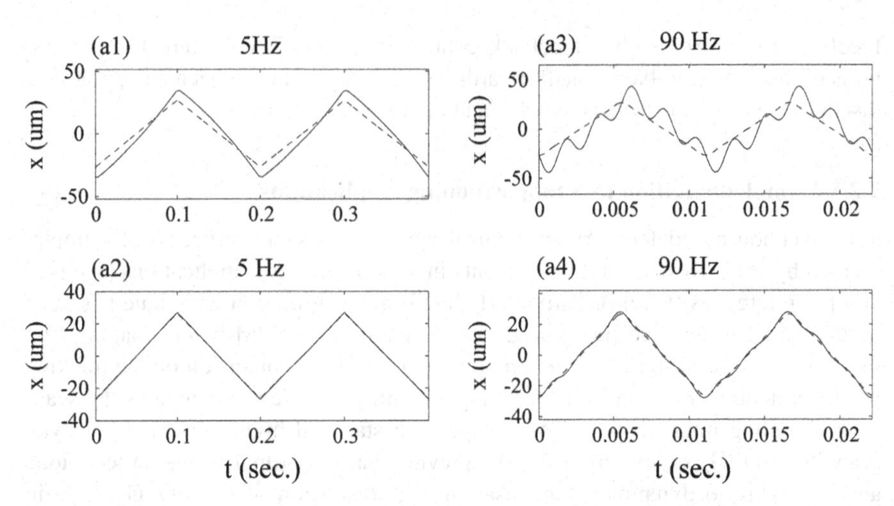

Fig. 9.10 The experimental tracking results by using (*top row*) the dc-Gain method and (*bottom row*) the inversion-based feedforward-feedback control technique at (*left column*) low speed and (*right column*) high-speed, where the frequency is the rate to scan one period of the triangle trajectory

repetitively (as numerous patterns/structures usually need to be fabricated). Such a pre-known desired trajectory and repetitive operation render the iterative learning control (ILC) an ideal choice for high-speed PBN: the pre-specified desired trajectory can be fully utilized in the ILC approach for precision tracking – the ILC technique can be noncausal, and the repetition allows the system dynamics and its variation to be compensated for through iteration, which otherwise are difficult to model and account for. These advantages of ILC approach have been demonstrated recently in high-speed PBN applications [7, 74].

The ILC approach corrects the control input by using the tracking error from the priori trials through iterations (see Fig. 9.11). A linear ILC law can be generally described as

$$u_{k+1}(j\omega) = Q(j\omega)[u_{k,FF}(j\omega) + L(j\omega)(y_d(j\omega) - y_k(j\omega))] \tag{14}$$

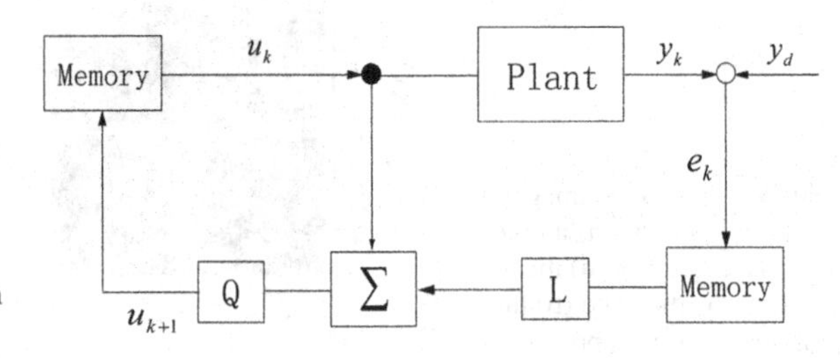

Fig. 9.11 The block diagram of iterative learning control

where $u_k(j\omega)$ and $y_k(j\omega)$ are the control input applied to the system and the output measured in the kth iteration, and $Q(j\omega)$ and $L(j\omega)$ are the ILC filters to be designed to compensate for adverse effects ($Q(j\omega)$) and the tracking error ($L(j\omega)$), respectively. As the adverse effects such as noise and system dynamics uncertainty tend to be in the high frequency range, the ILC filter $Q(j\omega)$ usually is designed to possess a low pass filter characteristic. Whereas the other ILC filter $L(j\omega)$ is designed to approximate the inverse of the system mapping $Q(j\omega)$ (Note that under the ideal scenario, i.e., the adverse effects such as noise and dynamics uncertainty can be ignored, exact tracking of the desired trajectory $y_d(j\omega)$ can be achieved with one step iteration by choosing $Q(j\omega) = 1$ and $L(j\omega) = G$). Particularly, the uncertainty in the system input-output mapping G can be accounted for in the design of the filter $L(j\omega)$ by using the notion of robust inversion, i.e., design $L(j\omega)$ to minimize the tracking error against the bound of the dynamics uncertainty [75, 76]. Similar ILC law can also be formulated in time-domain [77, 78].

It is advantageous to utilize ILC approaches in applications including PBN where repetitive operations are involved (e.g., numerous copies of a pattern/device need to be fabricated). We note that although the ILC approach per se, has been extensively studied [74, 79, 80]. In PBN applications, the desired trajectory $y_d(\cdot)$ is specified a priori by the given geometry path (contour) of the pattern/device to be fabricated along with the desired fabrication time, which renders the ILC approach an ideal choice – to achieve precision tracking in PBN applications. Specifically, the iterative mechanisms of ILC laws enable effectively compensation for the adverse effects described above (see Section 9.1.3) that are otherwise difficult to model accurately. Whereas such modeling errors inevitably limit the tracking precision of model-based control approaches (feedback or feedforward). Moreover, as the dynamics of PBN systems can vary in day-to-day operations due to effects such as thermal drift, probe ware and replacement, and variations of piezoelectric actuators, such constant dynamics variations are challenging to be accounted for by using other model-based control approaches without compromising the tracking performance – it is not realistic in PBN applications to redesign the controller every time when the dynamics variation becomes significantly large. On the contrary, such dynamics variations can be easily accounted for in the ILC framework – by updating the input through a few iterations.

9.2.4.1 Inversion-Based Iterative Control

The notion of system-inversion has also been introduced into the ILC framework [81–83], particularly for nanopositioning control [22, 31, 84–86]. The motivation is to fully utilize the available knowledge of system dynamics with the aim for rapid convergence and better tracking precision, particularly during high-speed. Note as discussed in Section 9.2.3, the use of system inverse is particularly important for nonminimum-phase systems.

Inversion-Based Iterative Control (IIC)

The IIC algorithm can be obtained by choosing the iterative filter $\rho(j\omega)$ in the general ILC law (14) as the inverse of the system dynamics model, i.e.,

$$u_{k+1}(j\omega) = u_k(j\omega) + \rho(j\omega)\hat{G}_m^{-1}(j\omega)[y_d(j\omega) - y_k(j\omega)] \tag{15}$$

where $G_m(j\omega)$ denotes the model of the system dynamics, and $\rho(j\omega) > 0$ is the frequency dependent iterative gain that is chosen to guarantee the convergence of the IIC law. Specifically, it can be shown [31] that the IIC law (15) converges at frequency ω if the phase error (defined as the difference between the phase of the "true" system dynamics $G(j\omega)$ and that of the model $G_m(j\omega)$) is less than $\pi/2$, and the iterative gain $\rho(j\omega)$ is chosen within the following range

$$0 < \rho(j\omega) < \frac{2\cos(|\Delta G(j\omega)|)}{\angle \Delta G(j\omega)}, \quad when\ |\Delta G(j\omega)| < \frac{\pi}{2}, \tag{16}$$

where $|\Delta G(j\omega)|$ and $\angle \Delta G(j\omega)$ are the magnitude error (defined as the ratio of the "true" dynamics gain with respect to the model gain, i.e., $|\Delta G(j\omega)| = |G(j\omega)|/|G_m(j\omega)|$) and the phase error at frequency ω, respectively. The advantage of the frequency-domain IIC law (15) is that the iterative coefficient $\rho(j\omega)$ is designed by utilizing the quantification of the modeling error (both the magnitude and the phase error), i.e., the iterative coefficient $\rho(j\omega)$ can be chosen towards rapid convergence (The optimal iterative coefficient to minimize the number of iterations is discussed in Ref. [72]). The range of iterative coefficient in Eq. (16) also implies that the tracking precision of the IIC law can be characterized by the tractable set of frequencies, i.e., the set of frequencies where the phase error is less than $\pi/2$, and the iterative control input should be set to zero at frequencies outside the tractable set. The quantification of the tractable set and the implementation of the IIC technique to nanopositioning control has been demonstrated in [22, 31, 85]. Note that during large-range, high-speed nanopositioning – as needed in high-speed PBN operations, both the vibration dynamics and the hysteresis effects can be excited, and result in even larger positioning error. As shown in Fig. 9.12, these two adverse effects of piezo-actuators on nanopositioning can be simultaneously removed by using the IIC technique.

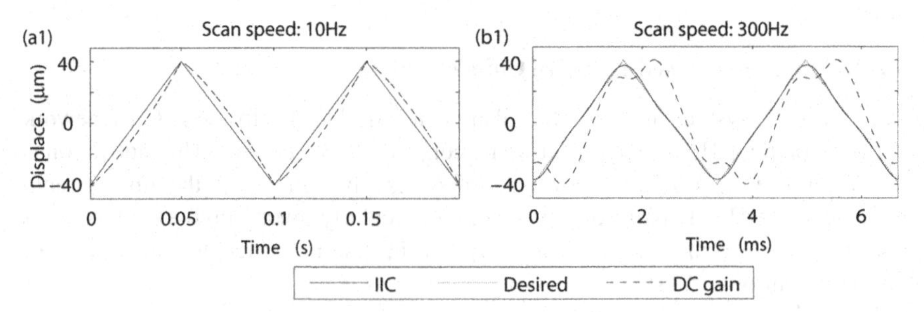

Fig. 9.12 The experimentally tracking results of a piezotube scanner to a large-range desired triangle trajectory at both (*left*) low-speed and (*right*) high-speed [22]

9.2.4.2 Extension to the Inversion-Based Iterative Control (IIC)

Next we briefly review two recent extensions along the line of inversion-based ILC approach: (1) the model-less IIC (MIIC) [86] and (2) the multi-axis IIC (MAIIC) [87] technique. The MIIC technique is proposed to eliminating the modeling process involved in the IIC procedure (to obtain the model $G_m(j\omega)$), while the MAIIC technique is developed to extend the IIC approach from single-input-single-output (SISO) systems to multi-input-multi-output (MIMO) systems.

Note that the convergence rate of the IIC law (15) depends on the quality of dynamics model $G_m(j\omega)$ (i.e., the size of the magnitude and the phase errors), whereas the modeling process itself can be time consuming and prone to errors (For example, the sweep-sine method has been widely used to model the system dynamics). Therefore, it is very much desirable, from practical viewpoint, to avoid the modeling procedure with no compromise to the convergence rate and tracking precision (Note that although ILC methods such as the classical PD-type of ILC laws [80, 88] also do not require system dynamics model, the performance of these simple ILC methods is not comparable to the IIC technique). Motivated by this need, the MIIC technique [86] has been developed. The key idea is to, instead of using a fixed dynamics model $G_m(j\omega)$, update the model of the system dynamics by using the measured input and output data along the iterations, i.e., by using the following replacement in the IIC law (15)

$$\frac{u_k(j\omega)}{y_k(j\omega)} \to G_m^{-1}(j\omega) \tag{17}$$

at frequencies where $y_k(j\omega)$ is not too small. Thus, by setting the iterative coefficient $\rho(j\omega) = 1$ (as the phase error is eliminated up to the noise level of the system), the IIC law (15) is reduced to

$$u_{k+1}(j\omega) = \frac{u_k(j\omega)}{y_k(j\omega)} y_d(j\omega) \tag{18}$$

when $y_k(j\omega) \neq 0$. Such a replacement not only eliminates the modeling procedure, but also improves both the convergence rate and the tracking precision at high-speed as the inverse of the system dynamics now is much more accurate and up to date. The MIIC technique has been applied to achieve high-speed nanopositioning in broadband nanomechanical property measurement of soft polymers [89] and PBN [7] as well. Figure 9.13 compares the star pattern fabricated by using the MIIC technique to that by using the dc-Gain method (i.e., no compensation). The efficacy of the MIIC technique for PBN applications is clearly demonstrated. Particularly note that the fabrication of the dashed-line star pattern required precision tracking in all *x*-*y*-*z* axes.

The IIC technique is also extended to the MAIIC method [87] for multi-axis motion control in the presence of cross-axis dynamics coupling. We note such multi-axis positioning systems widely exist in many areas including the multi-axis piezo-actuation systems used in many PBN processes. The goal of the extension is to

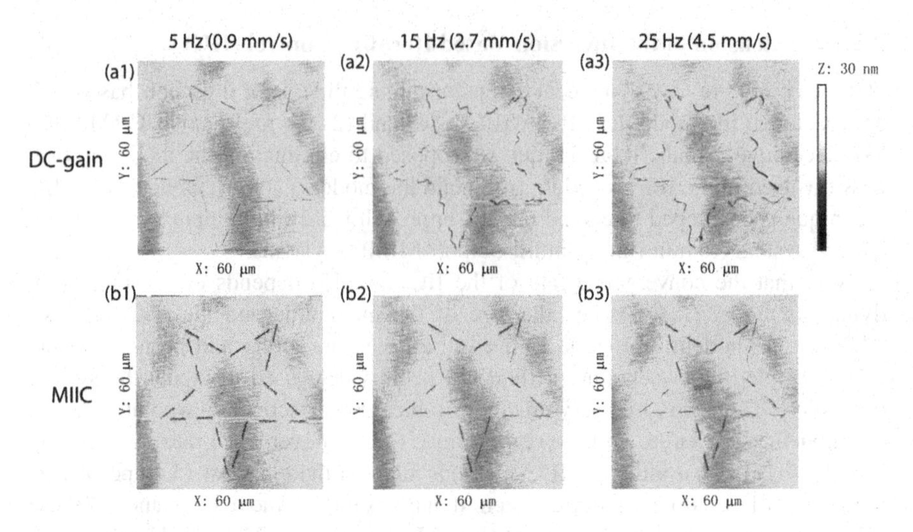

Fig. 9.13 The *star* patterns via mechanical scratching on a gold-coated silicon sample by using (**a**) the dc-Gain method and (**b**) the MIIC technique [7]

maintain both the simplicity and the performance of the IIC technique for such MIMO systems, which is obtained by decoupling the control of each axis in the MAIIC algorithm as follows,

$$\hat{U}_k(j\omega) = \hat{U}_{k-1}(j\omega) + \rho(j\omega)G_{I,m}^{-1}(j\omega)(\hat{Y}_d(j\omega) - \hat{Y}_{k-1}(j\omega)) \tag{19}$$

where $G_{I,m}$ is a diagonal matrix with the diagonal elements the model of the diagonal subsystems

$$G_{I,m}(j\omega) = \text{diag}[G_{11,m}(j\omega),\ G_{22,m}(j\omega),\ \ldots,\ G_{nn,m}(j\omega)], \tag{20}$$

$\rho(j\omega)$ is the frequency-dependent iteration coefficient matrix, i.e.,

$$\rho(j\omega) = \text{diag}\left[\rho_1(j\omega),\ \rho_2(j\omega),\ \ldots,\ \rho_n(j\omega)\right], \tag{21}$$

and $\hat{U}_k(j\omega)$, $\hat{Y}_k(j\omega)$ are the vector of input and output of each axis in the kth iteration, respectively (see Fig. 9.14).

It can be shown [87] that the convergence of the MAIIC law (19) (i.e., simultaneous precision tracking in multiple axes) can be achieved provided that the combined cross-axis coupling effect is less influential than the direct actuation on each axis, and the iterative coefficient $\rho_p(j\omega)$ for each axis p, is chosen by

$$0 < \rho_p(j\omega) < \frac{2}{|\Delta G_{pp}(j\omega)|}\left[\frac{\cos\angle\Delta G_{pp}(j\omega) - C_p(j\omega)}{1 - C_p(j\omega)^2}\right] \tag{22}$$

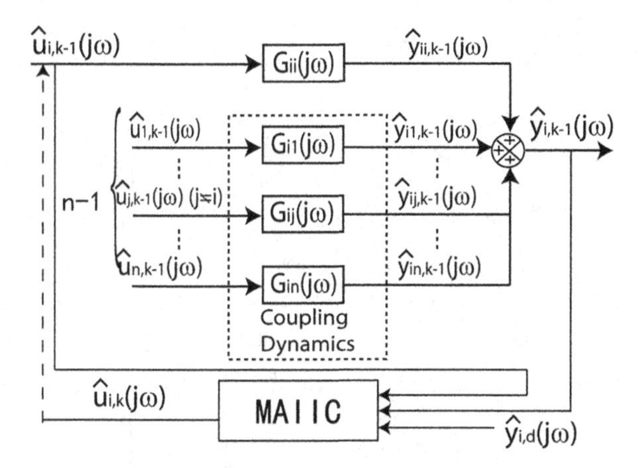

Fig. 9.14 The block diagram of the MAIIC law for the *i*th-axis output

where $\left|\Delta G_{pp}(j\omega)\right|$ and $\angle\Delta G_{pp}(j\omega)$ are the magnitude error and the phase error of the direct actuation subdynamics in the p-axis (defined similarly to the SISO case), and $C_p(j\omega)$ is the combined cross-axis coupling effect on the p-axis.

The efficacy of the MAIIC technique in multi-axis nanopositioning control has been illustrated through experiments. Figure 9.15 compares the tracking of a pyramid shape of trajectory in 3-D by using the MAIIC technique with that by using the IIC technique and that by using the DC-gain method. The DC-gain result clearly showed that both the dynamics and hysteresis and the cross-axis coupling effects are pronounced, and the cross-axis dynamics-coupling effect became dominant in the tracking results with the IIC technique where the control input for each axis was obtained by applying the IIC algorithm to the tracking of each single axis alone, and the obtained three inputs for x-, y- and z-axis respectively, were applied simultaneously. On the contrary, the MAIIC technique effectively removed both the dynamics-hysteresis and the cross-axis coupling effects (see Fig. 9.15). The performance of the MAIIC technique for multi-axis nanopositioning in PBN applications can be more directly evaluated from Fig. 9.16, where for demonstration purpose, two Chinese characters ("Na Mi", meaning nanometer) were fabricated by mechanical scratching a gold-coated silicon sample with the use of the above two techniques (the DC-gain and the MAIIC methods). The distortions in the DC-gain result are clearly seen. Particularly note that not only the shape of the two characters was distorted, but also did two extra lines cross each other appear between the two characters. These extra cross-lines were caused by the lateral-to-vertical coupling caused probe oscillations in the vertical direction. On the contrary, such extra cross-lines did not occur in the two characters fabricated by using the MAIIC technique – a clear demonstration of the efficacy of the MAIIC technique in compensating for the cross-axis coupling effect in multi-axis nanopositioning.

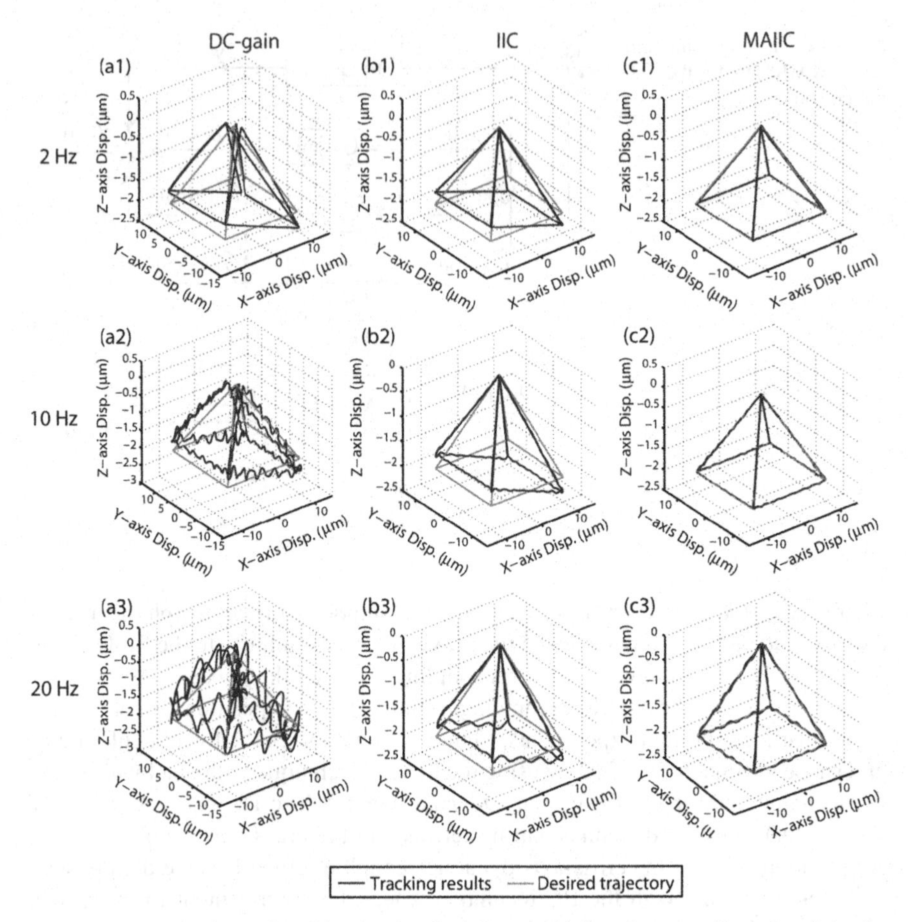

Fig. 9.15 The experimental tracking results of a 3-D pyramid shape by using (*1st column*) the dc-Gain method, (*2nd column*) the IIC method, and (*3rd column*) the MAIIC method at three different fabrication rate, where the fabrication rate (in Hz) is defined by the frequency to traverse all the *four side triangles* once in order [87]

9.2.5 Feedforward-Feedback Two-Degree-of-Freedom Control

In the feedforward-feedback two-degree-of-freedom (2DOF) framework (see Fig. 9.17), the advantages of feedback control (e.g., robustness against uncertainties and disturbances) and those of feedforward control (e.g., ease of use system dynamics model for tracking precision) are combined and arrive at an overall better control performance.

It has been well recognized in the control community (e.g., [90–92]) that the 2DOF approach is superior to the feedback control alone in achieving both tracking precision and robustness against disturbances and system uncertainties. The general approach is to first design a feedback controller to satisfy the regulation requirements (e.g., internal stability, attenuation of disturbances/noise effects) then, second, design a causal stable feedforward controller to improve the tracking

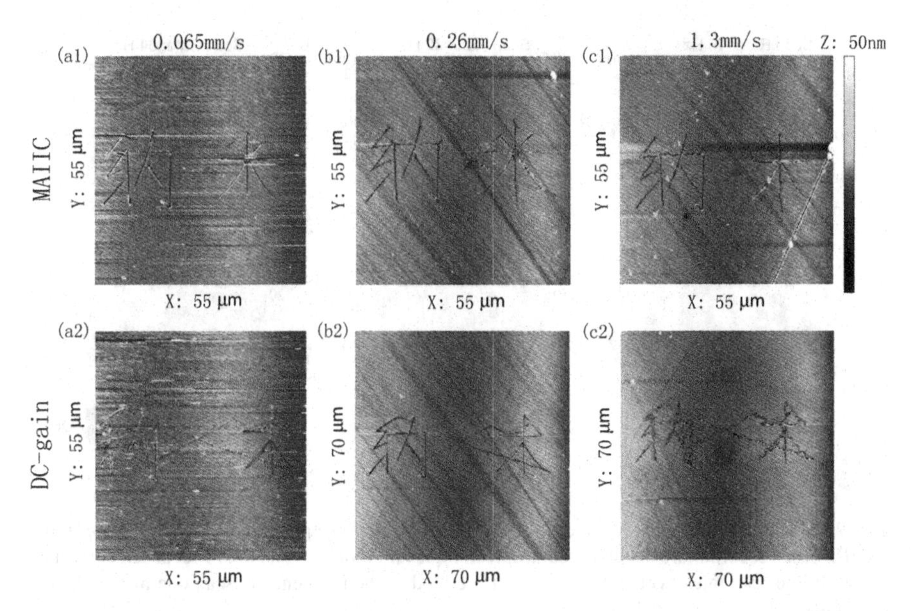

Fig. 9.16 The *two Chinese characters* "Na Mi" (meaning: nanometer) that were fabricated by using (*1st column*) the MAIIC technique with (*2nd column*) the dc-Gain method at three different fabrication speed

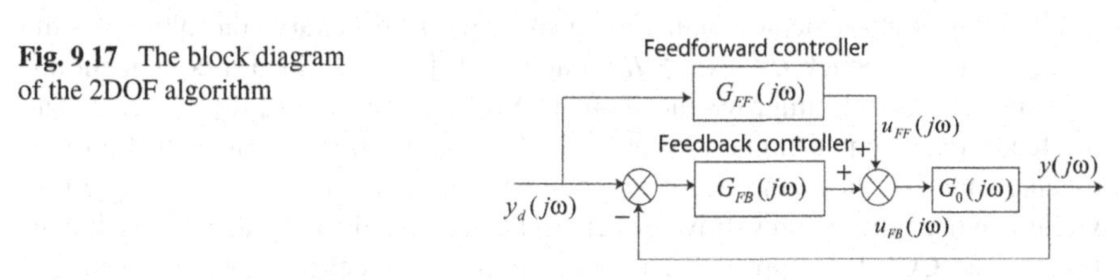

Fig. 9.17 The block diagram of the 2DOF algorithm

performance [90, 91] by using, for example, optimal control techniques (e.g., [57, 91]). Alternatively, the feedforward controller can be designed first with the goal of minimizing the feedforward tracking error upon given bound of system dynamics uncertainty, then the feedback controller can be designed to complement the feedforward controller so that it can improve the tracking precision as well as the closed-loop regulation properties. The idea is to fully utilize the a priori knowledge of the system dynamics in the feedforward path by designing a noncausal feedforward controller with guaranteed tracking precision, which, in turn, can be utilized to shape the feedback controller design. This idea has been explored recently in the so called *robust-inversion* 2DOF approach [72]. The efficacy of the 2DOF approach to nanopositioning applications has been demonstrated through experiments. The 2DOF controller design based on the H_∞ approach has been illustrated in [93, 94] for the lateral scanning in AFM imaging applications. The robust-inversion-based 2DOF control for achieving high-speed trajectory tracking has been illustrated in [72].

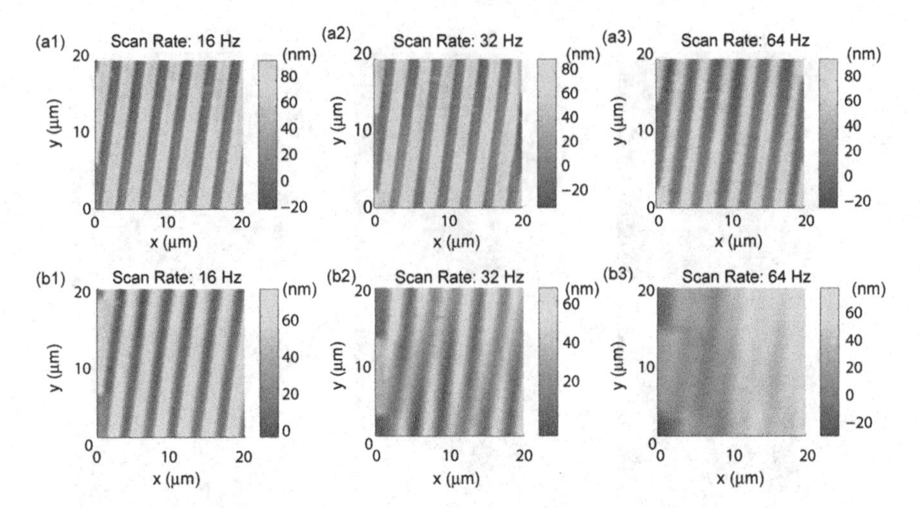

Fig. 9.18 The AFM images (contact-mode) of a calibration sample obtained by using (*top row*) the IIC-based feedforward-feedback control technique and (*bottom row*) the PI feedback control at different line scan rates, where the scan rate is defined as the frequency to scan one line forth and back [74]

Finally, the 2DOF framework can be combined with the ILC approach where the ILC law is implemented online to generate the feedforward control input – the *current-cycle-feedback ILC (CCF-ILC)* approach [95–97]. Such a combination – by using ILC to (1) improve the input to the feedback loop [98] or (2) update the feedforward input online [76, 99] – integrates the strengths of both controller together: the ILC law can effectively compensate for the repetitive tracking errors while leaving other nonrepetitive effects to be accounted for by the feedback controller. The CCF-ILC approach has been applied to precision tracking in micro- and nano-scale applications in [49]. Moreover, the CCF-ILC approach can also be extended for slow-varying repetitive operations where the desired trajectory is varying from one iteration to the other. In [75], the allowable variation of the desired trajectory (so that the convergence of the ILC law is guaranteed) is quantified for given tracking precision. As shown in Fig. 9.18, by implementing this approach to the tracking of the sample topography (i.e., the z-axis positioning) during the SPM imaging (contact-mode), the imaging speed can be substantially increased.

9.3 Summary of Discussions

The recent development of control techniques for nanopositioning needs in PBN processes has been reviewed. The needs for nanopositioning in PBN process and particularly, in high-speed operations towards high-throughput PBNs are discussed. The challenges involved in nanopositioning in general and PBN operations in specific are clarified. Three major control technique developments, the robust control approach, the stable-inversion feedforward control approach, and the iterative learning control, have been reviewed. In particular, the advantages and limits of each approach have been comparatively discussed. Experimental results of applying these control techniques to trajectory tracking in nanopositioning, scanning probe microscope imaging, and PBN operations have been presented.

Nanopositioning is central to transforming PBN technology from the proof-of-concept stage to large scale implementations. Even though the use of piezoelectric actuators provides us nanopositioning capability, challenges due to existing adverse effects need to be addressed to meet the needs in PBN operations, particularly when the fabrication is at high-speed. These adverse effects include the vibration dynamics, the nonlinear hysteresis behavior of piezoelectric actuators, the changes of the system dynamics characteristics, and disturbances due to the variations of the environment and/or the material properties. Control techniques are thereby, needed to compensate for these adverse effects towards precision motion in PBN applications, where the key is to achieving high-speed PBNs with no loss of operation range. Whereas the loss of operation range occurs when utilizing only the increase of piezoactuator bandwidth for high-speed PBNs. High-speed PBNs are particularly crucial in applications such as rapid prototyping at nanoscale and pattern-guided self-assembly nanofabrication approach. Moreover, although the parallel-probe approach is promising in arriving at high-throughput nanofabrication, issues related to the nonuniformity among the probes render the practical implementation of the parallel-probe approach to the distant future. Therefore, current developments of control techniques are geared towards high-speed nanopositioning.

The development of control techniques for PBN applications should (1) not only effectively tackle the adverse effects on nanopositioning, (2) but also judiciously exploit the available knowledge of the PBN system and operation. Viewed from these two criteria, the PID control technique widely used in the commercial nanopositioning systems is inefficient in both accounting for the adverse effects during high-speed, large-range motion and utilizing the known system-operation knowledge. Robust H_∞ control improves over PID control in adverse effects compensation and is effective for nanopositioning in general. The knowledge of the desired trajectory known in many PBN operations, however, is not fully utilized in the notion of robust H_∞ control. The development of the stable-inversion technique provides a systematic and efficient framework to utilize the knowledge of both the dynamics and the desired trajectory. As a feedforward approach, however, the performance of the stable-inversion technique can be sensitive to the variations of the PBN system and operation conditions. The system-operation variation effect is effectively compensated for in the iterative learning control approach that also fully uses the a priori knowledge of the desired trajectory when compensating for the adverse effects during PBN operations. The performance of iterative learning control can be further improved by combining the ILC input with feedback control loop in the feedforward-feedback 2DOF control framework. Experimental results are presented to facilitate the discussion of the above control techniques for PBN applications.

Further development of control techniques should tackle emerging challenges in enhancing the fabrication quality, throughput, and functionality of PBN technology.

One of such challenges is to develop an efficient control strategy to accurately capture and then compensate for the nonuniformity among the probes in parallel-probe approach, particularly as the probes need to be individually addressed to maintain the fabrication quality. More advanced control techniques also need to be developed to explore the nonlinearity of the probe-sample interaction and thereby, better manipulate the probe-sample interaction. More precise control of the probe-sample interaction is the key to the high quality of the PBN fabricated patterns/devices. Furthermore, with the development of active probes or coordinated multiple probes, it is now possible to achieve more than 3-DOF motion of the probe relative to the sample, which in turn, provides a more efficient approach to fabricate arbitrary 3-D devices at nanoscale – A significant advantage of PBN technology over other nanofabrication approaches. Control techniques, thereby, need to be developed for ensuring the precision motion over multiple DOFs (and the coordination of multiple probes). Finally, the development of SPM technology for imaging and material characterization can also be introduced to PBN and arrive at rapid evaluation of the quality of the fabricated patterns/devices online.

Acknowledgments The authors would like to thank co-workers Dr. Ying Wu, Dr. Kyong-soo Kim, and Ms. Yan Yan, for their contributions (as referred in the writing). The authors also would like to thank Prof. Zhiqun Lin from Iowa State University and Dr. Chanmin Su from the Bruker-Nano Instrument Inc. for their help in sample preparation (Lin) and SPM instrumentation (Su), respectively. The research was funded through NSF Grants No. CMMI 0624597, DUE 0632908, and CAREER-award CMMI-1066055.

References

1. Tseng, A. A., Notargiacomo, A., & Chen, T. P. (2005). "Nanofabrication by scanning probe microscope lithography: A review". *Journal of Vacuum Science and Technology, 23*, 877–894.
2. Gates, B. D., Xu, Q., Stewart, M., Ryan, D., Willson, C. G., & Whitesides, G. M. (2005). "New approaches to nanofabrication: Molding, printing, and other techniques". *Chemical Review, 105*, 1171–1196.
3. Wouters, D., & Schubert, U. S. (2004). "Nanolithography and nanochemistry: Probe-related patterning techniques and chemical modification for nanometer-sized devices". *Angewandte Chemie, 43*, 2480–2495.
4. Geissler, M., & Xia, Y. (2004). "Patterning: Principles and some new developments". *Advanced Materials, 16*, 1249–1269.
5. Garcia, R., Martinez, R. V., & Martinez, J. (2005). Nano-chemistry and scanning probe nanolithographies". *Chemical Society Review, 35*, 29–38.
6. Xie, X., Chung, H., Sow, C., & Wee, A. (2006). "Nanoscale materials patterning and engineering by atomic force microscopy nanolithography". *Materials Science and Engineering, 54*, 1–48.
7. Yan, Y., Zou, Q., & Lin, Z. (2009). "A control approach to high-speed probe-based nanofabrication". *Nanotechnology, 20*, 175301-1–175301-11.
8. Huang, S. M., Hong, M. H., Lu, Y. F., Lukyanchuk, B. S., Song, W. D., & Chong, T. C. (2002). "Pulsed-laser assisted nanopatterning of metallic layers combined". *Journal of Applied Physics, 91* (5), 3268–3274.
9. Chimmalgi, A., Choi, T. Y., Grigoropoulos, C. P., & Komvopoulos, K. (2003). "Femtosecond laser aperturless near-field nanomachining of metals assisted by scanning probe microscopy". *Applied Physics Letters, 82* (8), 1146–1148.

10. Avouris, P., Martel, R., Hertel, T., & Sandstrom, R. (1998). "AFM-tip-induced and current-induced local oxidation of silicon and metals". *Applied Physics A: Materials Science & Processing, 66*, s659–s667.
11. Fontaine, P. A., Dubois, E., & Stievenard, D. (1998). "Characerization of scanning tunneling microscopy and atomic force microscopy-based techniques for nanolithography on hydrogen-passivated silicon". *Journal of Applied Physics, 84* (4), 1776–1781.
12. Tseng, A. A., Notargiacomo, A., & Chen, T. P. (2005). "Nanofabrication by scanning probe microscope lithography: A review." *Journal of Vacuum Science & Technology B: Microelectronics and Nanometer Structures, 23*, 877–894.
13. Vettiger, P., Despont, M., Drechsler, U., Durig, U., Haberle, W., Lutwyche, M. I., et al. (2000). "The 'millipede': More than one thousand tips for future afm data storage". *IBM Journal of Research, 44*, 323–340.
14. Vettiger, P., Cross, G., Despont, M., Drechsler, U., Durig, U., Gotsmann, B., et al. (2002). "The 'millipede' – Nanotechnology entering data storage". *IEEE Transactions on Nanotechnology, 1*, 39–55.
15. Huo, F., Zheng, Z., Zheng, G., Giam, L. R., Zhang, H., & Mirkin, C. A. (2008). "Polymer pen lithography". *Science, 321*, 1658–1661.
16. Khizroev, S., & Litvinov, D. (2004). "Focused-ion-beam-based rapid prototyping of nanoscale magnetic devices." *Nanotechnology, 15*, R7–R15.
17. Watt, F., Bettiol, A. A., Kan, J. A., Teo, E. J., & Breese, M. B. (2005). "Ionbeam lithography and nanofabrication: A review". *International Journal of Nanoscience, 4*, 269–286.
18. Lindsey, J. S. (1991). "Self-assembly in synthetic routes to molecular devices biological principles and chemical perspectives: A review". *New Journal of Chemistry, 15*, 153–180.
19. Park, C., Yoon, J., & Thomas, E. L. (2003). "Enabling nanotechnology with self assembled block copolymer patterns". *Polymer, 44*, 6725–6760.
20. Croft, D., Shedd, G., & Devasia, S. (2001). "Creep, hysteresis, and vibration compensation for piezoactuators: Atomic force microscopy application". *ASME Journal of Dynamic Systems, Measurement and Control, 123* (1), 35–43.
21. Zou, Q., Leang, K. K., Sadoun, E., Reed, M. J., & Devasia, S. (2004). "Control issues in high-speed AFM for biological applications: Collagen imaging example". *Special Issue on "Advances in Nano-technology Control", Asian Journal Control, 8*, 164–178.
22. Wu, Y., & Zou, Q. (2007). "Iterative control approach to compensate for both the hysteresis and the dynamics effects of piezo actuators". *IEEE Transactions on Control Systems Technology, 15*, 936–944.
23. Leang, K. K., & Devasia, S. (2005). "Design of hysteresis-compensating iterative leaning control for atomic force microscopes". *Mechatronics, 16*, 141–158.
24. Moheimani, S. O. (2008). "Invited review article: Accurate and fast nanopositioning with piezoelectric tube scanners: Emerging trends and future challenges". *Review of Scientific Instruments, 79* (7), 071101.
25. Devasia, S., Eleftheriou, E., & Moheimani, S. O. (2007). "A survey of control issues in nanopositioning". *IEEE Transactions on Control Systems Technology, 15*, 802–823.
26. Abramovitch, D., Andersson, S., Pao, L., & Schitter, G. (2007). "A tutorial on the control of atomic force microscope". *Proceedings of American Control Conference* (pp. 3499–3502). New York City, NY.
27. Devasia, E., Eleftheriou, S., & Moheimani, S. O. (2007). "A survey of control issues in nanopositioning". *IEEE Transactions on Control Systems Technology, 15* (5), 802–823.
28. Requicha, A. A. (2003). "Nanorobots, NEMS and nanoassembly". *Proceedings of the IEEE, 91* (11), 1922–1932.
29. Rifai, O. E., & Youcef-Toumi, K. (2001). "Coupling in piezoelectric tube scanners used in scanning probe microscopes". *American Control Conference* (pp. 3251–3255). Arlington, VA.
30. Maess, J., Fleming, A. J., & Allgower, F. (2008). "Simulation of dynamics-coupling in piezoelectric tube scanners by reduced order finite element analysis". *Review of Scientific Instruments, 79* (1).

31. Tien, S., Zou, Q., & Devasia, S. (2005). Control of dynamics-coupling effects in piezo-actuator for high-speed AFM operation. *IEEE Transactions on Control Systems Technology, 13* (6), 921–931.
32. Wu, Y., Shi, J., Su, C., & Zou, Q. (2009). "A control approach to cross-coupling compensation of piezotube scanners in tapping-mode atomic force microscope imaging". *Review of Scientific Instruments, 80* (4).
33. Ando, T., Uchihashi, T., Kodera, N., Miyagi, A., Nakakita, R., Yamashita, H., et al. (2006). "High-speed atomic force microscopy for studying the dynamic behavior of protein molecules at work". *Japanese Journal of Applied Physics, 45* (3B), 1897–1903.
34. Ando, T., Kodera, N., Takai, E., Maruyama, D., Saito, K., & Toda, A. (2001). "A high-speed atomic force microscope for studying biological macromolecules". *Proceedings of the National Academy of Sciences of the USA, 98* (22), 12468–12472.
35. Frenken, J. W., Oosterkamp, T. H., Hendrisken, B. L., & Bost, M. J. (2005). "Pushing the limits of SPM". *Materials Today, 8* (5), 20–25.
36. Rosta, M. J., Crama, L., Schakel, P., Tol, E. v., Velzen-Williams, G. B., Overgauw, C. F., et al. (2005). "Scanning probe microscopes go video rate and beyond". *Review of Scientific Instruments, 76*, 53710–53718.
37. Humphris, A. D., Hobbs, J. K., & Miles, M. J. (2003). "Ultrahigh-speed scanning near-field optical microscopy capable of over 100 frames per second". *Applied Physics Letters, 83* (1), 6–8.
38. Schitter, G., Astrom, K., DeMartini, B., Thurner, P., Turner, K., & Hansma, P. (2007). "Design and modeling of a high-speed AFM-scanner". *IEEE Transactions on Control Systems Technology, 15* (5), 906–915.
39. Fleming, A., Kenton, B. J., & Leang, K. K. (2010). "Bridging the gap between conventional and video-speed scanning probe microscopes". *Ultramicroscopy, 110*, 1205–1211.
40. Clayton, G. M., Tien, S., Leang, K. K., Zou, Q., & Devasia, S. (2009). "A review of feedforward control approaches in nanopositioning for high-speed spm". *ASME Journal of Dynamic Systems, Measurement and Control, 131*, 061101-1–061101-19.
41. Ando, Y., Ikehara, T., & Matsumoto, S. (2007). "Development of three-dimensional microstages using inclined deep-reactive ion etching". *Journal of Microelectromechanical Systems, 16* (3), 613–621.
42. Yong, Y. K., Aphale, S. S., & RezaMoheimani, S. O. (2009). "Design, identification, and control of a flexure-based xy stage for fast nanoscale positioning". *IEEE Transactions on Control Systems Technology, 8* (1), 46–54.
43. Li, Y., & Bechhoefer, J. (2007). "Feedforward control of a piezoelectric flexure stage for AFM". *Review of Scientific Instruments, 78*, 013702.
44. Sebastian, A., Gannepalli, A., & Salapaka, M. (2007). "A review of the systems approach to the analysis of dynamic-mode atomic force microscopy". *IEEE Transactions on Control Systems Technology, 15* (5), 952–959.
45. Franklin, G. F., & Powell, A. E.-N. (2002). *Feedback Control of Dynamic Systems* (4th ed.). Englewood Cliffs, NJ: Prentice Hall.
46. Ogata, K. (1990). *Modern Control Engineering* (2nd ed.). Englewood Cliffs, NJ: Prentice-Hall.
47. Liu, G.-Y., Xu, S., & Qian, Y. (2000). "Nanofabrication of self-assembled monolayers using scanning probe lithography". *Accounts of Chemical Research, 33* (7), 457–466.
48. Bourne, K., Kapoor, S. G., & DeVor, R. E. (2010). "Study of a high performance AFM probe-based microscribing process". *ASME Journal of Manufacturing Science and Engineering, 132* (3), 030906-1–030906-10.
49. Barrett, R. C., & Quate, C. F. (1991). "Optical scan-correction system applied to atomatic force micropscopy". *Review of Scientific Instruments, 62*, 1393–1399.
50. Schitter, G., Menold, P., Knapp, H., Allgower, F., & Stemmer, A. (2001). "High performance feedback for fast scanning atomic force microscopes". *Review of Scientific Instruments, 72* (8), 3320–3327.

51. Salapaka, S., Sebastian, A., Cleveland, J. P., & Salapaka, M. V. (2002). "High bandwidth nano-positioner: A robust control approach". *Review of Scientific Instruments, 73* (9), 3232–3241.
52. Yan, Y., Wu, Y., Zou, Q., & Su, C. (2008). "An integrated approach to piezo actuators positioning in high-speed AFM imaging". *Review of Scientific Instruments, 79*, 073704–073712.
53. Skogestad, S., & Postlethwaite, I. (2005). *Multivariable Feedback Control: Analysis and Design* (2nd ed.). New York: Wiley.
54. Zhou, K., Doyle, J., & Glover, K. (1995). *Robust and Optimal Control.* New York: Prentice Hall.
55. Dullerud, G. E., & Paganini, F. G. (2001). *A Course in Robust Control Theory: A Convex Approach.* Berlin: Springer.
56. Dong, J., Salapaka, S. M., & Ferreira, P. M. (2008). "Robust control of a parallel-lkinematic nanopositioner". *ASME Journal of Dynamic Systems, Measurement and Control, 130*, 041007-1–041007-15.
57. Schitter, G., Stemmer, A., & Allgower, F. (2003). "Robust 2dof-control of a piezoelectric tube scanner for high-speed atomic force microscopy". *American Control Conference* (pp. 3720–3725). Denver, CO.
58. Sebastian, A., & Salapaka, S. (2005). "Design methodologies for robust nanopositioning". *IEEE Transactions on Control Systems Technology, 13* (6), 868–876.
59. Sahoo, R. D., Sebastian, A., & Salapaka, M. V. (2003). "Transient-signal-based sample-detection in atomic force microscopy". *Applied Physics Letters, 83* (26), 5521–5523.
60. Lee, C., & Salapaka, S. M. (2009). "Robust broadband nanopositioning: Fundamental trade-offs, analysis, and design in a two-degree-of-freedom control framework". *Nanotechnology, 20* (3), 035501.
61. Barth, J. V., Costantini, G., & Kern, K. (2005). "Engineering atomic and molecular nanostructures at surfaces". *Nature, 437*, 671–679.
62. Pantazi, A., Sebastian, A., Cherubini, G., Lantz, M., Pozidis, H., Rothuizen, H., et al. (2007). "Control of mems-based scanning-probe data-storage devices". *IEEE Transactions on Control Systems Technology, 15* (5), 824–841.
63. Pantazi, A., Sebastian, A., Pozidis, H., & Eleftheriou, E. (2005). "Two-sensor-based h_∞ control for nanopositioning in probe storage". *IEEE Conference on Decision and Control* (pp. 1174–1179). Seville, Spain.
64. Stein, G. (2003). "Respect the unstable". *IEEE Control System Magazine, 23* (4), 12–25.
65. Isidori, A. (1995). *Nonlinear Control Systems* (3rd ed.). London: Springer.
66. Zou, Q., & Devasia, S. (1999). "Preview-based stable-inversion for output tracking of linear systems". *ASME Journal of Dynamic Systems, Measurement and Control, 121*, 625–630.
67. Zou, Q., & Devasia, S. (2004). "Preview-based optimal inversion for output tracking: Application to scanning tunneling microscopy". *IEEE Transaction on Control Systems Technology, 12*, 375–386.
68. Zou, Q. (2009). "Optimal preview-based stable-inversion for output tracking of nonminimum-phase linear systems". *Automatica, 45*, 230–237.
69. Devasia, S., Chen, D., & Paden, B. (1996). "Nonlinear inversion-based output tracking". *IEEE Transactions on Automatic Control, 41* (7), 930–942.
70. Devasia, S., & Paden, B. (1998). "Stable inversion for nonlinear nonminimum phase time-varying systems". *IEEE Transactions on Automatic Control, 43*, 283–288.
71. Zou, Q., & Devasia, S. (2007). "Preview-based stable-inversion for output tracking of nonlinear nonminimum-phase systems: The VTOL example". *Automatica, 43* (1), 117–127.
72. Wu, Y., & Zou, Q. (2009). "Robust inversion-based 2-DOF control design for output tracking: Piezoelectric-actuator example". *IEEE Transactions on Control System Technology, 17*, 1069–1082.
73. Croft, D., McAllister, D., & Devasia, S. (1998). "High-speed scanning of piezo-probes for nano-fabrication". *ASME Journal of Manufacturing Science and Engineering, 120*, 617–622.

74. Helfrich, B. E., Lee, C., Bristow, D. A., Xiao, X. H., Dong, J., Alleyne, A. G., Salapaka, S. M., & Ferreira, P. M. (2010). "Combined h_∞-feedback control and iterative learning control design with application to nanopositioning systems". *IEEE Transactions on Control Systems Technology, 18* (2), 336–351.
75. Wu, Y., & Zou, Q. (2009). "An iterative based feedforward-feedback control approach to high-speed atomic force microscope imaging". *ASME Journal of Dynamic Systems, Measurement and Control, 131*, 061105-1–061105-9.
76. Wu, Y., Zou, Q., & Su, C. (2009). "A current cycle feedback iterative learning control approach for AFM imaging". *IEEE Transactions on Nanotechnology, 8* (4), 515–527.
77. Kavli, T. (1991). "Frequency domain synthesis of trajectory learning controller for robot manipulators". *Journal of Robotic Systems, 9* (5), 663–680.
78. Moore, K. L., Dahleh, M., & Bhatacharyya, S. P. (1992). "Iterative learning control: A survey and new results". *Journal of Robotic Systems, 9* (5), 563–594.
79. Moore, K. L., & Xu, J.-X. (2000). "Special issue on iterative learning control". *International Journal of Control, 73* (10).
80. Ahn, H.-S., Chen, Y., & Moore, K. L. (2007). "Iterative learning control: Brief survey and categorization". *IEEE Transactions on Systems, Man, and Cybernetics-Part C: Applications and Reviews, 37* (6), 1099–1121.
81. Ghosh, J., & Paden, B. (2002). "A pseudoinverse-based iterative learning control". *IEEE Transactions on Automatic Control, 47*, 831–836.
82. Ghosh, J., & Paden, B. (2000). "Nonlinear repetitive control". *IEEE Transactions on Automatic Control, 45*, 949–954.
83. Ghosh, J., & Paden, B. (1999). "Pseudo-inverse based iterative learning control for nonlinear plants with disturbances". *IEEE Conference on Decision and Control* (vol. 11, pp. 5206–5212). Phoenix, AZ.
84. Tien, S., Zou, Q., & Devasia, S. (2004). "Iterative control of dynamics-coupling effects in piezo-based nano-positioners for high-speed AFM". *Proceedings of the 2004 IEEE International Conference on Control Applications* (pp. 711–717). Taipei, Taiwan.
85. Kim, K., Zou, Q., & Su, C. (2008). "A new approach to scan-trajectory design and track: AFM force measurement example". *ASME Journal of Dynamic Systems, Measurement and Control, 130*, 051005-1–051005-10.
86. Kim, K.-S., & Zou, Q. (2008). "A model-less inversion-based iterative control technique for output tracking: Piezo actuator example". *ACC* (pp. 2710–2715). Seattle, WA.
87. Yan, Y., Wang, H., & Zou, Q. (2010). "A decoupled inversion-based iterative control approach to multi-axis precision positioning: 3-d nanopositioning example". *American Control Conference* (pp. 1290–1295). Baltimore, MD.
88. Arimoto, S., Kawamura, S., & Miyazaki, F. (1986). "Convergence, stability, and robustness of learning control schemes for robot manipulators". *Proceedings of the International Symposium on Robot Manipulators on Recent Trends in Robotics: Modeling, Control and Education* (pp. 307–316). Albuquerque, NM.
89. Xu, Z., Kim, K., Zou, Q., & Shrotriya, P. (2008). "Broadband measurement of ratedependent viscoelasticity at nanoscale using scanning probe microscope: Poly(dimethylsiloxane) example". *Applied Physics Letters, 93* (13), 133103.
90. Youla, D. C., & Bongiorno, J. J., Jr. (1985). "A feedback theory of two-degree-of-freedom optimal wiener-hopf design". *IEEE Transactions on Automatic Control, 30* (7), 652–665.
91. Prempain, E., & Postlethwaite, I. (2001). "Feedforward control: A full-information approach". *Automatica, 37* (1), 17–28.
92. Yaesh, I., & Shaked, U. (1991). "Two-degree-of-freedom h_∞-optimization of multivariable feedback systems". *IEEE Transaction on Automatic Control, 36* (11), 1272–1275.
93. Schitter, G., Stemmer, A., & Allgower, F. (2004). "Robust two-degree-of-freedom control of an atomic force microscope". *Asian Journal of Control, 6* (2), 156–163.
94. Lee, C., & Salapaka, S. M. (2009). "Robust broadband nanopositioning: Fundamental trade-offs, analysis, and design in a two-degree-of-freedom control framework". *Nanotechnology, 20* (3), 035501.

95. Goh, C., & Yan, W. (1996). "An h_∞ synthesis of robust current error feedback learning control". *ASME Journal of Dynamic Systems, Measurement and Control, 118*, 341–346.
96. Doh, T.-Y., Moon, J.-H., Jin, K., & Chung, M. (1999). "Robust iterative learning control with current feedback for uncertain linear systems". *International Journal System Science, 30* (1), 39–47.
97. Owens, D., & Munde, G. (2000). "Error convergence in an adaptive iterative learning controller". *International Journal of Control, 73*, 851–857.
98. Longman, R. (2000). "Iterative learning control and repetitive control for engineering practice". *International Journal of Control, 73* (10), 930–954.
99. Roover, D. d., & Bosgra, O. (2000). "Synthesis of robust multivariable iterative learning controllers with application to a wafer stage motion system". *International Journal of Control, 73* (10), 968–979.

Chapter 10
Scanning Probe Based Nanolithography and Nanomanipulation on Graphene

Pasqualantonio Pingue

Abstract Alternative lithographic techniques, in particular those based on scanning probe microscopy, have shown a great potential for fabricating nanostructures using various material and allowing high spatial resolution, alignment capabilities and high-resolution imaging during the different lithographic steps. More specifically, atomic force microscope (AFM) and scanning tunneling microscope (STM) have been in the recent past employed to image and modify at nanometer scale a new carbon material discovered in 2004 and called *graphene*, a single layer of carbon atoms arranged in a honeycomb crystal lattice. In this chapter a review of recent results obtained by scanning probe based nanofabrication on graphene nanostructures is presented. It is focused in particular on nanomanipulation, local anodic oxidation (LAO), electrochemical or thermal-stimulated desorption, static or dynamic ploughing as well as other AFM and STM based techniques on imaging, lithography and spectroscopy.

Keywords Scanning probe microscopy · Scanning tunneling microscopy · Atomic force microscopy · Scanning probe spectroscopy · Nanomanipulation · Nanolithography · Local anodic oxidation · Nanomaterial · Nanomechanic · Graphene · Two-dimensional crystals · Nanoribbon

Abbreviations

AFM	Atomic force microscope
dI/dU, dI/dV	Differential conductance
FLG	Few layer graphene
GNR	Graphene nano ribbons
GO	Graphene oxide
GO_{epi}	Graphene oxide epitaxial
I-V	Current-Voltage
KPM	Kelvin probe microscopy
LAO	Local anodic oxidation
LDOS	Local density of states
NEMS	Nano electro mechanical systems

P. Pingue (✉)
Laboratorio NEST, Scuola Normale Superiore, I-56127 Pisa, Italy
e-mail: p.pingue@sns.it

A.A. Tseng (ed.), *Tip-Based Nanofabrication*, DOI 10.1007/978-1-4419-9899-6_10,

SEM	Scanning electron microscope
SGM	Scanning gate microscopy
SLG	Single layer graphene
SPL	Scanning probe lithography
SPM	Scanning probe microscopy
SPN	Scanning probe nanomanipulation
SPS	Scanning probe spectroscopy
STM	Scanning tunneling microscope
TCNL	Thermo chemical nano lithography
TERS	Tip enhanced Raman scattering
UHV	Ultra high vacuum

10.1 Introduction

The "rise of graphene" [1] in Physics started in 2004 when A. Geim and K. Novoselov at Manchester University [2, 3] demonstrated that it was possible to isolate a single carbon layer of atoms having a honeycomb lattice structure, starting from bulk graphite samples and performing mechanical exfoliation by "scotch tape" method. With patience and practice, this simple method results in high-quality single layer graphene (SLG) flakes, both in terms of high electron mobility, low crystal defect density and easily more than 100 μm^2 in area size. Andrei Geim and Kostya Novoselov received the 2010 Nobel Prize in Physics for their extraordinary scientific work.

Atomic force microscopy (AFM) technique resulted to be very important for both performing imaging on those 2D crystallites and demonstrating that only a *single* layer was deposited on the substrate (typically a 300 nm thick SiO_2 layer thermally grown on Si chip, where the optical visibility of graphene is improved by interference effects [4]). In fact, in their case the presence of a folded region in a graphene flake allowed researchers to measure, trough the AFM topographic images, the differential layer step height, thus avoiding any artifacts due to different tip-substrate interactions on graphene and SiO_2 and allowing demonstration of a single atom thick carbon structure presence. In fact, this differential step height, matching the interlayer distance in the corresponding 3D crystals (0.34 nm) helped to distinguish between double or few-layer crystals and true single sheets. In Fig. 10.1, left panel, the historical first AFM image of a graphene flake reported in literature is shown, and the folded edge of the graphene layer together with its differential step height illustrated.

One of the first attempts to deposit SLG flakes on a substrate were also made by T. W. Ebbesen and H. Hiura [5] and by R. S. Ruoff's group [6] using another kind of mechanical exfoliation technique based on SPM nanomanipulation. In both cases, a silicon probe of a SPM was employed as tool to rubber graphite surface or etched pillars in order to obtain very thin layers (called "graphite origami", by analogy with Japanese traditional paper folding). Imaging was performed by scanning

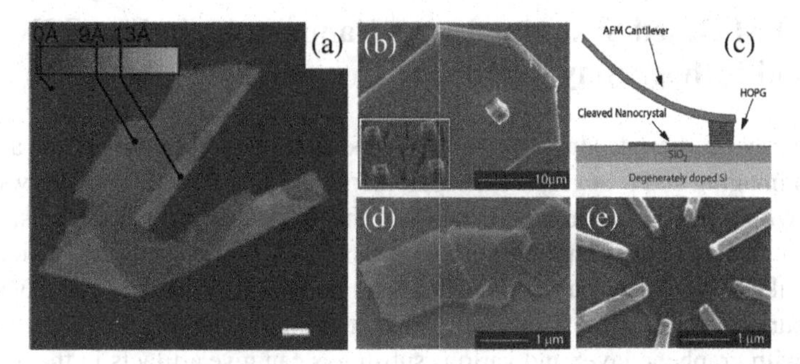

Fig. 10.1 *Left panel*: (**a**) Single layer graphene visualized by AFM (scale bar: 1 μm) on top of an oxidized Si wafer (300 nm of thermal grown SiO_2) (from Novoselov et al. [2]). *Right panel*: (**b**) Scanning electron microscope image of an HOPG crystallite mounted on a microcantilever. *Inset*: bulk HOPG surface patterned by masked anisotropic oxygen plasma etching; (**c**) schematic drawing of the microcleaving process; (**d**) thin graphite samples cleaved onto the SiO_2/Si substrate; (**e**) a typical mesoscopic device fabricated from a cleaved graphite sample (from Zhang et al. [7])

electron microscope (SEM) and AFM, but no evidence of single graphene sheets was found. The thinnest slab observed at that time in reference [6] was 98 nm thick. An improvement was performed by P. Kim and co-workers at Columbia University [7] by using graphite pillars mounted on a tipless AFM cantilever. In this way they successfully transferred thin graphite samples (as thin as 10 nm, or ~30 graphene layers) directly onto a SiO_2/Si substrate for subsequent device fabrication and electronic transport measurements characterization. In the right panel of Fig. 10.1, a cartoon of the SPM-based deposition technique together with the SEM images of the cantilevered graphitic pillar and of the obtained thin graphite layers are reported.

In the case of STM, the imaging and modification of the graphite surface by STM has a more than 20-year-old history [8, 9]. Therefore, we could say that scanning probe microscopy (SPM) techniques are strictly linked to graphene from the beginning of their rising and vice versa.

In this review we would like to focus on the SPM imaging techniques that have been applied to graphene in the last few years and on the scanning probe lithographic and nanomanipulation techniques that have been employed to modify graphene layers on nanometer scale, which have allowed studying their peculiar physical, chemical and mechanical properties. Scanning probe lithography (SPL) and nanomanipulation (SPN) were developed just after the discovery of the two major microscopy techniques within the SPM family: scanning tunneling microscopy (STM) [10] and atomic force microscopy (AFM) [11]. Their experimental applications in many fields have been recently reviewed [12, 13] and their contribution to the advancement of nanofabrication techniques in terms of resolution, low costs and flexibility highlighted [14]. We will show how SPL and SPN have been applied to graphene during the last few years, evaluating separately both STM and AFM-based techniques, and finally discussing perspective developments and new application fields.

10.2 AFM-Based Scanning Probe Nanomanipulation (SPN) and Lithography (SPL)

As discussed in the introduction, AFM has been employed first of all as a tool to perform imaging and topographic measurements on SLG flakes. AFM allows determining with high resolution the thickness of the exfoliated graphene layers, as well as the roughness of the graphene surface when SLGs are deposited on various substrates (like SiO_2, mica, quartz or Si_3N_4) and its characteristic rippled, buckled or folded surface. As already mentioned above, the interaction of silicon cantilevered probes with graphene layer and various substrates can give artifacts in the measurements of the relative step-height. Accordingly, intermittent contact-mode imaging parameters must be carefully chosen in order to minimize these effects and to obtain a reliable step-height value [15]. In Fig. 10.2 (top panel) SLG images and step

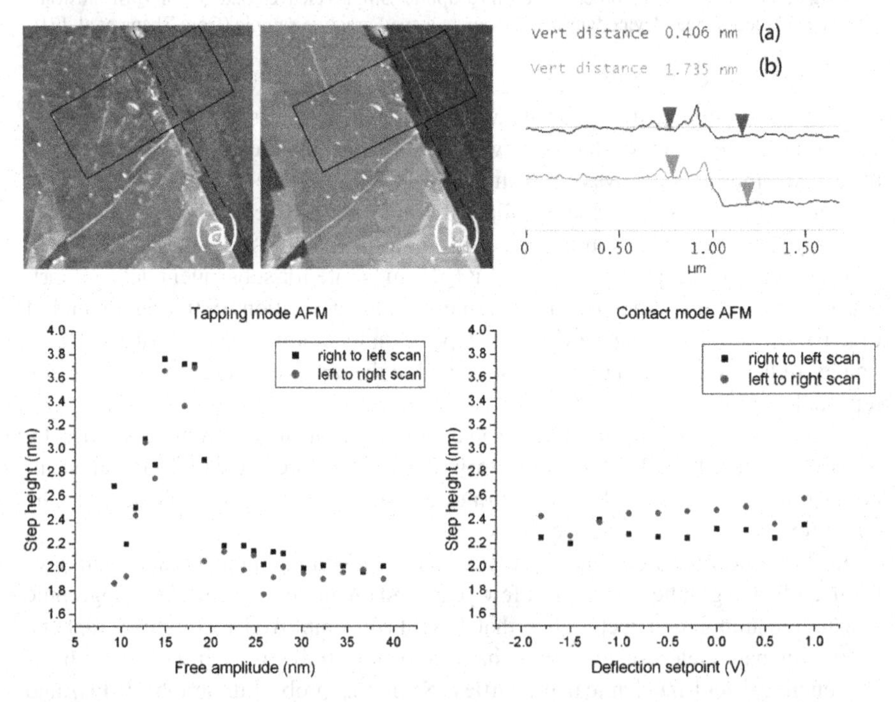

Fig. 10.2 *Top panel*: (**a**, **b**) intermittent contact mode AFM images of the regions of a single layer flake (each image is 2.5 μm × 2.5 μm). The images were acquired using a constant amplitude setpoint and two different free amplitudes, one higher (**a**) and another one lower (**b**) than the amplitude at which unstable imaging occurs on the SLG. In the first case the setpoint is in the repulsive regime for both silicon dioxide and SLG, while in the latter case imaging is in the attractive regime for SLG. Averaged profiles (obtained inside the *black marked area*) taken on each image show the increase in thickness when measurements are not performed in the repulsive regime. *Bottom panel*: intermittent contact mode AFM measurements of a few layer graphene flake step height at various free amplitudes (*left graph*). Contact mode AFM measurement of the same flake, using various deflections of the cantilever, i.e. changing the contact force (*right graph*). Changing the contact force does not have any effect on the step height, but the left to right and right to left scans differ by about 0.2 nm (from Nemes-Incze et al. [15])

heights obtained by AFM in different intermittent-contact conditions and parameters are reported, together with a comparison between step height values obtained in intermittent-contact mode and in standard contact mode (bottom panel) on few layer graphene flake. Notice how the step-height value can be strongly-dependent on the attractive or repulsive AFM measurement regime.

Rippling of SLGs surface has been invoked to explain thermodynamic stability of free-standing graphene sheets [16] and AFM-based imaging techniques have been employed to study this phenomenon in detail. Recently, Heinz and co-workers [17] have demonstrated by AFM imaging the fabrication of SLG that are flat down to the atomic level, performing scotch tape deposition on mica instead of on standard SiO_2 substrate (see Fig. 10.3). In general, AFM-based techniques are low-throughput imaging tools but remain essential piece of instrument for research of graphene, avoiding any modifications of the electronic properties of the imaged SLG flakes, as for example can happen when using other high-spatial resolution microscopy like scanning electron microscope (SEM) [18].

SPM imaging obtained by our research group on graphene deposited on SiO_2 by PDMS transfer technique [19–21], demonstrated "blistering" of the SLG surface. In Fig. 10.4 topography (a), phase (b) and 3D imaging (c) show suspended membranes in correspondence of small residual particles on the SiO_2 surface and, more in general, the presence of a blistered topography. Notice how phase

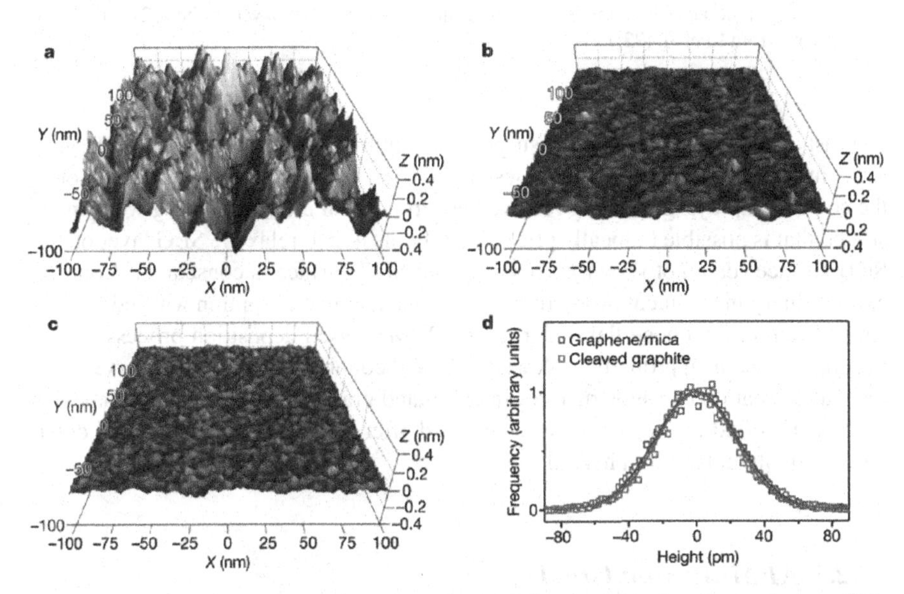

Fig. 10.3 Three-dimensional representations of the AFM topographic data for graphene on SiO_2 (**a**) and on mica (**b**) substrates. (**c**) AFM image of the surface of a cleaved kish graphite sample. Images **a**, **b** and **c** correspond to 200 nm × 200 nm areas and are presented with the same height scale. (**d**) Height histograms of the data in (**b**) and in (**c**) as *squares*. The histograms are described by Gaussian distributions (*solid lines*) with standard deviations of 24.1 and 22.6 pm, respectively (from Lui et al. [17])

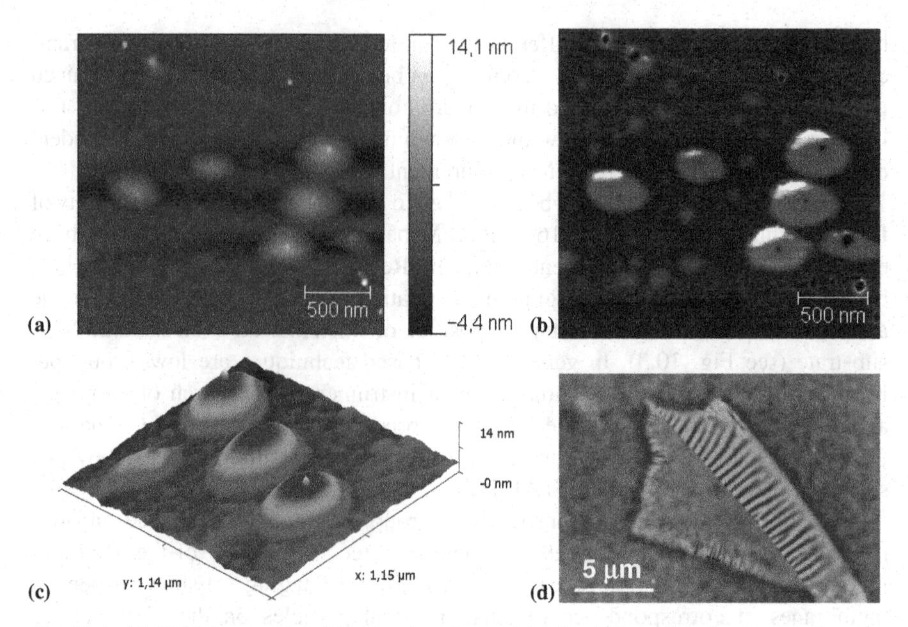

Fig. 10.4 (**a**) AFM topography, (**b**) phase imaging and (**c**) three-dimensional representation of a blistered SGL surface deposited by PDM transfer imprint on SiO_2 substrate (from Gohler et al., submitted); (**d**) AFM topographic image of the graphene buckling after annealing procedure. The height of the single-layer is 1 nm. The height of the connecting few-layer is about 2.5 nm. Scale bar is 5 μm (from Li et al. [22])

contrast enhances the presence of suspended membranes, (represented by light gray areas in the color scale) and how these regions have in some cases a black spot in the center. Employing spatially resolved electron-beam induced etching we demonstrated that is possible to locally prick a membrane and to relax the SLG layer on the SiO_2 surface (data not shown). This a typical SLG surface is consequent from the extraordinary mechanical properties of graphene and in our opinion was induced by stress-strain on graphene flakes during the PDMS-based deposition process. These peculiar mechanical properties allowed Z. Li and collaborators [22] to observe spontaneous formation of nanostructures in exfoliated graphene flakes: periodic buckles as those reported in Fig. 10.4d were obtained after annealing and quick cool down thermal treatments of graphene flakes.

10.2.1 AFM-SPN on Graphene

Both AFM-based SPN and SPL techniques applied on graphene have been developed in the past on bulk graphite (as already seen) and on other carbon-based materials such as nanotubes [23], amorphous carbon layer [24] or diamond films [25]. Those methods have been extended only recently also to SLGs. As mentioned in the introduction, we could consider "graphite origami" [5, 6] and the "microscopic cleaving process" based on an AFM cantilever [7] (see Fig. 10.1b) as first examples of SPN on few layer graphene flakes. After those pioneering works, SPN has been principally employed to study the mechanical properties of graphene: accurate mechanical characterization at very high spatial resolution can be only achieved by SPN and nanoindentation techniques. More in detail, McEuen [26] and Van der Zant [27] studied and measured by AFM force vs distance curves the mechanical spring constants of stacks of graphene sheets suspended over trenches in silicon dioxide, extracting the Young's modulus and the tension in those single-atom thick membranes. In Fig. 10.5 we report the schematics of the two experiments together with some results both related to force vs distance curves (Fig. 10.5, left panel) first on membrane elastic constant measured in function of the spatial position (Fig. 10.5, right panel) and then on the photolithographed trench. Measurements of the elastic properties and intrinsic breaking strength of monolayer graphene membranes have been recently obtained by J. Hone and coworkers [28] by nanoindentation

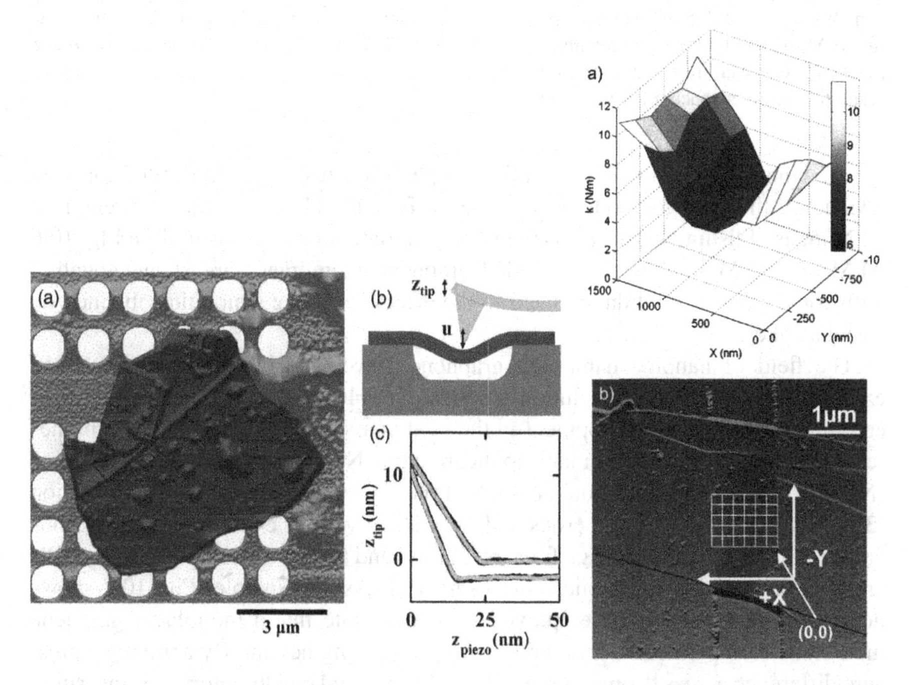

Fig. 10.5 *Left panel*: (**a**) An AFM height image of a suspended flake (~69 layers). (**b**) Schematic overview of the method used to determine the local compliance of the flake. (**c**) Two linear force-distance *curves* (offset for clarity) taken on the flake shown in (**a**). The approaching (*gray*) and retracting (*black*) parts of the curves lie on top of each other. The *bottom curve* is taken on an unsuspended part of the flake, while the *top curve* is taken on a suspended part (from Poot et al. [27]). *Right panel*: (**a**) Surface plot of the spring constant of a suspended graphene sheet vs the location of the AFM tip. (**b**) An amplitude AFM micrograph of the suspended sheet. Each data point was taken at the intersection of the grid located on the suspended portion of the graphene. The trench etched into the silicon dioxide is seen as a vertical stripe (from Frank et al. [26])

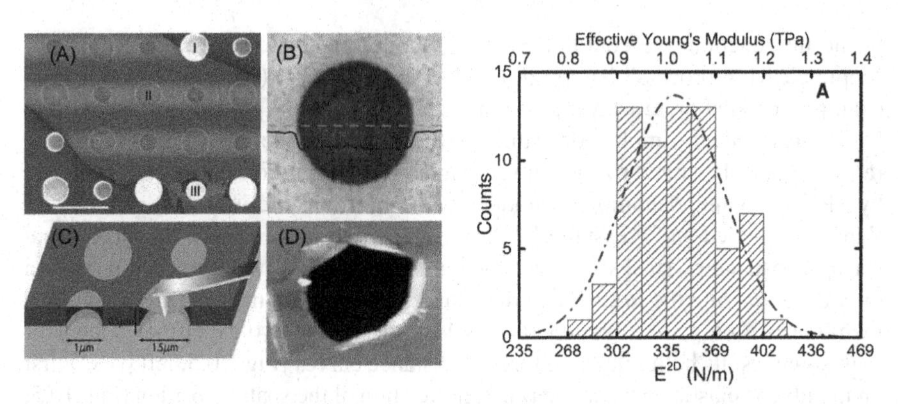

Fig. 10.6 *Left panel*: Images of suspended graphene membranes. (**a**) Scanning electron micrograph of a large graphene flake spanning an array of *circular holes* 1 and 1.5 μm in diameter. Area I shows a *hole* partially covered by graphene, area II is fully covered, and area III is fractured from indentation. Scale bar, 3 μm. (**b**) Noncontact mode AFM image of one membrane, 1.5 μm in diameter. The *solid line* is a height profile along the *dashed line*. The step height at the edge of the membrane is about 2.5 nm. (**c**) Schematic of nanoindentation on suspended graphene membrane. (**d**) AFM image of a fractured membrane. *Right panel*: Histogram of elastic stiffness. *Dashed line* represents Gaussian fit to data. The effective Young's modulus was obtained by dividing by the graphite interlayer spacing (from Lee et al. [28])

experiments with an atomic force microscope. Their results demonstrated for SLG a value for the Young's modulus the order of E = 1.0 TPa and breaking strength of 42 N/m, establishing graphene as the strongest material ever measured. In Fig. 10.6 AFM and SEM images of suspended graphene membranes are shown together with some experimental data related to the effective Young's modulus obtained on various flakes.

The field of nanomechanics on graphene, developed by SPN technique and exploiting the high AFM resolution in terms of applied forces and positioning, will certainly have a great development in the next years because graphene membranes represent an ideal system for many application like Nano ElectroMechanical System (NEMS) [29–31], gas filtration, or single-molecule trapping and chemical detection [32]. SLG's mechanical properties and stability in various experimental conditions can be studied in details, at high force resolution and at nanometer scale only by SPN and nanoindentation techniques based on SPM. As an example, Fig. 10.7 shows how AFM imaging can be employed to demonstrate that a monolayer graphene membrane is impermeable to standard gases, including helium. By applying a pressure difference across the membrane, P. L. McEuen and collaborators [33] measured both the elastic constant and the mass of a single layer of graphene. Their pressurized graphene membrane was defined by the Authors "the world's thinnest balloon" and provided a unique separation barrier between two distinct regions that was only one atom thick.

Moreover, SPN is starting to obtain interesting results in other classes of two-dimensional crystallites, in order to study their peculiar properties as topological insulators, including graphene oxide (GO) single layer flakes, the latter easily

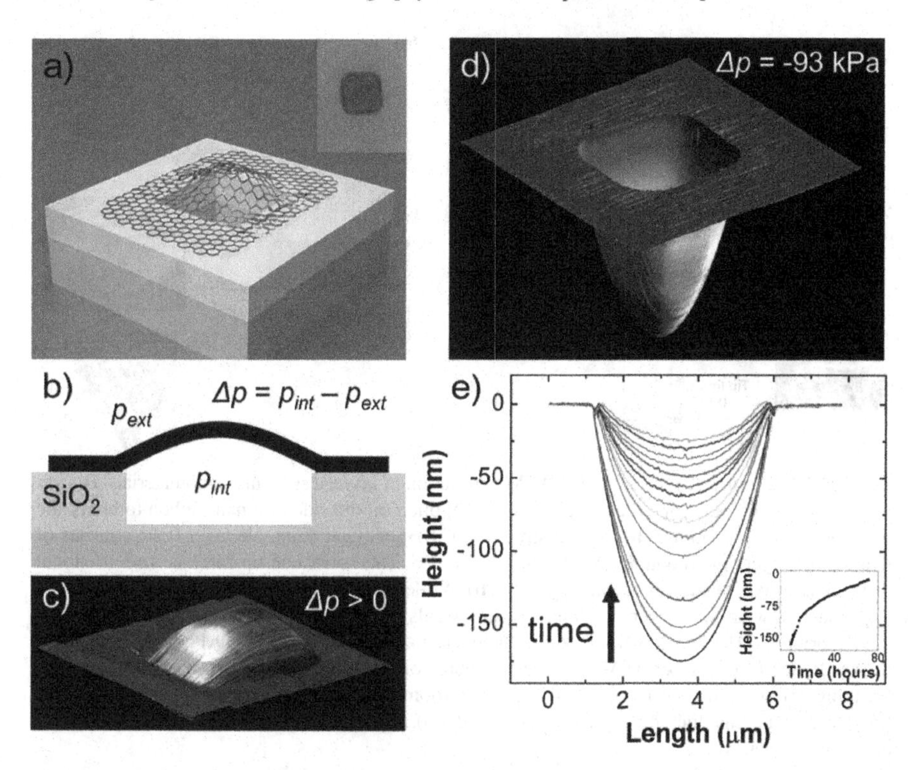

Fig. 10.7 (**a**) Schematic of a graphene sealed microchamber. (*Inset*) optical image of a single atomic layer graphene drumhead on 440 nm of SiO_2. The dimensions of the microchamber are 4.75 μm × 4.75 μm × 380 nm. (**b**) Side view schematic of the graphene sealed microchamber. (**c**) Tapping mode atomic force microscope (AFM) image of a ~ 9 nm thick many layer graphene drumhead with $\Delta p > 0$. The dimensions of the *square microchamber* are 4.75 μm × 4.75 μm. The upward deflection at the center of the membrane is z ~ 90 nm. (**d**) AFM image of the graphene sealed microchamber of (**c**) with $\Delta p \sim -93$ kPa across it. The minimum dip in the z direction is 175 nm. (**e**) AFM line traces taken through the center of the graphene membrane of (**a**). The images were taken continuously over a span of 71.3 h and in ambient conditions. (*Inset*) deflection at the center of the graphene membrane vs time. The first deflection measurement (z ~ 175 nm) is taken 40 min after removing the microchamber from vacuum (from Bunch et al. [33])

obtained by chemically-induced exfoliation methods starting from bulk graphite samples [34, 35]. As example, in Fig. 10.8, we report a recent experiment by Yi Cui and co-workers of SPN on Bi_2Se_3, where and AFM tip has been employed to separate different layers in nanoribbons and a single quintuple layer was obtained [36].

When the SPN produces a controlled and permanent modification of the nanostructure, we can use the term scanning probe lithography (SPL). In the final part of this section we would like to summarize the recent results of AFM-SPL on graphene.

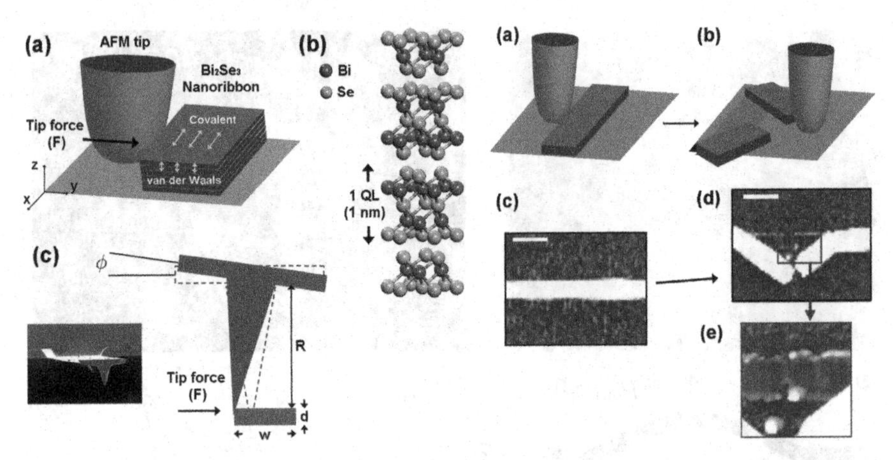

Fig. 10.8 *Left panel*: (**a**) Schematic of AFM exfoliation of layered structure nanomaterial–Bi_2Se_3 nanoribbon. Horizontal tip force (y-direction) is applied on the side of a nanoribbon to break the in-plane covalent bonding. (**b**) Crystal structure of Bi_2Se_3. One quintuple layer (QL) consists of five atomic layers of Bi and Se (Se–Bi–Se–Bi–Se). (**c**) The horizontal tip force is from torsional displacement (φ) of the tip. *Right panel*: (**a, b**) While the tip force breaks covalent bonding of upper layers, weak interlayer bonding, van der Waals bonding, is also broken in the middle of the nanoribbon. As a result, several bottom layers remained intact while the AFM tip breaks most of the Bi_2Se_3 layers. (**c–e**) AFM topographic image of a Bi_2Se_3 nanoribbon (d ~50 nm) before breaking (**c**) and after the breaking (**d**), and the zoom-in image (**e**). The bottom layers are not broken apart and maintain the same thickness (~2 nm). All scale bars indicate 500 nm (from Hong et al. [36])

10.2.2 AFM-SPL on Graphene

When the applied force between the AFM tip and the graphene flakes is increased along a certain line in order to permanently modify its surface, the typical results it is not a precise cut but a movement or a crumbling in a rather uncontrolled fashion due to the mechanical robustness of the graphene and to the relatively low interacting force of the flake with the SiO_2 surface. Some authors performed deposition by mechanical exfoliation on GaAs substrate (having a low-roughness surface compared to SiO_2 substrate) to increase this interaction and to mechanically structuring the flakes at nanometer scale [37]. In Fig. 10.9 are shown their main results, as a cut on a multilayer graphene flake (top panel) and a nanoscale peeling of another one (bottom panel), both obtained by AFM-SPL. Other authors manipulated and modified graphene directly on SiO_2 substrate, studying in this way its interaction with it [38], or performed nanomanipulation on ridges of few-layer epitaxial graphene grown on 4H-SiC (see Fig. 10.10a, b) [39]. Finally, recent results by Hua Zhang and co-workers [40], obtained on single layers of graphene oxide (GO) on SiO_2 substrates, seem to demonstrate that mechanical scratching by AFM results easier respect to SLG (as reported in Fig. 10.10c), as could be theoretically argued by the fact that GO has weakened chemical bonds compared to graphene.

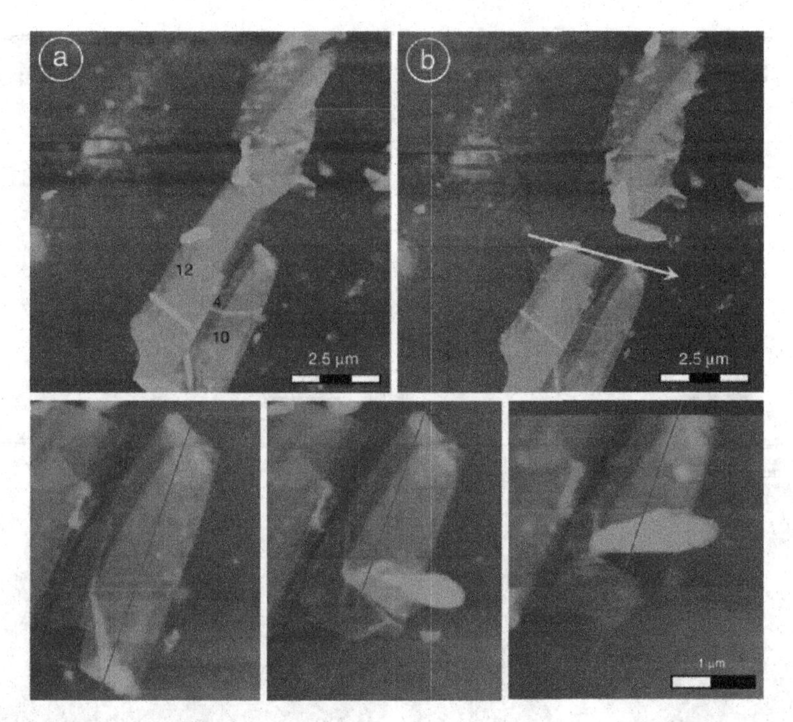

Fig. 10.9 *Top panel*: "Brute-force" mechanical manipulation of few-layer graphene flakes on GaAs (**a**, **b**). The number of layers is depicted in the figure. The flake in (**a**) is approached from the left with the AFM tip. The flake rips apart and rolls up to the top. *Bottom panel*: AFM micrographs after three subsequent nano-peeling steps (*left* to *right*) of a few layer graphene flake (from Giesbers et al. [37])

When a bias is applied between the AFM tip and the sample, a new kind of SPL can be done and a different modification of the surface can be obtained. More in detail, in the case of SLG flake where metallic ohmic contacts have been fabricated by standard e-beam lithography and in presence of a water bridge between the scanning probe tip and the sample, a process like the local anodic oxidation (LAO) [41–44] can be observed, as schematized in the cartoon of Fig. 10.11a. The strong electric field present between the AFM tip and the sample and the presence of a water meniscus, induces an electrochemical oxidation of carbon at room temperature, creating volatile carbon oxide molecules (CO_x) and therefore a spatially-controlled modification of the graphene surface, as observed from A. J. M. Giesbers and collaborators [37]. They found that this technique works only if the scan line is started at the edge of a graphene sheet (see Fig. 10.11b). This behavior was related to the fact that edge termination of graphene by, as example, hydrogen atoms can substantially facilitate the initial oxidation process. Another possible explanation was found in the hydrophobic character of graphene sample that allows at the water meniscus to be stable only moving the tip from the SiO_2 substrate towards the edge of the flakes. With this direct lithographic technique,

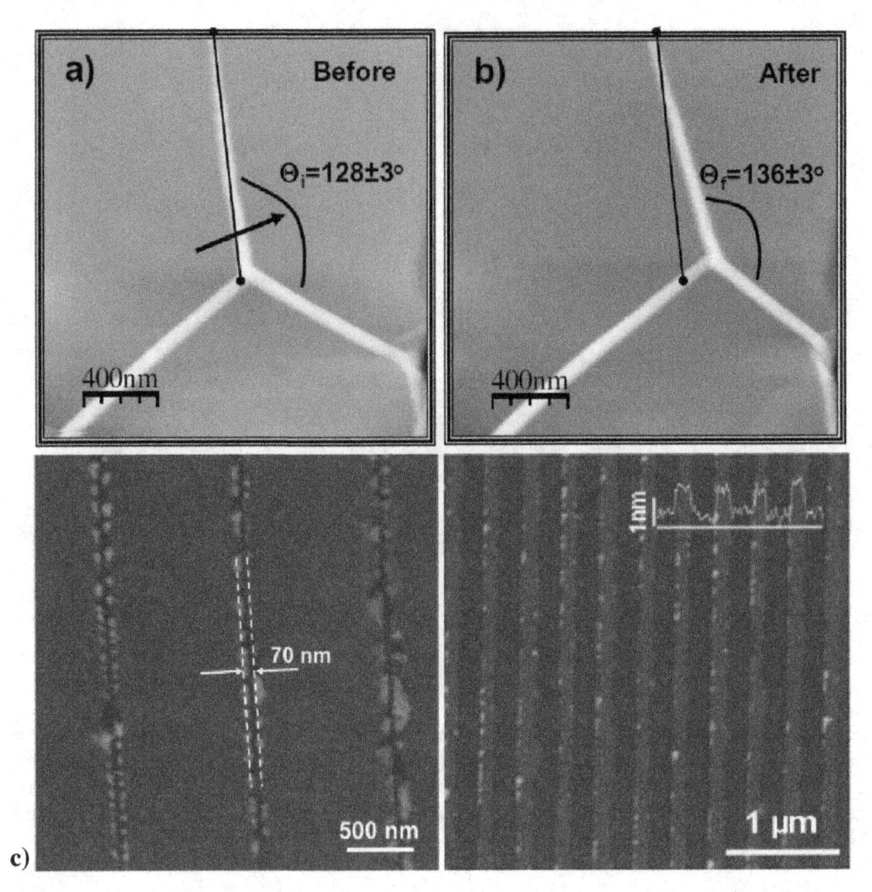

Fig. 10.10 In (**a**), an intermittent contact AFM image of epitaxial graphene grown on SiC substrate, obtained under ambient conditions is shown. Observe the presence of an interconnected ridge network, connected by ridge nodes. A *black line* in the upper half of the image serves to locate the ridge node. The *black arrow* shows the direction of a contact mode scan performed during a nanomanipulation experiment. In (**b**), the resulting AFM image after nanomanipulation is shown. The ridge node is displaced by ~200 nm. The subtended ridge angle also increases by ~8° Prakash et al. [39]). In (**c**) AFM images of nanogaps having various sizes and fabricated by mechanical scratching in graphene oxide flakes deposited on Si/SiO_2 substrate (from Gang et al. [40])

several fabrication steps can be eliminated, such the resist process necessary for the photo or e-beam lithography, and the capability to control the oxidation process at the nanometer scale together with the sample monitoring and imaging during and after the lithographic procedure can be finally achieved. Many complex patterns on SLG have been obtained in this way and transport measurements on graphene nanostructure accomplished. L. P. Rokhinson and coworkers [45] demonstrated LAO lithographic resolution on graphene nanoribbons down to 25 nm of width and the capability to pattern complex structures as Ahronov-Bohm nanorings (see Fig. 10.12). Interestingly, on the same work was also demonstrated the LAO process

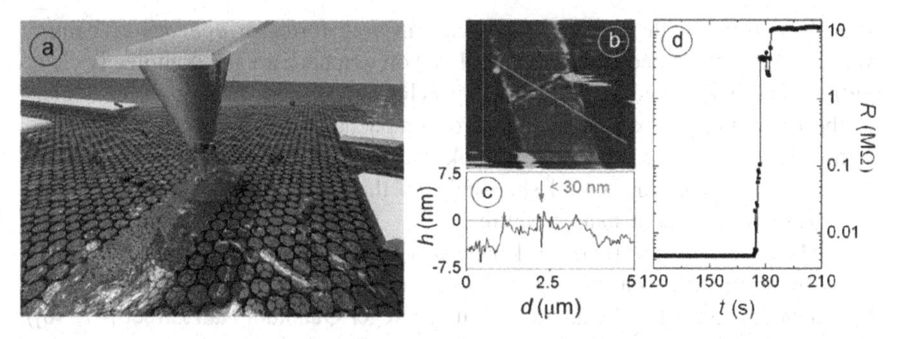

Fig. 10.11 Local anodic oxidation (LAO) with an AFM. (**a**) Artistic impression of the oxidation procedure: The AFM-tip oxidizes the graphene, in the presence of a water layer, into carbon-based oxides that escape from the surface. (**b**) AFM micrograph of an oxidized groove into a graphene ribbon. (**c**) Cross-section of the micrograph along the *gray line*, showing the oxidized trench. (**d**) The resistance of the ribbon, monitored during oxidation, increases dramatically as it is oxidized in two pieces (from Giesbers et al. [37])

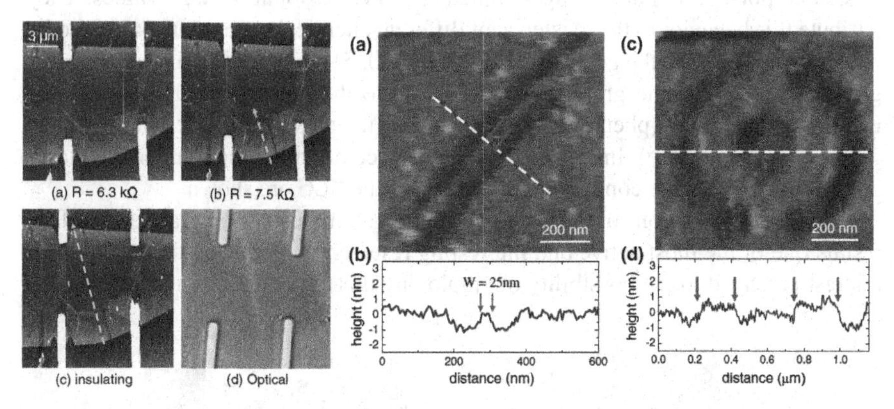

Fig. 10.12 *Left panel*: (**a**) AFM image of an uncut graphene flake (thickness ~5 nm). The *four white bars* in the picture are the metal contacts. The two-terminal resistance was 6.3 kΩ. (**b**) A trench was cut from the edge to the middle of the flake, along the direction indicated with the *dashed arrow*. The resistance increased to 7.5 kΩ. (**c**) The trench was cut through, electrically insulating the left and right parts of the flake. (**d**) Optical microscope image of the same flake with trenches. *Right panel*: (**a**) AFM image of a nanoribbon fabricated on a graphene flake with thickness ~1 nm. The width and length of the ribbon are 25 and 800 nm, respectively. (**b**) Height profile along the dashed line in (**a**). (**c**) A nanoring (inner radius ~160 nm, outer radius ~380 nm) patterned on a graphene flake. Two long trenches, not shown in the picture, were subsequently drawn from the circumference of the ring outward to the edges of the flake to electrically isolate the ring device. (**d**) Height profile along the *dashed line* in (**c**) (from Weng et al. [45])

on flakes electrically floating and it was described that, depending on the lithographic parameters and conditions, either trenches or bumps can be generated on the graphene surface. Researchers attributed bumps to partial oxidation of the graphene, with oxygen incorporated into the graphene lattice. They also verified that written oxidation lines originating either from the edge or from the middle of the flake (in contrast with reference [37]), and in both forms of trenches or bumps. At the moment these peculiar aspects of LAO on graphene are not completely clear. Further experiments and theoretical analysis have to be performed in order to clarify the field-induced oxidation kinetics on graphene and the water meniscus role in this electrochemical process. We think that in the next future high spatial resolution Raman scattering could shed light on the local chemical composition of LAO-patterned graphene nanostructures.

T. Machida and coworkers [46] patterned by LAO techniques both SLG Hall bars as nanoribbons, measuring peculiar electronic properties of Dirac fermions in these nanostructures (as the anomalous half-integer Quantum Hall Effect [47, 48]) and demonstrating the capability to perform by SPL energy band gap engineering in graphene nanoribbons (see Fig. 10.13).

SPL based on field-induced modification of SLG has been also studied by F. Pérez-Murano's Group at Barcelona [49] on a graphene layer fabricated by epitaxial growth on SiC substrate. In this case Authors, contrarily to the previous reported examples, were able to perform patterning both at positive as negative tip-to-sample polarization and using significantly lower absolute voltage values. They attribute this behavior to the existence of different chemical phenomena: a "cathodic oxidation" (including the generation of H_2O_2) [50, 51] or an SPM – induced hydrogenation process of the graphene surface, due to the presence of H ions created inside the water layer (phenomenon enhanced by the presence of free-charges at the graphene/SiC interface). In Fig. 10.14 the presence of patterning at opposite biases and their effects on the conducting properties of the SLG are shown.

The capability to control the density of hydrogen molecules on graphene surface results one of the most active and interesting research on graphene field. The great interest is related to the possibility of employing this extraordinary new material as

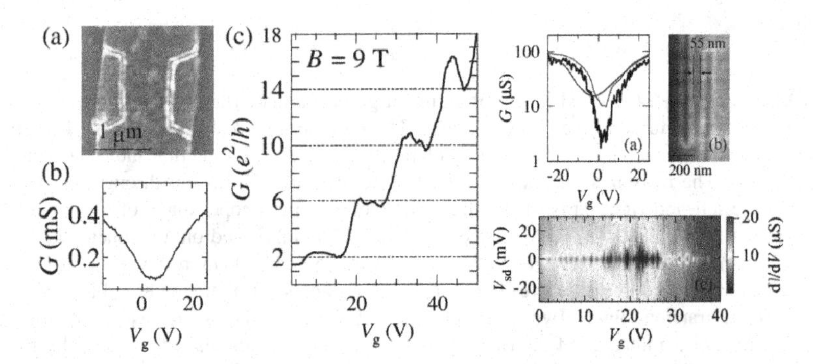

Fig. 10.13 *Left panel*: (**a**) AFM image of the *bar-shaped device* fabricated in single-layer graphene. (**b**) Two-terminal conductance G as a function of the gate-bias voltage V_g measured at T = 4.2 K in a zero magnetic field. (**c**) G as a function of V_g measured at T = 4.2 K and B = 9 T. *Right panel*: (**a**) Two-terminal conductance G as a function of the gate-bias voltage V_g measured at T = 300, 77, and 4.2 K in a zero magnetic field. (**b**) AFM image of a 55-nm-wide (W) and 800-nm-long (L) nanoribbon formed in single layer graphene. (**c**) *Color plot* of the differential conductance dI/dV of the 77-nm-wide and 600-nm-long nanoribbon as a function of the gate-bias voltage V_g and the source-drain bias voltage V_{sd} (from Masubuchi et al. [46])

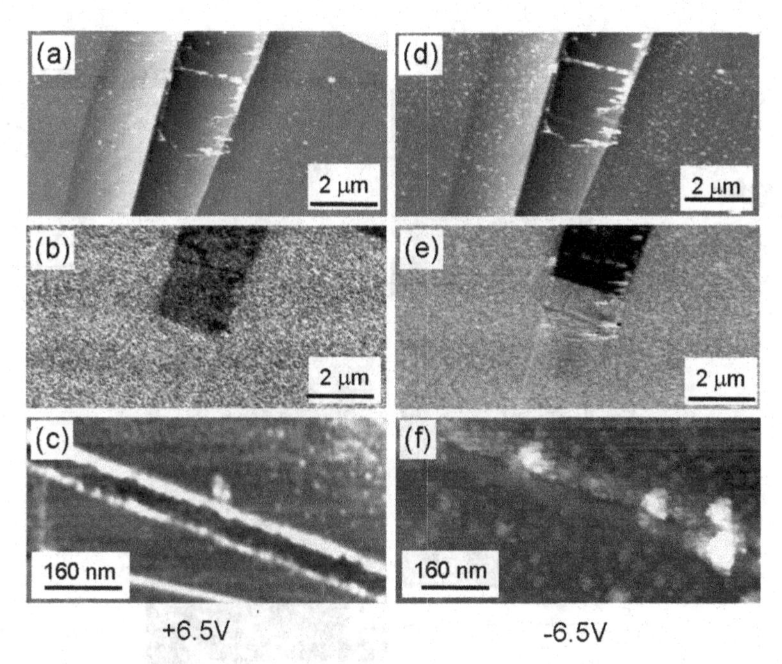

Fig. 10.14 (**a**, **c**, **d**, and **f**) Topographic and (**b**, **e**) electrical force dynamic AFM images of a graphene ribbon on a SiC substrate contacted with a gold electrode located at the top of the scanning area. The images are taken after scanning same areas in contact mode in order to remove surface contamination. Images (**a–c**) are taken after creating a *line* while polarizing the graphene layer at 6.5 V. Images (**d–f**) are taken after creating a *line* while polarizing the graphene layer at –6.5 V. Electrical force microscopy images (**b**, **e**) show that *both lines* effectively disconnect the graphene ribbon (from Rius et al. [49])

an efficient and safe hydrogen storage system or to use SPM as a tool to fabricate sculpted graphene nanoribbons [52] by hydrogen desorption of graphane, the fully hydrogenated graphene layer [53–55].

To conclude this section on AFM-based SPL we would like to underline the results obtained by E. Riedo and collaborators employing Thermo Chemical Nano Lithography (TCNL) technique [56]. A heated AFM tip produces a reduction chemical process in selected regions of both single layers of isolated graphene oxide (GO) and large-area GO films formed by on-chip oxidation of epitaxial graphene (GO_{epi}) grown on SiC wafer, as shown in Fig. 10.15a, b. Researchers demonstrated in this way the possibility to use SPL as a tool to directly create nanometric conductive paths on top of GO surface, employing conductive AFM to perform the spatially resolved current imaging of the nanoribbons. Similar results employing an electrochemical reduction tip-induced technique on GO flakes have also been recently obtained by P. Samorì, V. Palermo and colleagues [57]. In this paper Authors demonstrated that the electrochemical reduction process involve the presence of water at the tip-GO interface. In presence of a negative bias applied to the sample, water meniscus is oxidized, thereby generating hydrogen ions that take part in reducing

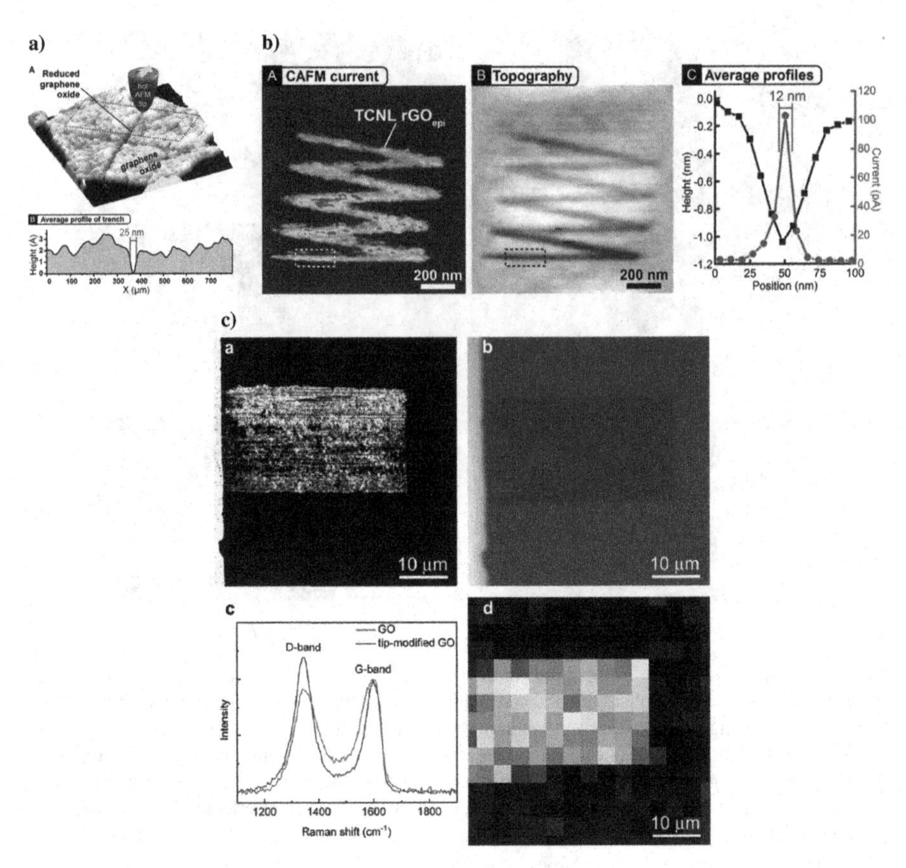

Fig. 10.15 (**a**) Local thermal reduction of a single-layered graphene oxide flake. (*A*) Topography of a cross shape of reduced GO formed after an AFM tip heats the contact to 330°C scanned across the GO sheet at 2 μm/s. (*B*) The averaged profile of the trench outlined in (*A*) shows that the width (FWHM) of the *line* can be as narrow as 25 nm. (**b**) Current and topographical images. (*A*) Room-temperature AFM current image (taken with a bias voltage of 2.5 V between tip and substrate) of a zigzag-shaped nanoribbon fabricated by TCNL on GO_{epi} at $T_{heater} \sim 1,060$°C with a linear speed of 0.2 μm/s and a load of 120 nN. (*B*) Corresponding topography image taken simultaneously with (*A*). (*C*) Averaged profiles of current and height of the cross sections that are indicated as dashed lines in (*A*) and (*B*) (from Wei et al. [56]). (**c**): (**a**) C-AFM current map of a tip-modified rectangular region of a few-layer GO film and a top-contact gold electrode (*left side*; current range ~50 nA). (**b**) Optical image of the same region showing contrast between the modified and unmodified GO. (**c**) Raman spectra obtained on modified and unmodified GO (normalized to the G-band peak). (**d**) Raman map representing the spatial dependence of the D-band/G-band intensity ratio, demonstrating the spatially-resolved reduction process in the GO surface (from Mativetsky et al. [57])

the GO surface [58]. In Fig. 10.15c we report conductive AFM, optical microscopy and micro Raman spectroscopy measurements that have been employed to characterize the spatially-resolved reduced GO region. While this type of research is still in its early days, it is possible that this new GO reduction technique will be extended so that arrays of AFM tips rapidly write circuits at high rates across graphene or

GO wafers, as thermo mechanical lithography did on polymers trough the millipede project [59].

AFM employed in conductive mode either to oxidize the graphene or to perform current imaging on reduced graphene oxide, as has been illustrated by few examples in this section, allows to smoothly switching toward the second part of this review where a tunnel current passing through a metallic tip is employed as feedback for imaging. Therefore, Scanning Tunneling Microscope (STM) and the results obtained on graphene both for imaging as for nanolithography and spectroscopy employing this atomic resolution microscopy technique, will be the argument of the following section.

10.3 STM-Based Scanning Probe Lithography (SPL) and Spectroscopy (SPS)

Scanning Tunneling Microscopy (STM) has long been used to observe the electronic topography of graphite and the "scotch tape" method was employed many times by researchers in the past to clean the surface of graphite before STM imaging (unfortunately throwing out the tape with the graphene flakes attached on it). In standard experiments, only three carbon atoms of the six present in the graphene lattice were visible in STM imaging due to the AB stacking of carbon layers on graphite. On the contrary, the STM imaging on single layer graphene on top of SiO_2 substrate should reveals the six atoms lattice cell structure, as confirmed by Flynn and co-workers in UHV-STM performed on mechanical exfoliated sample contacted by metallic electrodes (see Fig. 10.16, top panel) [60]. In this work the high crystal quality of these samples on relatively large scale (tens of nanometers size) was also demonstrated, together with the corrugation of the surface due to the underlying SiO_2 substrate (see Fig. 10.16, bottom panel). Other authors [61], performing STM measurements in ambient conditions, demonstrated indirectly for the first time that graphene surface primarily follows the underlying morphology of SiO_2 and thus it does not have intrinsic, independent corrugations. See Fig. 10.17 for typical large area or high resolution STM images of graphene on SiO_2 and for the height-height correlation function of the graphene sheet and SiO_2 surface. In particular, the reader should observe the low density of defects in the crystal structure and the presence of both triangular and hexagonal patterns in the high-resolution images (Fig. 10.17, left panel b and d) obtained by the same STM tip. Authors observed lattice spacing consistent with the graphene lattice, and the appearance of both triangular and hexagonal lattice in the image was attributed to the presence of strong spatially dependent perturbations which interact with graphene electronic states, probably related to the observed film curvature and/or the charge traps on the SiO_2 surface.

Another peculiar characteristic of STM imaging is the possibility to perform local spectroscopy by varying the value of the applied bias necessary to obtain the feedback tunneling current or by modulating it in order to extract current-voltage (I-V) or differential conductance vs voltage (dI/dU or dI/dV) data at different locations of the sample surface. This STM-based spectroscopy (STM-SPS) results very useful

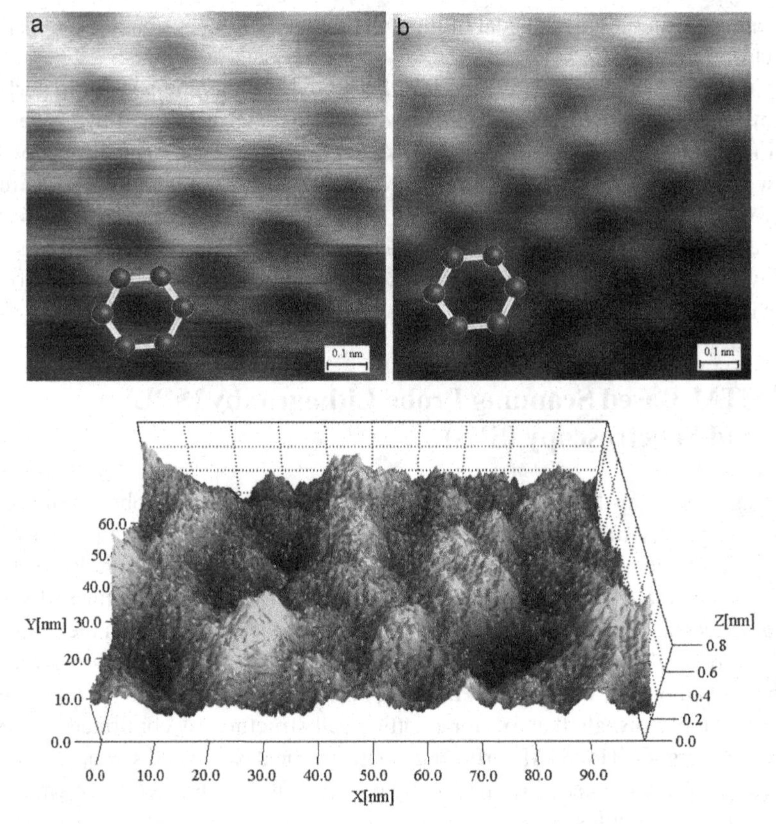

Fig. 10.16 *Top panel*: STM topographic images of different regions of the graphene flakes. The images were obtained with $V_{bias} \sim 1$ V (sample potential), I ~1 nA, and a scan area of 1 nm^2. A model of the underlying atomic structure is shown as a guide to the eye. (**a**) Image from a single layer of graphene: a honeycomb structure is observed. (**b**) Image of the multilayer portion of the sample: the characteristic three-for-six STM image of the surface of bulk graphite is observed. *Bottom panel*: Stereographic plot of a large-scale (100 × 62 nm) STM image of a single-layer graphene film on the silicon dioxide surface. The STM scanning conditions were $V_{bias} \sim 1$ V (sample potential) and I ~ 0.6 nA. The 0.8-nm scale of the vertical (Z) coordinate is greatly enlarged to accentuate the surface features (from Stolyarova et al. [60])

in the case of graphene in order to obtain information about the local electronic density or density of states (LDOS) and to study in particular the effects of oxygen or hydrogen absorption on its surface.

A recent work performed by G. Li et al. [62] demonstrated the possibility to perform low temperature and high magnetic field scanning tunneling microscopy and spectroscopy of graphene flakes on bulk graphite that exhibit the structural and electronic properties of graphene decoupled from the underlying substrate. They were thus able to find "true graphene" on top of bulk graphite samples and they also demonstrated the high quality of these decoupled single layers from electronic transport point of view. In Fig. 10.18 some topographic STM images of the SLG are

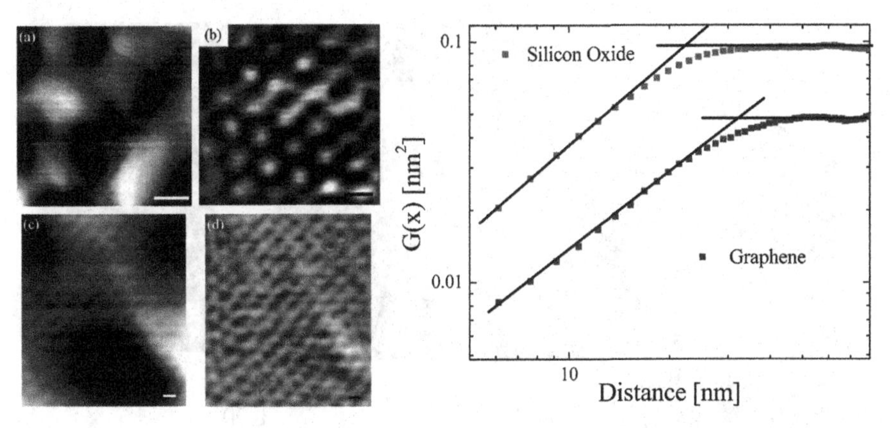

Fig. 10.17 *Left panel*: (**a**) A typical large-area STM image of graphene on top of SiO_2: peak-to-peak height variation of the image is approximately 2.5 nm. $V_{sample} = 1.1$ V and $I_{tunnel} = 0.3$ nA. The scale bar is 2 nm. (**b**) Atomically resolved image of a graphene sheet. $V_{sample} = 1.0$ V and $I_{tunnel} = 24$ pA. The scale bar is 2.5 Å. (**c**) STM image of another area. The scale bar is 2.5 Å. $V_{sample} = 1.2$ V and $I_{tunnel} = 0.33$ nA. (**d**) A high-pass filtered image of the large area scan shown in (**c**). Both triangular and hexagonal patterns are observed. The orientations of the *triangle* and *hexagons* are same. The scale bar is 2.5 Å. *Right panel*: The height-height correlation function of the graphene sheet and SiO_2 surface. The *lines* are fits to the large and small length behaviors (power-law and constant, respectively), and the point of intersection indicates the correlation length. This analysis is performed by selecting data from an acquired image showing both graphene and SiO_2 surfaces. Therefore, the tip morphology is the same for *both curves* and the tip-related artifact effect does not contribute to the analysis (from Ishigami et al. [61])

shown, while Fig. 10.19 reports the related STM-SPS data. Pronounced peaks in the tunneling spectra develop with increasing field, revealing a Landau level sequence that provides a direct way to identify graphene and to determine the degree of its coupling to the underlying substrate. The Fermi velocity and the quasiparticle lifetime, obtained from the positions and width of the Landau peaks, provide access to the electron-phonon and electron-electron interactions.

The presence of decoupled graphene layers on top of bulk graphene is interesting from many points of view. Among those, the fact that a honeycomb structure was already observed many times in the past by STM imaging on bulk graphite and many years before the graphene discovery. At that time this "anomalous" structure was attributed to multiple-tip artifacts [63], as other atypical STM images of graphite. Probably, we will never know if this interpretation of experimental data was right or if graphene was imaged for the first time by STM more than 23 years ago.

From STM imaging point of view, many interesting results have been obtained on graphene directly grown on SiC [64, 65] or on metals like Ni, Cu, Ru, Rh, Pt, Ir [66–68]. More in details, Moiré patterns effects on imaging have been shown and very interesting spatially-resolved spectroscopy (SPS) data obtained in different configurations and growth conditions. In Fig. 10.20 (left panel) an epitaxially- grown graphene monolayer on top of Ru(0001) surface is reported. It shows a periodic rippled structure and correlated charge inhomogeneities in the charge distribution. Real space measurements by scanning tunneling spectroscopy revealed the existence of

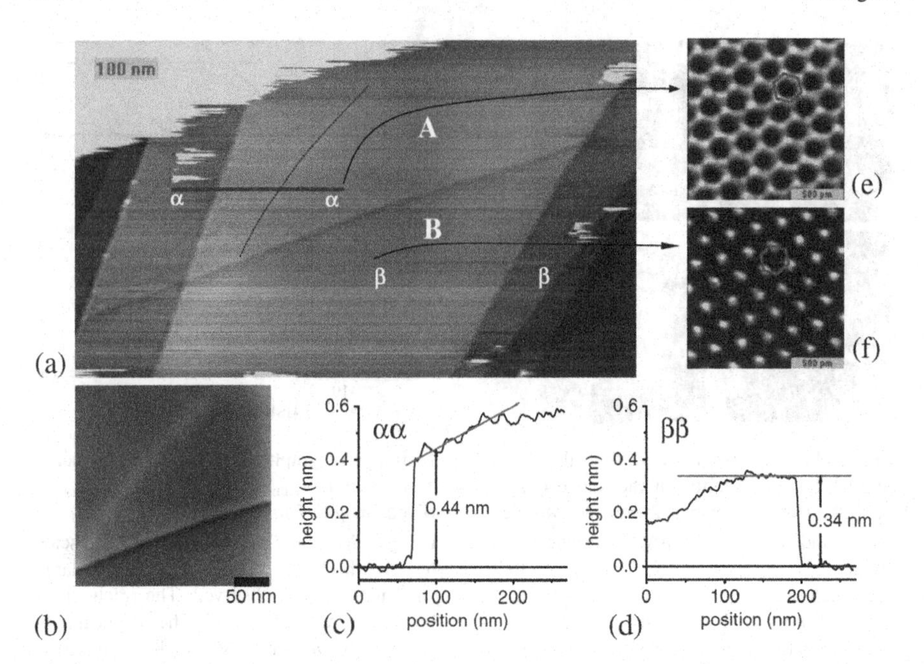

Fig. 10.18 Topography of the graphene layer isolated by extended defects on a graphite surface. (**a**) Large area topography. Two underlying defects are seen: a long ridge that runs diagonally under the top two layers and a fainter one under the first layer (*dashed line*). The long ridge separates a region with honeycomb structure (region *A*) above it, (**e**), from one with triangular structure (region *B*) below, (**f**). *Two arrows* mark positions where atomic-resolution images were taken. (**b**) High resolution image where the fainter ridge is visible. (**c**) Crosssectional cut along line in (**a**) showing that the separation between the top and second layer is larger than the equilibrium value (0.34 nm) near the fainter ridge. (**d**) Cross-sectional cut along line showing that the height of an atomic step far from the ridges is comparable to that in Bernal-stacked graphite. (**e, f**) Atomic-resolution image showing honeycomb structure in region *A* (atoms visible at all 6 *hexagon* vertices) and triangular structure in *B* (atoms seen only at 3 vertices corresponding to only one visible sublattice). A coherent honeycomb structure is seen over the entire region A. Set sample bias voltage and tunneling current were 300 mV and 9 pA for (**a**), 300 mV and 49 pA for (**b**), 200 mV and 22 pA for (**f**), 300 mV and 55 pA for (**e**) (from Li et al. [62])

electron pockets at the higher parts of the ripples, as also predicted by a simple theoretical model (see Fig. 10.20, right panel) [69]. In another interesting work, strained graphene nanobubbles were created by in situ growth of sub-monolayer graphene films in ultrahigh vacuum on a clean Pt(111) surface and imaged by STM as shown in Fig. 10.21 [70]. In this interesting work Authors were able to demonstrate by STM spectroscopy how these strained nanostructures, self-assembled on graphene, exhibit Landau levels that form in the presence of strain-induced pseudo–magnetic fields greater than 300 Tesla.

The presence of strain on graphene layers and how it can induce spatial modulations in the local conductance of single-layer graphene on SiO_2 substrates, has been also demonstrate by STM imaging and spectroscopy by M.L. Teague and colleagues [71]. They found that strained graphene exhibits parabolic, U-shaped conductance

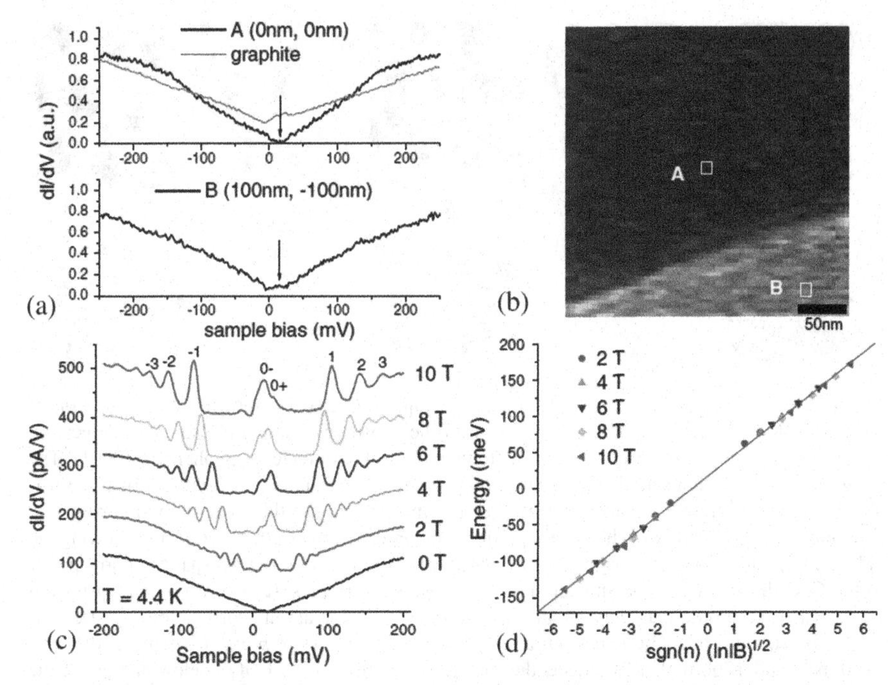

Fig. 10.19 Spectroscopic evidence for decoupled graphene layer. (**a**) Zero-field spectroscopy taken in regions *A* and *B* marked by *squares* in (**b**). *Top panel* shows that the tunneling conductance for decoupled graphene (region *A*) is *V shaped* and vanishes at the Dirac point (DP) (marked by the *arrow*). A typical spectrum for graphite is included for comparison. *Bottom panel* shows that in the more strongly coupled layer (region *B*) the differential conductance does not vanish at the DP. (**b**) Differential conductance (dI/dV) map at the DP energy as marked by *arrows* in (**a**). The *scanned area* is the same as in Fig. 10.1b. dI/dV vanishes in the *dark region* but is finite in the *bright region*. (**c**) Field dependence of tunneling spectra in region *A* showing a single sequence of Landau levels (LL). The peaks are labeled with LL index n. Each spectrum is shifted by 80 pA/V for clarity. (**d**) LL energy showing square-root dependence on level index and field. The *symbols* correspond to the peaks in (**c**), and the *solid line* is a fit. From the slope we obtain $v_F = 0.79 \times 10^6$ m/s (v_F is the Fermi velocity) and from the intercept $E_D = 16.6$ mV (E_D is the energy at the Dirac point) indicating *hole* doping. The tunneling junctions parameters were set at a bias voltage of 300 mV and tunneling current of 20 pA for (**a**) and (**c**) and 53 pA for (**b**). The amplitude of the ac bias voltage modulation was 5 mV for all spectra in the figure (from Li et al. [62])

versus bias voltage spectra rather than the V-shaped spectra expected for Dirac fermions, whereas V-shaped spectra are recovered in regions of relaxed graphene. Strain maps derived from the STM studies further revealed direct correlation with the local tunneling conductance.

But STM can also be employed to directly perform high resolution lithography (STM-SPL). This application is particularly important in the case of graphene nanostructures, where the STM capability to modify them at the atomic scale (and not just at the nanometer one as in the AFM case) provide a new tool to explore the peculiar transport and optical characteristics of this carbon-based material. L.P. Biró and collaborators demonstrated the extraordinary lithographic resolution achievable

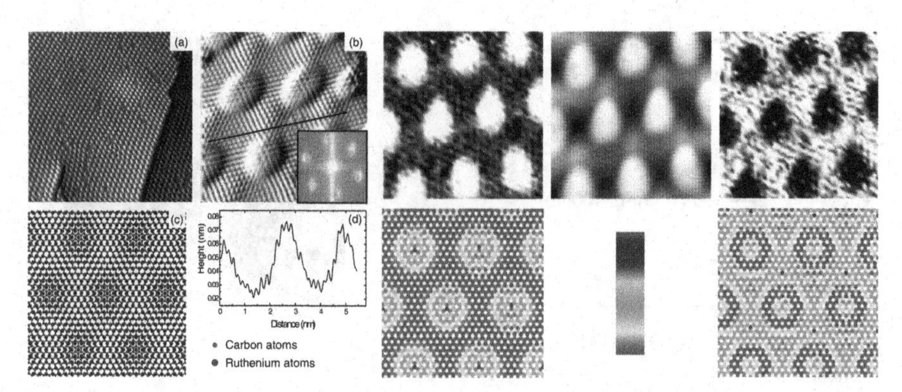

Fig. 10.20 (**a**) 76 nm × 76 nm STM image of graphene/Ru(0001) showing the decoration of a screw dislocation and a monoatomic step from the substrate. There are also some defects on the rippled structure. (**b**) 6.5 nm × 6.5 nm atomically resolved image of graphene/Ru(0001). The image was taken with a sample bias voltage of Vs = −4.5 mV and a tunnel current of It = 3 nA. The image is differentiated along the X direction in order to see the weak atomic corrugation superimposed to the ripples. The *inset* reproduces the Fourier transform of the image showing the (11 × 11) periodicity of the rippled graphene layer. The *larger hexagonal* pattern corresponds to the C–C distances and the smaller spots to the periodic ripples. (**c**) Corresponding structural model. (**d**) Line profile marked with an *arrow* in *panel* (**b**). The atomic corrugation in these conditions is around 5 pm. The *upper left* and *right images* in the *right panel* are maps of dI/dV at –100 meV and +200 meV and reflects the spatial distribution of the LDOS below and above the Fermi level, respectively, for an extended graphene layer on Ru(0001). The *central image* shows the topographic image recorded simultaneously. The *lower panel* of the *right panel* shows the corresponding calculations for the spatially resolved LDOS for a (11 × 11) periodically corrugated graphene layer (from Vázquez de Parga et al. [69])

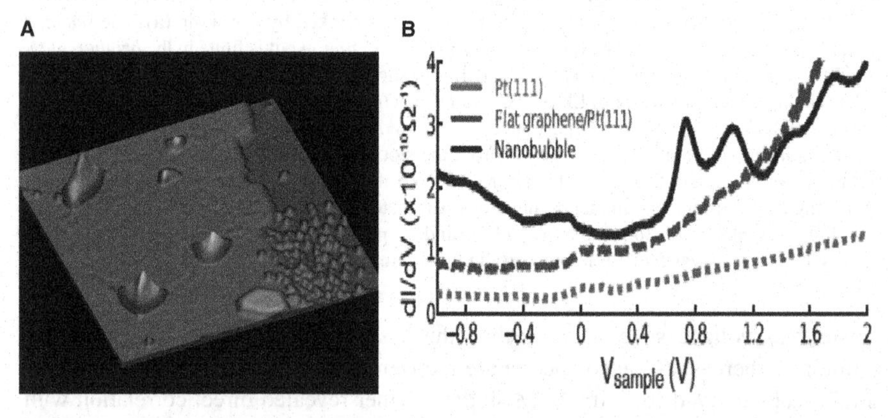

Fig. 10.21 STM images and STS spectra taken at 7.5 K. (**a**) Graphene monolayer patch on Pt(111) with four nanobubbles at the graphene-Pt border and one in the patch interior. Unreacted ethylene molecules and a small hexagonal graphene patch can be seen in the lower right ($I_{tunneling}$ = 50 pA, V_{sample} = 350 mV, 3D z-scale enhanced 4.6×). (**b**) SPS spectra of bare Pt(111), flat graphene on Pt(111) (shifted upward by 3×10^{-11}/Ohm), and the center of a graphene bubble (shifted upward by 9×10^{-11}/Ohm). V_{mod} = 20 mV (from Levy et al. [70])

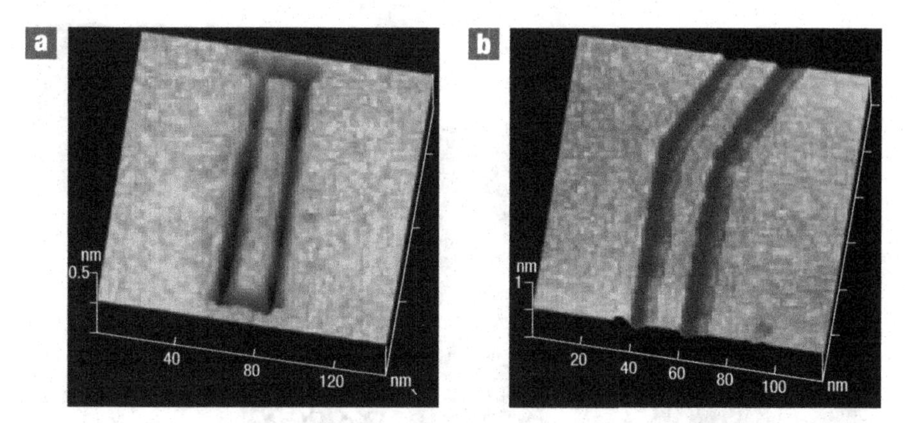

Fig. 10.22 Graphene nanostructures patterned by STM lithography. (**a**) 3D STM image of a 10-nm-wide and 120-nm-long graphene nanoribbon (GNR). (**b**) An 8-nm-wide 30-degree GNR bent junction connecting an armchair and a zigzag ribbon (from Tapasztó et al. [72])

by STM-SPL fabricating graphene nanoribbons [72] on top of bulk graphite sample. The mechanism employed to directly cut just the first graphene layer is not at the moment well clear. Authors suppose that most likely the breaking of carbon–carbon bonds by field-emitted electrons combined with the electron-transfer-enhanced oxidation of the graphene is responsible for the etching, as observed in the past in the case of carbon nanotubes [73]. Nevertheless, they achieved impressive lithographic results (see Fig. 10.22) combining the STM feature of surface modification with atomic-resolution imaging in order to engineer nanostructures with almost atomically precise structures and predetermined electronic properties. In fact, in this case the atomic resolution allowed fabrication of ribbons having 10–15 nm of width together with imaging of their crystallographic orientation. A topographic superstructure due to the interference of the electrons scattered at the irregularities of the edges, was also evidenced (see Fig. 10.23, left panel). Moreover, STM-SPS demonstrated the presence of a gap (0.5 eV) in the energy band structure of a 2.5 nm wide graphene nanoribbon (see Fig. 10.23, right panel): Authors attributed this huge energy gap to the confinement effect induced by nanofabrication. This high-resolution STM-SPL based approach to patterning graphene opens the route for allowing the room-temperature operation of GNR-based electronic devices.

Another interesting application of STM based SPL is the one studied by N. P. Guisinger and coworkers [74]. In their case, it was demonstrated the reversible and local modification of the electronic properties of graphene by hydrogen passivation and subsequent electron-stimulated hydrogen desorption with a scanning tunneling microscope tip. As shown in Fig. 10.24 (left panel a, b), STM is able to distinguish between clean graphene surface grown on SiC substrate from the hydrogenated one both from their different topographies as from their local electronic structures, as evidenced by the corresponding dI/dU data (left panel c, d). Hydrogen desorption can be also locally induced from the graphene surface in a controllable way (Fig. 10.24, left panel e), leaving behind pristine graphene clean surface

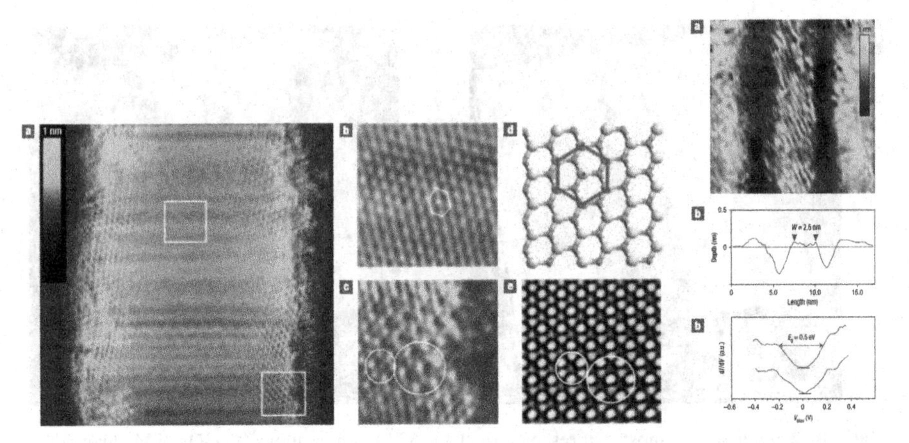

Fig. 10.23 *Left panel*: (**a**) Atomic-resolution STM image (20×20 nm^2, 1 nA, 200 mV) of a 15-nm-wide GNR displaying an atomically flat and defect-free structure. The *color scale bars* encode the height of the imaged features. (**b**, **c**) Magnified images of the defect-free lattice taken at the centre of the ribbon (**b**) and position-dependent superstructures near the edges (**c**). (**d**) Identification of crystallographic orientation from the triangular lattice observed in atomic-resolution STM images of HOPG-supported GNR. (**e**) Theoretical STM image of the superstructures at the edges of the ribbon. *Right panel*: (**a**) STM image (15×15 nm^2, 1 nA, 100 mV) of a 2.5-nm-wide armchair GNR. The *color scale bars* encodes the height of the imaged features. (**b**) Average line-cut of the STM image revealing the real width of the ribbon. (**c**) Representative STS spectra taken on the narrow ribbon showing an energy gap of about 0.5 eV (zero DOS marked by *horizontal lines*) (from Tapasztó et al. [72])

layer. This interesting technique has been finally exploited at room temperature to demonstrate high resolution lithographic capabilities, as reported in Fig. 10.24 (right panel a, b): lines and complex structure have been fabricated down to 5 nm of width. Notice how this STM-based patterning results stable at room temperature. STM-SPS was finally employed to find the characteristic lateral dimensions of the hydrogen-desorbed regions where the electronic structure changes its properties, demonstrating in this very elegant way the quantum-mechanic effects of lateral confinement at nanometer scale induced by hydrogen desorption on graphane.

10.4 Conclusions and Perspectives

In this review we have tried to show how graphene discovery and rising have been up to now strictly related to scanning probe based imaging techniques but also to the capability of SPM to modify its surface at nanometer scale. Many examples of SPN, SPL and SPS techniques applied to graphene have been reported together with some interesting properties of this novel carbon material that have been underlined by SPM characterization. Mechanical properties, electronic transport characteristics, and topographical information have been obtained and lithographic capabilities demonstrated at nanometer scale. From the research works reviewed in this new and fascinating field it results that research on graphene has to thank

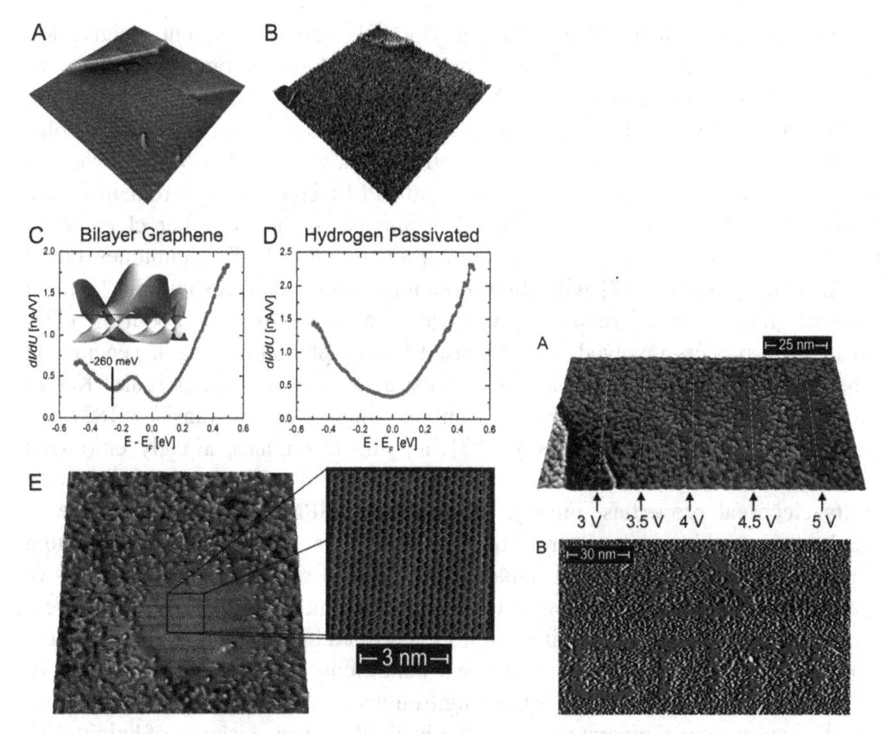

Fig. 10.24 *Left panel*: (**a**) This STM image (100 nm × 100 nm, sample bias 0.3 V, 0.1 nA) shows a cleanly prepared graphene surface that has been epitaxially grown on 4H:SiC(0001). (**b**) The surface is fully saturated with hydrogen at room temperature, as illustrated in this STM image (100 nm × 100 nm). (**c**) SPS is utilized to measure this characteristic dI/dU *curve* over cleanly prepared bilayer graphene. The Dirac point of the bilayer graphene is shifted from the Fermi level to a value of –260 meV. This shift is inherent to epitaxial graphene on SiC and is illustrated by the inset, where graphene's E(kx, ky) dispersion relation is plotted with an added plane representing the shifted Fermi level. (**d**) Similar dI/dU measurements were made on the hydrogen-saturated surface and do not show the same characteristic features indicating that the graphene's electronic structure has been modified. (**e**) This image shows a patterned *box* of graphene. When the bias is increased to +4.5 V, the hydrogen easily desorbs from the surface in a controllable way leaving behind pristine graphene, as illustrated in the atomic resolution image of the *inset*. *Right panel*: This STM image shows an increasing graphene pattern width as a function of increasing positive sample bias. The tip velocity and set point current were held constant. (**b**) To illustrate the level of control over the graphene writing, Authors patterned their institutional University logo and initials with a line width down to 5 nm (from Sessi et al. [74])

SPM techniques as unique tools for high resolution characterization. Moreover, an important issue related to SPM imaging techniques is also their gentleness in avoiding any modification of graphene properties during the topographic or spectroscopic analysis.

We have to mention that other methods as optical imaging and spectroscopy (e.g. Raileigh [75] and Raman [76, 77] scattering) have been employed on graphene characterization and strongly contributed to the present understanding of their extraordinary physical-chemical properties. Furthermore, electron beam lithography (EBL), nano imprint lithography (NIL) and subsequent reactive ion etching processes are at the moment the primarily employed lithographic techniques for the patterning of SLG samples. Nevertheless, we think that in the next future SPM, SPL and SPN will remain unique and flexible tools to perform high resolution imaging, patterning and nanomanipulation techniques of SLG-based devices in basic research. We believe that graphene and SPM techniques will remain linked together and that they will take mutual advantage of their peculiar characteristics. More in details, tip enhanced Raman scattering (TERS) techniques based on AFM imaging [78, 79] will allow obtaining high spatial resolution chemical maps of graphene nanostructures, giving information for example about the electrochemical process involved in LAO-based lithography and, more in general, on functionalized or strained SLG-based nanostructures at nanometer scale. Kelvin Probe Microscopy (KPM) [80, 81], Scanning single-electron transistor [82] and Scanning Gate Microscopy (SGM) [83] imaging techniques, already employed on SLG nanostructures, will allow to imaging and modify at nanometer scale theirs electrical properties. Finally, we think that SPM technique will have in graphene a new kind of substrate to perform an innovative class of interesting experiments and that SPM will remain a high spatial resolution and non-invasive tool to perform imaging on new synthesized graphene nanostructures. As recent intriguing example, AFM imaging on SLG deposited on mica has opened the possibility to study water adlayer in ambient conditions. Hydrophobic properties of graphene substrate allowed demonstrating trough direct imaging that submonolayers of water at room temperature form atomically flat, faceted islands of height 0.37 ± 0.02 nm, in agreement with the height of a monolayer of crystalline ice [84], as shown in Fig. 10.25. These kinds of studies could also have important repercussion

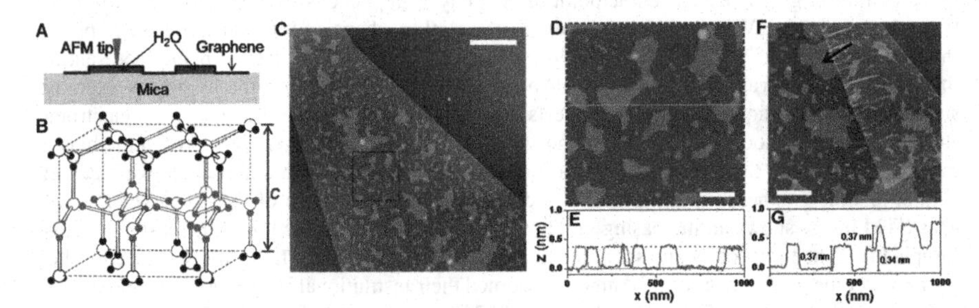

Fig. 10.25 Graphene visualizes the first water adlayer on mica surface at ambient conditions. (**a**) A schematic of how graphene locks the first water adlayer on mica into fixed patterns and serves as an ultrathin coating for AFM. (**b**) The structure of ordinary ice. *Open balls* represent O atoms, and smaller, *solid balls* represent H atoms. A single puckered bilayer is highlighted with *gray* in the central region. Interlayer distance is c/2 = 0.369 nm when close to 0°C. (**c**) AFM image of a monolayer graphene sheet deposited on mica at ambient conditions. (**d**) A close-up of the *blue square* in (**c**). (**e**) Height profiles along the *gray line* in (**d**) and from a different sample. The *dashed line* indicates z = 0.37 nm. (**f**) AFM image of another sample, where the edge of a monolayer graphene sheet is folded underneath itself. The *arrow* points to an island with multiple 120° corners. (**g**) The height profile along the *red line* in (**f**), crossing the folded region. Scale bars indicate 1 mm for (**c**) and 200 nm for (**d**) and (**f**). The same height scale (4 nm) is used for all images (from Xu et al. [84])

on water meniscus role and stability on LAO experiments described in the previous "AFM-SPL on graphene" section.

Finally, atomically precise graphene nanoribbons of different topologies and widths grown on top of Au(111) and Ag(111) substrates have been successfully imaged after their chemical synthesis thanks to STM. The topology, width and edge periphery of the graphene nanoribbon products are defined by the structure of the precursor monomers employed in this bottom-up lithographic technique [85]. These monomers can be designed to give access to a wide range of different graphene nanoribbons but STM imaging results a unique tool able to verify their presence, distribution, topography and electronic characteristics at the atomic scale. In Fig. 10.26 different shapes of these graphene nanoribbons are reported, with beautiful high resolution images of their atomic structure.

In perspective, we believe therefore that SLG will be also employed as a reliable, mechanically stable, conductive substrate for SPM research, becoming a

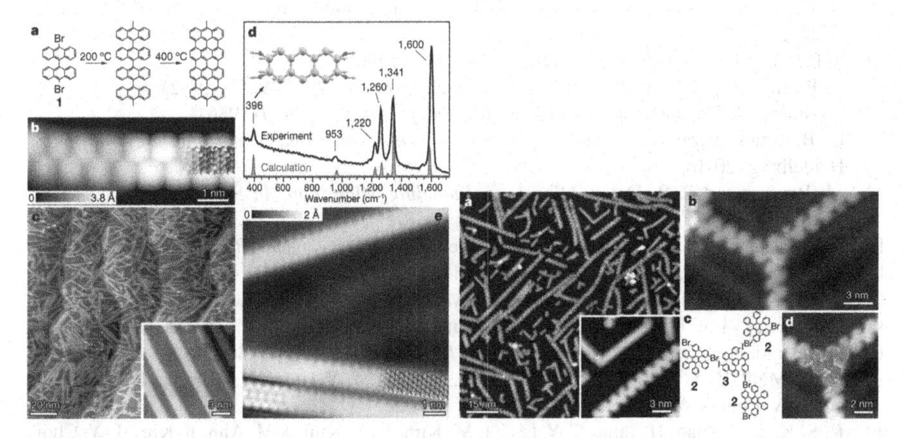

Fig. 10.26 *Left panel*: Straight GNRs from bianthryl monomers. (**a**) Reaction scheme from precursor 1 to straight N = 57 GNRs. (**b**) STM image taken after surface assisted C–C coupling at 200°C but before the final cyclodehydrogenation step, showing a polyanthrylene chain (*left*, temperature T = 5 K, voltage U = 1.9 V, current I = 0.08 nA), and DFT-based simulation of the STM image (*right*) with partially overlaid model of the polymer (*gray* carbon; *white* hydrogen). (**c**) Overview STM image after cyclodehydrogenation at 400°C, showing straight N = 57 GNRs (T = 300 K, U = 23 V, I = 0.03 nA). The *inset* shows a higher-resolution STM image taken at 35 K (U = 21.5 V, I = 0.5 nA). (**d**) Raman spectrum (532 nm) of straight N = 57 GNRs. The peak at 396 cm^{-1} is characteristic for the 0.74 nm width of the N = 57 ribbons. The *inset* shows the atomic displacements characteristic for the radial-breathinglike mode at 396 cm^{-1}. (**e**) High-resolution STM image with partly overlaid molecular model of the ribbon (T = 5 K, U = 20.1 V, I = 0.2 nA). At the *bottom left* is a DFT-based STM simulation of the N = 57 ribbon shown as a greyscale image. *Right panel*: (**a**) STM image of coexisting straight N = 57 and chevron-type GNRs sequentially grown on Ag(111) (T = 5 K, U = 22 V, I = 0.1 nA). (**b**) Threefold GNR junction obtained from a 1,3,5-tris(4″-iodo-2′-biphenyl)benzene monomer 3 at the nodal point and monomer 2 for the ribbon arms: STM image on Au(111) (T = 115 K, U = 22 V, I = 0.02 nA). Monomers 2 and 3 were deposited simultaneously at 250°C followed by annealing to 440°C to induce cyclodehydrogentation. (**c**) Schematic model of the junction fabrication process with components 3 and 2. (**d**) Model (*gray* carbon; *white* hydrogen) of the colligated and dehydrogenated molecules forming the threefold junction overlaid on the STM image from (**b**) (from Cai et al. [85])

"standard sample" for AFM and STM high resolution imaging on various molecular compounds deposited on it, while SPM will increase its importance as characterization and modification tool for graphene-based Nano Electro-Mechanical system (NEMS) and electro-optical devices, the new frontier in the "rise of graphene".

References

1. A. K. Geim and K. S. Novoselov, *Nat. Mater.* **6**, 183 (2007).
2. K. S. Novoselov, D. Jiang, F. Schedin, T. J. Booth, V. V. Khotkevich, S. V. Morozov, and A. K. Geim, *PNAS* **102**, 10451 (2005).
3. K. S. Novoselov, A. K. Geim, S. V. Morozov, D. Jiang, Y. Zhang, S. V. Dubonos, I. V. Grigorieva, and A. A. Firsov, *Science* **306**, 666 (2004).
4. S. Roddaro, P. Pingue, V. Piazza, V. Pellegrini, and F. Beltram, *Nano Lett.* **7**, 2707 (2007).
5. T. W. Ebbesen and H. Hiura, *Adv. Mater.* **7**, 582 (1995).
6. X. K. Lu, M. F. Yu, H. Huang, and R. S. Ruoff, *Nanotechnology* **10**, 269 (1999).
7. Y. Zhang, J. P. Small, M. E. S. Amori, and P. Kim, *Appl. Phys. Lett.* **86**, 073104 (2005).
8. G. Binnig, H. Fuchs, Ch. Gerber, H. Rohrer, E. Stoll, and E. Tosatti, *Europhys. Lett.* **1**, 31 (1986).
9. S. Park and C. F. Quate, *Appl. Phys. Lett.* **48**, 112 (1986).
10. G. Binnig, H. Rohrer, C. Gerber, and E. Weibel, *Phys. Rev. Lett.* **49**, 57 (1982).
11. G. Binnig, C. F. Quate, and Ch. Gerber, *Phys. Rev. Lett.* **56**(9), 930 (1986).
12. B. Bhushan, *Scanning Probe Microscopy in Nanoscience and Nanotechnology*. Springer, Heidelberg (2010).
13. M. Bowker and P. R. Davies, *Scanning Tunneling Microscopy in Surface Science*. Wiley, Weinheim (2010).
14. A. A. Tseng, A. Notargiacomo, and T. P. Chen, *J. Vac. Sci. Technol. B* **23**, 877 (2005).
15. P. Nemes-Incze, Z. Osvath, K. Kamaras, and L. P. Biró, *Carbon* **46**, 1435, (2008).
16. A. Fasolino, J. H. Los, and M. Katsnelson, *Nat. Mater.* **6**, 858 (2007).
17. C. H. Lui, L. Liu, K. F. Mak, G. W. Flynn, and T. Heinz, *Nature* **462**, 339 (2009).
18. D. Teweldebrhan and A. A. Balandin, *Appl. Phys. Lett.* **94**, 013101 (2009).
19. M. A. Meitl, Z.-T. Zhu, V. Kumar, K. J. Lee, X. Feng, Y. Y. Huang, I. Adesida, R. G. Nuzzo, and J. A. Rogers, *Nat. Mater.* **5**, 33 (2006).
20. K. S. Kim, Y. Zhao, H. Jang, S. Y. Lee, J. M. Kim, K. S. Kim, J.-H. Ahn, P. Kim, J.-Y. Choi, and B. H. Hong, *Nature* **457**, 706 (2009).
21. G. F. Schneider, V. E. Calado, H. Zandbergen, L. M. K. Vandersypen, and C. Dekker, *Nano Lett.* **10**, 1912 (2010).
22. Z. Li, Z. Cheng, R. Wang, Q. Li, and Y. Fang, *Nano Lett.* **9**(10), 3599 (2009).
23. T. Hertel, R. Martel, and P. Avouris, *J. Phys. Chem. B* **102**, 910 (1998).
24. A. Wienssa and G. Persch-Schuy, *Appl. Phys. Lett.* **75**, 1077 (1999).
25. T. Banno, M. Tachiki, H. Seo, H. Umezawa, and H. Kawarada, *Diam. Relat. Mater.* **11**, 387 (2002).
26. I. W. Frank, D. M. Tanenbaum, A. M. van der Zande, and P. L. McEuen, *J. Vac. Sci. Technol. B* **25**(6), 2558 (2007).
27. M. Poot and H. S. J. van der Zant, *Appl. Phys. Lett.* **92**, 063111 (2008).
28. C. Lee, X. Wei, J. W. Kysar, and J. Hone, Science **321**, 385 (2008).
29. J. S. Bunch, A. M. van der Zande, S. S. Verbridge, I. W. Frank, D. M. Tanenbaum, J. M. Parpia, H. G. Craighead, and P. L. McEuen, *Science* **315**, 490 (2007).
30. S. Shivaraman, R. A. Barton, X. Yu, J. Alden, L. Herman, M. V. S. Chandrashekhar, J. Park, P. L. McEuen, J. M. Parpia, H. G. Craighead, and M. G. Spencer, *Nano Lett.* **9**, 3100 (2009).
31. V. Singh, S. Sengupta, H. S. Solanki, R. Dhall, A. Allain, S. Dhara, P. Pant, and M. M. Deshmukh, *Nanotechnology* **21**, 165204 (2010).

32. F. Schedin, A. K. Geim, S. V. Morozov, E. W. Hill, P. Blake, M. I. Katsnelson, and K. S. Novoselov, *Nat. Mater.* **6**, 652 (2007).
33. J. S. Bunch, S. S. Verbridge, J. S. Alden, A. M. van der Zande, J. M. Parpia, H. G. Craighead, and P. L. McEuen, *Nano Lett.* **8**(8), 2458 (2008).
34. S. Stankovich, D. A. Dikin, R. D. Piner, K. A. Kohlhaas, A. Kleinhammes, Y. Jia, Y. Wu, S. T. Nguyen, and R. S. Ruoff, *Carbon* **45**, 1558 (2007).
35. G. Eda, G. Fanchini, and M. Chhowalla, *Nat. Nanotechnol.* **3**, 270 (2008).
36. S. S. Hong, W. Kundhikanjana, J. J. Cha, K. Lai, D. Kong, S. Meister, M. A. Kelly, Z.-X. Shen, and Y. Cui, *Nano Lett.* **10**(8), 3118 (2010).
37. A. J. M. Giesbers, U. Zeitler, S. Neubeck, F. Freitag, K. S. Novoselov, and J. C. Maan, *Solid State Commun.* **147**, 366 (2008).
38. S. Eilers and J. P. Rabe, *Phys. Status Solidi B* **246**(11–12), 2527 (2009).
39. G. Prakash, M. L. Bolen, R. Colby, E. A. Stach, M. A. Capano, and R. Reifenberger, *New J. Phys.* **12**, 125009 (2010).
40. G. Lu, X. Zhou, H. Li, Z. Yin, B. Li, L. Huang, F. Boey, and H. Zhang, *Langmuir* **26**(9), 6164 (2010).
41. J. Dagata, J. Schneir, H. H. Harary, C. J. Evans, M. T. Postek, and J. Bennett, *Appl. Phys. Lett.* **56**, 2001 (1990).
42. H. C. Day and D. R. Allee, *Appl. Phys. Lett.* **62**, 2691 (1993).
43. R. Garcia, R. V. Martinez, and J. Martinez, *Chem. Soc. Rev.* **35**, 29 (2006).
44. A. A. Tseng, A. Notargiacomo, and T. P. Chen, *J. Vac. Sci. Technol. B* **23**, 877 (2005).
45. L. Weng, L. Zhang, Y. P. Chen, and L. P. Rokhinson, *Appl. Phys. Lett.* **93**, 093107 (2008).
46. S. Masubuchi, M. Ono, K. Yoshida, K. Hirakawa, and T. Machida, *Appl. Phys. Lett.* **94**, 082107 (2009).
47. K. S. Novoselov, A. K. Geim, S. V. Morozov, D. Jiang, M. I. Katsnelson, I. V. Grigorieva, S. V. Dubonos, and A. A. Firsov, *Nature* **438**, 197 (2005).
48. Y. Zhang, J. W. Tan, H. L. Stormer, and P. Kim, *Nature* **438**, 201 (2005).
49. G. Rius, N. Camara, P. Godignon, and F. Pérez-Murano, *J. Vac. Sci. Technol. B* **27**(6), 3149 (2009).
50. J. H. Ye, F. Pérez-Murano, N. Barniol, G. Abadal, and X. Aymerich, *J. Vac. Sci. Technol. B* **13**, 1423 (1995).
51. G. Abadal, F. Pérez-Murano, N. Barniol, and X. Aymerich, *Appl. Phys. A: Mater. Sci. Process.* **66**, S791 (1998).
52. V. Tozzini and V. Pellegrini, *Phys. Rev. B* **81**, 113404 (2010).
53. M. H. F. Sluiter and Y. Kawazoe, *Phys. Rev. B* **68**, 085410 (2003).
54. J. O. Sofo, A. S. Chaudhari, and G. D. Barber, *Phys. Rev. B* **75**, 153401 (2007).
55. D. C. Elias, R. R. Nair, T. M. G. Mohiuddin, S. V. Morozov, P. Blake, M. P. Halsall, A. C. Ferrari, D. W. Boukhvalov, M. I. Katsnelson, A. K. Geim, and K. S. Novoselov, *Science* **323**, 610 (2009).
56. Z. Wei, D. Wang, S. Kim, S. Y. Kim, Y. Hu, M. K. Yakes, A. R. Laracuente, Z. Dai, S. R. Marder, C. Berger, W. P. King, W. A. de Heer, P. E. Sheehan, and E. Riedo, *Science* **328**, 1373 (2010).
57. J. M. Mativetsky, E.Treossi, E. Orgiu, M. Melucci, G. P. Veronese, P. Samorì, and V. Palermo, *J. Am. Chem. Soc.* **132**, 14130 (2010).
58. M. Zhou, Y. L. Wang, Y. M. Zhai, J. F. Zhai, W. Ren, F. A. Wang, S. J. Dong, *Chem. Eur. J.* **15**, 6116 (2009).
59. P. Vettiger, M. Despont, U. Drechsler, U. Durig, W. Haberle, M. I. Lutwyche, H. E. Rothuizen, R. Stutz, R. Widmer, and G. K. Binnig, *IBM J. Res. Develop.* **44**(3), 323 (2000).
60. E. Stolyarova, K. T. Rim, S. Ryu, J. Maultzsch, P. Kim, L. E. Brus, T. F. Heinz, M. S. Hybertsen, and G. W. Flynn, *PNAS* **104**(22), 9209 (2007).
61. M. Ishigami, J. H. Chen, W. G. Cullen, M. S. Fuhrer, and E. D. Williams, *Nano Lett.* **7**(6), 1644 (2007).
62. G. Li, A. Luican, and E. Y. Andrei, *Phys. Rev. Lett.* **102**, 176804 (2009).

63. H. A. Mizes, S. Park, and W. A. Harrison, *Phys. Rev. B* **36**, 4491 (1987).
64. G. M. Rutter, J. N. Crain, N. P. Guisinger, T. Li, P. N. First, and J. A. Stroscio, *Science* **317**, 219 (2007).
65. F. Varchon, P. Mallet, L. Magaud, and J.-Y. Veuillen, *Phys. Rev. B* **77**, 165415 (2008).
66. J. Wintterlin and M.-L. Bocquet, *Surf. Sci.* **603**, 1841 (2009).
67. Y. Pan, H. Zhang, D. Shi, J. Sun, S. Du, F. Liu, and H. Gao, *Adv. Mater.* **21**(27), 2777 (2009).
68. J. Coraux, A. T. N'Diaye, C. Busse, and T. Michely, *Nano Lett.* **8**(2), 565 (2008).
69. A. L. Vázquez de Parga, F. Calleja, B. Borca, M. C. G. Passeggi, Jr., J. J. Hinarejos, F. Guinea, and R. Miranda, *Phys. Rev. Lett.* **100**, 056807 (2008).
70. N. Levy, S. A. Burke, K. L. Meaker, M. Panlasigui, A. Zett, F. Guinea, A. H. Castro Neto, and M. F. Crommie, *Science* **329**, 544 (2010).
71. M. L. Teague, A. P. Lai, J. Velasco, C. R. Hughes, A. D. Beyer, M. W. Bockrath, C. N. Lau, and N. C. Yeh, *Nano Lett.* **9**(7), 2543 (2009).
72. L. Tapasztó, G. Dobrik, P. Lambin, and L. P. Biró, *Nature Nanotech.* **3**, 397 (2008).
73. D. H. Kim, J. Y. Koo, and J. J. Kim, *Phys. Rev. B* **68**, 113406 (2003).
74. P. Sessi, J. R. Guest, M. Bode, and N. P. Guisinger, *Nano Lett.* **9**(12), 4343 (2009).
75. C. Casiraghi, A. Hartschuh, E. Lidorikis, H. Qian, H. Harutyunyan, T. Gokus, K. S. Novoselov, and A. C. Ferrari, *Nano Lett.* **7**(9), 2711 (2007).
76. A. C. Ferrari, J. C. Meyer, V. Scardaci, C. Casiraghi, M. Lazzeri, F. Mauri, S. Piscanec, D. Jiang, K. S. Novoselov, S. Roth, and A. K. Geim, *Phys. Rev. Lett.* **97**, 187401 (2006).
77. D. Graf, F. Molitor, K. Ensslin, C. Stampfer, A. Jungen, C. Hierold, and L. Wirtz, *Nano Lett.* **7**(2), 238 (2007).
78. R. M. Stöckle, Y. D. Suh, V. Deckert, and R. Zenobi, *Chem. Phys. Lett.* **318**, 131 (2000).
79. Y. Saito, P. Verma, K. Masui, Y. Inouye, and S. Kawata, *J. Raman Spectrosc.* **40**, 1434 (2009).
80. Y.-J. Yu, Y. Zhao, S. Ryu, L. E. Brus, K. S. Kim, and P. Kim, *Nano Lett.* **9**(10), 3431 (2009).
81. A. Verdaguer, M. Cardellach, J. J. Segura, G. M. Sacha, J. Moser, M. Zdrojek, A. Bachtold, and J. Fraxedas, *Appl. Phys. Lett.* **94**, 233105 (2009).
82. J. Martin, N. Akerman, G. Ulbricht, T. Lohmann, J. H. Smet, K. Von Klitzing, and A. Yacoby, *Nat. Phys.* **4**, 144 (2008).
83. M. R. Connolly, K. L. Chiou, C. G. Smith, D. Anderson, G. A. C. Jones, A. Lombardo, A. Fasoli, and A. C. Ferrari, *Appl. Phys. Lett.* **96**, 113501 (2010).
84. K. Xu, P. Cao, and J. R. Heath, *Science* **329**, 1188 (2010).
85. J. Cai, P. Ruffieux, R. Jaafar, M. Bieri, T. Braun, S. Blankenburg, M. Muoth, A. P. Seitsonen, M. Saleh, X. Feng, K. Muellen, and R. Fasel, *Nature* **466**, 472 (2010).

Chapter 11
Diamondoid Mechanosynthesis for Tip-Based Nanofabrication

Robert A. Freitas Jr.

Abstract Diamond mechanosynthesis (DMS), or molecular positional fabrication, is the formation of covalent chemical bonds using precisely applied mechanical forces to build nanoscale diamondoid structures via manipulation of positionally controlled tooltips, most likely in a UHV working environment. DMS may be automated via computer control, enabling programmable molecular positional fabrication. The Nanofactory Collaboration is coordinating a combined experimental and theoretical effort involving direct collaborations among dozens of researchers at institutions in multiple countries to explore the feasibility of positionally controlled mechanosynthesis of diamondoid structures using simple molecular feedstocks, the first step along a direct pathway to developing working nanofactories.

Keywords Carbon placement · Diamond · Diamondoid · DMS · Hydrogen abstraction · Hydrogen donation · Mechanosynthesis · Minimal toolset · Molecular manufacturing · Nanofactory · Nanofactory Collaboration · Nanopart · Positional assembly · Positional fabrication · tooltips

Abbreviations

AFM	Atomic force microscope
CPU	Central processing unit
CVD	Chemical vapor deposition
DFT	Density functional theory
DMS	Diamond mechanosynthesis
DNA	Deoxyribonucleic acid
MEMS	Microelectromechanical systems
NIST	National Institute of Standards and Technology (U.S.)
NMAB	National Materials Advisory Board (U.S.)
NNI	National Nanotechnology Initiative (U.S.)
NRC	National Research Council of the National Academies (U.S.)
ONR	Office of Naval Research (U.S.)
SEM	Scanning electron microscopy

R.A. Freitas (✉)
Institute for Molecular Manufacturing, Palo Alto, CA 94301, USA
e-mail: rfreitas@rfreitas.com

A.A. Tseng (ed.), *Tip-Based Nanofabrication*, DOI 10.1007/978-1-4419-9899-6_11,

SPM Scanning probe microscope
UHV Ultra-high vacuum

11.1 Positional Diamondoid Molecular Manufacturing

Complex molecular machine systems [1–6], including microscale robotic mechanisms comprised of thousands or millions of nanoscale mechanical components such as gears, motors, and computer elements, probably cannot be manufactured using the conventional techniques of self-assembly. As noted in the final report [7] of the 2006 Congressionally-mandated review of the U.S. National Nanotechnology Initiative by the National Research Council (NRC) of the National Academies and the National Materials Advisory Board (NMAB): "For the manufacture of more sophisticated materials and devices, including complex objects produced in large quantities, it is unlikely that simple self-assembly processes will yield the desired results. The reason is that the probability of an error occurring at some point in the process will increase with the complexity of the system and the number of parts that must interoperate." Error detection and correction requires a minimum level of complexity that cannot easily be achieved via thermodynamically-driven self-assembly processes.

The opposite of self-assembly processes is positionally controlled processes, in which the positions and trajectories of all components of intermediate and final product objects are controlled at every moment during fabrication and assembly. Positional processes should allow more complex products to be built with high quality and should enable rapid prototyping during product development. Pure positional assembly is the norm in conventional macroscale manufacturing (e.g., cars, appliances, houses) but is only relatively recently [8, 9] starting to be seriously investigated experimentally for nanoscale manufacturing. Of course, we already know that positional fabrication can work in the nanoscale realm. This is demonstrated in the biological world by ribosomes, which positionally assemble proteins in living cells by following a sequence of digitally encoded instructions (even though ribosomes themselves are self-assembled). Lacking this positional fabrication of proteins controlled by DNA-based software, large, complex, digitally-specified organisms would probably not be possible and biology as we know it would not exist. Guided self-assembly – another hybrid approach combining self-assembly and positional assembly – is also being investigated experimentally [10, 11].

The most important materials for positional assembly may be the rigid covalent or "diamondoid" solids, since these could potentially be used to build the most reliable and complex nanoscale machinery. Preliminary theoretical studies have suggested great promise for these materials in molecular manufacturing. The NMAB/NRC Review Committee recommended [7] that experimental work aimed at establishing the technical feasibility of positional molecular manufacturing should be pursued and supported: "Experimentation leading to demonstrations supplying ground truth for abstract models is appropriate to better characterize the potential for use of bottom-up or molecular manufacturing systems that utilize processes more complex than self-assembly." Making complex nanorobotic systems requires manufacturing techniques that can build a molecular structure by positional assembly [9, 12]. This will involve picking and placing molecular parts one by one, moving them along controlled trajectories much like the robot arms that manufacture cars on automobile assembly lines. The procedure is then repeated over and over with all the different parts until the final product is fully assembled inside a desktop nanofactory.

Technologies required for the atomically precise fabrication of diamondoid nanorobots in macroscale quantities at low cost requires the development of a new nanoscale manufacturing technology called positional diamondoid molecular manufacturing, enabling diamondoid nanofactories. Achieving this new technology over the next 1–3 decades will require the significant further development of multiple closely related technical capabilities: diamondoid mechanosynthesis, programmable positional assembly, massively parallel positional fabrication and assembly, and nanomechanical design.

11.2 Diamondoid Mechanosynthesis (DMS)

Mechanosynthesis, or molecular positional fabrication, is the formation of covalent chemical bonds using precisely applied mechanical forces to build atomically precise structures. Mechanosynthesis will be most efficient when automated via computer control, enabling programmable molecular positional fabrication of nanostructures of significant size. Atomically precise fabrication involves holding feedstock atoms or molecules, and separately a growing nanoscale workpiece, in the proper relative positions and orientations so that when they touch they will chemically bond in the desired manner. In this process, a mechanosynthetic tool is brought up to the surface of a workpiece. One or more transfer atoms are added to (Fig. 11.1), or removed from, the workpiece by the tool. Then the tool is withdrawn and recharged. This process is repeated until the workpiece (e.g., a growing nanopart) is completely fabricated to atomic precision with each atom in exactly the right place. The transfer atoms are under positional control at all times to prevent unwanted side reactions from occurring. Side reactions are also avoided using proper reaction design so that the reaction energetics avoid undesired pathological intermediate structures and atomic rearrangements.

The positional assembly of diamondoid structures, some almost atom by atom, using molecular feedstock has been examined theoretically [13–25] via computational models of diamondoid mechanosynthesis (DMS). DMS is the controlled addition of individual carbon atoms, carbon dimers (C_2), or single methyl (CH_3) and like groups to the growth surface of a diamond crystal lattice workpiece in an inert manufacturing environment such as UHV. Covalent chemical bonds are formed one by one as the result of positionally constrained mechanical forces applied at the tip of a scanning probe microscope (SPM) apparatus, usually resulting in the addition of one or more atoms having one or more bonds into the workpiece structure.

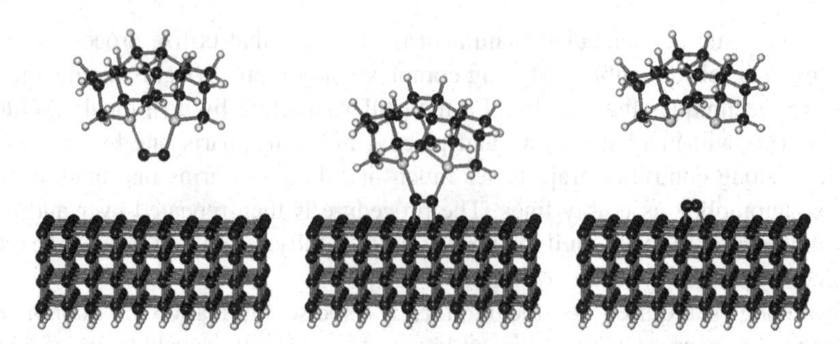

Fig. 11.1 Diamondoid mechanosynthesis: DCB6Ge dimer placement tool shown depositing two carbon atoms on a diamond surface (C = *black*, H = *white*, Ge = *yellow/gray*) [9].

Programmed sequences of carbon dimer placement on growing diamond surfaces in vacuo appear feasible in theory [20, 24].

11.2.1 Diamondoid Materials

Diamondoid materials include pure diamond, the crystalline allotrope of carbon. Among other exceptional properties, diamond has extreme hardness, high thermal conductivity, low frictional coefficient, chemical inertness, a wide electronic bandgap, and is the strongest and stiffest material presently known at ordinary pressures. Diamondoid materials also may include any stiff covalent solid that is similar to diamond in strength, chemical inertness, or other important material properties, and possesses a dense three-dimensional network of bonds. Examples of such materials are carbon nanotubes and fullerenes, atomically-precise doped diamond, several strong covalent ceramics such as silicon carbide, silicon nitride, and boron nitride, and a few very stiff ionic ceramics such as sapphire (monocrystalline aluminum oxide) that can be covalently bonded to pure covalent structures such as diamond. Of course, pure crystals of diamond are brittle and easily fractured. The intricate molecular structure of an atomically precise diamondoid product will more closely resemble a complex composite material, not a brittle solid crystal. Such products, and the nanofactory systems that build them, should be extremely durable in normal use.

11.2.2 Minimal Toolset for DMS

It is already possible to synthesize bulk diamond today. In a process somewhat reminiscent of spray painting, layer after layer of diamond is built up by holding a cloud of reactive hydrogen atoms and hydrocarbon molecules over a deposition surface. When these molecules bump into the surface they change it by adding, removing, or rearranging atoms. By carefully controlling the pressure, temperature, and the exact composition of the gas in this process – called chemical vapor deposition or CVD – conditions can be created that favor the growth of diamond on the surface. But randomly bombarding a surface with reactive molecules does not offer fine control over the growth process. To achieve atomically precise fabrication, the first challenge is to make sure that all chemical reactions will occur at precisely specified places on the surface. A second problem is how to make the diamond surface reactive at the particular spots where we want to add another atom or molecule. A diamond surface is normally covered with a layer of hydrogen atoms. Without this layer, the raw diamond surface would be highly reactive because it would be studded with unused (or "dangling") bonds from the topmost plane of carbon atoms. While hydrogenation prevents unwanted reactions, it also renders the entire surface inert, making it difficult to add carbon (or anything else) to this surface.

To overcome these problems, a set of molecular-scale tools must be developed that would, in a series of well-defined steps, prepare the surface and create hydrocarbon structures on a layer of diamond, atom by atom and molecule by molecule. A mechanosynthetic tool typically will have two principal components – a chemically active tooltip and a chemically inert handle to which the tooltip is covalently bonded. The tooltip is the part of the tool where site-specific single-molecule chemical reactions are forced to occur by the application of mechanical energy. The much larger handle structure is big enough to be grasped or positionally manipulated using an SPM or similar macroscale instrumentality. At least three types of basic mechanosynthetic tools (Fig. 11.2) have already received considerable theoretical (and some related experimental) study and are likely among those required to build molecularly precise diamond via positional control:

(1) *Hydrogen Abstraction Tools.* The first step in the process of mechanosynthetic fabrication of diamond might be to remove a hydrogen atom from each of two specific adjacent spots on the diamond surface, leaving behind two reactive dangling bonds. This could be done using a hydrogen abstraction tool [21] (Fig. 11.2a) that has a high chemical affinity for hydrogen at one end but is elsewhere inert. The tool's unreactive region serves as a handle or handle attachment

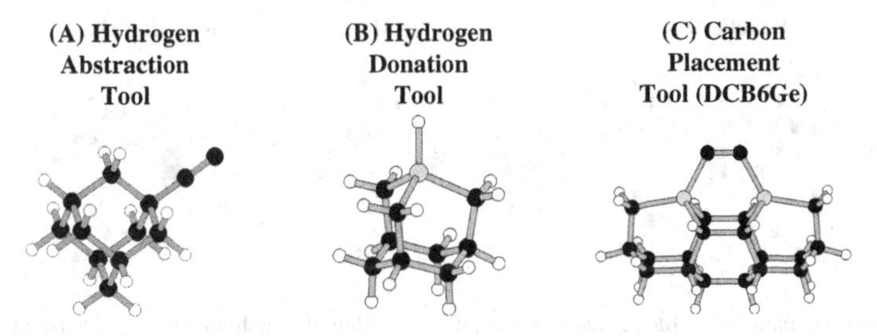

Fig. 11.2 Examples of three basic mechanosynthetic tooltypes that are required to build atomically precise diamond via positional control (C = *black*, H = *white*, Ge = *yellow/gray*) [24]. © 2007 Robert A. Freitas Jr. All Rights Reserved

point. The tool would be held by a high-precision nanoscale positioning device, initially perhaps a scanning probe microscope tip but ultimately a molecular robotic arm, and moved directly over particular hydrogen atoms on the surface. One suitable molecule for a hydrogen abstraction tooltip is the acetylene or "ethynyl" radical, comprised of two carbon atoms triple bonded together. One carbon of the two serves as the handle connection, and would bond to a nanoscale positioning device through a larger handle structure. The other carbon of the two has a dangling bond where a hydrogen atom would normally be present in a molecule of ordinary acetylene (C_2H_2), which can bond and thereby abstract a hydrogen atom from a workpiece structure. The environment around the tool would be inert (e.g., vacuum or a noble gas such as neon). The first extensive DMS tooltip trajectory analysis, examining a wide range of viable multiple degrees-of-freedom tooltip motions in 3D space that could be employed in a separate reaction sequence to recharge the ethynyl-based hydrogen abstraction tool (Fig. 11.3), was published in 2010 [25].

(2) *Carbon Placement Tools*. After the abstraction tool has created adjacent reactive spots by selectively removing hydrogen atoms from the diamond surface but before the surface is re-passivated with hydrogen, carbon placement tools may be used to deposit carbon atoms at the desired reactive surface sites. In this way a diamond structure can be built up on the surface, molecule by molecule, according to plan. The first complete tool ever proposed for this carbon deposition function is the "DCB6Ge" dimer placement tool [15] – in this example, a carbon (C_2) dimer having two carbon atoms connected by a triple bond with each carbon in the dimer connected to a larger unreactive handle structure via two germanium atoms (Fig. 11.2c). This dimer placement tool, also held by a nanoscale positioning device, is brought close to the reactive spots along a particular trajectory, causing the two dangling surface bonds to react with the ends of the carbon dimer. The dimer placement tool would then withdraw, breaking the relatively weaker bonds between it and the C_2 dimer and transferring the carbon dimer from the tool to the surface. A positionally controlled dimer could be bonded at many different sites on a growing diamondoid workpiece, in principle allowing the construction of a wide variety of useful nanopart

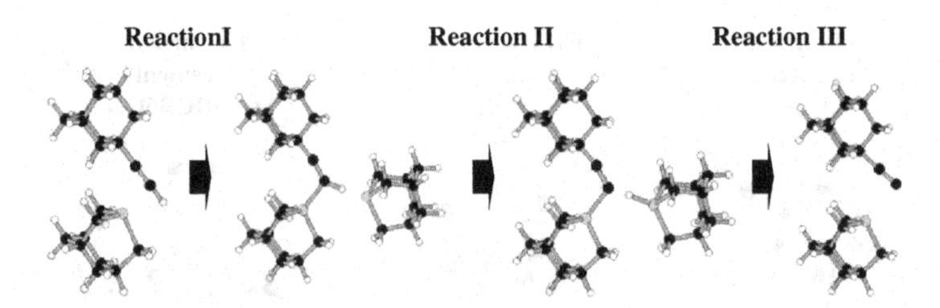

Fig. 11.3 Schematic of recharge reaction for 1-ethynyladamantane hydrogen abstraction tool (end product of Reaction III, structure at top) using sequence of three positionally controlled reactions involving two 1-germanoadamantane radical tooltips (C = *black*, H = *white*, Ge = *yellow/gray*) [25].

shapes. As of 2010, the DCB6Ge dimer placement tool remains the most studied [15, 17, 19, 20, 22, 24] of any mechanosynthetic tooltip to date, having had more than 150,000 CPU-hours of computation invested thus far in its analysis, and it remains the only tooltip motif that has been successfully simulated and theoretically validated for its intended function on a full 200-atom diamond workpiece surface [20]. Other proposed dimer (and related carbon transfer) tooltip motifs [13–15, 18, 22, 24] have received less intensive study but are also expected to perform well.

(3) *Hydrogen Donation Tools*. After an atomically precise structure has been fabricated by a succession of hydrogen abstractions and carbon depositions, the fabricated structure must be passivated to prevent additional unplanned reactions. While the hydrogen abstraction tool is intended to make an inert structure reactive by creating a dangling bond, the hydrogen donation tool [23] does the opposite (Fig. 11.2b). It makes a reactive structure inert by terminating a dangling bond by adding an H atom. Such a tool would be used to stabilize reactive surfaces and help prevent the surface atoms from rearranging in unexpected and undesired ways. The key requirement for a hydrogen donation tool is that it include a weakly attached hydrogen atom. Many molecules fit that description, but the bond between hydrogen and germanium is sufficiently weak so that a Ge-based hydrogen donation tool should be effective.

A 3-year study [24] representing 102,188 CPU hours of computing effort for the first time computationally analyzed a comprehensive set of DMS reactions and an associated minimal set of nine specific DMS tooltips that could be used to build basic diamond, graphene (e.g., carbon nanotubes), and all of the tools themselves including all necessary tool recharging reactions. The research defined 65 DMS reaction sequences incorporating 328 reaction steps, with 354 pathological side reactions analyzed and with 1,321 unique individual DFT-based (Density Functional Theory) quantum chemistry reaction energies reported. These mechanosynthetic reaction sequences range in length from 1 to 13 reaction steps (typically 4) with 0–10 possible pathological side reactions or rearrangements (typically 3) reported per reaction.

The first practical proposal for building a DMS tool experimentally was published in 2005 [19] and is the subject of the first diamond mechanosynthesis patent ever issued (in 2010) [19]. According to the original proposal, the manufacture of a complete "DCB6Ge" positional dimer placement tool would require four distinct steps: synthesizing a capped tooltip molecule, attaching it to a deposition surface, attaching a handle to it via CVD, then separating the tool from the deposition surface. An even simpler practical proposal for building DMS tools experimentally, also using only experimental methods available today, was published in 2008 as part of the aforementioned minimal toolset work [24]. Processes were identified for the experimental fabrication of a hydrogen abstraction tool, a hydrogen donation tool, and two alternative carbon placement tools (other than DCB6Ge). These processes and tools are part of the second mechanosynthesis patent ever filed and provide clear developmental targets for a comprehensive near-term DMS implementation program to begin working toward a more mature set of efficient, positionally controlled mechanosynthetic tools that can reliably build atomically precise diamondoid structures – including more DMS tools.

11.2.3 Experimental Activities to Date

The first experimental proof that individual atoms could be manipulated was obtained by IBM scientists in 1989 when they used a scanning tunneling microscope to precisely position 35 xenon atoms on a nickel surface to spell out the corporate logo "IBM". However, this feat did not involve the formation of covalent chemical bonds. One important step toward the practical realization of DMS was achieved in 1999 [26] with the first site-repeatable site-specific covalent bonding operation of a two diatomic carbon-containing molecules (CO), one after the other, to the same atom of iron on a crystal surface, using an SPM.

The first experimental demonstration of true mechanosynthesis, establishing covalent bonds using purely mechanical forces in UHV – albeit on silicon atoms, not carbon atoms – was reported in 2003 [27]. In this landmark experiment, the researchers vertically manipulated single silicon atoms from the Si(111)-(7×7) surface, using a low-temperature near-contact atomic force microscope to demonstrate: (1) removal of a selected silicon atom from its equilibrium position without perturbing the (7×7) unit cell, and (2) the deposition of a single Si atom on a created vacancy, both via purely mechanical processes. The same group later repeated this feat with Ge atoms [28]. By 2008, the Custance group in Japan [29] had progressed to more complex 2D structures fabricated entirely via mechanosynthesis using more than a dozen Si/Sn or Pb/In atoms [29], with a 12-atom 2D pattern created in 1.5 h ($\sim$450 s/atom). In late 2008 Moriarty's group at the University of Nottingham (U.K.) began a $3 million 5-year effort [30] employing a similar apparatus to produce 2D patterns using carbon atoms, to validate the theoretical DMS proposals of Freitas and Merkle [24]. If successful, Moriarty's work may lead to subsequent studies extending DMS from 2D to small 3D carbon nanostructures.

Positional control and reaction design are key for the success of DMS. Error correction will be difficult or even impossible in many cases, so each reaction must be executed precisely on the first attempt. To accomplish this, sequences of positionally controlled reaction steps must be chosen such that desired reactions are energetically favored, side reactions or other undesired reactions are energetically disfavored, and defect formation or unwanted lattice rearrangements are either energetically disfavored or blocked by sufficient barriers to prevent their occurrence. Reaction sequences for building diamond and fullerene nanostructures that appear to meet these requirements have been proposed theoretically [24] and are now being investigated experimentally [30].

An extensive bibliography of theoretical and experimental work on DMS is available at http://www.MolecularAssembler.com/Nanofactory/AnnBibDMS.htm.

11.3 Programmable Positional Fabrication and Assembly

After demonstration of basic DMS, the next major experimental milestone may be the mechanosynthetic fabrication of atomically precise 3D structures, creating readily accessible diamondoid-based nanomechanical components engineered to form desired architectures possessing superlative mechanical strength, stiffness, and strength-to-weight ratio. These nanoscale components may range from relatively simple diamond rings, rods and cubes [31] to more sophisticated "nanoparts" such as fullerene bearings [32–34], gears [35–37] and motors [38], composite fullerene/diamond structures [39], and more complex devices [13] such as diamondoid gears [40], pumps [40], and conveyors [41].

Atomically precise nanoparts, once fabricated, must be transferred from the fabrication site and assembled into atomically precise complex components containing many nanoparts. Such components may include gear trains in housings [42], sensors, motors, manipulator arms, power generators, and computers. These complex components may then be assembled, for example, into an even more complex molecular machine system that consists of many complex components. A micron-size medical nanorobot such as a respirocyte [1] constructed of such molecularly precise components may possess many tens of thousands of individual components, millions of primitive nanoparts, and many billions of atoms in its structure. The conceptual dividing line between fabrication and assembly may sometimes become blurred because in many cases it might be possible, even preferable, to fabricate nominally multipart components as a single part – allowing, for example, two meshed gears and their housing to be manufactured as a single sealed unit.

The process of positional assembly, as with DMS, can be automated via computer control as has been demonstrated experimentally in the case of individual atoms in the Autonomous Atom Assembly project sponsored by NIST and ONR [43], in the case of nanoscale objects in SEM [44, 45] and AFM-based [45–47] manipulation systems, and in the case of microscale parts in automated MEMS assembly [48, 49]. This capability will allow the design of positional assembly stations [46, 47] which receive inputs of primitive parts and assemble them in programmed sequences of steps into finished complex components. These components can then be transported to secondary assembly lines which use them as inputs to manufacture still larger and more complex components, or completed systems, again analogous to automobile assembly lines.

For nanofactories to be economically viable, we must also be able to assemble complex nanostructures in vast numbers – in billions or even trillions of finished units (product objects). Approaches under consideration include using replicative manufacturing systems or massively parallel fabrication, employing large arrays of scanning probe tips all building similar diamondoid product structures in unison, as in nanofactories [9, 13, 50]. This will require massively parallel manufacturing systems with millions of assembly lines operating simultaneously and in parallel, not just one or a few of them at a time as with the assembly lines in modern-day car factories. Fortunately, each nanoassembly production line in a nanofactory can, in principle, be quite small. Many millions of them should easily fit into a very small volume. Massively parallel manufacture of DMS tools, handles, and related nanoscale fabrication and assembly equipment will also be required, perhaps involving the use of massively parallel manipulator arrays or some other type of replicative system [50].

Reliability is an important design issue. The assembly lines of massively parallel manufacturing systems might have numerous redundant smaller assembly lines feeding components into larger assembly lines, so that the failure of any one smaller line cannot cripple the larger one. Arranging parallel production lines for maximum efficiency and reliability to manufacture a wide variety of products (possibly including error detection, error correction and removal of defective parts) is a major requirement in nanofactory design.

Computational tools for molecular machine modeling, simulation and manufacturing process control must be created to enable the development of designs for diamondoid nanoscale parts, components, and nanorobotic systems. These designs can then be rigorously tested and refined in simulation before undertaking more expensive experimental efforts to build them. Molecular machine design and simulation software is available [42] and libraries of predesigned nanoparts are slowly being assembled. More effort must be devoted to large-scale simulations of complex nanoscale machine components, design and simulation of assembly sequences and manufacturing process control, and general nanofactory design and simulation.

11.4 Nanofactory Collaboration

The NMAB/NRC Review Committee, in their Congressionally-mandated review [7] of the NNI, called for proponents of "site-specific chemistry for large-scale manufacturing" to: (1) delineate desirable research directions not already being pursued by the biochemistry community; (2) define and focus on some basic experimental steps that are critical to advancing long-term goals; and (3) outline some "proof-of-principle" studies that, if successful, would provide knowledge or engineering demonstrations of key principles or components with immediate value.

In direct response to these requirements, the Nanofactory Collaboration [9] is coordinating a combined experimental and theoretical effort to explore the feasibility of positionally controlled mechanosynthesis of diamondoid structures using simple molecular feedstock. The precursor to the Nanofactory Collaboration was informally initiated by Robert Freitas and Ralph Merkle in the Fall of 2000 during their time at Zyvex. Their continuing efforts, and those of others, have produced direct collaborations among 25 researchers or other participants (including 18 PhD's or PhD candidates) at 13 institutions in 4 countries (U.S., U.K., Russia, and Belgium), as of 2010. The Collaboration website is at http://www.MolecularAssembler.com/Nanofactory.

At present, the Collaboration is a loose-knit community of scientists and others who are working together as time and resources permit in various team efforts with these teams producing numerous co-authored publications, though with disparate funding sources not necessarily tied to the Collaboration. While not all participants may currently envision a nanofactory as the end goal of their present research (or other) efforts in connection with the Collaboration, many *do* envision this, and even those who do not currently envision this end goal have nonetheless agreed to do research in collaboration with other participants that we believe will contribute important advances along the pathway to diamondoid nanofactory development, starting with the direct development of DMS. While some work has been done on each of the multiple capabilities believed necessary to design and build a functioning nanofactory, for now the greatest research attention is being concentrated on the first key area: proving the feasibility, both theoretical and experimental, of achieving diamondoid mechanosynthesis. We welcome new participants who would like to help us address the many remaining technical challenges [51] to the realization of a working diamondoid nanofactory.

Acknowledgments The author acknowledges private grant support for this work from the Life Extension Foundation, the Kurzweil Foundation, and the Institute for Molecular Manufacturing.

References

1. R.A. Freitas Jr., Exploratory design in medical nanotechnology: a mechanical artificial red cell, *Artif. Cells Blood Subst. Immobil. Biotech.*, **26**, 411–430 (1998); http://www.foresight.org/Nanomedicine/Respirocytes.html
2. R.A. Freitas Jr., Nanodentistry, *J. Amer. Dent. Assoc.*, **131**, 1559–1566 (2000); http://www.rfreitas.com/Nano/Nanodentistry.htm
3. R.A. Freitas Jr., Clottocytes: artificial mechanical platelets, *IMM Report*, **18**, 9–11 (2000); http://www.imm.org/Reports/Rep018.html
4. R.A. Freitas Jr., Microbivores: artificial mechanical phagocytes using digest and discharge protocol, *J. Evol. Technol.*, **14**, 1–52 (2005); http://jetpress.org/volume14/Microbivores.pdf
5. R.A. Freitas Jr., Pharmacytes: an ideal vehicle for targeted drug delivery, *J. Nanosci. Nanotechnol.*, **6**, 2769–2775 (2006); http://www.nanomedicine.com/Papers/JNNPharm06.pdf
6. R.A. Freitas Jr., The ideal gene delivery vector: chromallocytes, cell repair nanorobots for chromosome replacement therapy, *J. Evol. Technol.*, **16**, 1–97 (2007); http://jetpress.org/v16/freitas.pdf
7. Committee to Review the NNI (National Nanotechnology Initiative), National Materials Advisory Board (NMAB), National Research Council (NRC), *A Matter of Size: Triennial Review of the National Nanotechnology Initiative*, The National Academies Press, Washington, DC (2006); http://www.nap.edu/catalog/11752.html#toc
8. T. Kenny, Tip-Based Nanofabrication (TBN), Defense Advanced Research Projects Agency (DARPA)/Microsystems Technology Office (MTO), Broad Agency Announcement BAA 07-59 (2007); http://www.fbo.gov/spg/ODA/DARPA/CMO/BAA07-59/listing.html
9. Nanofactory Collaboration Website (2010); http://www.MolecularAssembler.com/Nanofactory
10. J.D. Cohen, J.P. Sadowski, P.B. Dervan, Addressing single molecules on DNA nanostructures, *Angew. Chem. Int. Ed.*, **46**, 7956–7959 (2007).
11. J.H. Lee, D.P. Wernette, M.V. Yigit, J. Liu, Z. Wang, Y. Lu, Site-specific control of distances between gold nanoparticles using phosphorothioate anchors on DNA and a short bifunctional molecular fastener, *Angew. Chem. Int. Ed. Engl.*, **46**, 9006–9010 (2007).
12. R.A. Freitas Jr., Current status of nanomedicine and medical nanorobotics, *J. Comput. Theor. Nanosci.*, **2**, 1–25 (2005); http://www.nanomedicine.com/Papers/NMRevMar05.pdf

13. K.E. Drexler, *Nanosystems: Molecular Machinery, Manufacturing, and Computation*, Wiley, New York (1992).
14. R.C. Merkle, A proposed 'metabolism' for a hydrocarbon assembler, *Nanotechnology*, **8**, 149–162 (1997); http://www.zyvex.com/nanotech/hydroCarbonMetabolism.html
15. R.C. Merkle, R.A. Freitas Jr., Theoretical analysis of a carbon-carbon dimer placement tool for diamond mechanosynthesis, *J. Nanosci. Nanotechnol.*, **3**, 319–324 (2003); http://www.rfreitas.com/Nano/JNNDimerTool.pdf
16. J. Peng, R.A. Freitas Jr., R.C. Merkle, Theoretical analysis of diamond mechanosynthesis. Part I. Stability of C_2 mediated growth of nanocrystalline diamond C(110) surface, *J. Comput. Theor. Nanosci.*, **1**, 62–70 (2004); http://www.molecularassembler.com/Papers/JCTNPengMar04.pdf
17. D.J. Mann, J. Peng, R.A. Freitas Jr., R.C. Merkle, Theoretical analysis of diamond mechanosynthesis. Part II. C_2 mediated growth of diamond C(110) surface via Si/Ge-triadamantane dimer placement tools, *J. Comput. Theor. Nanosci.*, **1**, 71–80 (2004); http://www.MolecularAssembler.com/JCTNMannMar04.pdf
18. D.G. Allis, K.E. Drexler, Design and analysis of a molecular tool for carbon transfer in mechanosynthesis," *J. Comput. Theor. Nanosci.*, **2**, 45–55 (2005); http://e-drexler.com/d/05/00/DC10C-mechanosynthesis.pdf
19. R.A. Freitas Jr., A simple tool for positional diamond mechanosynthesis, and its method of manufacture. U.S. Provisional Patent Application No. 60/543,802, filed 11 February 2004; U.S. Patent Pending, 11 February 2005; U.S. Patent No. 7,687,146, issued 30 March 2010; http://www.freepatentsonline.com/7687146.pdf
20. J. Peng, R.A. Freitas Jr., R.C. Merkle, J.R. von Ehr, J.N. Randall, G.D. Skidmore, Theoretical analysis of diamond mechanosynthesis. Part III. Positional C_2 deposition on diamond C(110) surface using Si/Ge/Sn-based dimer placement tools, *J. Comput. Theor. Nanosci.*, **3**, 28–41 (2006); http://www.MolecularAssembler.com/Papers/JCTNPengFeb06.pdf
21. B. Temelso, C.D. Sherrill, R.C. Merkle, R.A. Freitas Jr., High-level *ab initio* studies of hydrogen abstraction from prototype hydrocarbon systems, *J. Phys. Chem. A*, **110**, 11160–11173 (2006); http://www.MolecularAssembler.com/Papers/TemelsoHAbst.pdf
22. R.A. Freitas Jr., D.G. Allis, R.C. Merkle, Horizontal Ge-substituted polymantane-based C_2 dimer placement tooltip motifs for diamond mechanosynthesis, *J. Comput. Theor. Nanosci.*, **4**, 433–442 (2007); http://www.MolecularAssembler.com/Papers/DPTMotifs.pdf
23. B. Temelso, C.D. Sherrill, R.C. Merkle, R.A. Freitas Jr., *Ab initio* thermochemistry of the hydrogenation of hydrocarbon radicals using silicon, germanium, tin and lead substituted methane and isobutane, *J. Phys. Chem. A*, **111**, 8677–8688 (2007); http://www.MolecularAssembler.com/Papers/TemelsoHDon.pdf
24. R.A. Freitas Jr., R.C. Merkle, A minimal toolset for positional diamond mechanosynthesis, *J. Comput. Theor. Nanosci.*, **5**, 760–861 (2008).
25. D. Tarasov, N. Akberova, E. Izotova, D. Alisheva, M. Astafiev, R.A. Freitas Jr., Optimal tooltip trajectories in a hydrogen abstraction tool recharge reaction sequence for positionally controlled diamond mechanosynthesis, *J. Comput. Theor. Nanosci.*, **7**, 325–353 (2010).
26. H.J. Lee, W. Ho, Single bond formation and characterization with a scanning tunneling microscope, *Science*, **286**, 1719–1722 (1999); http://www.physics.uci.edu/%7Ewilsonho/stm-iets.html
27. N. Oyabu, O. Custance, I. Yi, Y. Sugawara, S. Morita, Mechanical vertical manipulation of selected single atoms by soft nanoindentation using near contact atomic force microscopy, *Phys. Rev. Lett.*, **90**, 176102 (2003); http://link.aps.org/abstract/PRL/v90/e176102
28. N. Oyabu, O. Custance, M. Abe, S. Moritabe, Mechanical vertical manipulation of single atoms on the Ge(111)-c(2x8) surface by noncontact atomic force microscopy, *Abstracts of Seventh International Conference on Non-Contact Atomic Force Microscopy*, Seattle, Washington, DC, 12–15 September, 2004, p. 34; http://www.engr.washington.edu/epp/afm/abstracts/15Oyabu2.pdf

29. Y. Sugimoto, P. Pou, O. Custance, P. Jelinek, M. Abe, R. Perez, S. Morita, Complex patterning by vertical interchange atom manipulation using atomic force microscopy, *Science*, **322**, 413–417 (2008); http://www.sciencemag.org/cgi/content/full/322/5900/413
30. Nanofactory Collaboration, *Nanofactory Collaboration Colleague Awarded $3M to Conduct First Diamond Mechanosynthesis Experiments*, Nanofactory Collaboration press release, 11 August 2008; http://www.MolecularAssembler.com/Nanofactory/Media/PressReleaseAug08.htm
31. D. Tarasov, E. Izotova, D. Alisheva, N. Akberova, R.A. Freitas Jr., Structural stability of clean, passivated, and partially dehydrogenated cuboid and octahedral nanodiamonds up to 2 nanometers in size, *J. Comput. Theor. Nanosci.*, **8**, 147–167 (2011).
32. R.E. Tuzun, D.W. Noid, B.G. Sumpter, An internal coordinate quantum Monte Carlo method for calculating vibrational ground state energies and wave functions of large molecules: A quantum geometric statement function approach, *J. Chem. Phys.*, **105**, 5494–5502 (1996).
33. K. Sohlberg, R.E. Tuzun, B.G. Sumpter, D.W. Noid, Application of rigid-body dynamics and semiclassical mechanics to molecular bearings, *Nanotechnology*, **8**, 103–111 (1997).
34. D.W. Noid, R.E. Tuzun, B.G. Sumpter, On the importance of quantum mechanics for nanotechnology, *Nanotechnology*, **8**, 119–125 (1997).
35. D.H. Robertson, B.I. Dunlap, D.W. Brenner, J.W. Mintmire, C.T. White, Molecular dynamics simulations of fullerene-based nanoscale gears, in *Novel Forms of Carbon II*, C.L. Renschler, D.M. Cox, J.J. Pouch, Y. Achiba (Eds.), MRS Symposium Proceedings Series, Volume 349, pp. 283–288 (1994).
36. J. Han, A. Globus, R.L. Jaffe, G. Deardorff, Molecular dynamics simulations of carbon nanotube-based gears, *Nanotechnology*, **8**, 95–102 (1997).
37. A. Globus, C.W. Bauschlicher Jr., J. Han, R.L. Jaffe, C. Levit, D. Srivastava, Machine phase fullerene nanotechnology, *Nanotechnology*, **9**, 192–199 (1998).
38. R.E. Tuzun, D.W. Noid, B.G. Sumpter, Dynamics of a laser driven molecular motor, *Nanotechnology*, **6**, 52–63 (1995).
39. O.A. Shenderova, D. Areshkin, D.W. Brenner, Carbon based nanostructures: diamond clusters structured with nanotubes, *Mater. Res.*, **6**, 11–17 (2003); http://www.scielo.br/scielo.php?pid=S1516-14392003000100004&script=sci_arttext&tlng=en
40. T. Cagin, A. Jaramillo-Botero, G. Gao, W.A. Goddard III, Molecular mechanics and molecular dynamics analysis of Drexler-Merkle gears and neon pump, *Nanotechnology*, **9**, 143–152 (1998).
41. G. Leach, Advances in molecular CAD, *Nanotechnology*, **7**, 197–203 (1996).
42. M. Sims, Molecular modeling in CAD, *Machine Design*, **78**, 108–113 (2006).
43. NIST, Autonomous Atom Assembly (2004); http://cnst.nist.gov/epg/Projects/STM/aaa_proj.html
44. E.C. Heeres, A.J. Katan, M.H. van Es, A.F. Beker, M. Hesselberth, D.J. van der Zalm, T.H. Oosterkamp, A compact multipurpose nanomanipulator for use inside a scanning electron microscope, *Rev. Sci. Instrum.*, **81**, 023704 (2010).
45. M.-F. Yu, M.J. Dyer, G.D. Skidmore, H.W. Rohrs, X.K. Lu, K.D. Ausman, J. von Ehr, R.S. Ruoff, 3-Dimensional manipulation of carbon nanotubes under a scanning electron microscope, *Sixth Foresight Nanotechnology Conference*, November 1998; http://www.foresight.org/Conferences/MNT6/Papers/Yu/index.html
46. C.R. Taylor, K.K. Leang, Design and Fabrication of a Multifunctional Scanning Probe with Integrated Tip Changer for Fully Automated Nanofabrication, Paper 2617, *ASPE Annual Meeting*, 2008; http://www.leang.com/academics/pubs/LeangKK_2008_ASPE.pdf
47. K.K. Leang, C.R. Taylor, A novel multifunctional SPM probe with modular quick-change tips for fully automated probe-based nanofabrication, *Proceedings of the 2009 NSF Engineering Research and Innovation Conference*, Honolulu, HI, 22–25 June 2009.
48. K. Tsui, A.A. Geisberger, M. Ellis, G.D. Skidmore, Micromachined end-effector and techniques for directed MEMS assembly, *J. Micromech. Microeng.*, **14**, 542–549 (2004); http://dx.doi.org/10.1088/0960-1317/14/4/015

49. D.O. Popa, H.E. Stephanou, Micro- and meso-scale robotic assembly, *SME J. Manuf. Proc.*, **6**, 52–71 (2004).
50. R.A. Freitas Jr., R.C. Merkle, *Kinematic Self-Replicating Machines*, Landes Bioscience, Georgetown, TX (2004); http://www.MolecularAssembler.com/KSRM.htm
51. R.A. Freitas Jr., R.C. Merkle, Remaining technical challenges for achieving positional diamondoid molecular manufacturing and diamondoid nanofactories, Nanofactory Collaboration Website (2010); http://www.MolecularAssembler.com/Nanofactory/Challenges.htm

Chapter 12
Constraints and Challenges in Tip-Based Nanofabrication

Ampere A. Tseng

Abstract In the past decade, tip-based nanofabrication (TBN) has become a powerful technology for nanofabrication due to its low cost and unique atomic-level manipulation capabilities. While a wide range of nanoscale components, devices, and systems have been fabricated by TBN, this technology still faces a number of constraints and challenges, which can be categorized into three areas: repeatability (reliability), ability (feasibility), and productivity (throughput). This chapter reviews these constraints and discusses the challenges for potential approaches to circumventing them. First, the major TBN techniques and their recent advances are reviewed in brief. Then, specific approaches for enhancing its repeatability by using automated equipment, for increasing its ability by seeking strategies to create truly three-dimensional nanostructures, and for improving its productivity by parallel processing, speed increasing, and larger tips, are evaluated. Finally, a preliminary roadmap for the next several years and a recommendation of areas for future research and development are provided.

Keywords Atomic force microscopy · Atomic manipulation · Automation · Control · Dip-pen nanolithography · Dual sources · Hardware · Hot tip · Local anodic oxidation · Machining · Multiple probes · Multiresolution · Nanofabrication · Nanolithography · Nanostructure · Parallel processing · Piezodriver · Piezoelectric actuator · Photodetector · Scanning probe microscopy · Scanning near-field optical microscopy · Scanning tunneling microscopy · Scanning speed · Scratch speed · Scratching · Software · Tip · Tip based nanofabrication · Throughput

Abbreviations

AFM	Atomic force microscope/microscopy
AM	Atomic manipulation
APTS	Aminopropyl trimethoxysilane
BPL	Beam pen lithography
CAD/CAM	Computer-aided design/computer-aided manufacturing
CNC	Computer numerical control

A.A. Tseng (✉)
School for Engineering of Matter, Transport, and Energy, Arizona State University, Tempe, AZ 85287-6106, USA
e-mail: ampere.tseng@asu.edu

A.A. Tseng (ed.), *Tip-Based Nanofabrication*, DOI 10.1007/978-1-4419-9899-6_12,

DNA	Deoxyribonucleic acid
DPN	Dip-pen nanolithography
ECD	Electrochemical deposition
EFMT	Electric field-induced mass transfer
FM	Frequency-modulation
ITRS	International technology roadmap for semiconductors
KBS	Knowledge based software
LAO	Local anodic oxidation
MEMS	Micro electromechanical system
NFP	Nanofountain Probes
PCS	Probe control software
PDMS	Polydimethylsiloxane
PID	Proportional-integral-derivative
PMG	Phenolic molecular glass
RIE	Reactive ion etch
SAM	Self-assembled monolayer
SEM	Scanning electron microscope
SIMS	Secondary ion mass spectrometry
SNOM	Scanning near-field optical microscopy
SOI	Silicon-on-insulator
SPM	Scanning probe microscopy
SPP	Surface plasmon polariton
STM	Scanning tunneling microscopy
TBN	Tip-based nanofabrication
vdW	van der Waals

12.1 Introduction

The ever-shrinking trend of semiconductor devices and systems has driven the rapid development of nanofabrication so that these nanoscale devices and systems can fulfil their designed potential. The advances in nanofabrication have also facilitated and catalyzed the development of many newly emerging fields, including nanoplasmonics, nanophotonics, nanobiology, nanomedicine, and others. All of these newly emerging areas have novel functionalities and unique characteristics. For example, the development of nanoplasmonics heralded a list of possible breakthroughs, including realizing that a cloak of invisibility exists, with the potential to enhance surveillance and detection abilities [1, 2]; accomplishing a planar hyperlens that could render optical microscopes an order of magnitude more powerful and detect 45 nm particles in wafers or objects as small as DNA [3, 4]; and increasing the potential for energy harvesting through more efficient solar collectors [5, 6]. Similarly, nanophotonics holds great potential to revolutionize computer and consumer electronics that use light instead of electronic signals to process information [7] and to provide low-power, low-cost and high-speed telecommunication devices [8]. Furthermore, surmounting the nanoscale nature of DNA, viruses, and protein molecules, which put stringent nanometer size and precision requirements on many new generations of medical and biological devices [9, 10], would unlock great potential to provide cost-effective health care systems, including early disease diagnosis and monitoring, personalized medicine, protein and peptide delivery, nanorobots and nanoprobes, antibody therapeutics, and even cell repair [10, 11].

Despite all these important applications, the nanofabrication techniques adopted by the semiconductor industry are not cost-effective and feasible for the devices and systems in these emerging fields. For instance, the semiconductor-integrated circuits made by current methods of deep UV photolithography could be shrunk to a 32 nm feature size (half-pitch) with the help of resolution enhancement techniques such as high numerical aperture, off-axis illumination, phase-shift mask, optical proximity correction, and double processing as indicated in the recent edition of ITRS [12]. To including such enhancements, fabrication facilities and processes currently become orders of magnitude more complex and expensive, especially in the case of nanoscale optical masks, where costs can run upwards of a million dollars per mask. Such a state-of-the-art optical lithography approach is designated for high-volume production, in which highly sophisticated and carefully maintained fabrication equipment is employed.

The above mentioned applications in newly emerging areas are still in their infancy state where more development iterations or product revisions are necessary in order to lower nanofabrication costs and to make the processes more flexible and reconfigurable. This is particularly important for the prototyping of a small quantity of product. One enabling technology for low-quantity nanofabrication is Tip-Based Nanofabrication (TBN), which makes use of micro-cantilevers with functionalized nanoscale tips. These cantilevered nanotips, which are basically evolved from atomic force microscopy (AFM), are able to manipulate, modify, remove, and deposit materials for creating a wide range of nanoscale patterns and structures [13]. Many techniques based on the TBN or AFM platform have been developed with different degrees of similarity and success. The next great challenge for TBN is to build on the common tip based platform to expand the capacity of this technology to fabricate nanostructures with greater complexity, improved precision, better reliability, and higher performance at low cost.

In this chapter, the major constraints and challenges facing TBN will be presented. For the sake of clarity and relevance, most of the presentation and elaboration still focuses on AFM, which has become the most popular methodology in the TBN family adopted for nanofabrication. AFM can be performed in a wide range of environments, including in non-vacuum, air, and liquid solutions with different chemical and biological ingredients, and, indeed, is an excellent technique to illustrate the capabilities of TBN. Other popular techniques, such as scanning tunneling microscopy (STM) [14], dip-pen nanolithography (DPN), and scanning near-field optical microscopy (SNOM), will also be addressed. In DPN, the tip is loaded with chemical- or biological-molecular species that can react with the substrate to activate nanoscale chemical or biological reactions for creating different types of nanostructures [15]. In SNOM, the tip creates a localized photonic energy field and the resulting interaction between the tip and sample can trigger many different photonic related reactions, including electrostatic, optical, optochemical, optoelectrical, and optomagnetic [16].

In the subsequent sections, following a section on the basic principles of and recent advances in those major TBN techniques just mentioned, the constraints and challenges related to AFM automation and equipment precision are first presented with a recommendation for the appropriate direction of future research. Advances in patterning three-dimensional (3D) structures by TBN are then surveyed with an emphasis on the selection of proper techniques that have high potential for making small feature sizes with high precision in truly 3D fashions. Limitations on the throughput of TBN are also addressed with a focus on fruitful approaches for probe development, parallel processing, speed increasing, and larger tips. A wide range of nanostructures fabricated through a variety of TBN techniques are selected for illustrating advances in and limitations of these tip-based techniques. The information surveyed and illustrated is extrapolated to form the basis for a general assessment of the current constraints and future challenges facing TBN. In addition to this assessment of constraints and challenges, a table summarizing the TBN roadmap over the next several years and concluding remakes are finally presented.

12.2 Major Techniques in TBN

This section briefly reviews the current status of four major techniques used in TBN with the goal of introducing their general principles of, difference among, and recent advances in these techniques.

12.2.1 Atomic Force Microscopy (AFM)

AFM, introduced by Binnig et al. [17] in 1986, evolved from the scanning tunneling microscopy (STM) invented in 1982. Currently, AFM and STM are the two most important technologies in the scanning probe microscopy (SPM) family. AFM operates by measuring the attractive or repulsive forces between a tip and a sample, which vary according to the distance between the two. Since the tip is located at the free end of a flexible microcantilever, the attractive or repulsive forces cause the cantilever to deflect. Depending on the situation, the forces measured in AFM include van der Waals, capillary, chemical bonding, electrostatic, magnetic, among others. Typically, the deflection is measured using a laser beam reflected off the cantilever into an array of photodetectors as shown in Fig. 12.1.

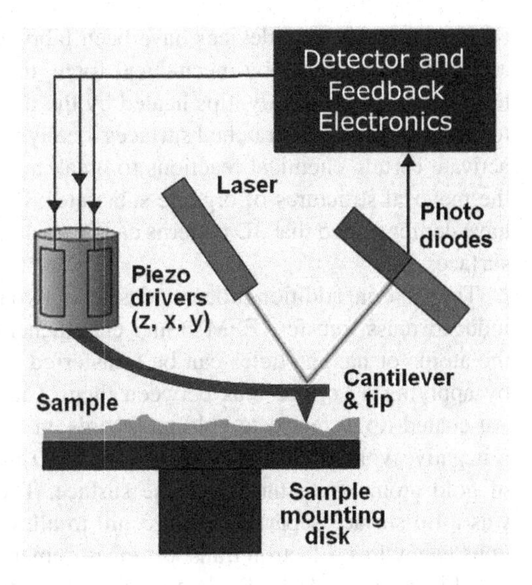

Fig. 12.1 Schematic of AFM apparatus

In the past two decades, due to its low cost and unique atomic-level manipulation capabilities, AFM has evolved from a versatile imaging instrument for atomic and molecular analyses to a vital tool for nanoscale material, component, and device fabrication [14, 18]. Many AFM fabrication techniques have been developed including material modification, material removal, and material addition, which constitute the three most important processes categorized by the taxonomy of manufacturing. Material modification consists of two major techniques: atomic manipulation and local anodic oxidation (LAO). The former is a technique to move individual atoms or molecules as well as nanoscale particles to specific positions to create a range of nanopatterns, with or without material property changes [19]. Oyabu et al. [20] used an AFM to perform lateral manipulation of an adsorbate (intrinsic Ge adatom) on Ge(111)-c(2×8) surfaces and demonstrated that the atom-by-atom creation of patterns at room temperature, composed of a few inherent atoms on a heterogeneous surface is possible. In LAO, nanoscale anodized oxide patterns can be formed by oxidizing the target sample surface through tip-induced anodic or chemical reactions, which are triggered by the high electric fields created by a bias voltage applied between a conducting tip and sample. Because of its inherent high reliability and nanoscale precision, the LAO created pattern can be used as a critical nanoscale element in many electronic devices and as a robust mask for subsequent etching or deposition in nanolithography [14, 18, 21]. The AFM patterned oxide (SiO_2), which can be etched out by a HF solution, has also been used as a sacrificial layer to define the stamp shape used for nanoimprinting [22].

In material removal, an AFM tip has been acted as a cutting tool to perform nanoscale machining process, in which nanoscale materials are directly scratched or machined by the tip loaded with mechanical forces. This material removing process is also known as mechanical scratching or AFM scratching and a wide range of nanostructures and devices have been fabricated in this fashion [14, 18, 23]. In addition to scratching by mechanical force, the AFM tip can also be loaded with thermal energy. Recently, tips heated by the thermal energy were utilized not only to melt or soften the scratched surfaces locally but also to provide thermal energy to activate certain chemical reactions to break the intermolecular bonds or to modify the material structures of organic substrates. Pires et al. [24] and Knoll et al. [25] have demonstrated that 3D patterns can be created by heated tips on many polymeric surfaces.

The material addition process consists of two major techniques: the electric field-induced mass transfer (EFMT) and electrochemical deposition (ECD). In EFMT, the atoms or nanoparticles can be transferred from the tip to the substrate surface by applying a voltage bias between them. Calleja et al. [26] used an oscillating Au-coated tip to fabricate gold nanowires on SiO_2 surfaces. The tip was biased at a negative voltage of 10–15 V with 0.5–5.0 ms pulses to facilitate the transport of gold atoms from the tip to the surface. The voltage was applied when there was a tip-surface separation of $\sim$3 nm to allow sequential pulses to form continuous nanowires. Electron transport measurements of the wires created show a clear metallic behavior. In ECD, the tip and the substrate act as two electrodes to form a nanoscale electrochemical cell that performs a chemical reaction. Agarwal et al. [27] demonstrated that biological molecules could be self-assembled on metallic surfaces using the ECD technique with an AFM tip operated in tapping mode. They demonstrated that histidine-tagged (his-tagged) peptides and proteins could be patterned on a nickel surface.

12.2.2 Scanning Tunneling Microscopy (STM)

An STM (scanning tunneling microscope) has typical resolutions of 0.1 nm in lateral movements and of 0.01 nm in the vertical or depth direction. With a resolution of 0.01 nm, individual atoms can routinely be imaged and manipulated on the surfaces of crystalline materials. An STM uses a sharp conducting tip with a bias voltage applied between the tip and a conducting target sample. When the tip is within the atomic range ($\sim$1 nm) of the sample, electrons from the sample begin to tunnel through the gap to the tip or vice versa, depending on the sign of the bias voltage, as shown in Fig. 12.2a. The exponential dependence of the distance between the tip and the target surface gives STM a remarkable sensitivity with sub-angstrom resolution in the vertical direction. Typically, tunneling current decreases by one order of magnitude as the gap distance is increased by 0.1 nm, which accounts for the extremely high vertical resolution of 0.01 nm. An STM is normally operated in an ultra high vacuum for high precision experiments, since vacuum provides the ideal barrier for tunneling. STM can be operated in either a constant current mode or a constant height mode. For a typical range of operating voltages from 10 mV to 1.0 V, the tip needs to be positioned close enough ($\sim$0.3–1.0 nm) to the surface in order for the tunneling current ($\sim$0.2–10 nA) to be measured conveniently. While STM

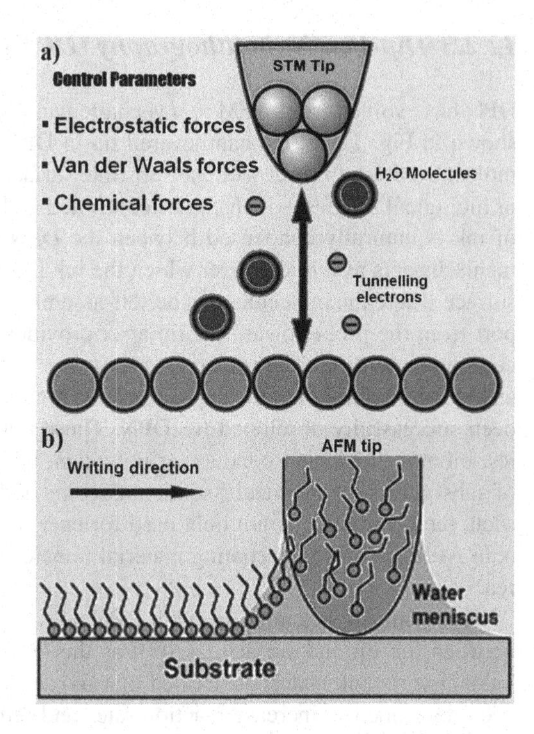

Fig. 12.2 Schematic of STM and DPN probes: (**a**) interaction of charged STM tip with conducting substrate, (**b**) molecular transport between ink coated DPN tip and substrate

can also be operated in a non-vacuum environment, including in air, water, and various other gases or liquids, these generally deteriorate its measurement sensibilities and lower its resolution. Further details on STM design and instrumentation can be found in a review book [28] while its applications to nanofabrication can be found in an article by Tseng et al. [14].

One of the most unique capabilities of STM in TBN is atomic manipulation (AM), which has been used to build model physical systems, such as quantum confined structures, magnetic nanostructures, and artificial molecules [29]. Until recently, because of its superior precision capabilities, STM was the sole technique that could be used for performing AM by moving one atom at a time [19]. Ternes et al. [30] have now developed an instrument that combines the strengths of AFM and STM to allow for the determination of the forces required to move a single atom on a surface. Their results indicate that for moving metal atoms on metal surfaces, the lateral force component plays the dominant role. Furthermore, measuring spatial maps of the forces during manipulation can yield the full potential energy landscape of the tip-sample interaction.

In general, AFM has a much broader potential and range of applications over STM, because AFM can be performed at room environment and concurrently loaded with mechanical, electrical, optical, and thermal energies on either conducting or non-conducting surfaces [14]. While STM can provide a higher resolution, it must be conducted with a conducting tip and target material, and prefers a high-vacuum environment.

12.2.3 Dip-Pen Nanolithography (DPN)

DPN has evolved from AFM to a versatile nanofabrication tool in its own right. As shown in Fig. 12.2b, the cantilevered tip in DPN is coated with a thin film of ink molecules that can react with the substrate surface to create a pattern with chemical or biological species, which can be controlled at the nanoscale level. A minute drop of ink is naturally condensed between the DPN tip and the substrate. The liquid meniscus acts as a bridge over which the ink molecules migrate from the tip to the surface where the molecules can be self-assembled or anchored. The capillary transport from the probe toward the tip apex provides a resupply of new molecules for continuous writing. The ink and substrate are chosen to ensure a chemical affinity and to favor adhesion of the deposited film. A number of different ink materials have been successfully developed for DPN. These materials include inorganics, organics, biomolecules, and conducting polymers, which are compatible with a variety of substrates such as metals, semiconductors, and functionalized surfaces [15]. As such, the tip in DPN is not only used for energy transfer to substrate surfaces as it is in AFM, but also for coating material transfer for further chemical or biological reactions.

The writing or scanning speed of DPN does depend mainly on the mass transport between the tip and surface, as well as the lead and down times for loading and unloading the ink materials. Heated tips and automated ink supplying systems have been developed to increase reaction rates and reduce loading and unloading times [15, 31]. Like AFM, efforts to develop multiprobe arrays with addressable ink wells are underway for DPN. Recent efforts combining DPN with AFM have made use of transparent, two-dimensional (2D) arrays of pyramid-shaped elastomeric tips or pens for large-area, high-throughput patterning of ink molecules [32]. Nevertheless, an automated ink delivery system for multi-probes remains one of the biggest challenges in DPN.

12.2.4 Scanning Near-Field Optical Microscopy (SNOM)

In recent years, the optical version of AFM, known as SNOM, has also shown great potential for fabricating various structures at the nanoscale. The major difference between SNOM and AFM is that the sharp cantilevered tip is replaced by a nanoscale light source/collector, or a scatterer. In an aperture SNOM, as shown in Fig. 12.3a, the probe tip is held very close to the sample and the evanescent light through the aperture is incident on the sample and excites the atoms at the surface to reradiate propagating waves. An apertureless probe, as shown in Fig. 12.3b, can also be used, in which a sharp tip without any apertures, similar to an STM or AFM tip, scans the sample within the scale of the near field. The sample is irradiated from above by far field light to interact with the near field induced by the tip to produce the evanescent field on the surface, which is sensed by a tip detector. Because the distance between the probe and the sample surface is much smaller than the

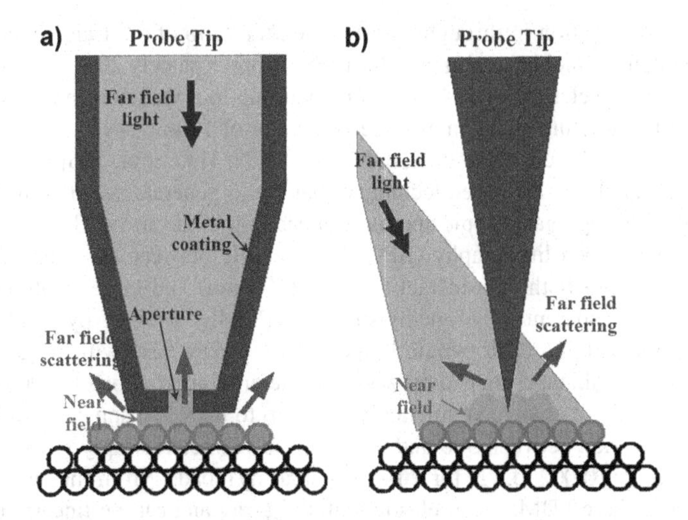

Fig. 12.3 Schematic of SNOM probes: (**a**) aperture tip, (**b**) apertureless tip

wavelength of the light source, SNOM works in the "near-field" and the spatial resolution of the optical excitation of materials is well below the far-field diffraction limit. In this way, the feature size of the resulting patterns fabricated by optical interactions between the tip and sample can be appropriately controlled at the nanoscale. A wide variety of energy forms, including electrostatic, optical, optochemical, optoelectrical, and optomagnetic, have been induced by SNOM-based techniques [16].

In general, the SNOM-based TBN techniques can overcome the diffraction limit of the light source, but they suffer from inherently low throughput and restricted scan area. Many recent efforts have focused on the development of multiple-SNOM probes to increase the throughput by parallel processing with these multiple or arrayed probes [33]. Zhang et al. [34] used e-beam lithography and subsequent pattern transfer by RIE (reactive ion etch) to fabricate a cantilevered array containing 16 aperture probes. This aperture SNOM probe array can be used for parallel processing to increase the throughput for fabrication or to reduce the scanning time for imaging. Each tip has a uniform aperture to accommodate the wavelength of a light source as short as 250 nm. Haq et al. [35] developed an array containing 16 SNOM-probes and demonstrated that their arrayed probes were capable of executing parallel chemical transformations at high resolution over macroscopic areas. Light beams generated by a spatial modulator or a zone plate array were coupled to an array of cantilever probes with Al-coated hollow tips. Each of the probes could be separately controlled with aperture-to-aperture spacing of 125 μm. They demonstrated the selective photodeprotection of nitrophenylpropyloxycarbonyl (NPPOC)-protected aminosiloxane monolayers on silicon dioxide surface and the subsequent growth of nanostructured polymer brushes by atom-transfer radical polymerization, as well as the fabrication of 70 nm structures in photoresist by the probe array immersed underwater. A variation in throughput ratio (i.e., the ratio of the transmitted intensity to the intensity incident upon the aperture) of approximately 20% was measured with light at a wavelength of 514 nm, corresponding to a variation in aperture size of 5 nm for the apertures with an averaged diameter of 100-nm.

By combining the concepts of SNOM with DPN, Huo et al. [36] developed a massively parallel SNOM-based technique that could generate patterns by passing 400-nm light through nanoscopic apertures at each tip in the array. Their technique, known as "beam pen lithography (BPL)," can toggle between near- and far-field distances, allowing both sub-diffraction limit (100 nm) and larger features to be generated. A transparent polydimethylsiloxane (PDMS) array of pyramid-shaped tips was fabricated and each pyramidal pen had a square base with edges several tens of μm in length, tapering to a tip with a diameter of approximately 100 nm. The aperture size did not change significantly from tip to tip (less than 10% variation), and could be varied between 500 and 5,000 nm in diameter, simply by controlling the contact force (0.002–0.2 N for a 1-cm^2 pen array) loaded from the back side of the tip array, where PDMS is an elastic material array and can be linearly or elastically deformed. Thus, in addition to many fabrication parameters, the pattern size or resolution could also be modulated by controlling the tip aperture size or contact force.

12.2.5 Tip-Based Nanofabrication (TBN) and Dual Sources

The nanoscale fabrication and imaging abilities of DPN and SNOM are essentially dictated by the nature of the cantilevered tip, similar to the principle applied in AFM. Most of the time, the technical advancements achieved using one of the four major techniques outlined above (AFM, STM, DPN, and SNOM) can be modified for implementation to the others without major difficulty. Based on the preceding overview of these four techniques, it is expected that more hybrid techniques combining or mixing the advantages of individual techniques and using multiple nanotips or microtips (or micro pens or stamps) will soon be developed to extend the patterning capabilities and/or increase the throughput. As a result, the technology used by these four techniques, as well as extended hybrid techniques, has recently been categorized as TBN and has become one of the major research thrust areas in nanofabrication [13].

In addition to mixing or combining different techniques, TBN has also been improved by loading additional energy sources on the tips. For example, as mentioned earlier, both electrical and thermal energy can be added to AFM and DPN tips to trigger or speed up certain chemical or physical reactions to modify (break or form) materials. If an AFM tip is loaded with an electrical bias, LAO, EFMT, or ECD can be performed, while a heated AFM tip can activate chemical reactions to break intermolecular bonds. The addition of thermal or electrical energy to DPN tip has also improved this technology. In DPN, the molecular inks need to be mobile under ambient conditions, which reduces the number of viable inks and complicates the dynamic control of deposition. This makes it difficult to prevent ink deposition when the tip and/or tip-sample meniscus are in contact with the substrate. Nelson et al. [37] demonstrated that a heated DPN can circumvent these limitations by utilizing a heated tip to deposit ink that is solid at room temperature but flows upon melting. Also, voltage can be applied to a DPN tip to generate a current between the tip and the substrate, which can trigger electrochemical reactions to pattern the substrate. Jegadesan et al. [38] illustrated that these local electrochemical reactions can be used to electropolymerize neutral polymer precursor substrates, forming protruding conductive polymeric features.

12.3 Equipment Precision and Automation

Two of the major challenges in the development of TBN are to increase its reproducibility (repeatability) and productivity (throughput). Towards these goals, extensive efforts have been made to implement both automation and precision strategies in an AFM-based system to reduce fabrication lead and down times as well as improve the product quality and reproducibility, especially patterning reliability and hardware resolution. This section discusses the nature and potential of these efforts. Subsequent sections will elaborate on other efforts, including creating truly 3D nanostructures by a more controllable fashion and using multiple probes for parallel-processing, enhancing the patterning domain through use of microscale tips or stamps, and increasing the writing speeds by better control strategy.

The quality and productivity of the nanostructures created by TBN depend greatly on the capacities of equipment and instruments used and the extent to which the tip can be precisely controlled. It is unavoidable that the TBN system be automated and that precision of control be in the sub-nanometers for dimension accuracy, and sub-nanonewtons for force accuracy. It is therefore a priority should be established to refine and enhance system hardware and software as well as its controllability and integration. The main components of a typical AFM system are probes, scanners (or drivers), displacement detective sensors, and electronics for signal management, as shown in Fig. 12.1. These components and their related subjects are discussed in this section.

12.3.1 Hardware Development and Control

Since their inception, AFM and STM have used piezoelectric tube drivers to move the tip or to scan the sample. Typically, in STM, a sharp metal tip is directly attached at the end of a piezotube driver (P_z) to control the height of the tip above the target surface, while another two piezotube drivers (P_x and P_y) are mounted or coupled with P_z to scan the tip in two lateral directions. In AFM, as shown in Fig. 12.1, the cantilever is directly mounted on the piezotube driver, P_z for vertical control, while the lateral piezotube drivers, P_x and P_y, can be coupled with P_z or with the sample mounting stage. Because of its simplicity in design and construction as well as its compatibility in integration, the piezotube driver with quartered electrodes has become the most popular scanner used in the commercially available SPMs for 3D positioning. A flexure guidance system incorporated with piezoelectric stacks for actuation has been developed to alleviate friction and stiction [39].

Positioning accuracy and speed of a piezodriver are limited due to certain properties inherent in piezoelectric materials, which mainly include hysteresis, creep, thermal drift, and vibrational dynamics. These inherent properties have detrimental effects on the positioning accuracy of the piezo driver. The displacement and motion of a piezo scanner can have a highly non-linear dependence on the applied voltage. For example, because of the hysteretic property, if a piezoelectric scanner is driven by a voltage amplifier, the resulting displacement can deviate from linear by as much as 15% between the forward and backward movements. The deviation due to creep normally occurs when the applied voltage to the piezo undergoes a sudden change and can result in significant loss in precision when positioning is required over extended periods of time, especially during slow operation of AFM for nanofabrication [40, 41]. The errors due to thermal drift are from the thermal expansion and contraction of the mechanical components of the system. Mokaberi and Requicha [42] showed that a Kalman filter could be effectively used to estimate the drift and found that a 1°C change in temperature could cause a 50 nm drift for a typical AFM operating in ambient temperatures. Consequently, the thermal drift deviation can be significant when the AFM nanofabrication is performed in the ambient temperature. If a heated tip is used, the thermal drift problem can amount to a substantial hurdle that requires a solution.

To cope with these nonlinearity errors or effects, in open-loop control, the controller can use a non-linear voltage or charge profile to drive the piezo scanner such that is moves linearly. Fleming and Leang [43] demonstrated that charge drives could reduce the error caused by hysteresis to less than 1% of the scan range. However, the charge profile is difficult to find and can vary greatly from one scanning condition to another. Often, the profile can be taught during calibration procedures using surface gratings with known height steps in the vertical direction and known pitch in the lateral dimensions. The open-loop control strategy has achieved only limited success and only works in scanning patterns similar to that of the calibration gratings. For example, some commercially available AFM software can partially compensate for the hysteresis effect by scanning only in the same direction [41, 44].

Control techniques have been developed and utilized to alleviate and diminish these detrimental effects as well as to improve the tip's moving or scanning speed [45]. The increase in tip speeds from certain control techniques to enable high-bandwidth nanopositioning will be presented in Section 12.5.3; only the basics of feedback control will be discussed in this section. In implementing closed-loop or feedback control, a feedback sensor to measure the displacement and to guide the scanner movement is critical. Integration of a displacement sensor into the piezo scanners or drivers of AFMs can be a delicate task. In AFM, three types of sensors, including laser-beam deflective, optical interferometric, and capacitive, have been adopted for measuring the z displacement or deflection of the cantilever [41]. Among them, the laser-beam deflective sensor is the most popular choice for displacement sensing, because its sensitivity and response time are usually better than the other two.

The laser-beam deflective sensor consists of a laser, photodiodes, and the required optical lens. For measuring the vertical displacement or movement, as shown in Fig. 12.1, a laser beam is directed by a prism onto the back of a cantilever near its free end and the reflected beam from the vertex of the cantilever is directed through a mirror onto a quad-photodiode detector, in which each diode is located in one of the four quadrants. The differential signal from the top and bottom photodiodes provides the AFM signal of the cantilever vertical or bending deflection, while the difference between the left and right diodes detects the cantilever twist due to torsion. The tip deflection or twist will change the direction of the reflected laser beam, changing the intensity difference between two selected photodiodes (feedback signal). For example, in one of the AFM operating modes called constant force mode, for topographic imaging or for any lithographic operation, in which the applied normal force is to be kept constant, a feedback circuit is used to modulate the voltage applied to the P_z driver to adjust the height of the driver, such that the cantilever's vertical deflection (given by the intensity difference between the top and bottom diodes) remains constant during scanning. The driver height variation is thus a direct measure of the surface topography of the sample. The AFM is carried out under a constant force in this mode.

Since it is difficult to use laser-beam deflective sensors to measure lateral displacements, capacitive sensors have emerged as the most popular sensors for lateral displacement measurements in AFMs [28, 41]. The main drawback to use this type of sensor is the noise. If the capacitive sensor is used to generate a frequency response, the root-mean-square noise can be measured to be about 20 pm/(Hz)$^{1/2}$ [45]. Over a bandwidth of 10 kHz, this can yield a noise of 2 nm, which is clearly inadequate for atomic level manipulations. Limiting the bandwidth to 30 Hz would reduce the sensor noise to a level in sub-nm precision that is sufficient for AFM, but this significantly limits the speed of the scanner. Consequently, in most AFMs, the lateral displacement or movements are open-loop controlled. The patterning errors caused by an AFM using open-loop control in the lateral directions can easily be larger than 10 nm [46], which is unacceptable in most TBN operations. Consequently, to increase fabrication speed, the noise levels generated by capacitive sensors must be reduced.

Apart from the noise problem, the cross-coupling issue, which exists among the three-axes of piezotube scanners, also needs to be addressed. Experimental results confirm that this cross-coupling could be substantial. As indicated by Bhikkaji et al. [47], a cross-coupling of 20 dB at low frequencies is normal, and this effect could be even more severe at the resonance frequency of the scanner. This effect becomes more significant if the driver is used in the high-speed scanning regime. Additionally, the accuracy of the nanopositioning in AFMs and other TBN applications is limited by the uncertainties (variations) in the system dynamic behaviors ubiquitously that exist in almost all mechanical systems. To combat these adverse effects on nanopositioning control for AFM applications, including TBN, advanced model-based control techniques have been developed and significant improvement of AFM performance in various applications has been achieved [48]. It is expected that both the positioning accuracy and the bandwidth of piezoelectric tube scanners can be substantially improved by employing model-based control and estimation techniques together with a deeper understanding of the properties of piezoelectric materials. More discussion on positioning control will be presented in the throughput section.

To reduce the noise caused by increasing the bandwidth or scanning speed and the effect from cross-coupling of multi-axis piezo drivers, Moheimani [45] has made a list of recommendations for utilizing advanced feedback control and estimation techniques as well as innovative ideas and methods that have emerged in other fields, e.g., smart structures. By combining these techniques, significant improvement in the operating conditions of a scanner based on a capacitive-based feedback sensor can be achieved. This will, in turn, translate into faster and more accurate AFMs. However, real-time implementation of these methods often requires access to fast digital signal processing hardware and the requisite software. For small piezo tubes, with the first resonance frequency in the tens or hundreds of kHz, such implementation remains quite challenging. Analog control implementation could be an answer, although one would need to address a significant number of issues that arise with such implementation.

Note that the imaging capability of AFM can monitor the geometry or morphology of the nanostructure created in situ. The patterning errors caused by adverse factors, such as noise, piezo creep, piezo drift and dynamics, can be compensated to a certain degree by modifying the processing parameters, depending on the requirement. However, since this approach is not only unreliable but also time consuming, it is not recommended for further investigations for TBN.

12.3.2 Software Development and Automation

In TBN operations, software is as important as the hardware. Software for TBN can include many features or modules. Software provides not only the interface and (software) drivers enabling the user to operate or to communicate with the hardware and external instruments, but also the instructions for the hardware to perform the specified nanofabrication tasks, such as atomic manipulation, LAO, mechanical scratch or tip-induced chemical reaction and deposition. Initially, AFM software was implemented to acquire images with sub-nanometer resolution with certain abilities to perform data acquisition, image processing, and position and movement controls. Development of these general abilities has largely been parallel to evolution in computer technologies and has enabled AFM to measure different types of tip-sample interaction and to become a powerful analytical tool for nanotechnology.

The importance of software for AFM has been recognized by both commercial and non-commercial organizations, many of whom have developed their own software and file formats for different purposes over the past 20 years [49]. A freeware called WSXM, which can read most AFM file formats, is a good example of general AFM software development [50]. WSXM contains a lithography module that enables the user to program and implement additional algorithms or control abilities. Two major purposes of this type of modules are to move and load the tip by providing the link to drive the hardware and to set the appropriate range of values for each of the major processing parameters so the equipment can perform the required tasks concurrently.

One of the requirements for TBN software is to implement a computer-aided design/computer-aided manufacturing (CAD/CAM) feature to enhance the design and patterning capabilities of TBN, or specifically AFM, so that the tip motion and fabrication steps can be controlled by a computer. An AFM integrated with a CAD/CAM system is similar to a machine tool equipped with a CAD/CAM system, also known as a CNC (computer numerical control) machine tool, which is operated by programmable commands encoded on a linked personal computer (PC). Since the 1950s, the CNC machine tool has revolutionized the manufacturing process. Cruchon-Dupeyrat et al. [51] integrated CAD and microcircuit design software into an AFM control system for guiding the AFM to rapidly produce self-assembled monolayers (SAM) to have the designed nanopatterns. They demonstrated that the AFM system equipped with CAD/CAM software could be used to create a 327-nm pie-shaped pattern with a "cowboy loop" around the pie-pattern by nanografting of octadecanethiol in a hexanethiol matrix. A nanografted script with the same thiol SAM was also created by this AFM system. Johannes et al. [52] coupled CAD software with a custom-built three-axis AFM system to generate a 3D pattern. Several micro-sized LAO characters on Si substrates were created to illustrate the capability of the system developed.

Some knowledge-based software (KBS) systems have also been developed for advising and guiding users on how to direct and manipulate the AFM tip and the related processing parameters so as to perform certain required fabrication steps. In atomic manipulation, a software system called probe control software (PCS) has been developed to monitor the underlying phenomena during manipulation processes [53]. It was successfully tested for the manipulation of Au nanoparticles on poly-L-lysine coated mica substrate. An AFM operated in dynamic mode assisted by PCS was used to study the manipulation process by analyzing the simultaneously recorded cantilever amplitude and deflection. The results show that the contact force between the tip and the nanoparticles is responsible for the onset motion. Moreover, utilizing the PCS for colloidal Au nanoparticles on an (aminopropyl) trimethoxysilane (APTS) coated silicon substrate, more complex manipulation was performed, including building a simple 3D pyramidal structure and rotating and translating a linked two-particle structure [53–55]. For investigating the atomic manipulation on an insulator surface, Nishi et al. [56] introduced thermal drift compensation software for better approaching the target atom to enhance tip manipulation capabilities.

Zhao et al. [57] also developed a computer-aided AFM system for automated manipulation of nanoparticles to form designed patterns with applications in nanodevice prototyping.

In AFM scratching, Robinson et al. [58] developed photolithographic simulation software to direct AFM to perform nanoscratching for mask repair and 2D shape reconstruction. Repair results were shown for various processes to highlight the relative strengths and weaknesses of the system developed. Advances in repair dimensional precision and overall imaging performance were also demonstrated.

We expect to see that KBS continuously advances to provide instructions and advices to users for conducting and manipulating AFM equipment to create the required products, as well as to perform the required fabrication steps. With KBS, TBN will become more powerful and versatile.

12.4 Three-Dimensional (3D) Fabrication

In the real world, all objects are 3D. In the engineering world, the construction of truly 3D structures remains a persistent challenge. A large variety of 3D nano- and micro-structures, such as optical lenses, photonic crystals, scanning probes, and gears/racks has been studied and produced [59–61]. This section briefly reviews TBN efforts in 3D patterning and elaborates on associated challenges.

To create real 3D nano- and micro-structures, automated equipment or a CAD/CAM equipped system is a prerequisite. The appropriate relationships among the major processing parameters and the geometrical dimensions must also be available. Since efforts in AFM automation were reviewed in the preceding section, this section addresses different issues. Three approaches – material removal, material modification, and material addition – are selected to illustrate 3D patterning schemes. For the first approach, a process where the tip is imposed with force or force with heat, such as AFM scratching is presented. For the second one, a tip charged with electric bias to perform LAO is examined. For the third approach, material addition, the bottom-up schemes by atomic manipulation and molecular self assembly are reviewed.

12.4.1 Patterning by Material Removal

AFM scratching techniques with a heated or unheated tip can be used to make 3D structures. The scratching approach using an unheated tip is presented first, followed by a description and discussion of the techniques utilizing a hot tip.

With an automated or CAD/CAM equipped AFM, either the tip or workpiece can be moved or rotated in 3D and a curvilinear pattern of grooves can be scratched with a singlepass or overlapping multiple passes. Yan et al. [62] coupled an AFM with a commercial piezo-driven stage to perform 3D scratching of a 1-μm thick copper film, which was deposited on a Si substrate. A few 3D structures were scratched layer-by-layer (or slice by slice). Depending on the accuracy required, the 3D profile of the object was sliced into a number of layers, and each layer was treated as a 2D contour. In scratching a smooth or curved surface, the pitch size, which dictates the smoothness and depth of the scratched pattern, is extremely critical. Theoretically, it can be found that the ratio of the pitch to the curvature radius of the tip should be smaller than or equal to 1.274. The pitch (or pixel pitch) of the overlapped grooves is the distance between two adjacent parallel-grooves scratched. Yan et al. [62] reported that the scratch depth increased by 30% when the feed or pitch is reduced to 66% of the original pitch size. Also, it is well known that in layered manufacturing or rapid prototyping, the geometric inaccuracy or the size of the "stair-step" errors is in the same order as the size of the pitch or feed [63].

Efforts have also been made to modify AFM equipment for scratching 3D or curvilinear patterns. Bourne et al. [64] assembled an AFM based scratching system that could hold an AFM probe at varying angles relative to the workpiece and permit the deflections of the AFM cantilever to be measured through a displacement sensor. This system or assembly combined an AFM probe with a five-axis microscale stage, which had a resolution of 20 nm and can achieve high scratching speeds. Since the workpiece could be rotated by 360°, the assembly can scratch curvilinear patterns of grooves. Grooves with lengths of 82 mm but depths of only a few hundred nm, using a single tool pass at scratching speeds as high at 417 μm/s, were demonstrated. The authors also observed that groove formation involved significant chip formation, while plowing occurred in particular at low load levels [64].

Heated tips can also be used to induce local chemical reactions for tearing covalently bonded polymers. However, breaking a primary chemical bond by a heated tip at very fast time scales is not easy because of the large energy barriers of covalent bonds. Instead, Pires et al. [24] chose a special resist material, called phenolic molecular glass (PMG), in which organic molecules are bound by hydrogen-bonds. These H-bonds can still provide sufficient stability to the material for lithographic processing, but are weak enough to be efficiently activated thermally by the hot tip. Instead of heating the tip continuously, Pires et al. demonstrated that a 3D pattern could be written by simultaneously applying a force and a temperature pulse for several μs. The force pulse pulled the tip into contact while the temperature pulse heated the tip and triggered the breaking or patterning process of the PMG resist. Uniform depth of nanogrooves was created in the resist upon single exposure events, in which the groove depth was controlled as a function of the applied temperature and force. A microscale replica of the Matterhorn Mountain with a resolution of 15 nm into a 100-nm thick PMG film was created. The replica was created by 120 steps of layer-by-layer scratches, resulting in a 25-nm tall structure [24].

Since the depth of scratching could be dictated by the applied force and tip temperature (or processing temperature), a nanostructure with a specific depth profile could be engraved or scratched by controlling the tip force or temperature. Knoll et al. [25] scribed grooves with various or curved depths by changing the tip normal force to generate 3D polymeric nanostructures in single layers. The patterning depth can vary from pixel to pixel by controlling the applied tip force. Figure 12.4 shows that a 3D world nano-map was scratched into a self-amplified

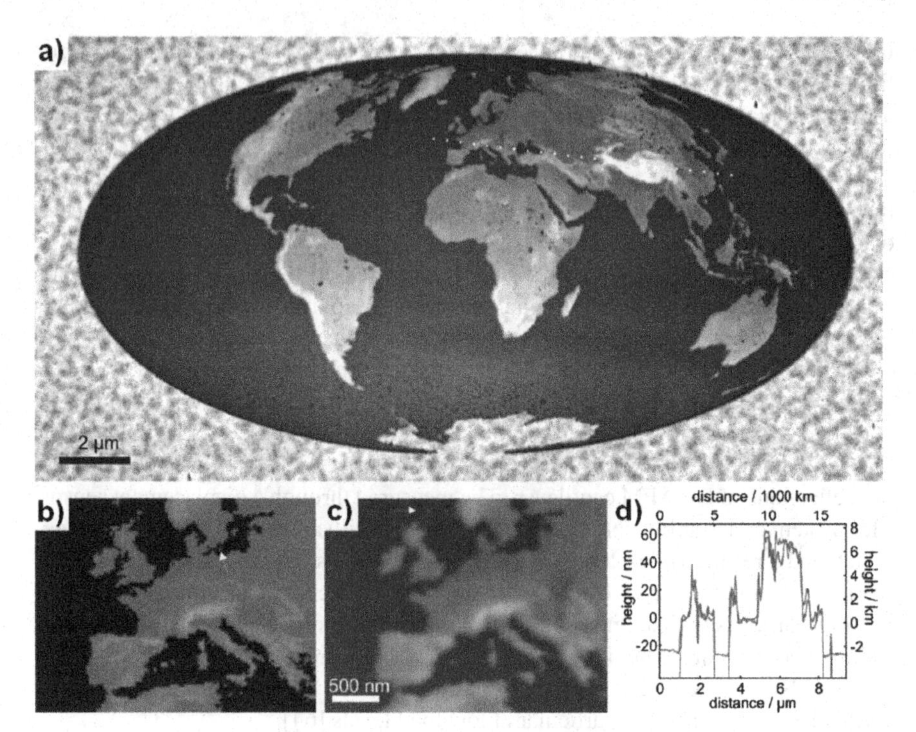

Fig. 12.4 Scratching 3D nanomap on 250-nm thick polyphthalaldehyde polymer film: (**a**) AFM image of digitized 3D world map. A Si tip heated to a temperature of 700°C with a force pulse of 14 μs was used in scratching. The relief is composed of 5×10^5 pixels with a pitch of 20 nm. (**b**) Programmed bitmap of sub-area, (**c**) imaged relief the sub-area in **b**. The *white arrows* indicate positions with features of 1 (in **b**) and 2 (in **c**) pixel width in the original bitmap. The 2 pixel wide features were reproduced to a resolution of ~ 40 nm for the patterning process. (**d**) Cross-section profiles along the *dotted line* shown in (**a**), for the original data and the relief reproduction. The cross-section cuts, from left to right, through the Alps, the Black Sea, the Caucasian mountains, and the Himalayas (reprinted with permission from [25] by Wiley-VCH Verlag GmbH & Co. KGaA)

depolymerization polymer using a heated tip with approximately 40 nm lateral and 1 nm vertical resolution. A Si tip heated to a temperature of 700°C with a force pulse duration of 14 μs was used in the process. The nano map created is composed of 5×10^5 pixels with a pitch of 20 nm. The full map pattern was written within one layer at a pixel rate of 60 μ/s or a writing speed of 0.3 mm/s; the total patterning time amounted to 143 s. The digital geological-elevation profile of the world image of Fig. 12.4a was obtained from U.S. Geological Survey and was transformed into a digital 3D force profile by providing the depth-force scratching relationship of the phthalaldehyde polymer, while the writing temperature was kept constant. In scratching or writing the 3D nanomap, a depth of 8 nm corresponds to 1 km of real-world elevation. Figure 12.4b, c show a comparison of a sub-area in the programmed bitmap and the imaged relief, respectively. Figure 12.4d shows the cross-section profiles along the dotted line shown in Fig. 12.4a. The cross-section cuts, from left to right, through the Alps, the Black Sea, the Caucasian mountains, and the

Himalayas. Interestingly, the material also exhibits good RIE resistance, enabling the direct pattern-transfer into silicon substrates with a vertical amplification of six.

12.4.2 Patterning by Local Anodic Oxidation (LAO)

In LAO, the shape, especially the height and width, of the oxide line induced can be controlled by adjusting the processing parameters, such as the imposed bias, tip scanning speed, and the relative humidity of environment. Roughly speaking, the height and width of the oxide line grow at a rate that depends on the negative bias linearly and on the scanning speed in an inverse logarithmic fashion [21]. Based on a large number of theoretical and experimental studies, software has been developed to provide the guidelines for AFM oxidation, which include the parameters for scanning mode, scanning path, and scanning speed, overlapping pitch, pulse duration, applied voltage, ambient humidity, tip geometry, and tip material [13]. By controlling these operating parameters, the oxide patterns can have not only the desired shape (or contour) but also the desired variation. These patterns can be truly 3D and be used as grayscale masks for making intricate 3D nanostructures, similar to those used for photolithography or laser machining of micro-optical devices.

By varying the applied bias voltage, Fernandez-Cuesta [65] could control the thickness of the oxide grown and demonstrated that gray oxide masks for pattern transferring could be fabricated to make nanoimprinting stamps by subsequent wet etching. With a similar approach, Chen et al. [66] developed a technique for making gray-scale oxide mask by controlling the variation of LAO thickness or height and reported that convex, concave, and arbitrarily shaped Si microlenses with diameters as small as 2 μm could be fabricated. An AFM image of convex lens with diameters of 2, 3, and 4 μm reported by Chen et al. is shown in Fig. 12.5a, which demonstrates that one can fabricate different sizes of microlens in single layers. Figure 12.5b shows a concave microlens with a diameter of 4.5 μm. Figure 12.5c shows a single convex microlens with a diameter of 4.5 μm for comparison; the oxidation tip speed was identical 4 μm/s. Figure 12.5d displays the AFM profiles of the microlenses shown in Fig. 12.5b, c. The upper curves of each profile set (concave and convex) are oxide profiles that are already multiplied by a factor of 30, which is corresponding to the pattern transfer ratio, while the lower curves of each profile set are the Si microlens final profiles from RIE. Note that a concave microlens can be used as a mold to press a resist capping on the substrate for pattern replication. It is more difficult to fabricate a concave microlens with the conventional processes of photoresist reflow, microjet fabrication, or photosensitive glass. In contrast, the combination of AFM oxidation and RIE anisotropic etching is a good way to fabricate both convex and concave microlenses. Since the pixel size and pitch could be controlled within the order of tens of nanometers with desired gray-scale levels, the LAO has the unique capability to create arbitrarily designed microlens structures with exquisite precision and resolution.

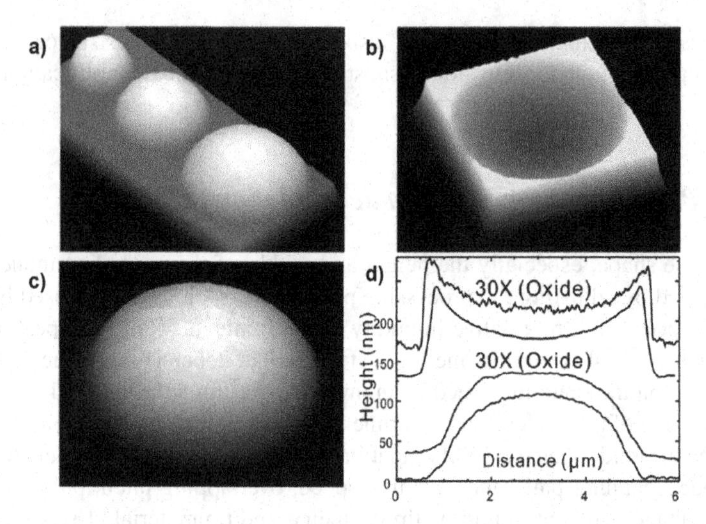

Fig. 12.5 AFM images of Si microlenses fabricated by AFM gray-scale oxidation with RIE dry etching: (**a**) microlenses with diameters of 2, 3, and 4 μm, (**b**) concave microlens with a diameter of 4.5 μm, (**c**) convex microlens with a diameter of 4.5 μm, (**d**) AFM measured profiles of microlenses shown in (**b**) and (**c**) before and after RIE (reproduced with permission from [66] by The Optical Society of America)

12.4.3 Patterning by Atomic Manipulation (AM)

One of the most unique capabilities of STM or AFM is its ability to manipulate atoms and molecules. By direct manipulation of atoms and molecules, a 3D nanostructure with atomic precision can be built. In fact, atomic manipulation is the most basic fabrication approach, since, in principle, all 3D structures or substances can be built one atom by one atom or one molecule by one molecule, if the building speed can be made sufficiently fast.

Until recently, because of its superior precision capability, STM was the sole technique used for performing AM by moving one atom at a time [19]. In STM, atomic resolution is possible by detecting a current of electrons quantum-mechanically tunneling through the vacuum gap between a voltage-biased metallic tip and a conductive surface. This tunneling current, I_t, has a monotonic exponential dependence on the tip–surface separation, z_d, and can be represented by a one-dimensional approximation [67]:

$$I_t = C_I \exp(-2\kappa z_d) \qquad (1)$$

where C_I is a proportional constant and κ is the decay constant for the wave functions in the gap barrier. Typically, I_t is of the order of pico- to nano-amperes, which can be measured with ordinary instrument. For a typical work function of 4 eV, $\kappa = 0.1/\text{nm}$, which implies that I_t decreases by an order of magnitude when the gap z_d is increased by only 0.1 nm. If I_t is keeping within 2%, then the gap z_d remains constant within 0.001 nm. This fact is the basis for a STM to achieve a resolution less than 0.01 nm [67].

The pioneering work demonstrating the ability of positioning single atoms on a metallic surface was conducted by Eigler and Schweizer [68] in 1990. Experiments were performed using an STM at cryogenic temperatures in UHV (ultrahigh vacuum), in which 35 xenon atoms were positioned on a Ni(110) surface to spell out the IBM logo. Since then the majority of the manipulations of atoms or molecules were conducted with STM at low temperatures. STM manipulation has been used to build model physical systems, such as quantum confined structures, magnetic nanostructures, artificial molecules, and computation with individual molecules [29]. STM manipulation of atoms and molecules has also been used to create various nanostructures, which can be characterized and modified in situ by using the tunneling current for rotations, diffusional jumps, vibrational excitations, desorption, and dissociation. As demonstrated by Rieder et al. [69], by tuning the voltage into the energy levels of specific vibrations or electronic levels, new opportunities for making molecular engines and switches become possible. Iancu et al. [70] found that two conformations of isolated single TBrPP-Co molecules on a Cu(111) surface can be manipulated or switched without altering their chemical composition by applying +2.2 V voltage pulses from a STM tip at 4.6 K. As a result, two different Kondo temperatures, which can act as a molecular switch, are obtained by this single molecular switching mechanism. Grill and Moresco [71] presented several examples of molecular wire-electrode systems, where single molecules were placed in contact in a controlled fashion. The associated electronic contact can be characterized using the additional contribution to the tunneling current and also using the influence on the electronic states of the electrode and the molecule. Changed chemical structures of the molecule resulted in different shapes and dimensions of electrodes can lead to a variety of contact configurations and molecular wire-electrode electronics.

In AFM, atomic manipulation has been based on the ability in detecting the interaction forces between the apex atoms of the tip and the atoms at the surface. These interatomic forces, F_{ts}, involve a variety of long- and short-range forces, most commonly being van der Waals (vdW) interactions, which are always present in handling nano objects. As a result, the vdw force is also sometimes used loosely as a synonym for the totality of atomic or intermolecular forces. The vdW force is a long-range electromagnetic force between permanent or induced temporary dipoles. The force is caused by fluctuation in the electrical dipole moment of atoms or molecules and their mutual polarization. The attractive potential (p_a) of this vdW, also known as London force, can be approximated as

$$p_a(z) = -A_1/z^6 \tag{2}$$

where A_1 is the interaction constant defined by London [72] and dependent on the polarization and ionization potential associated and z is the distance between the centers of the paired atoms. In general, the above equation is effective only up to several tens nm. When the interactions are too far apart the dispersion potential, p_a, decays faster than $1/z^6$; this is called the retarded regime [73].

The potential of the repulsion interaction (p_r) between neutral atoms or molecules can be approximated as

$$p_r(z) = B_1/z^n \tag{3}$$

B_1 is a proportional constant and the exponent n is higher than 10 for most of materials [74]. Repulsive forces between atoms and molecules arise mainly from Pauli or ionic repulsion. As a result, this repulsion potential is also called Pauli repulsion. If n = 12, the above potential can incorporate with the attractive potential interaction shown in Eq. (2) to form the well-known Lennard–Jones potential, p_{lj}:

$$p_{lj}(z) = 4p_0\left[\left(\frac{z_0}{z}\right)^{12} - \left(\frac{z_0}{z}\right)^{6}\right] \tag{4}$$

where p_{lj} is the Lennard–Jones potential including both the repulsive and the attractive terms, p_o is the depth of the potential well, z_o is the potential equilibrium distance at which $p_{lj}(z_o) = 0$. Based on the potential p_{lj} presented above, the corresponding Lennard–Jones force f_{lj}, can be found as

$$f_{lj}(z) = -\frac{dp_{lj}}{dz} = \frac{12p_o}{z}\left[\left(\frac{z_e}{z}\right)^{12} - \left(\frac{z_e}{z}\right)^{6}\right] \tag{5}$$

where z_e is the force equilibrium distance equal to $2^{1/6}z_o$. At $z = z_e$, $f_{lj}(z_o) = 0$; thus p_{lj} becomes minimum and equals $-p_o$. The minimum f_{lj} or maximum attractive force can be found to be $-2.3964\,p_o/z_o$ at $z = (13/7)^{1/6}z_e$ or $z = (26/7)^{1/6}z_o$.

Figure 12.6 shows the effect of separation distance (z) variations on the atomic interactive potentials of p_{lj} and force of f_{lj} for the atom pair of N–N and Au–Au. In fact, the atomic interactions of a wide range of atoms are between these two atomic pairs, since N is the most abundant gas in the atmosphere while Au is the most common precious metal. As shown, the potential well of Au–Au atoms is

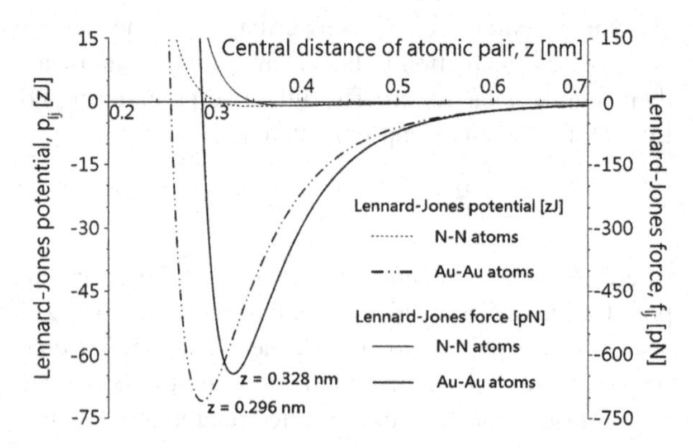

Fig. 12.6 Interatomic potentials and forces of N–N and Au–Au atom pairs (courtesy of Professor Ampere A. Tseng of Arizona State University)

approximately 65 times larger than that of N–N atoms, while the magnitude of the interatomic force of Au–Au atoms is 646.6 pN and about 75 times bigger than that of N–N atoms (8.542 pN). The maximum attractive forces occur at 0.328 and 0.388 nm for Au–Au and N–N, respectively.

As illustrated in Fig. 12.6, these attractive interatomic forces are between pN and nN with the separation distance between the tip and surface atoms being less than 1 nm. These pN or nN interatomic forces are difficult to be recognized by AFM contact mode. As the resolution and sensitivity of AFM is improved with time, these interatomic forces recently become detectable by AFM. Especially, by operating the AFM in dynamic mode under the frequency-modulation (FM), these interaction forces can be very precisely identified and quantified [29, 30]. In this dynamic mode FM, the cantilever is oscillated at near resonance while keeping the oscillation amplitude constant, and the forces acting on the AFM tip are detected as changes in the cantilever resonant frequency and phase. In fact, the signal change of the resonant frequency of an oscillating cantilever can be greatly amplified as a result of the interaction change. The resonant angular frequency is changed from ω_o to ω_e due to the influence of the attractive interaction force change can be found as

$$\omega_e = \sqrt{k_e/m} = \sqrt{\frac{k - (dF_{lj}/dz)}{m}} = \omega_o\sqrt{1 - \frac{dF_{lj}}{kdz}} \tag{6}$$

where ω_0 is the original resonant angular frequency of the cantilever and dF_{lj}/dz is the interaction force change with respect to the gap change (z), in which the cantilever is oscillating in the z direction. Here, F_{lj} is the interaction force and equal to the sum of the Lennard-Jones forces (Eq. 5) of all atoms or molecules involved [75]. If only two atoms are considered, F_{lj} becomes f_{lj}, which can be directly calculated from Eq. (5). Also, k_e and k are the original and effective spring constant of the cantilever, respectively.

If z is larger than z_e, k_e becomes smaller than k due to the attractive force of F_{lj} and ω_e too becomes smaller than ω_o as shown in the above equation. If the cantilever is vibrated at the frequency ω_d (a little larger than ω_0) where a steep slope can be found for a curve representing free space frequency vs. amplitude, the amplitude change (ΔA) at ω_d becomes very large even with a small change of intrinsic frequency caused by atomic attractions. Therefore, the amplitude change measured in ω_d reflects the distance change (Δz) between the tip and the surface atoms. If the change in the effective resonance frequency, ω_e, resulting from the interaction between the surface atoms and the tip, or the change in amplitude (ΔA) at a given frequency (ω_d) can be measured, the dynamic mode feedback loop can then compensate for the distance change between the tip and the sample surface. By maintaining a constant amplitude (A_0) and distance (z_d), dynamic mode can measure the topography of the sample surface by using the feedback mechanism to control the z-scanner movement following the measurement of the force gradient represented in Eq. (6).

Note that, at closest tip–surface distances, i.e., in the region $z < z_e$, the repulsive forces become significant and can seriously hamper atomic resolution by either blurring the short-range interaction or by causing the tip to come into hard contact with the surface. In the past several years, the appropriate control of the distance between the atom at the oscillating tip and the surface atoms to detect the tiny interatomic forces has compelled the AFM community to advance its instrumentation, including reducing the detrimental effect by thermal drift, as reported by Morita et al. [76]. They also indicated that the development of a technique that could arbitrarily construct designed molecules from individual atoms using observation, chemical identification and manipulation of atoms might enable us to construct complex molecules and high polymers from multiatom species in the future and the assembly of molecules using two-atom species, which was expected to be achieved in 2020, was the first step in this direction.

12.4.4 Patterning by Self-Assembly

Self-assembly using biomolecules, such as ligands, peptides, proteins, DNA, and viruses, has been employed to form ordered structures, from self-assembled monolayers (SAMs) to intricate 3D self-folding structures. Chung et al. [77] developed a platform to organize discrete molecular elements and nanostructures into deterministic patterns on surfaces with the potential to form a real 3D nanostructure. Three phases with the aid of AFM were involved. In their technique, two-step nanografting was first used to create patterns of SAMs to drive the organization of virus particles that had been either genetically or chemically modified to bind to the SAMs. Virus-SAM chemistries were described that provided irreversible and reversible binding, respectively. In the second phase, the SAM patterns were applied as affinity templates that were designed to covalently bind oligonucleotides to the SAMs and that were selected for their ability to mediate the subsequent growth of metallic nanocrystals. In the final phase, the liquid meniscus, which had been condensed at the AFM tip-substrate contact, was used as a physical tool to modulate the surface topography of a water soluble substrate and to guide the hierarchical assembly of Au nanoparticles into nanowires. These three phases have the potential to provide a general route toward development of a generic platform for 3D molecular and materials organizations.

Additionally, the inherent self-assembly properties of deoxyribonucleic acid (DNA) are highly programmable and versatile. In biotechnology, DNA is not only the most fundamental building block for biological systems, but also a kind of promising molecule as nano-lead to build or connect nano-devices due to its stable linear structure and certain conductivity. Eventually, by proper manipulation of biomolecules or DNA using the bottom-up approach, 3D structures should be able to be built or self assembled. Researchers have studied the AFM manipulation of single molecules by capturing and placing them at specific reactive sites on a substrate. The inability to discern a particular molecule in a solution or a particular position on a molecule becomes most conspicuous when handling DNA. Since genetic information of DNA is recorded as a linear sequence of bases, the position of the base has an essential meaning. The AFM based technique, which has the capability to control the position and confirmation of individual molecule, can overcome these difficulties. A recent study indicated that with a proper surface treatment of the substrate, the AFM tip could pick a single DNA molecule with reasonable success rate at 75% [78]. As an extension, a new single DNA was allowed to hybridize with the picked DNA, and conjugated with the picker DNA by use of a ligase. Various applications of the tip for manipulating single molecules and preparing new nanomaterials are envisaged.

The AFM tip has also been realized as a nanomanipulation tool to perform space-resolved modification of the DNA structure, also known as molecular surgery, including varied manipulating modes such as "cutting", "pushing", "folding", "kneading", "picking up", "dipping". Hu et al. [79] aligned DNA strands on a solid substrate to form a matrix of 2D networks and used an AFM tip to cut the DNA network in order to fabricate fairly complex artificial patterns. The AFM tip was also utilized to push or manipulate the dissected DNA strands to form spherical nanoparticles and nanorods by folding up the DNA molecules into ordered structures in air. Beyond its biological importance, short DNA sequences are of increasing interest as excellent and versatile building blocks for 3D nanostructures, in material science and nanotechnology. These structures with molecular surgery could be in a more deterministic form and can leverage the strengths of TBN to enable the construction of precisely patterned 3D structures and building blocks for self assembly.

Although the natural structures are all self-assembled 3D forms, the general process in creating true desirable 3D nanostructures, which can be self-assembled in a controllable fashion, have not well-developed yet because the interaction processes, the rules controlling yield, and the fault tolerance in self-assembly are still not fully understood [61]. Nevertheless, with the high-resolution imaging capability, a nanoscale tip can access and manipulate to arbitrary positions on any one of the molecules, it is expected that this nanotip can not only stimulate, catalyze, or initiate but also manipulate, regulate and direct the self-assembling process. As compared to TBN, the possibility using other tools in regulating or controlling the final form of the self-assembled structures is relatively remote. Thus, the tip should play a pivotal role in the control of the final geometry of the nanostructures self-assembled and have a wide impact on nano and DNA technology.

12.5 Challenges in TBN Throughput

One of the major challenges in the development of TBN is to increase throughput. Efforts and perspectives on the implementation of automated or CNC-based AFM systems for reducing fabrication lead and down times and for improving product quality were reviewed in the precedent sections. In this section, three approaches that have been used for improving throughput will be introduced and assessed. These approaches are: using multiple probes, increasing writing speeds, and enhancing tip size. Each of these approaches with an introduction on probe development will be separately presented.

12.5.1 Probe Development

In general, a tip is characterized by its material properties and dimensions, including apex radius, cross-sectional shape, height, aspect ratio, hardness, and stiffness. In AFM, the tip is integrated with or attached to a microscale cantilever. Thus an AFM probe consists of the tip and the cantilever and is distinguished by its stiffness or spring constant and its resonant frequency. The tip and cantilever are typically fabricated from the same material when mass production or integrated fabrication processes are required. Both the tip and the cantilever can be realized in distinct ways: direct fabrication by etching and indirect fabrication by molding. As indicated by Santschi et al. [80], material deposition and milling by focused ion beams have also been used to fabricate different types of tips.

Efforts have been made to design and fabricate special probes other than those for AFM imaging. In AFM machining or scratching, the microscale cantilever is normally under relatively larger applied loads (several hundreds nN or larger) than those specified for imaging or other AFM-based fabrication processes. The scratching probes normally possess higher spring constants (higher stiffness) and higher resonance. Ashida et al. [81] developed a diamond tipped cantilever with a stiffness of 820 N/m, or about 1,000 times greater than that of a typical cantilever used for imaging. Kawasegi et al. [82] also fabricated several Si cantilevers with a spring constant on the order of 500 N/m. These cantilevers can allow for a normal load on the order of 500 μN, and the scratched depth on a Si surface of up to 100 nm. To increase their wear-resistance tips, those tips used for scratching are normally coated with diamond or made of diamond-like materials.

As mentioned earlier, many heated tips have been developed to perform thermochemical nanolithography [24, 25, 83], in which the tips are normally heated by a resistive heating elements embedded in the cantilever. The heating elements are frequently made of doped Si with different shapes and resistivities because they can be easily integrated into a Si cantilever by using a SOI wafer for fabrication [84, 85]. Because of its chemical stability and uniform temperature coefficient of resistivity, Pt has also been used as the heating element. Chiou et al. [86] demonstrated that a Si tip can be heated by an integrated Pt heater up to 120°C to scratch an 800-nm wide groove on a PMMA substrate.

In DPN, the AFM-type tip is required to be re-coated with inks from time to time in writing relatively large or complex patterns. Recoating is time-consuming and tedious, since it needs ink replenishment and tip-position realignment. To have continuous and uniform ink feeding, Kim et al. [31] and Taha et al. [87] developed aperture tips, which were integrated with a micro reservoir, to provide continuous or reliable supply of feeding ink. In fact, two types of aperture based probes, which were also known as Nanofountain Probes (NFP), were developed for DPN. The first one was similar to a micropipette-based probe in which the aperture is constructed at the apex of a hollow pyramidal tip utilizing the back of the tip as a reservoir [87, 88]. The second type had a volcano tip, equipped with a integrated microchannel and an on-chip reservoir, as shown in Fig. 12.7, in which an ink solution is fed into the reservoir and driven by capillary action through the

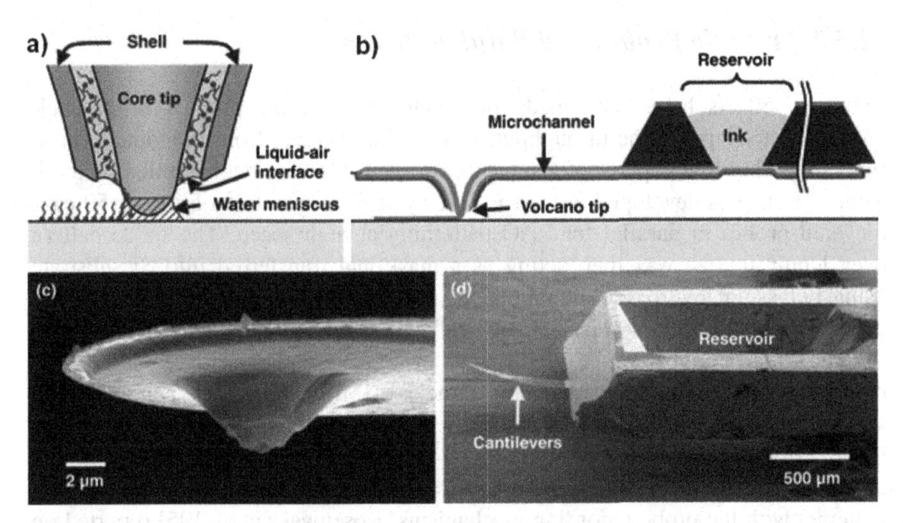

Fig. 12.7 Schematic of volcano tip for DPN: (a) writing mechanism of volcano tip, where molecular ink fed from reservoir forms liquid-air interface at annular aperture of the tip, (**b**) ink from reservoir is delivered to the tip via capillary forces, (**c**) SEM image of volcano tip, (**d**) SEM image of volcano dispensing system (reproduced with permission from [31] by Wiley-VCH Verlag GmbH & Co. KGaA)

microchannel to the volcano tip to form a liquid-air interface around the volcano core as indicated by Kim et al. [31]. Since the NFP tip involves a few of microfluidic components, it is expected that to maintain appropriate ink flowability in these components can be a problem. The probe may not be able sustain an extended period of operation, if the aperture diameter is not large enough. On the other hand, if the tip diameters are too large, these NFP probes can cause resolution problems in applications. Further improvement in the NFP probes to have a better nanoscale resolution capability should be one of the priorities for making it a true nanofabrication tool.

In SNOM applications, aperture tips are also preferred because the photonic near-field generated between the tip and sample becomes more controllable. However, the amount of the supplied photon energy or laser through the aperture is limited by its internal diameter. The tip can be heated up to a few hundred or even thousand degrees, eventually to melt and destroy the tip, if the diameter of the aperture is too small (much less than 100 nm) or the aperture tip is not appropriately designed. Many efforts have been dedicated to making more powerful SNOM probes increasing neither its energy supply nor its diameter [16]. Recently, Wang et al. [89] modified a hollow SNOM probe by adding nanogratings on the tip to transport and concentrate localized surface plasmonic polariton (SPP) wave without losing its imaging resolution. By adding the nanogratings, which consists of concentric plasmonic resonance grooves on the metallic sidewalls of SNOM aperture, the power throughput is increased at over 530 times comparing with single aperture probe with 405 nm source and 100 nm diameter aperture size.

12.5.2 Multiple Probes and Parallel Processing

Extensive efforts have been made in developing multiple probes for parallel-processing to improve the throughput. Approaches ranging from individual multi-function probes to independently activated array probes have been applied [90, 91]. Minne et al. [92] developed an expandable system to operate an array of 50 cantilevered probes in parallel for LAO patterning at high speed. The oxide pattern over 1 cm^2 in size was then acting as a mask and transferred into Si substrate using potassium hydroxide etching. IBM has extended this multiple-probe concept to develop a data-storage system, called Millipede, which uses large arrays (64×64) of AFM-type cantilevered tips to write, read, and erase data on very thin polymer films. The parallel indentation tracks of Millipede can be achieved with spacing of 18 nm between tracks and 9 nm within a track, and depth of 1 nm leading to a storage density of more than 1 Tbit/in^2 [93, 94]. The Millipede-like system capability had also been enhanced by integrating MEMSs (micro electromechanical systems) with the probe recording mechanisms. Rosenwaks et al. [95] reported on parallel AFM-based writing in ferroelectric domains of $LiNbO_3$ and $RbTiOPO_4$, in which the writing speed of ferroelectric domain-based devices is limited by both the physical processes in a single domain and the velocity of the tip. Hagleitner et al. [96] and Yang et al. [97] have developed tiny MEMS-based actuators and sensors to concurrently operate different probe arrays (up to 72×72) to achieve recording density at terabit per square inch and fast data manipulability or accessibility. These efforts on enhancing the parallel AFM-based recording technology should provide building blocks for the development of massively parallel system for AFM lithography.

In the fabrication of multiple probes, not only are the height and shape of each tip important, the dimensions and alignments of the multi-cantilevers also play crucial roles. These geometric factors have consequences in particular for variation in probe properties such as deflection, compliance, and resonant frequency. For example, during scratching, the approach angle for a cantilevered single probe to a planar substrate is not a critical issue. However, in the case of a multiple parallel probe arrangement, the approach angle becomes critical in the plane parallel to the surface. The alignment of the cantilevers and the approach of the array determine which tip comes in contact with the surface first. Often, the two outermost probes are intentionally made longer to serve as the adjustment or pilot probes [80].

Similar to AFM, DPN multiple probes have also developed for parallel processing to increase throughput. In DPN, the tip is mainly used for ink material transfer to activate chemical or biological reactions on the substrate surfaces, unlike an AFM tip that is utilized for energy transfer between the tip and substrate. Thus the characteristics of DPN, including patterning procedure are different from that of AFM, so that the parallel processing strategies developed for AFM cannot be directly applied to DPN. For example, the process of the molecule-based inking in DPN is less sensitive to tip-substrate contact force as compared to AFM [98]. By demonstrating the parallel processing of DPN has Salaita et al. [99] developed a 2-D 55,000-pen array of AFM cantilevers for parallel patterning molecules across 1-cm^2 substrate areas

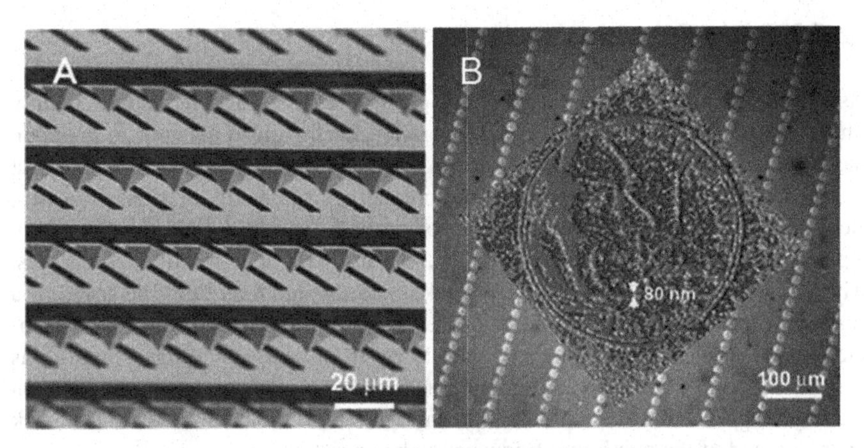

Fig. 12.8 Multiple DPN probe: (**a**) SEM image of part of a 2D 55,000-pen array, (**b**) AFM image of miniaturized replica of five-cent coin generated by depositing 1-octadecanethiol on gold-on-SiOx substrate followed with wet etching, where background is optical image of representative region of the substrate on which approximately 55,000 duplicates were generated (reproduced with permission from [99] by Wiley-VCH Verlag GmbH & Co. KGaA)

at sub-100-nm resolution. The 55,000-pen array as shown in Fig. 12.8a involves a passive-pen array, where each pen is a duplication tool and a single-tip feedback system is needed. The passive-pen array is relatively easier to be implemented and operated than the active pens, in which each pen in the array can be independently actuated. Using this 55,000-pen array a miniaturized replica of the face of the five-cent coin is generated by depositing 1-octadecanethiol on a gold-on-SiOx substrate followed with chemical etching. Figure 12.8b shows the miniaturized replica where the associated background is an optical micrograph of a representative region of the substrate on which approximately 55,000 duplicates are created [99]. On the other hand, Bullen and Liu [100] have developed an active-pen array in which each pen is electrostatically actuated. Active-pen arrays offer the benefit of multiple shape and multiple ink flexibilities. Also, utilizing such arrays, one can prepare many nanostructures of differing feature size and shape in a single experiment. Although the active-pen array has been developed, much work needs to be done to increase the operation flexibility and probe density of the array.

Indeed, researchers have used multi-probe TBN to write arrays of simple geometries, which should be a potentially useful tool for creating more complex nanoscale devices. Certainly, as comparing to the single probe system, it is more complicated in designing multiple probes. Multiple requirements including the software and hardware compatibilities, array architecture, control strategy, and instrumentation involved in a multiple tip array should be taken into consideration. Furthermore, the pattern uniformity and reliability of a large number of probes represent the additional challenges in using this kind of multiprobe in lithography. Nevertheless, parallel processing is the right approach for increasing the throughput. Efforts to satisfy these requirements or challenges should be encouraged.

12.5.3 Probe Speed Increasing

As shown in Fig. 12.1, the tip is moved by piezo actuators. Thus, its speed is mainly limited by the bandwidth of feedback operation for the piezo actuator to maintain the set value for the tip–sample interaction force and by the maximum frequency at which the actuator can be driven without producing unwanted vibrations or noises. Because of the inherent properties in piezoelectric materials, the noise generated in a piezo actuator includes hysteresis, creep, thermal drift, and vibrational dynamics, which limit the bandwidth of the feedback operation and thus limit the speed of a scanner. Efforts to reduce the noise caused by increasing the bandwidth or scanning speed as well as to cope with the effects from cross-coupling of multi-axis piezo drivers have been detailed in Section 12.3.1. This section will only focus on the direct approaches to extend the allowable limits of the bandwidth of feedback operation and of the maximum frequency to be operated.

Clayton et al. [101] reviewed the control approaches that enable higher bandwidths to increase the operating speed of AFMs. In particular, they recommended that inversion-based control could find the feedforward input needed to account for the positioning dynamics and, thus, achieve the required bandwidth as well as precision. Similar efforts in Japan reported by Morita et al. [76] had underway to enhance the resonant frequency of piezo actuators beyond their natural resonant frequencies and to manipulate the driving signal to extend the bandwidth of the mechanical transfer function. This is a kind of inverse transfer function compensation and can create an approximate inverse-transfer function for an arbitrary transfer function [76]. In a different approach, Kodera et al. [102] developed a dynamic PID (proportional-integral-derivative) controller that could change its gain parameters automatically, depending on the tip–sample interaction, and could eliminate the sharp dependence of the feedback bandwidth on the setpoint. With all those recent developments, the feedback bandwidth had reached ~70 kHz. Several studies had reported that fast AFM can be operated at a rate of 30 ms/frame or 30 frame/s [103, 104]. This rate gives the AFM to have real-time panning and zooming capabilities and to be comparable to a typical e-beam lithography system.

Moreover, the feedback bandwidth is also determined by the delays that occur in various steps in the feedback loop. In dynamic mode, a cantilever is normally oscillated close to its resonant frequency. Since it takes at least a half cycle of the oscillation to read the amplitude signal even with the fastest detection technique, the feedback bandwidth can never exceed 1/4 (in practice ~1/8) of the cantilever's resonant frequency. In general, the resonant frequency must be very high, while its spring constant should be kept as small as possible, which requires small cantilevers. However, in many TBN processes, a small cantilever may have adverse effects on its capability.

The writing motion tends to introduce dynamic or vibration-caused errors in the AFM-probe positioning and scanning. By reducing the dynamic-caused errors, Fantner et al. [105] used an additional piezo actuator to impose a simultaneous displacement in the opposite direction to counterbalance the impulsive or vibration forces generated by the motion of the AFM piezo scanner. Kodera et al. [106]

developed a piezo actuator with an actively damping structure and its resonant vibration can be greatly reduced and at the same time enhances the response speed of the piezo actuator. They used a mock scanner or actuator whose transfer function is similar to that of the real system. However, since the capacity of available piezo-actuators is limited, the bandwidth of the z-scanner is limited to ~150 kHz.

Furthermore, by using a spiral scan instead of the normal raster scan, Mahmood and Moheimani [107] reported that because the spiral scan can be produced by using single frequency cosine and sine signals with slowly varying amplitudes to the x-axis and y-axis of an AFM scanner, respectively, the single tone input signals can be applied. As a result, the scanner can be moved at higher speeds without exciting the mechanical resonance of the device and with relatively small control efforts. They demonstrated that high-quality images can be generated at scan frequencies well beyond the raster scans [107]. Based on a spiral AFM, Hung [108] found that the time to complete an imaging cycle could be reduced from 800 to 314 s by using spiral scans instead of the line-by-line scan, without sacrificing the image resolution. Since the spiral AFM can be directly applied for performing many AFM based nanofabrication activities without any further modification, it is expected that the tip writing speed or AFM throughput can be improved by using spiral AFM.

In a different approach, Carberry et al. [109] developed control software called "multitouch interface" to increase the scanning speed by producing intuitive and responsive control environment to shorten the user interface time. They demonstrated this by scanning around two chromosomes in water.

In industrial applications, a high-speed AFM can be very useful. It can be practically used to scan over a large surface, such as 300-mm wafers. The cantilever and deflection detector have to be miniaturized to gain high resonant frequency. Self-sensing and self-actuation cantilevers appear to be ideal for such a miniaturized cantilever scanner. Although this type of cantilever is already available, its resonant frequency in air (~ 60 kHz) is not sufficiently high for high speed scanner [80, 110]. Thus, the realization of a small enough self-sensing and self-actuation cantilever may take years.

12.5.4 Micro/Macro Tips and Stamps

In principle, it should be straightforward that a micro- or macro-scale tip can be used to perform the activities by a nanoscale tip, including AFM scratching, LAO, and deposition of various materials. Certainly, the processing or operation parameters for a micro or macro tip may be different from those of a nanoscale tip. Ashida et al. [81] applied a microscale diamond-coated tip to mechanically scratch Si substrates and found that material removal rate could be higher but the resulted resolution or surface roughness was deteriorated. Dinelli et al. [111] used a focused ion beam (FIB) to mill an AFM tip to form a microscale square-shaped stamp for imprinting (indenting) and a nanoscale tip (~250 nm in diameter) for imaging. The cantilever used had a spring constant of ~38 N/m. They demonstrated that the square shape of the stamp could be faithfully imprinted on a polystyrene film sample, while various imprinted features could be obtained by changing normal loads and dwell times.

Many reserch goups, including Hoeppener et al. [112], Shirakashi [113] and Cavallini et al. [114], have applied microscale tip or stamp to perform LAO for making oxide patterns, which have been used successfully as the masks for the subsequent etching process. Farkas et al. [115] used both nanoscale tips and microscale stamps for local oxidation of iron- and Group IV-metal thin films and no major differences were found on the underlying kinetics of the induced metal oxidation process. They also observed that if nitrogen was incorporated into the metal film during LAO, the induced oxide growth process could be dramatically enhanced as compared with that of single-crystal silicon.

The use of micro or macro scale tips can cover larger surface area and speed up the fabrication process, but can also lose the nanoscale resolution normally associated with a nanoscale tip. To increase throughput without scarifying the nanoscale resolution, a multi-resolution probe, which can be a cluster of probes equipped with different scales of tips to respectively perform different functions, should be developed to satisfy different precision and tolerance requirements in patterning nanostructures having multiscale precision requirements. This type of multi-resolution tools has been available for many macroscale fabrication processes [63]. Also, to be a vital nanofabrication tool, it is imperative to be able to produce a functional device, which may contain both nanoscale and macroscale components. These components may come about as a combination of nanoscale and traditional micro- or macro-fabrication processes. Therefore, the ability of the integration of nanofabricated components with subsequent manufacturing steps and the consolidation of nanostructures into micro/macroscopic structures becomes extremely critical.

Figure 12.9 shows a multi-resolution probe designed by Tseng [116], in which different tips can be conveniently changed during operation. As shown, the multi-resolution probe consists of six major parts: self-actuation probes, multiple probe holder, position-sensitive detector (PSD), piezo-driven sample stage,

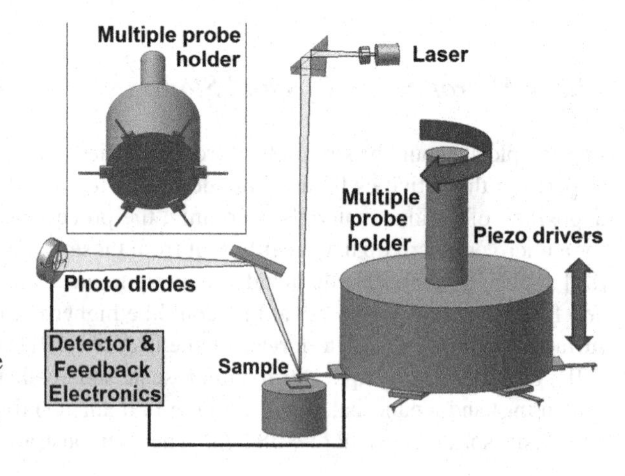

Fig. 12.9 Schematic of atomic force fabricator with multiple probe holder, equipped with 6 self-actuation probes shown in the insert (courtesy of Professor Ampere A. Tseng of Arizona State University)

feedback electronics, and software. The probe, called Multi-head Atomic Force Fabricator (MAFF), which is equipped with six self-actuation probes to alleviate the bandwidth limitation caused by the massive piezo-actuators. Each probe is designed to have different levels of stiffness to perform one type of nanofabrication processes, although more than six probes can be mounted on the holder. The deflections of these microcantilevers are measured by a laser deflection technique, which has the large geometrical amplification of the deflection of cantilever leading to a high sensitivity. Contact forces and force gradients are measured for the static and dynamic modes, respectively, during writing.

Santschi et al. [80] had provided a good review on the subject of self-sensing for AFM-type probes, while Morita et al. [76] recommended that a photo-thermal device should be a good candidate for self-actuation. Note that many of self-sensing devices, such as piezoresistive, capacitive, and piezoelectric, can be reversely used as self-actuators. Since the noise can significantly degrade the resolution and sensitivity of the device involved, the good discussion on the origins of noise related to cantilever structures provided by Santschi et al. [80] could be very informative for those who are interested in design the self-sensing or self-actuating elements. Note that, most of these self-sensing or self-actuating elements require more sophisticated microfabrication or MEMS processes for the integration of these elements into the cantilevers. As mentioned earlier, Bullen and Liu [100] utilized microfabrication techniques to develop electrostatically actuated probes to perform the dip-pen nanolithography. Takami et al. [90] and Wang et al. [91] had also successfully applying the MEMS technique to fabricate independently driven multiple-tip probe to perform multifunctional TBN.

12.6 Roadmap for Tip-Based Nanofabrication

The well-known International Technology Roadmap for Semiconductors (ITRS) has implied that a roadmap is an assessment of an industry's technological requirements and identifies options for overcoming roadblocks to achieving the industry's common goals [12]. It analyzes the technological challenges and needs facing its industry in the near future. In this vein, the preliminary roadmap presented in this section assesses the needs and challenges facing the TBN community. Recently, since the nanofabrication industry has become extremely competitive, this roadmap will help the industry better understand future technological advancements in TBN, especially for enhancing existing facilities to improve product performance while reducing fabrication cost. This roadmap will also help ensure that technological advancements in TBN can satisfy the needs of the nanotechnology community and it can serve as a guideline for future research in TBN [13, 76].

Several roadmaps related to TBN have been reported, including the DARPA requirements and rationales for advancing TBN technology [117], the 2006 Japanese AFM roadmap [76], the Battelle Roadmap for Productive Nanosystems [118], as well as a TBN review article by Tseng et al. [13]. They are adopted as the basis for the updated roadmap presented in this section. Through the use of functionalized AFM cantilevers and tips, DARPA required establishing a controlled environment for automated parallel fabrication of individual nanostructures with control over position, size, shape, and orientation at the nanometer scale. This capability should also include the ability to fabricate devices with controlled differences and the ability for in-situ detection of the nanostructure. The roadmap reported by Morita et al. [76] outlined the future prospects for AFM in three areas: atom manipulation in different environments, high resolution AFM, and high-speed AFM. These prospects were given from the viewpoint of Japanese groups that had been involved in these three areas. The Battelle roadmap pointed the way for strategic research in fabricating a wider range of materials and products with atomic precision to meet many of the greatest global challenges.

The resulting performance metrics are summarized in Table 12.1, which specifically gauges the TBN capabilities like the position accuracy, size variation, manipulation ability, energy species imposed to the tip, heterogeneity ability, multiple-tip fabrication rate, tip shape variation, and in-situ tip-height sensing. Regarding the first metric, it is essential to control the position accuracy of the nanostructures with

Table 12.1 Suggested roadmap for tip-based nanofabrication

Metric	Now	2013	2015	2017
Feature position accuracy [nm]	80	40	20	10
Feature size variation [% of dimension]	15	10	2	1
Atomic manipulation	Atom manipulation in all kinds of fluids at RT	Assembly of atomic devices using 3 atom species	Assembly of atom clusters using 3 atom species	Assembly of molecules with single atom species
Energy species imposed in tip	Dual source in single tip	Dual source in single tip	Triple source in single tip	Triple source in multiple tip
Heterogeneity in parameter: size, shape, or orientation	Two values of one parameter	Five values of two parameters	Continuous control over two parameters	Continuous control over three parameters
Fabrication rate or speed (no. of structure/min/tip with array tips)	1/min/tip Single tip	5/min/tip Five-tip array	50/min/tip Thirty-tip array	100/min/tip Fifty-tip array
Tip shape variation [% of dimension with required operations]	Height < 10%, radius < 20% 10^2 operations	Height < 5%, radius < 10% 10^3 operations	Height < 2%, radius < 4% 10^5 operations	Height < 1%, radius < 2% 10^6 operations
In-situ tip height sensing [nm]	20	10	5	2

respect to each other and with respect to preexisting features on a substrate within a certain number of nm. The metrics are set in the range in which the position accuracies achieved are 80, 40, 20, and 10 nm in 2011 (current ability), 2013, 2015 and 2017, respectively. This metric can be verified by using AFM to detect the position of reference markers on the substrate prior to, during, and after the fabrication. The size accuracy or dimension variation is critical since the ability to fabricate a specific structure with a specific dimension dictates the use of the unique dimension-dependent properties in nanodevices. The size control capability is absent in many existing TBN methods. This second metric requires the nanostructures to be fabricated with controlled accuracy or variations with respect to the characteristic dimension of the nanostructures in parameters. For example, in fabricating an array of 1,000-nm long nanowires, the length variation among the nanowires fabricated should be controlled within 10% of the characteristic dimension (1,000 nm) in 2013, i.e., 100 nm.

The current atomic manipulation ability can mechanically move a single atom in room temperature, air, and liquid environment. In the near future, the assembly of atom clusters and atom devices with novel functions from single to multi-atom species at RT will become important. Furthermore, the development of a technique that can arbitrarily construct designed molecules from individual atoms using observation, chemical identification and manipulation of atoms is the most difficult and challenging issue for atom manipulation. Such a technology may enable us to build complex molecules and high polymers from multi-atom species in the future. Currently, the tip can be loaded with single or dual energy sources, which include mechanical electrical, optical, chemical, and biological. The tip with multiple sources should enhance the capability of all TBN involved. Basically imposing multiple sources concurrently on the tip can not only alleviate the limitations of using a single source but also generate new abilities to handle special materials and new fabrication process, such as one where a bottom-up scheme is integrated with a top-down procedure.

The heterogeneity ability is the capability to fabricate nanostructures that are intentionally different from one to the next. It is understood that many nanodevices cannot be built from arrays of nanostructures with identical parameters (size, shape, and orientation). Currently, one of the three parameters, size, shape, and orientation can be controlled within the metric set for the size accuracy at two specified occasions, while, in 2015, the two structure parameters should be continuously controlled within the specified accuracy, i.e., 1% of the characteristic dimension. To achieve this metric, local tuning of the structure parameters can probably be through the local control of the fabrication environment. The metric on the fabrication rate is to demonstrate that a controlled nanofabrication technology can scale to useful quantities of throughput. At present, one simple nanostructure can be best fabricated at 1 min/tip by assuming that a single tip is used. In 2013 and 2015, the metrics are, respectively, five and 60 nanostructures in 1 min/tip with the assumption that linear arrays of 5 and 30 tips are required. The thresholds for this metric are set with the understanding that automation and parallel operation will eventually be required, and that these can be difficult to implement.

The metric on the tip shape variation is a requirement for the stability of the geometry and characteristics of the tools. For tip-based methods, the size and shape of the tip is likely to be important for maintaining control of the shapes and positions of the manufactured nanostructures. These metrics are set by two geometric parameters and one wear parameter: height, radius, and the number of operations in assuming that the tip is a conical-shape with a spherical end. Different tip geometries should still be parameterized by an overall "height" and the radius of curvature of the end. The currently tip life for performing LAO or imaging can reach 100 min for continuous operation at the best. The metric for the wear parameter implicitly requires continuous operation for 4 h in 2013, 7 h in 2015 and 10 h of operation in 2017. The metrics are set in assuming a growing understanding of tip wear and reliability, and a better method for limiting wear. For TBN, the ability to detect height above the surface is considered to be essential in the control of the environment near the tip. During operation, at least some of the time, the tip is out of contact with the substrate and in-situ tip detection of this separation is necessary. Currently, the ability for the in-situ tip height sensing should be within 20 nm. The metrics for the in-situ tip sensing become 10, 5 and 2 nm in 2013, 2015, and 2017, respectively. This in-situ sensing should be integrated with the arrays to enable use during the automated fabrication.

The eight metrics discussed above provide the basis for the formation of the preliminary TBN roadmap shown in Table 12.1. Undoubtedly, the metrics set forth in the TBN roadmap are extremely challenging. However, it is believed that, by focusing all TBN efforts on the metrics or topics listed, collective and innovative approaches can be developed within several years with the ability to fabricate nanostructures, such as nanowires, nanorods, nanotubes, nano-graphene structures and quantum dots, with nanometer-scale control over the size, orientation, and position of each nanostructure. With this ability, a wide range of nano-scale structures should be possible for the first time [13, 76, 118]. It is understood that a roadmap should be built by the consensus of leaders in industry, universities and government, and thus have the benefit of collective input beyond what any single person, company, university or government agency might provide. The roadmap presented is extremely preliminary, but it is a first step in this consensus building process.

12.7 Concluding Remarks

The technology of tip-based nanofabrication (TBN) was reviewed with an emphasis on recent progress and future challenges. Four major techniques under the family of TBN, atomic force microscopy (AFM), scanning tunneling microscopy (STM), dip-pen nanolithography (DPN), and scanning near-field optical microscopy (SNOM), were studied and their capabilities for material modification, deposition, and removal, as well as for other lithographic activities, were discussed. In particular, major advances in these techniques were presented to illustrate the respective feasibilities and potentials of the technique considered. The information gathered in this study was adopted for the assessment of the constraints and challenges facing the TBN community, and has led to possible solutions for future development and to a preliminary roadmap for TBN technology over the next several years.

Without a doubt, TBN looks powerful and promising, but major challenges remain in realizing its full potential. Three areas of constraints and challenges, including repeatability (reliability), ability (feasibility), and productivity (throughput), were highlighted. The rationale and specific approaches for enhancing the repeatability by automation were first introduced. The hardware and software developments required for achieving automation with required precision in TBN were specifically elaborated. The ability to create truly 3D nanostructures was then assessed with a focus on seeking the best strategy to develop 3D-engineered nanostructures in a controlled fashion.

Low throughput is the challenge common to all TBN techniques. Three approaches to improving throughput – operating multiple-tip in parallel, increasing the tip writing speed (rate), and using micro/macro scale tips – were assessed. The first one refers to many multiple-tip systems, which are similar to the IBM Millipede systems and have been developed for all four TBN techniques considered. However, in operating multiple tips in parallel, retaining the ability to control each tip independently still poses a constraint on the number of tips in a system. A more versatile control strategy to handle a great number of tips should be developed. Much effort has also been dedicated to increasing writing speed. Several studies have reported that fast AFMs have been developed to make TBN comparable to fast e-beam lithography systems. Approaches to developing even faster writing speed were proposed and discussed in some detail. Furthermore, many micro/macro scale tips have been developed to manage larger surface areas at reasonable writing speeds. It is also believed that a multiresolution TBN system equipped with both nanoscale and microscale tips, as well as loaded with single or multiple sources should be developed to provide multilevel accuracy and functional requirements in patterning different types of nanostructures.

Indeed, in TBN, the tip can initiate physical, chemical and biological changes through mechanical, thermal, electrical, magnetic, and/or optical interactions and can pattern nanostructures by material removal, modification, or deposition. However, with the tip is loaded with multiple energy sources, the resulting process becomes more complicated and more parameters must be correctly controlled. In many cases, slight changes in certain parameters, which might normally be insignificant when the tip is loaded with a single source, can substantially impact the quality and properties of the nanostructures created by multiple-source methods. As a result, it is expected that the challenges for multiple-source processes will be even greater than those for single-source processes. Therefore, the relevant phenomena, especially those related to the newly developed processes using combined or hybrid techniques need to be carefully investigated. Otherwise appropriate and useful TBN techniques, including hardware and software, cannot be developed. Moreover, TBN equipment may need to be redesigned and new processing strategies may need to be developed to address new challenges posed by combining various TBN techniques.

In the coming decade, TBN is expected to become an essential fabrication tool in many areas of science and technology and to have a revolutionary impact on every aspect of the manufacturing industry. On the thirtieth anniversary of the invention of scanning probe microscopy, it is especially meaningful to have this article that reviews the current accomplishments and constraints of TBN, that elaborates future challenges and developments of its applications to nanofabrication, and that provides an associated roadmap to the next decade in TBN. Finally, it is worth noting out that the purpose of this article was not to present an exhaustive dissertation of the constraints and challenges in this broad field, but rather to help the reader appreciate the essential potentials and major trends in TBN.

Acknowledgements The author would like to acknowledge the support of Pacific Technology, LLC of Phoenix (USA) and the National Science Council (ROC) under Grant No. NSC99-2811-E-007-014 in funding a University Chair professorship at National Tsing Hua University (NTHU) in Hsinchu, Taiwan, where the author spent a semester in preparation of this manuscript in 2010. The author is grateful to Professors Wen-Hwa Chen, Hung Hocheng, and Chien-Chung Fu of NTHU for their hospitality and encouragement during the author's stay in Hsinchu. Very special thanks are to Professor Jun-ichi Shirakashi of Tokyo University of Agriculture and Technology (Japan), Dr. Andrea Notargiacomo of CNR-IFN (Italy), Dr. Luca Pellegrino of CNR-SPIN (Italy), Professor Thomas W. Kenny of Stanford University (USA), Professor Ari Requicha of the University of Southern California (USA) and Dr. John Dagata of National Institute of Standards and Technology (USA) for their useful information and fruitful discussions. The author is thankful for the assistance provided by Ms Yu-Shan Huang and Mr. Gwo J. Wu of NTHU in preparing this manuscript.

References

1. H. A. Atwater, The promise of plasmonics, *Sci. Am.*, **296** (4), 56–63 (2007).
2. V. M. Shalaev, Transforming light, *Science*, **322**, 384–386 (2008).
3. A. A. Tseng, C. D. Chen, C. S. Wu, R. E. Diaz, M. E. Watts, Electron-beam lithography of microbowtie structures for next generation optical probe, *J. Microlith., Microfab. Microsyst.*, **1**, 123–135 (2002).
4. I. I. Smolyaninov, Y.-J. Hung, C. C. Davis, Magnifying superlens in the visible frequency range, *Science*, **315**, 1699–1701 (2007).
5. S. Pillai, K. R. Catchpole, T. Trupke, M. A. Green, Surface plasmon enhanced silicon solar cells, *J. Appl. Phys.*, **101**, 093105 (2007).
6. K. Nakayama, K. Tanabe, H. A. Atwater, Plasmonic nanoparticle enhanced light absorption in GaAs solar cells, *Appl. Phys. Lett.*, **93**, 121904 (2008).
7. P. N. Prasad, *Nanophotonics*, Wiley, New York, 2004.
8. J. B. Pendry, D. Schurig, D. R. Smith, Controlling electromagnetic fields, *Science*, **312**, 1780–1782 (2006).
9. H. Craighead, Future lab-on-a-chip technologies for interrogating individual molecules, *Nature*, **442**, 387–393 (2006).
10. V. Wagner, A. Dullaart, A.-K. Bock, A. Zweck, The emerging nanomedicine landscape, *Nat. Biotechnol.*, **24**, 1211–1217 (2006).
11. R. A. Freitas, Jr., Current status of nanomedicine and medical nanorobotics, *J. Comput. Theor. Nanosci.*, **2**, 1–25 (2005).
12. *The International Technology Roadmap for Semiconductors (ITRS)*, 2009 ed., www.itrs.net/links/2009ITRS/Home2009.htm (2010).
13. A. A. Tseng, S. Jou, A. Notargiacomo, T. P. Chen, Recent developments in tip-based nanofabrication and its roadmap, *J. Nanosci. Nanotechnol.*, **8**, 2167–2186 (2008).
14. A. A. Tseng, A. Notargiacomo, T. P. Chen, Nanofabrication by scanning probe microscope lithography: a review, *J. Vac. Sci. Technol. B*, **23**, 877 (2005).

15. K. Salaita, Y. Wang, C. A. Mirkin, Applications of dip-pen nanolithography, *Nat. Nanotechnol.* **2**, 145–155 (2007).
16. A. A. Tseng, Recent developments in nanofabrication using scanning near-field optical microscope lithography, *Optical Laser Technol.*, **39**, 514–526 (2007).
17. G. Binnig, C. F. Quate, C. Gerber, Atomic force microscope, *Phys. Rev. Lett.*, **56**, 930–933 (1986).
18. A. A. Tseng, L. Pellegrino, J.-I. Shirakashi, Nanofabrication using atomic force microscopy, in *Encyclopedia of Nanoscience and Nanotechnology*, chapter 287, 2nd ed., by H. S. Nalwa, American Scientific, Valencia, CA, 2012.
19. A. A. Tseng, Z. Li, Manipulations of atoms and molecules by scanning probe microscopy, *J. Nanosci. Nanotechnol.*, **7**, 2582–2595 (2007).
20. N. Oyabu, Y. Sugimoto, M. Abe, O. Custance, S. Morita, Lateral manipulation of single atoms at semiconductor surfaces using atomic force microscopy, *Nanotechnology*, **16**, S112–S117 (2005).
21. A. A. Tseng, T.-W. Lee, A. Notargiacomo, T. P. Chen, Formation of uniform nanoscale oxide layers assembled by overlapping oxide lines using atomic force microscopy, *J. Micro/Nanolith. MEMS & MOEMS*, **8,** 043050 (2009).
22. I. Fernandez-Cuesta, X. Borrise, F. Perez-Murano, Atomic force microscopy local andic oxidation of thin Si_3N_4 layers for robust prototyping of nanostructures, *J. Vac. Sci. Technol. B.*, **24**, 2988 (2006).
23. A. A. Tseng, J. Shirakashi, S. Nishimura, K. Miyashita, A. Notargiacomo, Scratching properties of nickel-iron thin film and silicon using atomic force microscopy, *J. Appl. Phys.*, **106**, 044314 (2009).
24. D. Pires, J. L. Hedrick, A. De Silva, J. Frommer, B. Gotsmann, H. Wolf, M. Despont, U. Duerig, A. W. Knoll, Nanoscale three-dimensional patterning of molecular resists by scanning probes, *Science*, **328**, 732–735 (2010).
25. A. W. Knoll, D. Pires, O. Coulembier, P. Dubois, J. L. Hedrick, J. Frommer, U. Duerig, Probe-based 3-D nanolithography using self-amplified depolymerization polymers, *Adv. Mater.*, **22**, 3361–3365 (2010).
26. M. Calleja, M. Tello, J. Anguita, F. García, R. García, Fabrication of gold nanowires on insulting substrates by field-induced mass transport, *Appl. Phys. Lett.*, **79**, 2471 (2001).
27. G. Agarwal, R. R. Naik, M. O. Stone, Immobilization of histidine tagged proteins on nickel by electrochemical dip pen nanolithography, *J. Am. Chem. Soc.*, **125**, 7408–7412 (2003).
28. C. J. Chen, *Introduction to Scanning Tunneling Microscopy*, 2nd ed., Oxford University, New York, NY, 2008.
29. O. Custance, R. Perez, S. Morita, Atomic force microscopy as a tool for atom manipulation, *Nat. Nanotechnol.*, **4**, 803–810 (2009).
30. M. Ternes, C. P. Lutz, C. F. Hirjibehedin, F. J. Giessibl, A. J. Heinrich, The force needed to move an atom on a surface, *Science*, **319**, 1066–1069 (2008).
31. K. H. Kim, N. Moldovan, H. D. Espinosa, A nanofountain probe with sub-100 nm molecular writing resolution, *Small*, **1**, 632–635 (2005).
32. L. Huang, A. B. Braunschweig, W. Shim, L. Qin, J. K. Lim, Matrix-assisted dip-pen nanolithography and polymer pen lithography, *Small*, **6**, 1077–1081 (2010).
33. S. S. Choi, J. T. Ok, D. W. Kim, M. Y. Jung, M. J. Park, Modeling of a nanoscale oxide aperture opening for a NSOM probe, *J. Kor. Phys. Soc.* **45**, 1659–1663 (2004).
34. Y. Zhang, K. E. Docherty, J. M. R. Weave**,** Batch fabrication of cantilever array aperture probes for scanning near-field optical microscopy, *Microelectronic Eng.*, **87**, 1229–1232 (2010).
35. E. ul Haq, Z. Liu, Y. Zhang, S. A. A. Ahmad, L.-S. Wong, S. P. Armes, J. K. Hobbs, G. J. Leggett, J. Micklefield, C. J. Roberts, J. M. R. Weaver, Parallel scanning near-field photolithography: the Snomipede, *Nano Lett.*, **10**, 4375–4380 (2010).
36. F. Huo, G. Zheng, X. Liao, L. R. Giam, J. Chai, X. Chen, W. Shim, C. A. Mirkin, Beam pen lithography, *Nat. Nanotechnol.*, **5**, 637–640 (2010).

37. B. A. Nelson, W. P. King, A. R. Laracuente, P. E. Sheehan, L. J. Whitman, Direct deposition of continuous metal nanostructures by thermal dip-pen nanolithography, *Appl. Phys. Lett.*, **88,** 033104 (2006).
38. S. Jegadesan, P. Taranekar, S. Sindhu, R. C. Advincula S. Valiyaveettil, Electrochemically nanopatterned conducting coronas of a conjugated polymer precursor: SPM parameters and polymer composition, *Langmuir*, **22,** 3807–3811 (2006).
39. S. S. Aphale, B. Bhikkaji, S. O. R. Moheimani, Minimizing scanning errors in piezoelectric stack-actuated nanopositioning platforms, *IEEE Trans. Nanotechnol.*, **7**, 79–90 (2008).
40. I. Mayergoyz, *Mathematical Models of Hysteresis*, Springer, New York, NY 1991.
41. E. Meyer, H. J. Hug, R. Bennewitz, *Scanning Probe Microscopy*, Springer, Heidelberg, Germany, 2004.
42. B. Mokaberi, A. G. Requicha, Drift compensation for automatic nanomanipulation with scanning probe microscopes, *IEEE Trans. Autom. Sci. Eng.*, **3**, 3 (2006).
43. A. J. Fleming, K. K. Leang, Charge drives for scanning probe microscope positioning stages, *Ultramicroscopy*, **108**, 1551–1557 (2008).
44. I. A. Mahmood, S. O. R. Moheimani, Making a commercial atomic force microscope more accurate and faster using positive position feedback control, *Rev. Sci. Instrum.*, **80**, 063705 (2009).
45. S. O. R. Moheimani, Invited review article: accurate and fast nanopositioning with piezoelectric tube scanners: emerging trends and future challenges, *Rev. Sci. Instrum.*, **79**, 071101 (2008).
46. H. Kuramoch, K. Ando, T. Tokizaki, M. Yasutake, F. Perez-Murano, J. A. Dagata, H. Yokoyama, Large scale high precision nano-oxidation using an atomic force microscope, *Surf. Sci.*, 566–568, 343–348 (2004).
47. B. Bhikkaji, M. Ratnam, A. J. Fleming, S. O. R. Moheimani, *IEEE Trans. Control Syst. Technol.*, **5**, 853(2007).
48. Y. Yan, Q. Zou, Z. Lin, A control approach to high-speed probe-based nanofabrication, *Nanotechnology*, **20**, 175301(2009).
49. P. Zahl, M. Bierkandt, S. Schröder, A. Klust, The flexible and modern open source scanning probe microscopy software package GXSM, *Rev. Sci. Instrum.*, **74**, 1222 (2003).
50. I. Horcas, R. Fernández, J. M. Gómez-Rodríguez, J. Colchero, J. Gómez-Herrero, A. M. Baro, WSXM: a software for scanning probe microscopy and a tool for nanotechnology, *Rev. Sci. Instrum.*, **78**, 013705 (2007).
51. S. Cruchon-Dupeyrat, S. Porthun, G.-Y. Liu, Nanofabrication using computer-assisted design and automated vector-scanning probe lithography, *Appl. Surf. Sci.*, **175–176**, 636–642 (2001).
52. M. S. Johannes, J. F. Kuniholm, D. G. Cole, R. L. Clark, Automated CAD/CAM-based nanolithography using a custom atomic force microscope, *IEEE Trans. Autom. Sci. Eng.*, **3**, 236–239 (2006).
53. B. Mokaberi, A. A. G. Requicha, Drift compensation for automatic nanomanipulation with scanning probe microscopes, *IEEE Trans. Autom. Sci. Eng.*, **3**, 199–207 (2006).
54. A. A. G. Requicha, D. J. Arbuckle, B. Mokaberi, J. Yun, Algorithms and software for nanomanipulation with atomic force microscopes, *Int. J. Robotics Res.*, **28**, 512–522 (2009).
55. R. Resch, C. Baur, A. Bugacov, B. E. Koel, A. Madhukar, A. A. G. Requicha, P. Will, Building and manipulating three-dimensional and linked two-dimensional structures of nanoparticles using scanning force microscopy, *Langmuir*, **14**, 6613–6616 (1998).
56. R. Nishi, D. Miyagawa, Y. Seino, I. Yi, S. Morita, Non-contact atomic force microscopy study of atomic manipulation on an insulator surface by nanoindentation, *Nanotechnology*, **17**, S142–S147 (2006).
57. W. Zhao, K. Xu, X. Qian, R. Wang, Tip based nanomanipulation through successive directional push, *ASME J. Manuf. Sci. Eng.*, **132**, 0309091 (2010).

58. T. Robinson, A. Dinsdale, M. Archuletta, R. Bozak, R. White, Nanomachining photomask repair of complex patterns, in *Photomask Technology 2008, Proceedings of SPIE*, Vol. 7122, The International Society for Optical Engineering, 2008.
59. V. K. Parashar, A. Sayah, M. Pfeffer, F. Schoch, J. Gobrecht, M. A. M. Gijs, Nano-replication of diffractive optical elements in sol-gel derived glasses, *Microelectronic Eng.*, **67–68**, 710–719 (2003).
60. A. A. Tseng, Recent developments in micromachining of fused silica and quartz using excimer lasers, *Phys. Status Solidi A*, **204**, 709–729 (2007).
61. T. G. Leong, A. M. Zarafshar, D. H. Gracias, Three-dimensional fabrication at small size scales, *Small*, **6**, 792–806 (2010).
62. Y. Yan, T. Sun, Y. Liang, S. Dong, Investigation on AFM-based micro/nano-CNC machining system, *Int. J. Mach. Tools Manuf.*, **47**, 1651–1659 (2007).
63. A. A. Tseng, M. Tanaka, Advanced deposition techniques for freeform fabrication of metal and ceramic parts, *Rapid Prototyping J.*, **7**, 6–17 (2001).
64. K. Bourne, S. G. Kapoor, R. E. DeVor, Study of a high performance AFM probe-based microscribing process, *ASME J. Manuf. Sci. Eng.*, **132**, 030906 (2010).
65. I. Fernandez-Cuesta, X. Borrise, F. Perez-Murano, Atomic force microscopy local oxidation of silicon nitride thin films for mask fabrication, *Nanotechnology*, **16**, 2731–2737 (2005).
66. C.-F. Chen, S.-D. Tzeng, H.-Y. Chen, S. Gwo, Silicon microlens structures fabricated by scanning-probe gray-scale oxidation, *Opt. Lett.*, **30**, 652–654 (2005).
67. P. K. Hansma, J. Tersoff, Scanning tunneling microscopy, *J. Appl. Phys.*, 61, R1–R23 (1987).
68. D. M. Eigler, E. K. Schweizer, Positioning single atoms with a scanning tunnelling microscope, *Nature*, **344**, 524–526 (1990).
69. K.-H. Rieder, G. Meyer, F. Moresco, K. Morgenstern, S.-W. Hla, J. Repp, M. Alemani, L. Grill, L. Gross, M. Mehlhorn, H. Gawronski, V. Simic-Milosevich, J. Henzl, K. F. Braun, S. Foelsch, L. Bartels, Force induced and electron stimulated STM manipulations: routes to artificial nanostructures as well as to molecular contacts, engines and switches, *J. Phys.: Conf. Series*, **19**, 175 (2005).
70. V. Iancu, A. Deshpande, S.-W. Hla, Manipulating Kondo temperature via single molecule switching, *Nano Lett.*, **6**, 820–823 (2006).
71. L. Grill, F. Moresco, Contacting single molecules to metallic electrodes by scanning tunnelling microscope manipulation: model systems for molecular electronics, *J. Phys. Condens. Matter.*, **18**, S1887–S1908 (2006).
72. F. London, Zur Theorie und Systematik der Molekularkrafte (On the theory and systematic of the molecular forces), *Z. Phys*, **63**, 245–279 (1930).
73. J. N. Israelachvilli, *Intermolecular and Surface Forces*, 2nd ed., Academic, London, 1991.
74. M. N. Magomedov, The calculation of the parameters of the Mie–Lennard-Jones potential, *High Temperature*, **44**, 513–529 (2006) (Translated from *Teplofizika Vysokikh Temperatur*, **44**, 518–533, 2006).
75. A. A. Tseng, Atomic interactions in nanofabrication, *J. Nanoscience and Nanotechnology*, (in press).
76. S. Morita, H. Yamada, T. Ando, Japan AFM roadmap 2006, *Nanotechnology*, **18**, 084001 (2007).
77. S. W. Chung, A. D. Presley, S. Elhadj, S. Hok, S. S. Hah, A. A. Chernov, M. B. Francis, B. E. Eaton, D. L. Feldheim, J. J. Deyoreo, Scanning probe-based fabrication of 3D nanostructures via affinity templates, functional RNA, and meniscus-mediated surface remodeling, *Scanning*, **30**, 159–171 (2008).
78. D. Kim, N. K. Chung, J. S. Kim, J. W. Park, Immobilizing a single DNA molecule at the apex of AFM tips through picking and ligation, *Soft Matter*, **6**, 3979–3984 (2010).
79. J. Hu, Y. Zhang, H. B. Gao, M. Q. Li, U. Hartmann, Artificial DNA patterns by mechanical nanomanipulation, *Nano Lett.*, **2**, 55–57 (2002).

80. C. Santschi, J. Polesel-Maris, J. Brugger, H. Heinzelmann, Scanning probe arrays for nanoscale imaging, sensing and modification, in *Nanofabrication: Fundamentals and Applications*, ed. by A. A. Tseng, pp. 65–126, World Scientific, Singapore, 2008.
81. K. Ashida, N. Morita, Y. Yoshida, Study on nano-machining process using mechanism of a friction force microscope, *JSME Int. J. Ser. C*, **44**, 244–253 (2001).
82. N. Kawasegi, N. Takano, D. Oka, N. Morita, S. Yamada, K. Kanda, S. Takano, T. Obata, K. Ashida, Nanomachining of silicon surface using atomic force microscope with diamond tip, *ASME J. Manuf. Sci. Eng.*, **128**, 723–729 (2006).
83. R. Szoszkiewicz, T. Okada, S. C. Jones, T. D. Li, W. P. King, S. R. Marder, E. Riedo, High-speed, sub-15 nm feature size thermochemical nanolithography, *Nano Lett.*, **7**, 1064–1069 (2007).
84. B. W. Chui, T. D. Stowe, Y. S. Ju, K. E. Goodson, T. W. Kenny, H. J. Mamin, B. D. Terris, R. P. Ried, D. Rugar, Low-stiffness silicon cantilevers with integrated heaters and piezoresistive sensors for high-density AFM thermomechanical data storage, *IEEE J. Microelectromech. Syst.*, **7**, 69–78 (1998).
85. D. W. Lee, T. Ono, M. Esashi, Electrical and thermal recording techniques using a heater integrated microprobe, *J. Micromech. Microeng.*, **12**, 841–848 (2002).
86. C. H. Chiou, S. J. Chang, G. B. Lee, H. H. Lee, New fabrication process for monolithic probes with integrated heaters for nanothermal machining, *Jpn. J. Appl. Phys., Part 1*, **45**, 208–214 (2006).
87. H. Taha, R. S. Marks, L. A. Gheber, I. Rousso, J. Newman, C. Sukenik, A. Lewis, Protein printing with an atomic force sensing nanofountain pen, *Appl. Phys. Lett.*, **83**, 1041 (2003).
88. A. Meister, S. Jeney, M. Liley, T. Akiyama, U. Staufer, N. F. de Rooij, H. Heizelmann, Nanoscale dispensing of liquids through cantileverd probes, *Microelectron. Eng.*, **67–68**, 644–650 (2003).
89. Y. Wang, Y.-Y. Huang, X. Zhang, Plasmonic nanograting tip design for high power throughput near-field scanning aperture probe, *Opt. Express*, **18**, 14004–14011 (2010).
90. K. Takami, M. Akai-Kasaya, A. Saito, M. Aono, Y. Kuwahara, Construction of independently driven double-tip scanning tunneling microscope, *Jpn. J. Appl. Phys. Part 2*, **44**, L120–L122 (2005).
91. X. F. Wang, C. Liu, Multifunctional probe array for nano patterning and imaging, *Nano Lett.*, **5**, 1867–1872 (2005).
92. S. C. Minne, J. D. Adams, G. Yaralioglu, S. R. Manalis, A. Atalar, C. F. Quate, Centimeter scale atomic force microscope imaging and lithography, *Appl. Phys. Lett.*, **73**, 1742–1744 (1998).
93. P. Vettiger, G. Cross, M. Despont, U. Drechsler, U. Durig, B. Gotsmann, W. Haberle, M. A. Lantz, H. E. Rothuizen, R. Stutz, G. K. Binnig, The "millipede"-nanotechnology entering data storage, *IEEE Trans. Nanotechnol.*, **1**, 39–54 (2002).
94. H. Pozidis, W. Haberle, D. Wiesmann, U. Drechsler, M. Despont, T. R. Albrecht, E. Eleftheriou, Demonstration of thermomechanical recording at 641 Gbit/in^2, *IEEE Trans. Magn.*, **40**, 2531–2536 (2004).
95. Y. Rosenwaks, D. Dahan, M. Molotskii, G. Rosenman, Ferroelectric domain engineering using atomic force microscopy tip arrays in the domain breakdown regime, *Appl. Phys. Lett.*, **86**, 012909 (2005).
96. C. Hagleitner, T. Bonaccio, H. Rothuizen, J. Lienemann, D. Wiesmann, G. Cherubini, J. G. Korvink, E. Eleftheriou, Modeling, design, and verification for the analog front-end of a MEMS-based parallel scanning-probe storage device, *IEEE J. Solid-State Circuit.*, **42**, 1779–1789 (2007).
97. J. P. Yang, J. Q. Mou, N. B. Chong, Y. Lu, H. Zhu, Q. Jiang, W. G. Kim, J. Chen, G. X. Guo, E. H. Ong, Probe recording technology using novel MEMS devices, *Microsyst. Technol.* 13, 733–740 (2007).
98. A. Chad, C. A. Mirkin, Development of massively parallel dip-pen nanolithography, *ACS Nano*, **1**, 79–83 (2007).

99. K. Salaita, Y. Wang, J. Fragala, R. A. Vega, C. Liu, C. A. Mirkin, Massively parallel dip–pen nanolithography with 55,000-pen two-dimensional arrays, *Angewandte Chemie Inter. Ed.*, **45**, 7220 (2006).
100. D. Bullen, C. Liu, Electrostatically actuated dip pen nanolithography probe arrays, *Sens. Actuators A Physical*, **125**, 504–511 (2006).
101. G. M. Clayton, S. Tien, K. K. Leang, Q. Zou, S. Devasia, A review of feedforward control approaches in nanopositioning for high-speed SPM, *ASME J. Dyn. Syst., Meas. Control*, **131**, 061101 (2009).
102. N. Kodera, M. Sakashita, T. Ando, A dynamic PID controller for high-speed atomic force microscopy, *Rev. Sci. Instrum.*, **77**, 083704 (2006).
103. A. D. L. Humphris, M. J. Miles, J. K. Hobbs, A mechanical microscope: high-speed atomic force microscopy, *Appl. Phys. Lett.*, **86**, 034106 (2005).
104. L. M. Picco, L. Bozec, A. Ulcinas, D. J. Engledew, M. Antognozzi, M. A. Horton, M. J. Miles, Breaking the speed limit with atomic force microscopy, *Nanotechnology*, **18**, 044030 (2007).
105. G. E. Fantner, G. Schitter, J. H. Kindt, T. Ivanov, K. Ivanova, R. Patel, N. H. Anderson, J. Adams, P. J. Thurner, I. W. Rangelow, P. K. Hansma, Components for high speed atomic force microscopy, *Ultramicroscopy*, **106**, 881–887 (2006).
106. N. Kodera, H. Yamashita, T. Ando, Active damping of the scanner for high-speed atomic force microscopy, *Rev. Sci. Instrum.*, **76**, 053708 (2005).
107. A. Mahmood, S. O. R. Moheimani, Fast spiral-scan atomic force microscopy, *Nanotechnology*, **20**, 365503 (2009).
108. S.-K. Hung, Spiral scanning method for atomic force microscopy, *J. Nanosci. Nanotechnol.*, **10**, 4511–4516 (2010).
109. D. M. Carberry, L. Picco, P. G. Dunton, M. J. Miles, Mapping real-time images of high-speed AFM using multitouch control, *Nanotechnology*, **20**, 434018 (2009).
110. S. R. Manalis, S. C. Minne, A. Atalar, C. F. Quate, High-speed atomic force microscopy using an integrated actuator and optical lever detection, *Rev. Sci. Instrum.*, **67**, 3294–3297 (1996).
111. F. Dinelli, C. Menozzi, P. Baschieri, P. Facci, P. Pingue, Scanning probe nanoimprint lithography, *Nanotechnology*, **21**, 075305 (2010).
112. J.-I. Shirakashi, Scanning probe microscope lithography at the micro- and nano-scales, *J. Nanosci. Nanotechnol.*, **10**, 4486–4494 (2010).
113. S. Hoeppener, R. Maoz, J. Sagiv, Constructive micolithography: electrochemical priting of monolayer template patterns extends constrictive nanolithography to the micrometer-millimeter dimension range, *Nano Lett.*, **3**, 761–767 (2003).
114. M. Cavallini, P. Mei, F. Biscarini, R. Garcia, Parallel writing by local oxidation nanolithography with submicrometer reslution, *Appl. Phys. Lett.*, **83**, 5286 (2003).
115. N. Farkas, R. D. Ramsier, J. A. Dagata, High-voltage nanoimprint lithography of refractory metal films, *J. Nanosci. Nanotechnol.*, **10**, 4423–4433 (2010).
116. A. A. Tseng, Multi-head atomic force fabricator, US patent pending.
117. T. W. Kenny, *Tip-Based Nanofabrication (TBN)*, BAA No. 07-59, US Defense Advanced Research Projects Agency, Arlington, VA, 2007.
118. UT-Battelle, *Productive Nanosystems, a Technology Roadmap*, Battelle Memorial Institute and Foresight Nanotech Institute (2007).

Index

A.A. Tseng (ed.), *Tip-Based Nanofabrication*, DOI 10.1007/978-1-4419-9899-6,

J

K

L

N

About the Editor

Ampere A. Tseng is a Professor of Engineering at Arizona State University (ASU). He received his Ph.D. in Mechanical Engineering from Georgia Institute of Technology in 1978 and has published more than 250 referred papers with nine US patents under his credentials. Dr. Tseng has edited more than ten technical monographs and has been an editor for more than ten different technical journals. Professor Tseng was a recipient of the Superior Performance Award of Martin Marietta Laboratories, RCA Service Award, Alcoa Foundation Research Award, and ASU 1999–2000 Faculty Award. He chaired the ASME Materials Division in 1991–1992 and was selected as an ASME fellow in 1995. Also, he chaired the 2000 NSF Workshop on Manufacturing of MEMS and 1st International Workshop on Tip-Based Nanofabrication in 2008 as well as he co-chaired 1992 International Conference on Transport Phenomena in Processing and the 1999 NSF USA-China Workshop on Advanced Machine Tool Research. Dr. Tseng has received more than three million dollars in research funding directly from government agencies and industries. In 1990, Professor Tseng managed to secure 12 million dollars from US Department of Energy to establish Center for Automation Technology and became its first center director at Drexel University. From 1996 to 2001, Dr. Tseng was the founding Director of the Manufacturing Institute at ASU and the nine million dollars' donation from Motorola to Manufacturing Institute in 1997 was the largest single gift in ASU history. Professor Tseng had been awarded a Nippon Steel fellowship to perform research in Japan in 1992, two National Science Council Professorships to conduct research in National Taiwan University and National Cheng Kung University in Taiwan. He also secured two grants from the US-Czech Science and Technology Program to perform research in Brno University of Technology, and one grant from Finland Tekes Program to conduct research with University of

Oulu. Professor Tseng was bestowed an Honorary Guest Professor by Shanghai Jiao Tong University in 2000 and by Beijing Tsinghua University and University of Science and Technology of China in 2003. Recently, Dr. Tseng was appointed as a University Chair Professor by National Tsing Hua University (Taiwan) and by Taiwan University of Science and Technology and granted as a Visiting Consultant by Singapore Nanyang Technological University.